Geophysical Hydrodynamics of Currents, Waves, and Ice in the Atlantic Arctic

Aleksey Marchenko · Eugene Morozov ·
Dmitry Frey · Nataliya Marchenko

Geophysical Hydrodynamics of Currents, Waves, and Ice in the Atlantic Arctic

Review and Research

Springer

Aleksey Marchenko
The University Centre in Svalbard
Longyearbyen, Norway

Eugene Morozov
Russian Academy of Sciences
Shirshov Institute of Oceanology
Moscow, Russia

Dmitry Frey
Russian Academy of Sciences
Shirshov Institute of Oceanology
Moscow, Russia

Nataliya Marchenko
Arctic Technology
The University Centre in Svalbard
Longyearbyen, Norway

ISBN 978-3-032-14121-7 ISBN 978-3-032-14122-4 (eBook)
https://doi.org/10.1007/978-3-032-14122-4

Researchers near the Paula Glacier in Spitsbergen (Courtesy of E. Karulin and E. Morozov)

This Springer imprint is published by the registered company Springer Nature Switzerland AG
The registered company address is: Gewerbestrasse 11, 6330 Cham, Switzerland

Preface

This is a review paper of long-term field and numerical research.

We accept the concept that Eastern Arctic is the Arctic region influenced by the Atlantic Ocean. The book summarizes our research in a wide region around Spitsbergen in the Barents, Greenland, and Norwegian seas as well as in Spitsbergen fjords. The book is based on our previous publications in journals and conference presentations and provides new data on the recent research of geophysical flows and waves in the region. Most of the results are based on field observations. Interpretation of field data is combined with numerical modeling of currents and waves.

Field measurements include moorings in the Nordic and Arctic seas, CTD and LADCP casts from ship and ice cover, and satellite beacons on ice floes and icebergs.

A review of researches on currents in the Nordic seas is given based on literature sources and measurements of the authors. Currents in the Fram and Denmark straits were analyzed based on data of long-term moorings and CTD/LADCP casts along sections in the straits. Difference regimes of overflows of Arctic waters over the transversal sill in the Denmark Strait were analyzed. A review is given about the currents in the Norwegian, Greenland, and Barents seas.

Internal waves have been studied based on field observations from ice and moorings. The influence of the critical latitude for tidal internal waves of frequency M_2 (period 12.4 h) is shown on the example of moored measurements. The critical latitude for internal waves M_2 is 74.5° N. At this latitude, inertial and tidal M_2 frequencies become equal; hence, internal tides cannot propagate over flat bottom. High amplitude internal tides were recorded in the Strait of Kara Gates based on moorings and numerical modeling.

We studied icebergs of the Barents Sea that are formed from outflow glaciers of Spitsbergen, Franz Josef Land, and Novaya Zemlya. The average mass of an iceberg was estimated at around one million tons. The general direction of iceberg drift corresponds to the direction of surface currents. Most of the icebergs formed by the outflow glaciers of Spitsbergen and Franz-Josef Land drift southwest following the East Spitsbergen current, and some icebergs may drift into the central regions of the Barents Sea under the influence of wind, waves, and drift ice.

Ice drift and deterioration of ice and icebergs in the study region have been studied using satellite observations of ice motion and signals from the beacons placed on ice. Measurements of ice displacement allowed us to record a tsunami wave generated by glacier motion into the sea in the fjord. We studied flexural gravity waves in the ice and surface manifestations of various waves in the ocean combined with their measurements in the water column (wind waves propagating from the ice-free region, seiches, internal waves). We also studied freshwater outflow from the glaciers descending into seawater in Spitsbergen and freezing of tidal seawater jets flowing under ice cover in a narrow strait.

Longyearbyen, Norway — Aleksey Marchenko
Moscow, Russia — Eugene Morozov
Moscow, Russia — Dmitry Frey
Longyearbyen, Norway — Nataliya Marchenko

Acknowledgements The book is based on the field and laboratory experiments performed at the University Center of Svalbard and on the research ships of the Shirshov Institute of Oceanology by the authors and on literature sources. The study was supported by the Russian Foundation for Basic Research (projects nos. 17-08-00085, 18-05-00283, 20-08-00246), and NWO-RFBR, project no. 047.017.2006.003, and by the Russian Science Foundation (project nos. 21-77-20004 (field measurements), 23-77-30001 (analysis and interpretation of long term data), 25-17-00309 (analysis of internal wave generation mechanisms)). Investigations within these projects have been continuing for many years.

The authors are grateful to numerous colleagues who participated in the field work, data analysis, and helped to perform observations in the ocean: Z. Kowalik, P. Chistyakov, E. Karulin, V. Markov, M. Karulina, A. Sakharov, S. V. Pisarev, A. V. Ivanov, K. V. Filchuk, I. V. Ryzhov, T. G. Elizarova, N. A. Diansky, S. V. Muzylev. We thank our colleagues who helped in the organization of expeditions: Captain L. V. Sazonov and I. Fer. Thanks go to colleagues who participated in data processing, analytical works, and preparation of the manuscript.

We also thank the crews of research vessel *Akademik Mstislav Keldysh* for their contribution to this research.

Competing Interests The authors have no competing interests to declare that are relevant to the content of this manuscript.

Contents

About the Authors

Aleksey Marchenko graduated in mechanics in 1982 from the Lomonosov Moscow State University and then received a Ph.D. in 1987 and a doctoral degree in 1997. He has background in continuum mechanics and fluid dynamics. His career developed at the Ishlinsky Institute for Problems in Mechanics and Prokhorov General Physics Institute of the Russian Academy of Sciences (Moscow, Russia, 1985–2006), and then at the University Centre in Svalbard (UNIS) (Longyearbyen, Norway), where he is Professor of Arctic Technology Department since 2006. His research and teaching focuses on ice physics and mechanics, wave processes in ice covered waters and ocean tides. He elaborated mathematical models and conducted numerical simulations, field and laboratory works. He organized and participated in the field works on drift ice in the Barents Sea and land fast ice in Spitsbergen fjords annually from 2007 to 2022. Since 2010 he is a member of the Norwegian Scientific Academy for Polar Research. Since 2005 he is a member of the National Committee on Theoretical and Applied Mechanics (Russia). From 2005 to 2022 he was a member of the International Committee of the International Conference on Port and Ocean Engineering Under Arctic Conditions (POAC).

Eugene Morozov graduated in oceanography in 1970 from the Moscow Institute of Physics and Technology, and then received a Ph.D. in 1975 and a doctoral degree in 1990. In 2025 he became corresponding member of the Russian Academy of Sciences. He is a field oceanographer and a specialist in the observations in the open ocean. His career developed at the Shirshov Institute of Oceanology in Moscow where he heads a laboratory. His research focuses on mesoscale dynamics, internal waves, and circulation. His research is also related to the Arctic and Antarctic oceanography. He participated in more than 50 oceanic expeditions all over the globe and was the leading scientist in many of them. From 2011 to 2015 he was the President of the International Association for the Physical Sciences of the Ocean (IAPSO).

Dmitry Frey graduated in applied physics and mathematics in 2014 from the Moscow Institute of Physics and Technology, and then received a Ph.D. in 2019. Since 2015, he has been working at the Laboratory of Hydrological Processes of the Shirshov Institute of Oceanology as a senior scientist. His research focuses on different aspects of ocean circulation including bottom gravity flows, western boundary currents, Arctic and Antarctic oceanography, and equatorial undercurrents. He is a field oceanographer and specialist in numerical modeling.

Nataliya Marchenko is a research associate at the University Centre in Svalbard (UNIS), located in Longyearbyen (78° N) Norway. She graduated from the Lomonosov Moscow State University (1988) and defended her Ph.D. thesis devoted to forest fire risk assessment, using the Geographical Information System (GIS) in 1992. Since 2006, Dr. N. Marchenko works at the Arctic Technology Department of UNIS and does research on sea ice, navigation, and risk assessment in Polar waters, laser scanning and GIS. Her scientific interests also include coastal dynamic and ecology.

Abbreviations

ADCP	Acoustic Doppler Current Profiler
ADT	Absolute Dynamic Topography
BSO	Barents Sea Opening
BSX	Barents Sea Exit
CTD-profiler	Conductivity-Temperature-Depth profiler
DSOW	Denmark Strait Overflow Water
DWBC	Deep Western Boundary Current
FBCO	Faroe Bank Channel Overflow
FSCBW	Faroe Shetland Channel Bottom Water
FZ	Fracture Zone
GEBCO	General Bathymetric Chart of the Oceans
INMOM	Institute of Numerical Mathematics Model
ISOW	Iceland Scotland Overflow Water
LADCP	Lowered Acoustic Doppler Current Profiler
LNADW	Lower North Atlantic Deep Water
LSW	Labrador Sea Water
MAR	Mid-Atlantic Ridge
MEIW	Modified East Icelandic Water
MIZ	Marginal Ice Zone
MNADW	Middle North Atlantic Deep Water
NADW	North Atlantic Deep Water
NAO	North Atlantic Oscillation
NSAIW	Norwegian Sea Arctic Intermediate Water
NSDW	Norwegian Sea Deep Water
PF	Polar Front
RFBR	Russian Foundation for Basic Research
RSF	Russian Science Foundation
SSH	Sea Surface Height
SST	Sea Surface Temperature
UML	Upper Mixed Layer
UNADW	Upper North Atlantic Deep Water

WOCE	World Ocean Circulation Experiment in the 1990s
WOD	World Ocean Database

Chapter 1
Currents and Tides in the Nordic Seas and Eastern Arctic

1.1 General Notes and Introduction

The Nordic Seas include the Greenland, Norwegian, and Iceland seas. The Barents Sea is adjacent to this large intermediate basin between two oceans, through which water, heat, and salt exchange occurs between the Atlantic Ocean and the Arctic Ocean. Sometimes this water basin is assumed as the region north of the Greenland-Scotland Ridge. The square of the basin of the Nordic Seas is approximately 25×10^6 km^2, which is less than 1% of the global ocean, but it plays a very important role in the interaction between the atmosphere and ocean; hence it has a pronounced effect on the European Climate. The climate of the region varies strongly from one season to another. The structure of bottom topography over this region is very diverse related to such a small area. Shallow shelves and steep continental slopes are located close to the deep basins.

The anomalies between sea surface temperatures and surface air temperatures in the Nordic Seas are the strongest in the world. This is a comparatively warm region because of the prevailing southwesterly winds and warm oceanic currents from the south. The mean temperatures in the region exceed by more than 15 °C the temperatures at the same latitudes of the Earth. This region plays the major role in the water mass formation of the Atlantic Ocean and it is an important pathway for water exchange between the North Atlantic and the Arctic Ocean.

Warm Atlantic waters transported by the North Atlantic Current pass through these seas warming the Arctic region and Europe. The densest Arctic waters overflow the thresholds in the Fram Strait and over the Greenland Ridge in the Denmark Strait between Iceland and Greenland. They also overflow the submarine sills between Iceland, Faroe Islands, and Shetland Islands and eventually form North Atlantic Deep Water. These passages are quite shallow. The depth of the sill in the Denmark Strait is 650 m and the one in the Faroe Channel is 850 m.

The most prominent topographic elevation in the Nordic Seas is the mid-oceanic ridge, which connects Spitsbergen and Iceland. Three main parts of this ridge system

A. Marchenko et al., *Geophysical Hydrodynamics of Currents, Waves, and Ice in the Atlantic Arctic*, https://doi.org/10.1007/978-3-032-14122-4_1

are known as the Kolbeinsey Ridge, the Mohn Ridge, and the Knipovich Ridge. The Greenland and the Norwegian seas are separated by the Mohn Ridge. The Greenland and Iceland seas are separated by the Jan Mayen Fracture Zone; the Aegir Ridge (Helgeland Ridge) is the boundary between the Norwegian and Iceland seas. A scheme of bottom topography is shown in Fig. 1.1.

Four large basins are recognized in the Nordic Seas. West of the mid-oceanic ridge, the Greenland Basin is the largest and deepest. The maximum depth is 3600 m. Sometimes a smaller but deep (3200 m) Boreas Basin is distinguished in the northern part of the Greenland Sea separated from the southern basin (Greenland Basin) by the Greenland Fracture Zone. In the Iceland Sea, the Iceland Plateau is an elevation between Iceland and Jan Mayen Island to the east of the Kolbeinsey Ridge and to the south of the Jan Mayen Fracture Zone. The Jan Mayen Ridge runs to the south from Jan Mayen Island.

Two deep basins are recognized in the Norwegian Sea: the Norwegian and Lofoten basins. The Norwegian Basin is the largest and deepest in the Nordic Seas. It is located

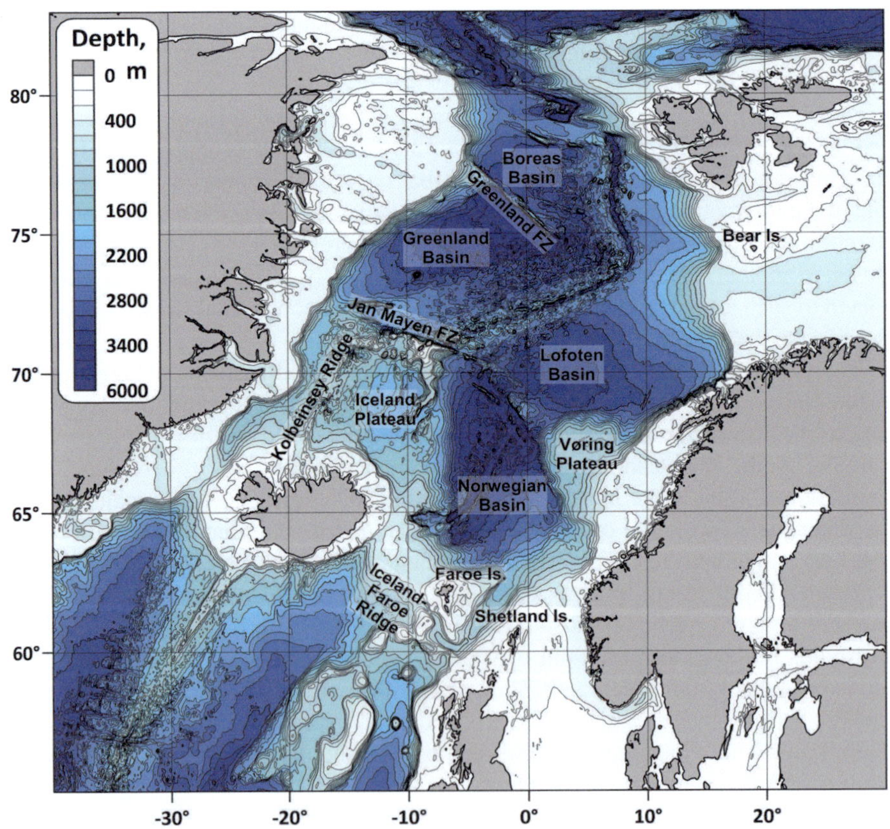

Fig. 1.1 A scheme of bottom topography of the Nordic Seas

north of the Iceland-Faroe Ridge. West of this basin is the Iceland Plateau and north of it is the Vøring Plateau and the continental slope of Norway. The maximum depth of the Norwegian Basin is 3600 m, but it is even deeper than 3800 m in a narrow trough (65°N, 5°W). The shallower and smaller Lofoten Basin is located north of the Norwegian Basin. The Mohn Ridge is the boundary between these basins.

The water masses of each of the Nordic Seas are characterized by diverse properties; hence, they influence the circulation in each of the basins assuming that bottom topography is also a factor that strongly influences surface and deep circulation. A scheme of surface circulation in the basins is shown in Fig. 1.2.

Cold dense waters of high salinity are formed in the Greenland Sea because of cold winters in the region close to Greenland and Arctic. Higher salinity of water is due to the inflow of high saline waters from the Atlantic Ocean that cools at high latitudes. The waters that flow to the Greenland Sea from the Arctic through the Fram Strait are also dense. However, these waters that actually form North Atlantic Deep Water are not the densest water that fills the bottom layers of the Atlantic Ocean.

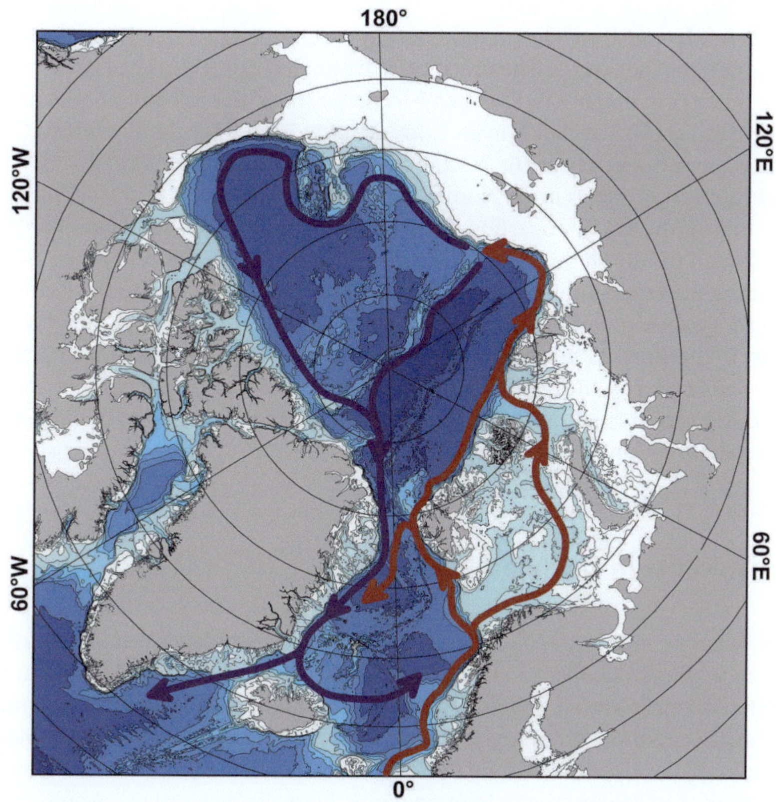

Fig. 1.2 Scheme of surface circulation in the Nordic Seas and Arctic Ocean based on many publications. The red arrows indicate warm currents from the North Atlantic (subsurface in the Arctic Ocean) and blue currents show cold water flowing from the Arctic Ocean

The bottom water formed in the Weddell Sea of the Antarctic is the densest because the sills that Antarctic Bottom Water overflows are much deeper than those around Iceland. Part of Arctic waters cannot overflow these sills and remains in the Arctic Ocean forever. The circulation of the Nordic Seas is cyclonic. Water exchange with the Atlantic Ocean in the upper layer occurs in the southeastern part of the Norwegian Sea and partly west of Iceland.

The continuation of the North Atlantic Current in the Norwegian Sea is the warm Norwegian Atlantic Current. The Norwegian Atlantic Current flows through the Faroe-Shetland Channel and then along the coast of Norway to the Arctic and divides into two branches, one flowing to Spitsbergen (the Spitsbergen Current) and the other branch flowing into the Barents Sea. A small portion of mixed Atlantic waters flows around the western coast of Iceland to the Iceland Sea as the North Icelandic Irminger Current. Another small portion of inflowing water from the south is the Faroe Current that flows across the Iceland-Faroe Ridge.

The outflow from the Arctic to the Nordic Seas occurs as deep overflows over the sills in the Denmark Strait and across the Iceland–Scotland Ridge. The total inflow of cold Arctic water to the Nordic Seas is estimated at approximately 3 Sv. The cold southward East Greenland Current enters the Greenland Sea through the Fram Strait from the Arctic. The transport is estimated at 1.3 Sv. The East Greenland Current transports much Arctic ice and icebergs that calved from the glaciers over the Greenland shelf. The Jan Mayen Current splits from the East Greenland Current near the eastern boundary of the Nordic Seas steered by bathymetry.

There are several gyres in the Nordic Seas. The subsurface waters leave the Nordic Seas to the south through overflows between Greenland and Scotland. The deeper waters flow to the south through the Denmark Straight and over the Faroe-Shetland Ridge along the Faroe-Bank Channel. The flow of Atlantic Water to the Barents Sea is strongly governed by topography, but the local wind field is also important. The transport to the Barents Sea based on difference estimates fluctuates near 1 Sv. The Barents Sea is a basin of formation of dense bottom water due to cooling and brine release during ice formation. All the properties of this region make it a "hot spot" as an extraordinary climatic warm anomaly.

1.2 Currents in the Norwegian Sea

The Norwegian Sea is an intermediate basin on the pathway of Atlantic water spreading to the Arctic Ocean. The most important current in the Norwegian Sea is the North Atlantic Current, which is an extension of the Gulf Stream. Two branches of the North Atlantic Current exist. These branches are the inner stable branch that follows along the continental slope (Norwegian Atlantic Slope Current) and variable western branch in deeper waters (Norwegian Atlantic Front Current). Approximately at a latitude of Tromsø (70° N), the Norwegian Atlantic Slope Current splits into a branch following the continental shelf and slope to the north towards Spitsbergen and another branch that follows the northern coastline of Norway and flows to the

Barents Sea. The Norwegian Coastal Current follows the coastline to the Barents Sea along the entire western coast of Norway closer to the coastal flows. The warm surface waters from the North Atlantic enter the Nordic Seas from the south. They cool in the Nordic Seas as they flow northward, submerge, and eventually become the Deep Western Boundary Current (DWBC) of North Atlantic Deep Water that flows in the deep-water layer to the South Atlantic.

The warm Norwegian Atlantic Current is an eastward branch of the North Atlantic Current. This current enters the Norwegian Sea north of Scotland and flows north-eastward along the coast of Norway and then enters the Barents Sea. The surface temperatures range from 8 °C in the south to 4 °C in the north; hence, the current exerts a warming influence on the climate of Northern Europe. It contributes to a mild climate in the region and rich fauna of the sea. The velocities of the main current stream are as high as 25 cm/s. The northern continuation of the Norwegian Atlantic Current is also called the West Spitsbergen Currents as it approaches Spitsbergen and then in the Fram Strait it loops into the East Greenland Current (Fig. 1.3 shows the schemes of currents based on different literature sources).

Falina et al. (2015) present estimates of Atlantic water inflow into the Nordic Seas. The inflow estimates in the summer months are as follows: 1 Sv through the Denmark Strait, 3–4 Sv through the Faroe Shetland Strait, and 3–4 Sv through the Shetland Strait (Fig. 1.4).

Mork and Blindheim (2000) analyzed hydrographic observations (1955–1996) over the Svinøy section to investigate variations in the Atlantic inflow to the Nordic Seas. This section is directed to the northwest from about 62°00′ N on the Norwegian coast to 64°40′ N at the prime meridian (Fig. 1.3). The Svinøy section crosses the eastern branch of the Norwegian Atlantic Current and crosses the Arctic Front farther offshore. This is the most important region for the inflow of Atlantic waters northward and the transport through this section is maximal among the other northward currents flowing to the Norwegian Sea. The observations show that two cores of high salinity are observed in summer, but in spring and winter the offshore core is less pronounced. The mean volume geostrophic transports of Atlantic water over the section are 4.5, 5.0, and 5.6 Sv in spring, summer, and winter, respectively. Interannual variations within ± 1.5–2.0 Sv in the geostrophic transport were recorded (Fig. 1.5).

Orvik et al. (2001) continued analysis of the Svinøy section data in 1996–1999. The measurements were supported by current meters on six moorings. Two clearly pronounced branches of the Norwegian Atlantic Current were revealed. The eastern branch appears as a narrow, topographically trapped, nearly barotropic, 30–50 km wide current, with a maximum velocity of 117 cm/s. The western branch is also about 30–50 km wide, and appears as an unstable frontal jet about 400 m deep with a maximum velocity of 87 cm/s. The observations between these two prominent branches show an eddy field with a recirculation to the southwest. Transport estimates from the current records in the eastern branch show an annual mean inflow of 4.2 Sv with variations between 2.0 and 8.0 Sv on a monthly time scale. The four-year current record shows an annual cycle with summer to winter variations in the proportion of 1–2 (Fig. 1.5).

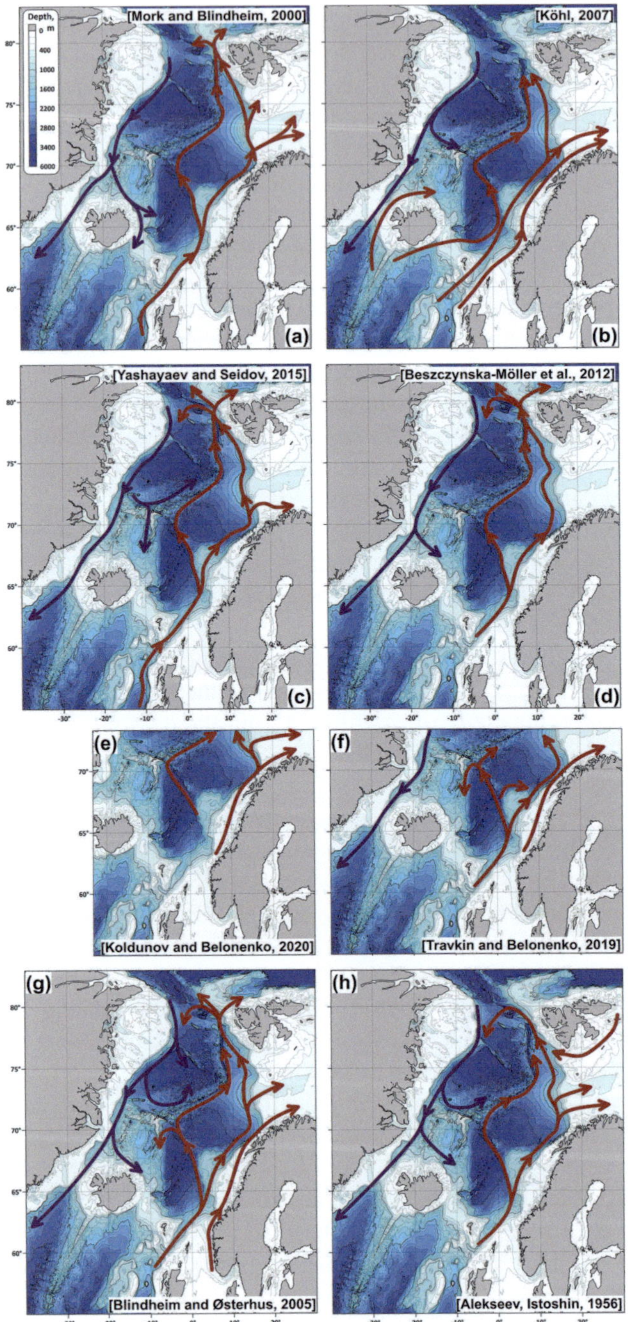
Depth, m
0
400
1000
1600
2200
2800
3400
6000
[Mork and Blindheim, 2000]
(a)
[Köhl, 2007]
(b)
[Yashayaev and Seidov, 2015]
(c)
[Beszczynska-Möller et al., 2012]
(d)
(e)
[Koldunov and Belonenko, 2020]
(f)
[Travkin and Belonenko, 2019]
(g)
[Blindheim and Østerhus, 2005]
(h)
[Alekseev, Istoshin, 1956]

◀**Fig. 1.3** Different schemes of currents in the Norwegian Sea based on publications **a** Mork and Blindheim (2000), **b** Köhl (2007), **c** Yashayaev and Seidov (2015), **d** Beszczynska-Möller et al. (2012), **e** Koldunov and Belonenko (2020), **f** Travkin and Belonenko (2019), **g** Blindheim and Østerhus (2005), **h** Alekseev and Istoshin (1956). Publications Koldunov and Belonenko (2020) and Travkin and Belonenko (2019) analyze only the currents in the Norwegian Sea; hence, we truncated the northern part of the figures

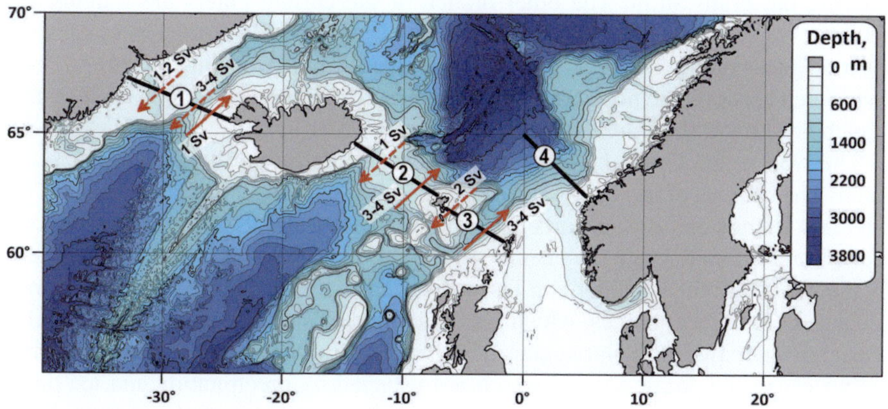

Fig. 1.4 Scheme of repeated oceanographic sections in the Denmark Strait (1), Faroe-Iceland Strait (2), Shetland Strait (3), and Svinøy section (4). Arrows show the direction and transport of Atlantic and Arctic waters; numerals indicate transport in Sv based on CTD/LDCP sections. The measurements were carried out by oceanographers of the Shirshov Institute of Oceanology, Russian Academy of Sciences, adopted from (Falina et al. 2015)

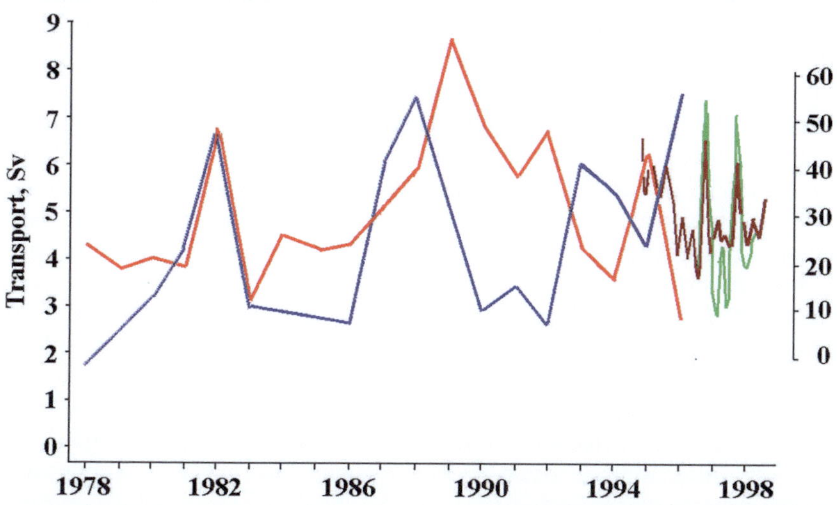

Fig. 1.5 Combined graph of geostrophic volume transports over the Svinøy section in 1978–1996; reference depth for zero velocity is 1000 m; March–April (blue), July–August (red) (Mork and Blindheim 2000); moored transport (green) and velocities (brown) in 1995–1998. (Orvik et al. 2001)

Gascard et al. (2004) investigated the Norwegian Atlantic Current based on three hydrographic sections across the Lofoten Basin in May 2000. The Norwegian Atlantic Current splits in the northern part of the Lofoten Basin midway between the Lofoten Islands and Bear Island. One branch of the current is progressing northward west of Bear Island (Bjørnøya) becoming the West Spitsbergen Current before entering the Fram Strait. The other branch of the current enters the Barents Sea between Norway and Bear Island and is heading towards the east. The estimate of the total geostrophic transports of the Norwegian Atlantic Current equal to 7.2 Sv is distributed between the northern branch heading to Spitsbergen (4.8 Sv) and eastern branch heading to the Barents Sea (2.4 Sv).

The Norwegian Coastal Current is the second intense northward flow of warm water. The sources of its water are partly from the North Atlantic Current east of the Shetland Islands and it transports part of the fresher Baltic Sea water. It also flows to the Barents Sea through the opening of the Barents Sea, a wide strait to the Arctic Ocean. This current is colder and less saline than the Norwegian Atlantic Current because of the other sources, which are cooler and less saline: the North and Baltic seas and water from the Norwegian fjords and rivers. It is warmer and saltier than the waters of the Arctic Ocean, which is freshened by precipitation and ice on its shores. Winter temperatures in the flow generally range between 2 and 5 °C. Mork (1981) writes that the Norwegian Coastal Current is referred to as the continuation of the Baltic Current mainly because of its low salinity. The transport of fresh water from the Baltic amounts to 450 km^3 per year but the contribution from other sources, mainly the southern North Sea and the Norwegian coast, is considerable.

Köhl (2007) writes that water masses in the Lofoten Basin are characterized by warm and saline Atlantic water in the upper 600–700 m. Salinities in this layer exceed 35 PSU, while the temperatures are over 4 °C. Norwegian coastal waters are dominated by two main water masses, water from the Norwegian Coastal Current and water from the North Atlantic Current (Atlantic water). As the Norwegian Coastal Current flows northward, its waters mix with the nearby flowing Norwegian Atlantic Current; hence, its salinity increases. The current is both wind-driven, entraining water along the Norwegian coast by southwesterly winds (forming elevation and thus pressure differences), and also geostrophically driven by its salinity distribution which forms density gradients.

Fer et al. (2020) analyzed the properties of the Norwegian Atlantic Slope Current based on the moored measurements at the Lofoten Escarpment during 14 months in 2016–2017 and revealed that the core of the current is located at depths from 200 to 650 m over the 650 and 1500 m isobaths. The strongest currents were recorded in winter (15 cm/s). High-frequency variability is characterized by fluctuations that reach 80 cm/s lasting from one to two weeks and extend down to 600 m. In contrast to observations over the Svinøy section (2° E, 63° N), the slope current is not barotropic and varies strongly with depth. The mean volume transport of Atlantic water is estimated at 2.0 ± 0.8 Sv with summer and winter mean values of 1.6 and 2.9 Sv, respectively. The baroclinic conversion from mean potential energy into eddy kinetic energy (e.g., baroclinic instability) over the Lofoten Escarpment can be substantial with volume-averaged values up to 2×10^{-4} Wm^3.

Yashayaev and Seidov (2015) analyzed the historic hydrographic data from the World Ocean Database to study the long-term variability of the Atlantic water throughflow over the Norwegian Sea to the western Barents Sea. Variability of water temperature, salinity, and density were analyzed along six hydrographic sections crossing the Atlantic water flow and coastal currents. The sections are located from the northern North Sea (69° N) throughout the Norwegian Sea to the Kola Section in the Barents Sea (33°30′ E) (Fig. 1.3).

Several water masses are distinguished in the Norwegian Sea. Coastal Water is located closer to the coast of Norway. It is transported by the Norwegian Coastal Current. This water originates in the Baltic Sea; water supply generally occurs from the rivers along the coast. The water is relatively fresh and strongly varies in temperature throughout the year. Warm Atlantic water of high salinity transported initially by the North Atlantic Current and its continuations occupies the upper 500 m of the open sea, which is determined by the depths of the thresholds over which this water enters the Norwegian Sea (Belonenko et al. 2014); the temperatures of this water are relatively stable throughout the year. Intermediate Water occupies the layer below the thermocline with Atlantic water; it is fresher than Atlantic water and colder. Cold and saline water occupies the deeper parts of the Lofoten Basin.

Koszalka et al. (2011) investigated circulation in the Norwegian Sea using drifters. The analysis of drifter trajectories confirmed the features of the general circulation in the Nordic Seas such as branches of the Norwegian currents, Lofoten Vortex, West Spitsbergen Current, and currents in the Greenland Sea. At the same time, they revealed a new feature of in the surface circulation, which are two anticyclonic recirculations at the Svinøy section. Their Lagrangian analysis of drifter trajectories shows that currents generally do not follow exactly the well-known schemes of circulations but represent meandering motion in the entire basin of the Nordic Seas (Fig. 1.6b). Averaging of the data by clustering drifters gives a mean pattern of the flows (Fig. 1.6a). Mean speeds based on drifters are shown in Fig. 1.7. The data were united in clusters with given number of drifters.

Poulain et al. (1996) analyzed the data of 107 surface drifters in 1991–1993 in the Norwegian and Greenland seas. The drifters revealed not only the general large-scale cyclonic transport of waters, but also cyclonic rotation of smaller-scale eddies in the Norwegian, Lofoten, and Greenland basins, and also over the Iceland Plateau. Jakobsen et al. (2003) analyzed the circulation in the surface layer of the Norwegian and Greenland seas based on the drift data of 387 buoys in the 1990s. Following Poulain et al. (1996), the authors report the existence of four large-scale cyclonic eddies.

Hátún et al. (2021) considered circulation gyre in the Norwegian Basin in the southern part of the Norwegian Sea. The Norwegian Sea Gyre (NSG) is a large body of Arctic Intermediate Water and deep dense overflow waters, which circulate counterclockwise within the Norwegian Basin. The strongest flow of the Norwegian Sea Gyre occurs downstream of a confluence of subarctic waters from the Iceland Sea and the Jan Mayen Ridge over steep bathymetry north of the Faroe slope. The Norwegian Sea Gyre circulation regulates key aspects of both the poleward Atlantic

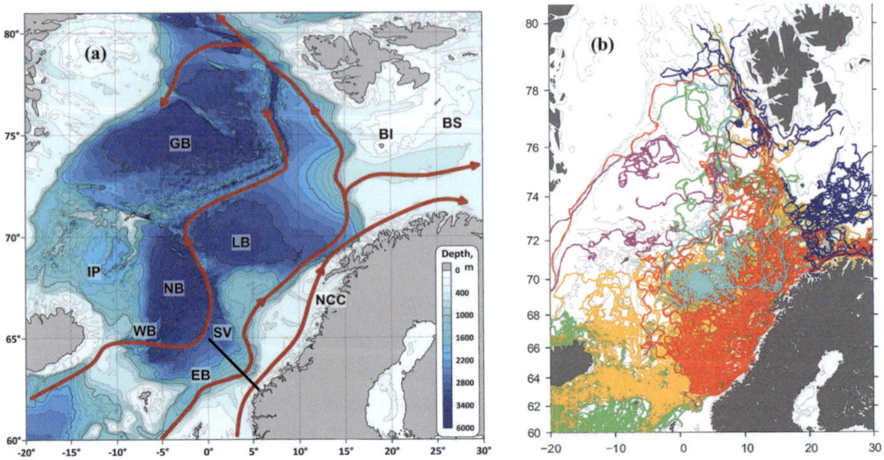

Fig. 1.6 **a** Main bathymetric features and pathways of Atlantic water flow in the Nordic Seas. BI is Bear Island; SV is Svinøy section marked with a black line. The two branches of the Norwegian Atlantic Current are the eastern branch (EB: Norwegian Slope Current) and the western branch (WB: Norwegian Front Current); and NCC is Norwegian Coastal Current; LB is Labrador Basin; NB is Norwegian Basin; GB is Greenland Basin; BS is Barents Sea; IP is Iceland Plateau. **b** Drifter trajectories colored according to the region of deployment: Svinøy (red), Lofoten Basin (cyan), Gimsøy and Barents Sea Opening (blue), Greenland Sea (pink), Iceland-Faroe-Ridge and Iceland Plateau (orange). Drifters advected from the North Atlantic are plotted in green Koszalka et al. (2011)

water currents and the equatorward near-bottom and mid-depth flows in the Norwegian Sea, which are the main arteries of the Meridional Overturning Circulation. The authors reveal close links between the Norwegian Sea Gyre circulation and: (1) the observed Faroe Bank Channel Overflow (FBCO) transport, (2) variable depth of the main thermocline separating Atlantic water from the underlying colder and denser subarctic water masses, and (3) satellite-derived sea-surface heights (SSH) in the southern Nordic Seas. In general, a strong Norwegian Sea Gyre and weak Faroe Bank Channel Overflow transport are associated with an uplifted thermocline and low sea surface height. Hátún et al. (2021) show that inflowing warm Atlantic water and returning dense overflow water at greater depths are intrinsically coupled. The connections involve both vertical undulations of the main interface, which separates the warm and cold waters, and the circulation strength of the Norwegian Sea Gyre.

Rossby et al. (2009a) present a 50-year hydrographic climatology of the Norwegian Sea analyzing water properties and their variation on isopycnal surfaces. They report that the warmest waters on each isopycnal surface are located in the northern part of the Lofoten Basin, close to the regions where the isopycnal outcrops, a clear indication of downward mixing of the warmer more saline waters on shallower isopycnals due to convective cooling at the surface. These saline waters are transported and shed from the Norwegian Atlantic Slope Current by means of anticyclonic eddies separated from the current due to a very high level of eddy activity.

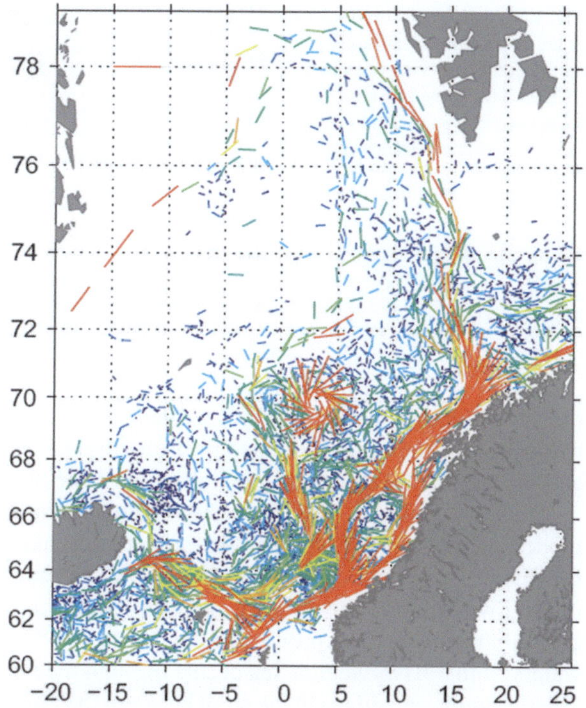

Fig. 1.7 Drifter-derived estimates of the mean velocity, color-coded by the speed magnitude. Estimates are obtained by clustering the drifter data in clusters with varying prescribed number of members $m = 12$ drifter days. The arrow in the upper left corner marks a speed of 30 cm/s. Adopted from (Koszalka et al. 2011)

The isopycnal analysis of Rossby et al. (2009a) shows that the water mass boundary between the waters of Arctic origin in the west and Atlantic waters in the east aligns tightly with the Jan Mayen, Mohn, and Knipovich ridges suggesting low cross-ridge exchange. Thus, the main mass and heat exchange between the eastern and western basins is limited to the northern and southern ends of the ridge system. Atlantic waters spread into the Greenland Sea through the Fram Strait and Artic waters propagate into the southern part of the Norwegian Sea just north of the Iceland-Faroe Ridge.

The authors of Belonenko et al. (2014) realized numerical model ECC02 for the region of the Norwegian Sea and plotted velocity vectors at different depths. A scheme of currents shown in Fig. 1.8 corresponds to the existing concept of the Norwegian Atlantic Current, which is a continuation of the North Atlantic Current and is a system of currents consisting of two branches covering the entire basin of the Norwegian Sea. The eastern branch, following the Norwegian continental slope, turns north near 70° N, 17° E, then deviates slightly to the west, while the other branch, directed to the north, splits at 68° N, 1° E into two new flows. One of them cyclonically, turns to the west at 69° N, 3° W and then heads south along the 6° W meridian with strengthening currents. The second, breaking away from the main branch at the bifurcation point, turns to the east and subsequently joins the coastal eastern branch with velocities reaching 4.5 cm/s, but this branch gradually weakens with increasing depth, and at 1000 m the maximum velocities are observed not along

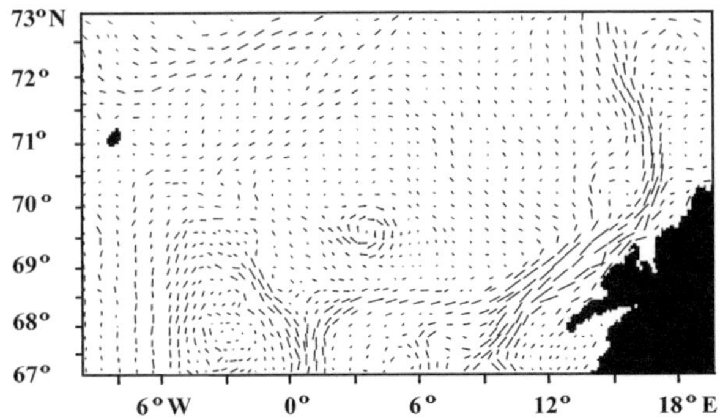

Fig. 1.8 Surface circulation in the Norwegian Sea based on the eddy-resolving model ECC02 with a resolution of 4 km (velocity scale is in cm/s). Adopted from Belonenko et al. (2014)

the edge of the shelf, but west of the branch. In general, it is clear that the cyclonic nature of the general circulation of the Lofoten Basin is due to its topographic features and the currents at all depths generally follow the bottom topography.

Thus, based on the data from the implementation of the ECC02 hydrodynamic model, the most important features of water circulation in the Lofoten Basin were identified: a cyclonic gyre spreading throughout the entire thickness of the transporting water along the isobaths, and the existence of an anticyclonic eddy in the center of the basin most pronounced at depths of 350–550 m. The outflow of bottom waters from the Norwegian Sea occurs through the Faroe Bank Channel overflowing several submarine ridges in the strait between the Shetland Islands and Faroe Islands. Eventually this flow becomes part of North Atlantic Deep Water flow.

The depths over the Iceland Faroe Ridge do not exceed 500 m, while the depth of the sill in the Faroe Bank Channel is approximately 850 m. This channel is a continuation of the Faroe-Shetland Channel, whose depths exceed 1000 m. The main flow of Arctic waters occurs through these channels located between the Faroes and Shetland Islands. A map of the bottom topography (Fig. 1.9) shows the Faroe–Shetland Channel narrowing into the Faroe Bank Channel. The bottom flow from the Norwegian Sea to the southwest follows the shallowing channel towards the southwest; then the channel makes a turn to the northwest and reaches a sill, which is 850 m deep. Then, the channel deepens into the basin of the North Atlantic. The flow of cold water is cascading down the slope into the North Atlantic. This flow occurs under the influence of pressure gradient, bottom friction, and the Earth's rotation. The overflow descends the Iceland-Faroe slope and mixes with overlaying water (Seim and Fer 2001; Saunders 2001; Mauritzen et al. 2005; Fer et al. 2010). Our analysis is based on several review publications.

Hansen and Østerhus (2007) write that bottom flow of water colder than 3 °C and salinity below 35.0 PSU occupies a layer below 500 m to the bottom. The overflow water is much colder and less saline than the upper Atlantic water.

The water in the Faroe-Shetland Channel originates from the waters north of the channel: Modified East Icelandic Water (MEIW), whose temperature and salinity values range within 1.0–3.0 °C and 34.7–34.9 PSU; Norwegian Sea Arctic Intermediate Water (NSAIW) with temperature and salinity variations from –0.5 to +0.5 °C

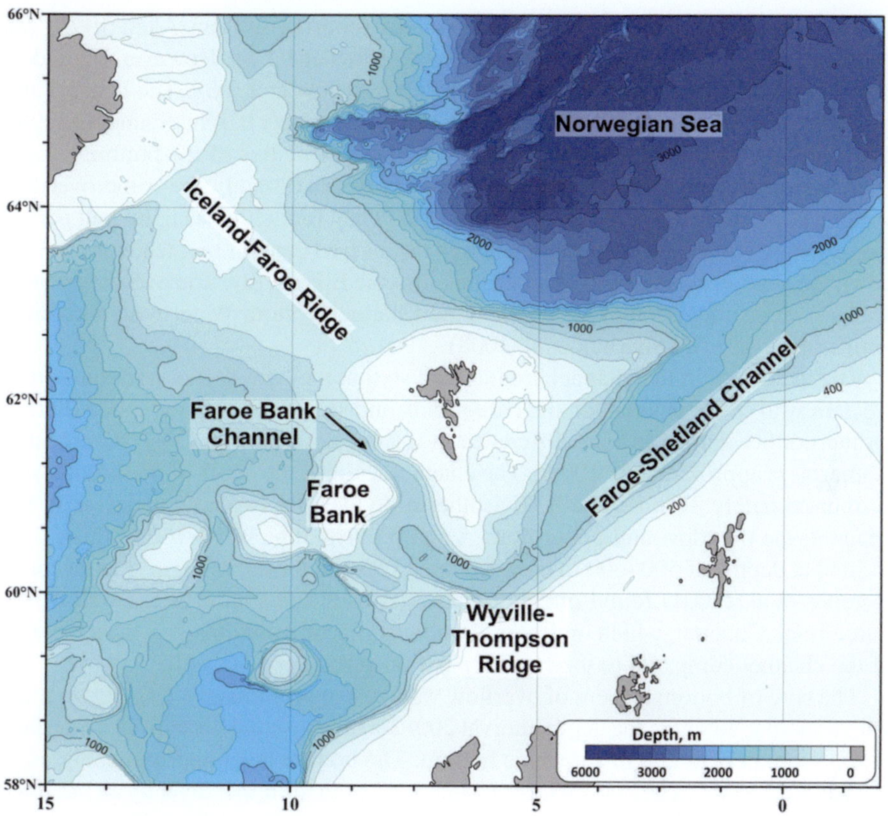

Fig. 1.9 Bathymetry of the region and the main passages for the bottom flows from the Norwegian Sea to the Atlantic Ocean

and 34.87–34.90 PSU; and Norwegian Sea Deep Water (NSDW) below the –0.5 °C potential isotherm whose salinity is approximately 34.91 PSU. The properties of the deepest water in the Faroe–Shetland Channel are slightly different from its origin in the Norwegian Sea. The properties of the two initial intermediate water masses MEIW and NSAIW are less pronounced in the channel, which is possibly caused by mixing. Generally speaking, only the coldest and low saline waters of the Norwegian Sea basin reach the sill of the Faroe Bank Channel.

Hansen and Østerhus (2007) analyzed the ADCP data from moorings in the Faroe Bank Channel over a period of 1995–2005. The mean overflow transport based on the velocity field was estimated at 2.1 ± 0.2 Sv with seasonal and inter-annual variations of approximately 10%, but without any trend over the whole period. The bottom current transport with potential density exceeding 27.8 kg m^{-3} was 1.9 ± 0.3 Sv with average properties: potential temperature $\theta = 0.25$ °C, salinity $S = 34.93$ PSU.

These estimates exceed the previous values reported in (van Aken 1995; van Aken and de Boer 1995) who estimated the overflow at 1.1 Sv. However, (Saunders 1990) gives an estimate of cold water (<3 °C) overflow at 1.9 Sv. The dense water transport over the Wyville-Thomson Ridge is approximately 0.3 Sv (Saunders 1990). The depths of the ridge are approximately 600 m. Velocities of the bottom current over the ridge sometimes exceed 100 cm s^{-1}. Approximately the same overflow volume was estimated in (Sherwin et al. 2008). They found that a significant part of the overflow is channeled through a newly discovered canyon that leads southward down the southern flank of the Faroe Bank into the Ellett gully. The overflow waters are sometimes identified as Faroe Shetland Channel Bottom Water with a nominal temperature of 0 °C (Sherwin et al. 2008).

The isotherms in the channel are inclined so that they are deeper over the southwestern wall of the channel on the Faroe Bank and shallower over the Faroe Plateau on the northeastern side. The difference in the deepening of the 0°–7 °C potential isotherms is approximately 100 m. The channel width at the depth of these isotherms is approximately 16 km. The 3 °C isotherm divides the two water masses in the channel: the overflow and upper flow. A sharp pycnocline is 50–100 m thick. It is located at depths of 500–600 m between the 27.55 and the 27.95 kg m^{-3} isopycnals. Borenäs et al. (2001) found pitching of isotherms at the southwestern wall of the Faroe Bank Channel, which indicates close locations of isotherms 1–7 °C at the wall of the channel compared to their wider position at the opposite wall.

The core of bottom current of overflow waters is approximately 200–300 m thick with a sharp gradient in the depth interval 200–300 m above the bottom. The interface is approximately at 250 m above the bottom. The bottom layer approximately 70 m thick is well mixed. The bottom layer is separated from the upper layer by an interface 120–180 m thick. The typical speed of the overflow current is 50–100 cm s^{-1}, the maximum currents reach 135 cm s^{-1} when the current accelerates down the slope of the Faroe Bank Channel to the ocean (Fer et al. 2010). The mean direction of the flow along the Faroe Bank Channel at the sill is to the northwest (304°).

McKenna et al. (2016) analyzed propagation of Arctic waters to the North Atlantic and indicated that only Norwegian Sea Arctic Intermediate Water (NSAIW) and Norwegian Sea Deep Water (NSDW) propagate through the conduit of the Faroe Bank Channel to participate in the Atlantic thermohaline circulation. Mixing processes in the Faroe-Shetland Channel result in an exchange of properties between the inflow and outflow in the channel. The flow of colder Norwegian Sea Deep Water occurs closer to the Faroe Plateau, thus causing the previously mentioned inclination of isotherms.

Belonenko et al. (2014) analyzed the general circulation of the Norwegian Sea and tried to explain the contradiction in the previous interpretation of data. Some previous researchers reported about general cyclonic circulation in the Norwegian Sea and especially in the Lofoten Basin. At the same time, detailed measurements by other researchers (Ivanov and Korablev 1995a; Köhl 2007) revealed an anticyclonic vortex in the center of the Lofoten Basin with a center at 70° N, 2° E.

Köhl (2007) writes that the Lofoten Basin is a region of high mesoscale activity. This region is the major heat reservoir in the Nordic Seas and is therefore a region of strong atmosphere–ocean interaction, which is especially high in the Lofoten Basin (Poulain et al. 1996). A unique natural phenomenon is located in the center of the Lofoten Basin in the Norwegian Sea. It is a quasi-permanent anticyclonic Lofoten Vortex. Russian field hydrographic surveys in the 1970s and 1980s planned to resolve the synoptic scale of this anomalous region and revealed a quasi-permanent anticyclonic vortex (Ivanov and Korablev 1995a). Ivanov and Korablev (1995b) describe this vortex as topographically controlled circulation along *f/H* contours. The vortex is a doubly convex lens usually smaller than 50-km in horizontal scale, which intensifies in winter (Fig. 1.10). Köhl (2007) studies the generation mechanism and the conditions for the stability of a quasi-permanent vortex in the center of the Lofoten Basin using a high-resolution ocean circulation model and altimeter data. The vortex manifests itself by pronounced sea surface height. This quasi-permanent vortex is formed from anticyclonic eddies that separate from the eastern branch of the Norwegian Atlantic Current. The large-scale bottom depression of the Lofoten Basin plays a crucial role for trapping anticyclones into the deep basin and enabling the dynamical stability of the vortex. Strong heat loss occurs in winter. The cold water in the upper part of the vortex preconditions convection in the following winter. This process deepens the convection in the upper part of the vortex.

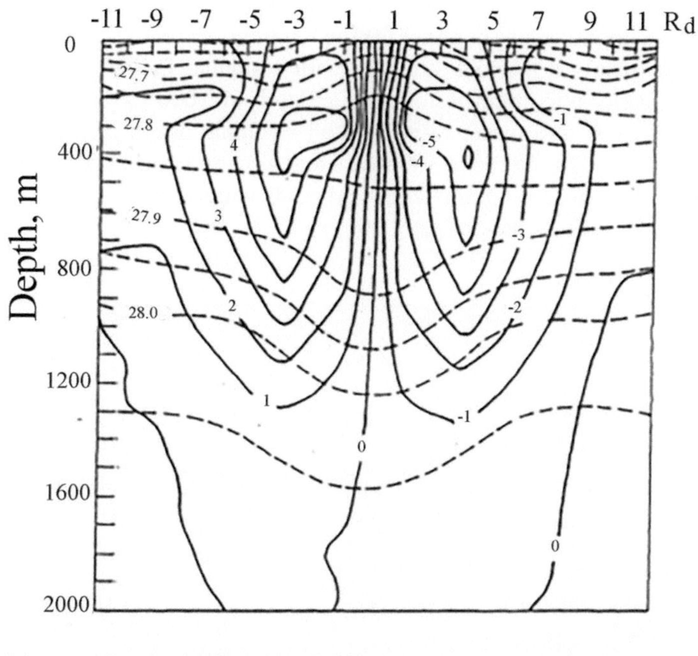

Fig. 1.10 Scheme of the distribution of potential density (sigma-t, dashed line) and geostrophic velocity relative to the 2000 m reference level (solid line, cm/s) over the mean vertical section across the Lofoten Vortex. R_d is the baroclinic radius of deformation, positive velocities are directed into the figure (Ivanov and Korablev 1995a, b)

Ivanov and Korablev (1995a, b) suggested that winter convection is the mechanism of the vortex generation. They suggested that the vortex originates primarily from anticyclones that were originally formed due to the instability and meandering of the Norwegian Atlantic Current and then propagated into the Lofoten Basin. Another mechanism of eddy formation was suggested in Bashmachnikov et al. (2018): the eddies are generated on the slopes of the basin when the current velocity is low. The role of convection according to the opinion of Köhl (2007) is mainly in setting the conditions in the upper part of the lens-like hydrographic structure. Rossby et al. (2009b) support the idea of Köhl (2007) that eddies separated from the Norwegian Atlantic Current maintain the energy of the Lofoten Vortex.

Travkin and Belononko (2019) write that the Lofoten Vortex is a remarkable dynamical phenomenon in the center of the Lofoten Basin. It is a quasi-permanent anticyclonic eddy, the core of which is represented by a lens of warm and saline water in the 300–1000 m depth with a horizontal scale of 100 km. The Lofoten Vortex extending to a depth of 3250 m was discovered during the hydrographic studies of the Arctic and Antarctic Research Institute in 1970s–1980s (Alexeev et al. 1991; Ivanov and Korablev 1995a, b), and it is currently one of the most studied vortexes in the global ocean, which is a strong quasi-stationary anticyclonic eddy in the center of the Lofoten Basin. A section of potential temperature along 69.8°N in August 2003 from (Köhl 2007) (Fig. 1.11) shows the double convex lens structure of the vortex with the center at about 600-m depth and a diameter of about 30 km. The vortex is mainly a thermal structure with a smaller haline contribution. The subsurface velocity maximum is found at a depth of ~600 m. High temperatures and salinities below 500 m are caused by anticyclonic origin of the eddy. Lower temperatures and salinities above 500 m are due to winter convection. The vortex is interacting with the surrounding eddies and is also advected by the mean cyclonic circulation in the Lofoten Basin (Köhl 2007).

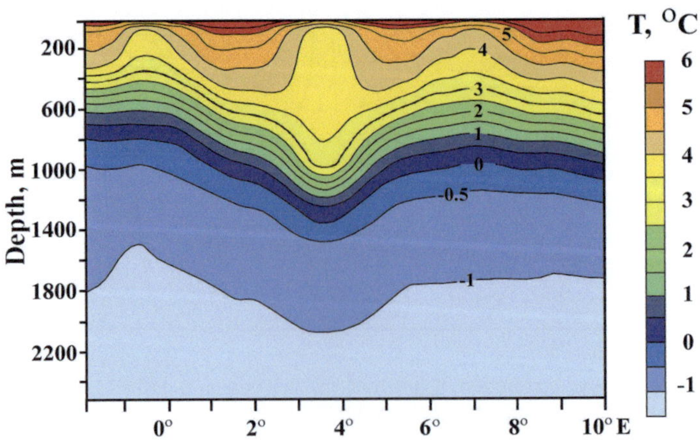

Fig. 1.11 Potential temperature along 69.8°N in August 2003, (Köhl 2007)

Raj et al. (2015) describe the quasi-permanent anticyclonic vortex on the basis of hydrographic and satellite observations. The authors analyze the vortex together with the main forcing mechanisms. The influence of the vortex on the hydrography of the Lofoten Basin is also shown. Atlantic Water in the Nordic Seas penetrate deep into the Lofoten Vortex and maintain the persistent existence of the vortex in the deepest part of the Lofoten Basin. They found its cyclonic drift and revealed seasonality in its eddy intensity with the maximum during late winter and minimum during late autumn. Two main characteristics of the Lofoten Vortex, which distinguishes it from other eddies in the Lofoten Basin are its location and persistence. Tracking of the center of the Lofoten Vortex confirms it to be persistently residing in the deepest part of the Lofoten Basin. Eddy merging processes are studied, and merging of eddies from the slope current is revealed to provide anticyclonic vorticity to the Lofoten Vortex.

The authors of Fedorov et al. (2019) use the Argo data, numerical modeling, and satellite altimetry to study the vortex. The numerical simulation is based on the Massachusetts Institute of Technology (MIT) ocean general circulation model (Marshall et al. 1997). The depth of the mixed layer based on modeling and Argo measurements can be as high as 800 m. Fedorov et al. (2019) report that winter deep convection is a necessary condition for its stability and is one of the main mechanisms of its existence because deep convection forms favorable conditions for its annual regeneration. Another mechanism that maintains high anticyclonic vorticity in the center of the basin is entrainment of mesoscale eddies that break away from the Norwegian Atlantic Current. The region of the Lofoten Vortex coincides with the region of the highest mixed layer depths; hence the existence of deep mixed layer proves conditions for deep convection and maintains the existence of an anticyclonic eddy in the center of the basin.

Fedorov et al. (2019) (Fig. 1.12) show the θ,*S*-curves for the profiles in the Lofoten Vortex and beyond it. It is clearly seen that the set of θ,*S*-curves is much less uniform outside the vortex compared to the profiles located inside the Lofoten Vortex. The differences and diversity of the θ,*S*-curves characterizing the profiles outside the Lofoten Vortex are not surprising, because these profiles were measured over the entire Lofoten Basin with the exception of the selected vortex basin.

Figure 1.13 shows the spatial distribution of the upper mixed layer (UML) depths based on the simulations using the MIT model over the season in January–April. One can see from Fig. 1.13 that the maximum upper mixed layer (UML) depths are recorded in the region of the Lofoten Vortex. The maximum UML depth based on the Argo data is located at approximately 3° E, and that based on the MIT model simulations is at 4° E. The UML depths of 500–600 m are predominantly located in the immediate vicinity of the Lofoten Vortex, where the greatest ocean bottom depths are also found. In the rest region of winter convection, the UML depth is small and amounts only to 100–300 m. Similar UML estimates were also obtained using the MIT model data (Fig. 1.13), but in general, it should be noted that they are somewhat overestimated compared to the UML estimates based on the Argo data. However, according to the MIT estimates, the region of maximum depths exceeding

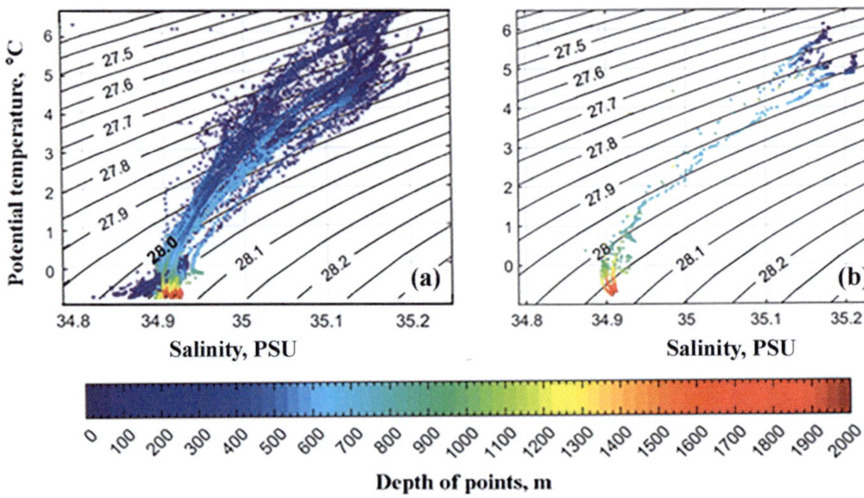

Fig. 1.12 Graphs of θ,S-curves of Argo profiles (April, 2002–2017) located outside and inside the Lofoten Vortex measured in April. The color shows the depth of the point on the profile (Fedorov et al. 2019)

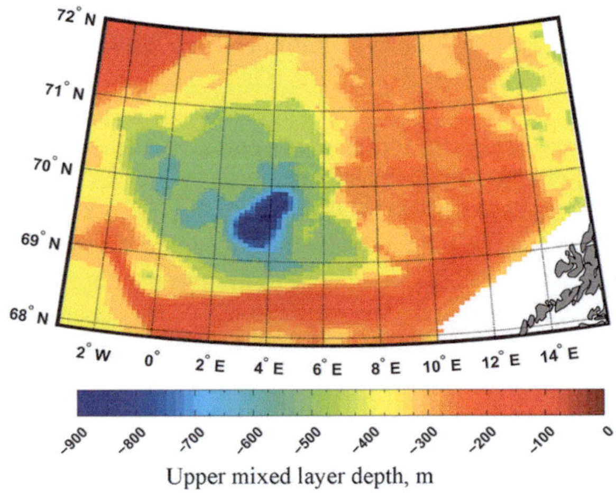

Fig. 1.13 Spatial distributions of maximum UML depths (m) based on the MIT Global Circulation Model data (profiles were constructed for each grid point) in 1993–2012 (Fedorov et al. 2019)

800 m is located approximately 1° to the east of the region of the maximum UML depths identified using Argo data.

The largest number of profiles with UML depths exceeding 300 m, based on the MIT estimates were observed in March, and the maximum of such measurements based on the Argo float data was in February. At the same time, the UML depths greater than 500 m were estimated only as 1% in March and April based on the MIT data in March, and according to the Argo float data they were measured also in March and April. From June to November deep convection is absent in the Lofoten

Basin. Unfortunately, the number of Argo profiles in this region is still insufficient to obtain a representative picture, from 2001 to 2017 only 12 profiles with UML depth exceeding 1000 m were observed.

Belonenko et al. (2021a) analyzed a series of numerical experiments to study the influence of bottom topography on the properties of the Lofoten Vortex. In the experiments they everywhere used the same lateral and surface boundary conditions, but with different bottom topography of the Norwegian Sea. They found that a bowl-shaped bottom topography is a necessary condition for the formation of the vortex; the existence of a topographic boundary (Helgeland Ridge) between the Lofoten Basin and the Norwegian Basin in the western part of the basin is crucial for the generation of the vortex because the eddies do not leave the Lofoten Basin towards the adjacent Norwegian Basin in the southwest. When the boundary between the basins is removed, a quasi-permanent anticyclonic vortex is formed in the Norwegian Basin.

Koszalka et al. (2011) and Søiland and Rossby (2013) write that the Lofoten Vortex is a small eddy, about 1/10 the scale of the Lofoten Basin. Analysis is based on 75 kHz vessel-mounted acoustic Doppler current profiler, CTD casts, and six RAFOS floats. The radius of its core of homogeneous water is 7–8 km (comparable to the local radius of deformation. The eddy reaches deeper than 1000 m. The measurements show that the core has a solid body rotation with a relative vorticity very close to the Coriolis parameter. The maximum orbital velocities are as high as 80 cm/s at 18 km radius. Yu et al. (2017) write based on 3-year measurements with gliders that the mean radius of the Lofoten Vortex is 18 $\pm$ 4 km and it propagates cyclonically with a mean speed of approximately 3–4 cm/s. The anticyclonic azimuthal peak velocity varies between 0.5 and 0.7 m/s located between 700 and 900 m depth.

Heat loss to the atmosphere in winter due to thermal convection clearly plays the key role in weakening the eddy over time. However, the float data do not show the eddy spinning down in winter; on the contrary, there is an indication the Lofoten Vortex may reintensify in the winter months through penetrative convection that deepens and sharpens the underlying pycnocline. The supply of heat is advected to the eddy by anticyclonic eddies separated from the eastern branch of the Norwegian Atlantic Current. The geostrophic transport relative to a 2500 m reference level is 25 Sv (Søiland and Rossby 2013). The authors also showed ADCP data when crossing the vortex. The velocity vectors measured using the onboard profiler are shown in Fig. 1.14. The authors also demonstrated rotation of float 931 in the vortex during 11 months since its deployment in the vortex at a radius of 18 km at the radius of maximum velocity. Figure 1.15 shows the 11-month trajectory of the float that looped with high velocity near the maximum radius for the first six months, after which moved to the periphery with slower velocity of the vortex and then escaped about two months later (Fig. 1.16). Temperature distribution across the vortex is shown in Fig. 1.16 based on CTD casts. The section shows a typical temperature distribution in an anticyclonic eddy.

The authors of Bashmachnikov et al. (2018) present two patterns of vertical structure of isopycnals in the Lofoten Vortex in the warm and cold seasons (Fig. 1.17). Typical warm season structure is characterized by a doubly convex lens. The upper

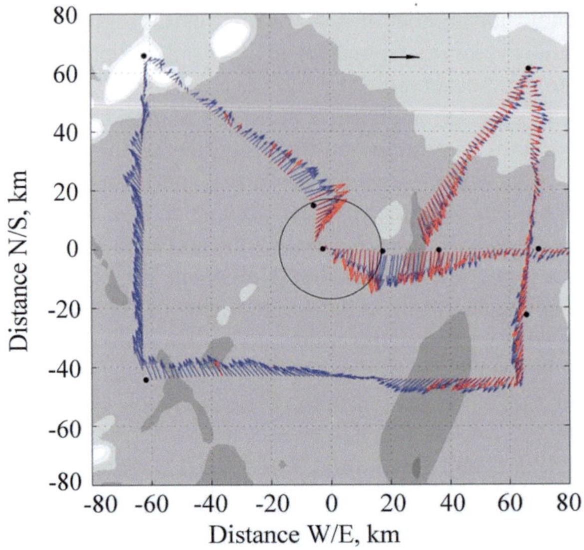

Fig. 1.14 SADCP velocity vectors along the ship track in the vortex region in July 2010 (blue vectors from 44 m depth, and red vectors from 700 m, respectively. The 17 km radius circle marks the maximum orbital velocities. The black dots indicate CTD stations and the vector in the top right indicates 0.5 m/s. Bathymetric contours in 50 m steps are shown in grey color from 3150 to 3350 m (light to dark). Origin in the figure is at 70.07N, 3.65E (Søiland and Rossby 2013)

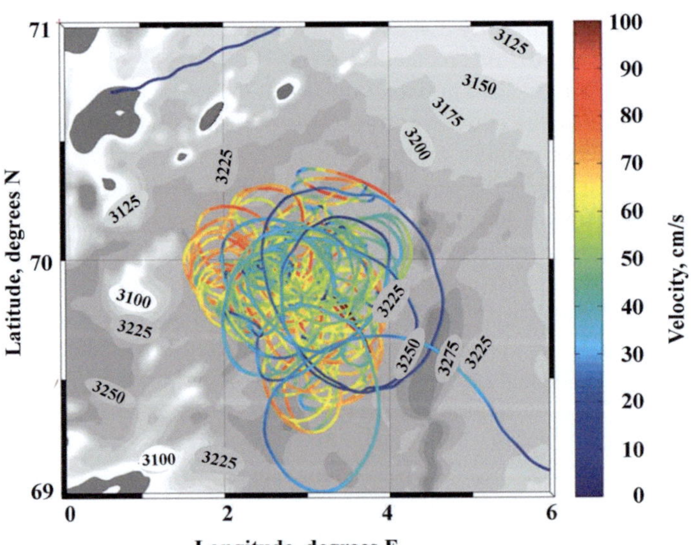

Fig. 1.15 Trajectory of float 931. Smoothed twice-daily tracking data is shown. Velocity of the float is shown in color. The color scale is on the right, cm/s. Bottom topography is in grey (Søiland and Rossby 2013)

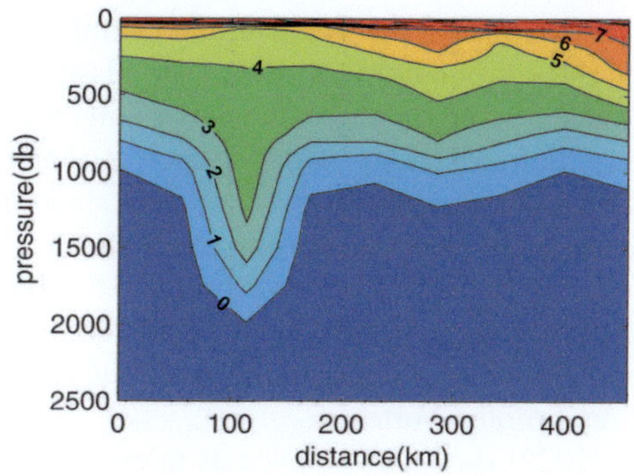

Fig. 1.16 Temperature section of nine equally spaced CTD casts (tics on the bottom axis) along 69.5° N from –0.92 to + 10.8 E (Søiland and Rossby 2013)

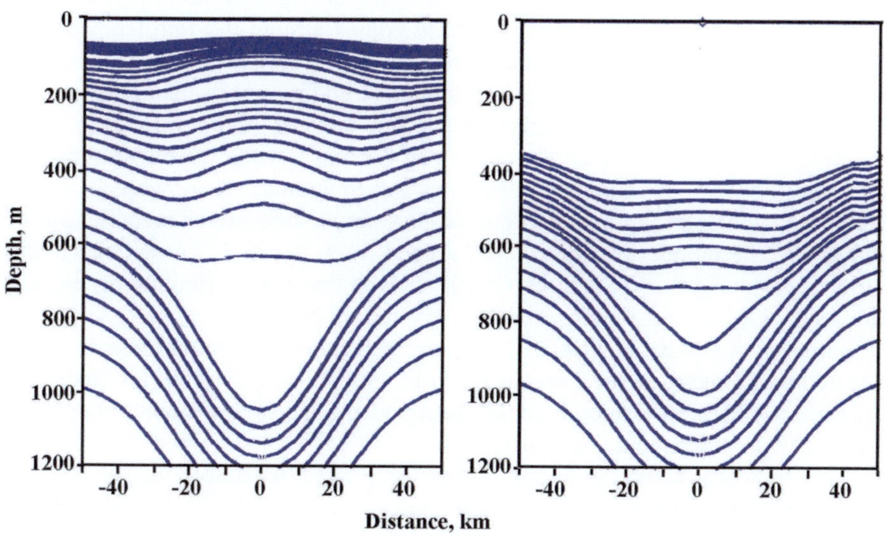

Fig. 1.17 Mean isopycnal depths (blue lines) across the Lofoten Vortex as a function of distance from the vortex axis, presented for two typical configurations of isopycnals for: **a** warm season, **b** cold season (when deep convection in the Lofoten Vortex core is detected) (Bashmachnikov et al. 2018)

isopycnals are turned upwards and those in the deeper layers turn to the bottom. In the cold season deep convection bends and displaces isopycnals down.

Bosse et al. (2019) analyzed the properties of the Lofoten Vortex. The vortex core, in nearly solid-body rotation, exhibits an unusually low vertical vorticity close to the local inertial frequency. Subsurface floats as deep as 800 m are trapped within the core for their entire deployment duration (up to 15 months). The potential vorticity is reduced in the core by two orders of magnitude relative to the surroundings, creating a barrier. In winter, this barrier weakens and lateral exchanges and heat flux between the eddy and the surroundings increase. The dissipation timescale for the eddy energy was estimated based on a simple energy budget as three years, during which wintertime convection seasonally modulates potential and kinetic energy. This conclusion is supported by Santeva et al. (2021); they show that the slope of the bottom and the increase in the steepness of the basin slopes to the southwest (the region of splitting of the Norwegian Atlantic Current into two branches) significantly increases the stability of the eddy. This relatively small variability of the topography forms a region of small gradients of background potential vorticity, coinciding with the mean position of the Lofoten Vortex.

Zhmur et al. (2021, 2022) show that isopycnals in cyclones and anticyclones are generally deformed in opposite directions. Cyclonic eddies bring isopycnals closer together, drawing the nearest outer isopycnals into the vortex core, while anticyclonic eddies move isopycnals apart. The latter leads to an increase in the Brunt-Väisälä frequency in the cyclone core and a corresponding decrease in the anticyclone.

Travkin and Belononko (2019) analyzed eddies around the Lofoten Vortex in 1993–2017. Seasonal variability of mesoscale eddies in the Lofoten Basin was manifested both in the seasonal change of number of eddies and in the corresponding change of their thermohaline characteristics. The number of anticyclones in winter and in spring was greater than the number of cyclones, while there was almost no difference between the numbers in summer and in autumn. There were about 235 cyclones in winter and spring; their number in summer and autumn increases almost up to 280. The number of anticyclones ranged from 262 to 277 based on satellite altimetry. Lundrigan and Demirov (2019) showed that the distribution of the eddy radii in the Lofoten Basin ranges between 19 and 25 km.

Gordeeva et al. (2021) write that only one percent of the total mesoscale eddies detected by altimetry in the Lofoten Basin in 1993–2017 (>11,550) are long-lived (lifetime > 35 days). Cyclonic eddies surround the quasi-permanent anticyclonic Lofoten Vortex forming a shield around it.

Zinchenko et al. (2019) analyzed tracks of long-lived (lifetime more than 35 days) individual 120 cyclonic and 210 anticyclonic eddies in 1993–2017. They showed based on satellite measurements that two predominant systems of mesoscale eddy formation exist in the Lofoten Basin. This fact may reflect two different mechanisms of eddy generation. The first one is the separation of eddies from the Norwegian Current during its meandering. The second one is the direct generation of eddies inside the Lofoten Basin. There are three distinct regions of eddy generation in the frontal zone of the Norwegian Current, from where eddies move to the west and northwest, forming three main trajectories. Anticyclonic eddies dominate in the region of the Lofoten Vortex. They rotate intensively within the Lofoten Vortex in the anticyclonic gyre and interact obviously with the Lofoten Vortex by merging. Cyclonic eddies are localized in the vicinity of two points with centers of 69.5° N, 4.0° E, and 70.0° N, 2.5° E.

Raj et al. (2016) investigated mesoscale eddies in the Lofoten Basin based on the satellite altimetry data (1995–2013) together with Argo profiling floats and surface drifter data and found that mesoscale eddies in the Lofoten Basin are predominantly generated and reside locally. Abundance of eddies is observed in two main locations, over the deep depression in the western Lofoten Basin, and along the slope current in the east. Anticyclonic eddies live longer than cyclonic eddies in the Lofoten Basin. Long lived anticyclones reside mainly in the western Lofoten Basin.

Sandalyuk et al. (2020) analyzed the 3D structure of eddies in the Lofoten Basin. Anticyclones in the Lofoten Basin are warmer and saltier in their southwestern sector compared to their northeastern side. Cyclones are colder and less salty in their south-eastern side. It is likely that this feature of mesoscale eddies in the Lofoten Basin reflects the properties of the large-scale temperature and salinity gradients in the sea, which are strongly affected by the warm and salty flow of the Norwegian Atlantic Slope Current.

The authors of Belonenko et al. (2021b) used a model of simulated Lagrangian particles and GLORYS12V1 reanalysis in June 2016 to track the vortex pair consisting of the large quasi-permanent Lofoten Vortex and a cyclone generated nearby the Lofoten Vortex. The cyclone was actually part of a shield around the Lofoten Vortex. This structure formed an uneven vortex pair. The cyclone moved around the Lofoten Vortex passing a distance of 350 km along its periphery with a speed of 14 cm/s, which is comparable to the orbital velocity of the Lofoten Vortex during the study period. The orbital velocity of the satellite cyclone was 15 cm/s.

Volkov et al. (2013) analyzed satellite altimetry data and revealed a cyclonic wavelike propagation of the synoptic-scale sea surface height anomalies around the center of the Lofoten Basin. They identified dipole and quadrupole wavelike modes of propagation. The phase speed of the waves ranged from 2 to 10 km/day and their wavelength was approximately 500 km. The dispersion analysis suggests that the observed propagation is a manifestation of barotropic topographic Rossby waves.

Volkov et al. (2015) modeled generation of the Lofoten Vortex using high-resolution numerical experiments with the mean horizontal grid spacing of 18 km, 9 km, and 4 km. Based on the model simulations, the Lofoten Vortex did not form in the18-km experiment. The most realistic (compared to the observations) simulation of the vortex was provided by the 4-km experiment, which better reproduced eddy variability in the region. Thus, the experiment provided evidence of the importance of eddies in the formation and stability of the vortex. The model revealed how anticyclonic eddies that are usually stronger and more numerous in the basin than cyclonic eddies contribute to the intensification of the Lofoten Vortex. The Lofoten Vortex itself is not stationary and drifts anticlockwise within the basin bounded by approximately the 3250 m isobath. The analysis of the barotropic vorticity budget in the 4-km experiment shows that the advection of the relative vorticity gradient by the eddies is the main mechanism that drives the variability of the Lofoten Vortex.

This mechanism initially suggested in Köhl (2007) is confirmed in the paper (Trodahl et al. 2020) based on modeling. Observations report the existence of a long-lived anticyclonic vortex in the Lofoten Basin. Despite repeatedly recorded existence of intense anticyclones, the observations cannot firmly decide whether this is the signature of a single vortex or a succession of ephemeral vortices. A vortex persisting for decades requires some energy supply mechanism. Wintertime convection and vortex merging have been proposed candidates. The authors of (Trodahl et al. 2020) examine the Lofoten Vortex dynamics using a high-resolution regional ocean model. The model is initialized from a coarser state with a weak eddy field. The current of warm Atlantic water flowing along the Norwegian continental slope intensifies and sheds anticyclonic eddies that drift into the basin. Anticyclones shed from the boundary current propagate westward and make their way into the deepest parts of the basin along generally cyclonic routes, in agreement with earlier studies (Volkov et al. 2015). After half a year, anticyclones arrive at the center. Here, they interact with the Lofoten Vortex and some interactions lead to a coalescence. Thus, anticyclones provide the nucleus for a vortex that remains distinct throughout the simulation. Analyses show that this vortex is regenerated by repeated absorption and vertical stacking of lighter anticyclones. This compresses and intensifies the combined vortex taking into account potential vorticity conservation. The vortex becomes more vertically stratified and also losses some water mass during the process. Wintertime convection serves mainly to vertically homogenize and densify the vortex, rather than intensifying it. Further, topographic guiding of anticyclones shed from the continental slope is vital for the existence and reinvigoration of the Lofoten Vortex. In this scenario, the Lofoten Vortex is maintained through repeated merging events. Fluid remains gradually exchanged, although the vortex is identified as a persistent extremum in potential vorticity.

Motion of anticyclones from the Norwegian Atlantic Current to the center of the basin is facilitated by the large-scale bottom depression of the Lofoten Basin. This motion explains the selection of the position of the vortex near the center of the basin. Simulation by Köhl (2007) shows that eastern Norwegian Atlantic Current and Norwegian Coastal Current (NCC) are not stable and produce strong anticyclonic (and cyclonic) eddies.

Raj et al. (2020) report about the interaction between mesoscale eddies and gyre circulation in the Lofoten Basin based on satellite altimetry. The authors revealed distinct cyclonic drift of the anticyclonic and cyclonic eddies. A well-defined cyclonic drift pattern of eddies was found during the period when the gyre circulation of the basin was strengthened, while a clear cyclonic drift of eddies was absent during a weakened gyre. Analysis of barotropic energy conversion in the reanalysis data shows maximum transfer of energy from the eddies to the mean flow in the Lofoten Vortex region. The study shows that the gyre circulation when strengthened, receives more energy from the Lofoten Vortex and loses less energy to those eddies circulating around the Lofoten Vortex.

In addition to the mesoscale Lofoten Vortex another eddy of lesser scale exists in the Lofoten Basin. This is the Moskenstraumen maelstrom near the Lofoten archipelago, which is a tidal whirlpool, one of the strongest in the world. This whirlpool was described in old legends and in a widely known story by Edgar Poe "Descent into the Maelström" (1841), in which a sailor tells how he survived when his ship was drawn into the whirlpool and pulled down.

The Moskstraumen maelstrom is a strong whirlpool dangerous for navigation; it is located in the open sea in the Lofoten Islands off the coast of Norway at 67°48′ N, 12°50′ E between the Moskenesøya and Værøy islands in the southwestern part of the Lofoten Archipelago. A small island Mosken is in the middle of the strait (Gjevik

Fig. 1.18 Photo of the Lofoten Moskstraumen maelstrom

et al. 1997). The Moskstraumen maelstrom is caused by the topography of the seabed. Forcing of water motion is caused by the semidiurnal tide. The currents in the Lofoten Maelstrom are among the world's strongest tidal currents. The current is amplified by the shallow ridge between the Moskenesøya and Værøy islands approximately 8 km long. The speed of the current is as high as 32 km per hour. The vortex reaches the sea bottom (Fig. 1.18). The strait is 40–60 m deep, and is considerably shallower than the surrounding sea. The tide fills up the Vestfjord twice a day, and the difference in height between high and low tides can be up to 4 m at the coast.

The tidal wave moves into the Norwegian Sea. Near the Lofoten Islands, the sea level difference between the open sea and the basin between the islands and Norway (Vestfjorden) drives water between the islands. Gjevik et al. (1997) and Moe et al. (2002) developed a high-resolution depth-integrated tidal model with a horizontal step of 0.5 km. Thus, the model made possible to study the transition of the tide from a progressive wave over the outer shelf to standing wave in the fjords. The simulated amplitude and phase for the M_2 tide constituent (Fig. 1.19) agree with observations at tidal gauge stations. The model revealed enhanced tidal currents between the Lofoten Islands.

The difference in sea level northwest of the Lofoten Islands and southwest of the islands in Vestfjorden is as high as 20–30 cm. This sea level difference forces the current through the passages between the islands. A strong sea level gradient reaching 30 cm appears between the outer and inner line of the islands. This level forces a strong tidal current. The topography enhances the flow. The shelf is relatively broad south of Lofoten and narrow in the northern part. This causes scattering of the tidal wave propagating northward. Numerical simulations of the current field by B. Lynge shows the velocity field near the islands of Moskenesøya and Værøy (Fig. 1.20) (Lynge et al. 2010).

Since the difference in sea level is largest at high and low water the current would be strong around the time of high and low water. The Moskstraumen is at the maximum outflow (northwest) intensity 1–2 h after local high water and strongest inward (southeast) 1–2 h after local low water. This is different from the tidal currents

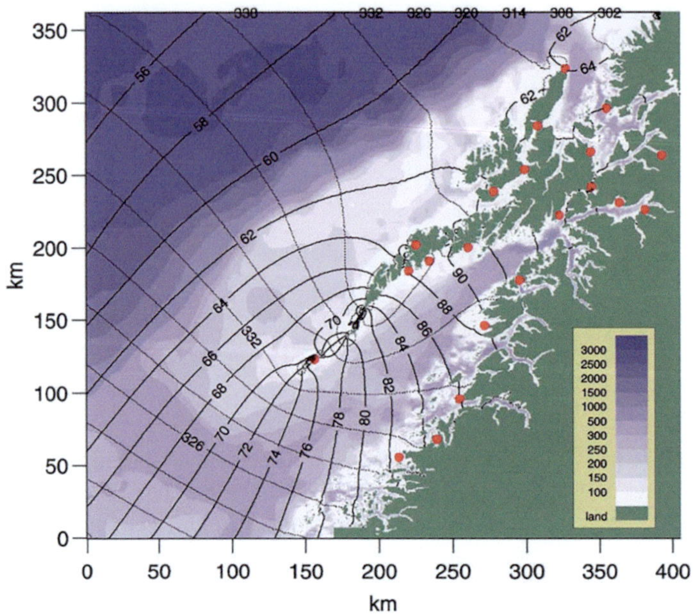

Fig. 1.19 The M_2 sea surface elevation. Solid lines show amplitude in cm with a 2 cm step and phase (broken lines, 2-degrees separation). A 2-degrees step corresponds to a time delay of the time equal to 4 min. Tidal stations are marked by (red) dots. Shading of blue colors shows depth with scale in legend (m) (Moe et al. 2002; Gjevik et al. 1997)

in enclosed fjords where usually the current is weak (slack water) around the time of high and low tide.

Whirlpools are formed caused by the tidal currents between the high and low tide, as the current reaches its maximum. These whirlpools are generated when the water encounters resistance with a landmass or water with different properties. As the current meets this resistance, smaller currents break off from the main flow and become their own independent rotating structures. These whirlpools reach proportions of up to 10 m in diameter and 4–5 m in depth, with speeds of up to 6 nautical miles per hour.

1.3 Bottom Circulation in the Norwegian Sea

Circulation in the Norwegian Sea has been studied in many publications. They are mostly related to the surface currents and intermediate depths. A review can be found in (Filyushkin et al. 2018). Our research deals with the bottom circulation in the Norwegian Sea related to the catastrophic sinking of nuclear submarine "Komsomolets" 30 years ago. There are several publications related to this catastrophe and

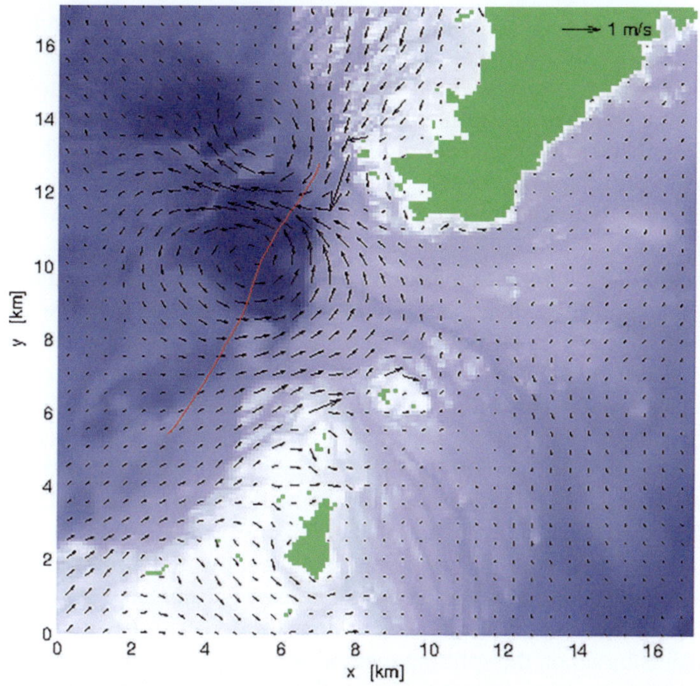

Fig. 1.20 Numerical simulations of the tidal velocities (https://www.farleia-forlag.no/moskstraumen/index_mosk_eng.html) (Lynge et al. 2010; Lynge, 2011)

spreading of currents in the region (Aleinik et al. 1999; Fomin et al. 1996; Aleinik et al. 2002; Lukashin and Scherbinin 2007a; Sidorova and Shcherbinin 2004).

Nuclear submarine "Komsomolets" sank in the Norwegian Sea on April 7, 1989 at a point about 180 nautical miles southwest of Bear Island at a depth of about 1690 m approximately at 73°44' N, 13°16' E (Oceanological Studies… Vinogradov 1996). It is a potential hazard to the environment due to the possible radioactive contamination from the reactor and the nuclear torpedo warheads. There were plans to salvage the submarine totally, to salvage the torpedo part, or to seal off the torpedo part of the submarine. However, the current decision to protect the nature from contamination was to leave "Komsomolets" where it is (Blindheim et al. 1994).

Dangerous radioactive components, which may be dissolved in seawater and spread in the sea, include cesium-137 and strontium-90. Plutonium, which is almost not dissolved in water would settle in the sediments and remain there for long. This research is an attempt to analyze how radioactive elements would spread if a leakage occurs. The study is based on numerical modeling and moored measurements of currents. Previous studies resulted in conclusions that the nuclear submarine is a minor radioactive pollution problem because the amount of radioactive material is relatively small, the wreck sank at a very deep location, and the amount of water for dilution is the whole sea (Blindheim et al. 1994). The possible radioactivity will spread in the water at great depths at isopycnic surfaces. Previous researches revealed that the currents in the region are not uniform but fluctuating (Fomin et al. 1996). Possible radioactive pollution would remain in the deep water and would be diluted slowly. It is not possible that polluted deep water would ascent to the surface from a depth of about 1700 m. The concentration of pollution in the water would not be high even if a leakage occurs. The residence time of the circulation system was estimated as a few centuries (Blindheim et al. 1994). In this analysis we follow (Morozov et al. 2019a; Diansky et al. 2021).

In this study we apply a numerical model to estimate the velocities of bottom currents in the region of catastrophe and compare the simulation results with the direct measurements of currents on moorings and previous results. We applied the Institute of Numerical Mathematics Ocean Model (INMOM) to simulate the circulation in the bottom layer of the Norwegian Sea (Diansky et al. 2002; Diansky 2013; Marchuk et al. 2005; Zalesny et al. 2012). This model is based on the system of the so-called primitive ocean hydrodynamics equations in spherical horizontal coordinates. We applied the hydrostatic and Boussinesq approximations and the vertical σ-coordinate system. The horizontal components of the velocity vector, potential temperature, salinity, and fluctuations of the sea surface height above the mean sea level are the prognostic variables of the model. Seawater density is calculated from the equation of state with the account for water compression (Brydon et al. 1999). The realization of the model is based on the splitting method with respect to the physical processes and, when possible, with respect to the spatial coordinates (Marchuk et al. 2005). Application of this method makes the model different from other well-known ocean models. The thermo-hydrodynamic equations of the model are written in a special symmetrized form so that it is possible to split the operator of the full problem into a set of simpler operators and construct spatial approximations of the relevant groups of summands (in different equations). All the split discrete problems satisfy the energy conservation law, which is valid for the initial differential problem. The simulation domain is between 00°00′ E and 40°00 E and between 69°00′ N and 80°00′ N, or 400 by 220 computational grid nodes. The vertical resolution was determined by the need to simulate the circulation in the bottom layer. The vertical coordinate of the INMOM sigma-model is determined by relation.

$$\sigma = (Z - \zeta)/(H - \zeta),$$

where, z is the traditional downward vertical coordinate, H is the ocean depth at the given point, and ζ is sea surface height relative to its unperturbed state. The application of sigma-levels is especially efficient for the simulation of the bottom circulation and rough bottom topography because it makes modeling bottom oceanic domains possible at significantly variable depths within one simulation (Diansky et al. 2002).

In our simulation, we selected 36 vertical levels; the model resolution increased with depth for better simulation of the bottom currents. The lower 15 sigma-levels were specified in the bottom layer. The distance between them was about 40 m at the depth of the location of the wreck; the vertical resolution depends on the ocean depth, which varied between the coastal regions and 4000 m. The ability of the INMOM model to reproduce circulation with a high spatial resolution under the condition of steep bottom topography is shown in (Diansky et al. 2013).

The bottom topography was taken from the GEBCO2013 database (https://www.gebco.net/data_and_products/gridded_bathymetry_data/). The horizontal resolution of the grid was 30 arc-second. The initial conditions for the temperature and salinity field were specified from the climatic monthly mean data of atlas (Locarnini et al. 2010). The initial velocity fields and sea surface height deviations were set zero (state of rest). These data were also used to specify the annual cycle of temperature and salinity at liquid boundaries. The heat and salt fluxes at the surface were supplemented by the relaxation additives to fit the model temperature and salinity values to the climatic data using the "nagging" method with a coefficient of ~50 m/yr (Diansky et al. 2002).

The results of simulations were compared with the available field observations at several moorings in the region. The measurements were also compared with the data of the CTD/LADCP profilers, which measure profiles of the thermohaline characteristics and current velocities, respectively.

The results of model simulations are shown in Fig. 1.21, which presents the vectors of currents in the bottom layer. The bottom currents were simulated for two seasons in January and July. Both charts show that bottom currents are quite low and generally there are no jets of high velocity in the entire sea except for the bottom stream of the bottom water from the Barents Sea flowing from the Bear Island Trough. This stream turns to the north around Spitsbergen. These conclusions are consistent with the conclusions based on the direct measurements in the Bear Island Trough in 2017 (Frey et al. 2017). A chart of velocity vectors in (Filyushkin et al. 2018) shows vectors of currents at a depth of 1500 m based on the ARGO data. It is worth noting that at this depth individual jets can flow into the Bear Island Trough. Hence, there are compensating bottom currents down the canyon. We also note that cold and dense waters are not formed in the Barents Sea, but they can be formed on the slopes of Spitsbergen.

We also analyzed the results of velocity measurements on moorings deployed in the vicinity of the wreck:

Mooring 1 at 73°42.0' N, 13°10.5' E. The Potok instrument was set at 1674 m over the ocean depth equal to 1687 m. The measurements started on July 27, 1995 and continued up to December 25, 1996.

Mooring 2 at 73°42.6' N, 13°06.0' E. The Potok instrument was set at 1700 m over the ocean depth equal to 1710 m. The measurements started on July 28, 1994 and continued up to July 3, 1995.

Time series of the zonal and meridional components of the currents near the bottom measured on mooring 2 are shown in Fig. 1.22.

Velocity components fluctuated with a period of a few days. The speed generally did not exceed 15 cm/s. The mean velocities over the entire period of measurements on both moorings were within 1 cm/s directed to the northeast. This is confirmed by the progressive vector diagram shown in Fig. 1.23 based on the measurements in August, 1995 at point of the wreck and mooring. Water particles displaced over

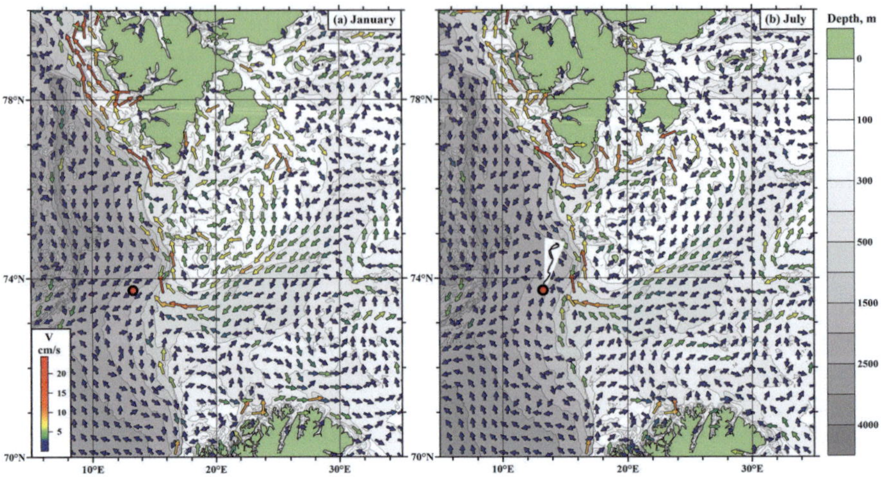

Fig. 1.21 Simulated velocities of the bottom circulation in the Norwegian Sea in January (**a**) and July (**b**). Location of the submarine on the bottom is indicated with a red dot. Progressive vector from Fig. 1.23 is superimposed on the chart in Fig. 1.21b. Adopted from Morozov et al. (2019a). Used with permission of the Russian Journal of Earth Sciences

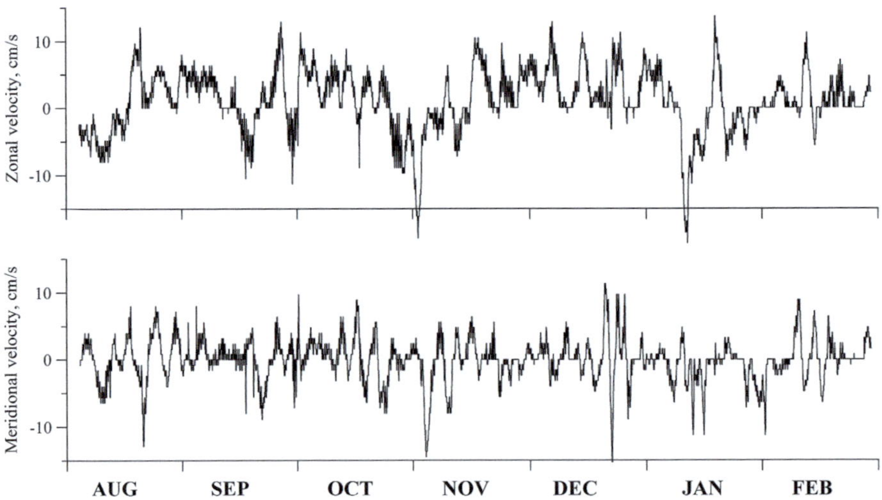

Fig. 1.22 Zonal and meridional components of bottom current velocity on the mooring located at 73°42.6' N, 13°06.0' E at a depth of 1700 m over the ocean depth equal to 1710 m. Adopted from Morozov et al. (2019a). Used with permission of the Russian Journal of Earth Sciences

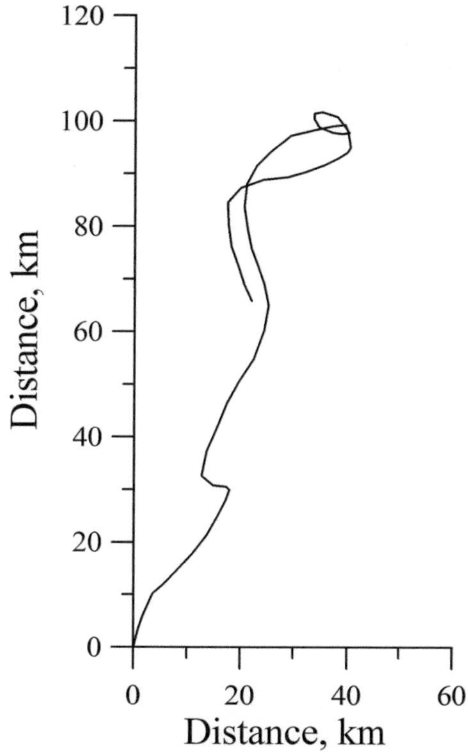

Fig. 1.23 Progressive vector diagram of currents in the bottom layer at mooring at 73°42.0' N, 13°10.5' E during August, 1995 at a depth of 1674 m over the ocean depth equal to 1687 m. Adopted from Morozov et al. (2019a). Used with permission of the Russian Journal of Earth Sciences

a distance not exceeding 100 km during one month. This conclusion coincides with the conclusions made by Aleinik et al. (2002).

Higher frequency fluctuations of velocity can be seen from the analysis of power spectra of these fluctuations. Graphs of power spectra of the zonal and velocity components of currents measured close to the bottom on mooring are shown in Fig. 1.24. A prominent semidiurnal tidal peak is clearly seen on both graphs. The periods of inertial and semidiurnal M_2 periods at this latitude does not make possible their separation. In addition, fluctuations with a period of 5–6 days give a contribution to the fluctuations of the mesoscale range, which can be caused by the internal dynamics of the ocean in this region.

Modeling simulation using the INMOM model with a high vertical and horizontal resolution in the bottom layers revealed that the currents in the bottom layer of the Norwegian Sea are not strong excluding the region of the continental slope of Spitsbergen where a jet of the Arctic waters flowing from the Bear Island Trough turns to the north. The fluctuations of currents with periods of 5–6 days and 12 h are several times greater than the mean currents in the region which are of the order of 1 cm/s. The results of the model simulation with a high-resolution model tuned to simulate the bottom currents confirm the results of previous research that the nuclear submarine on the bottom of the Norwegian Sea is a minor radioactive pollution problem because

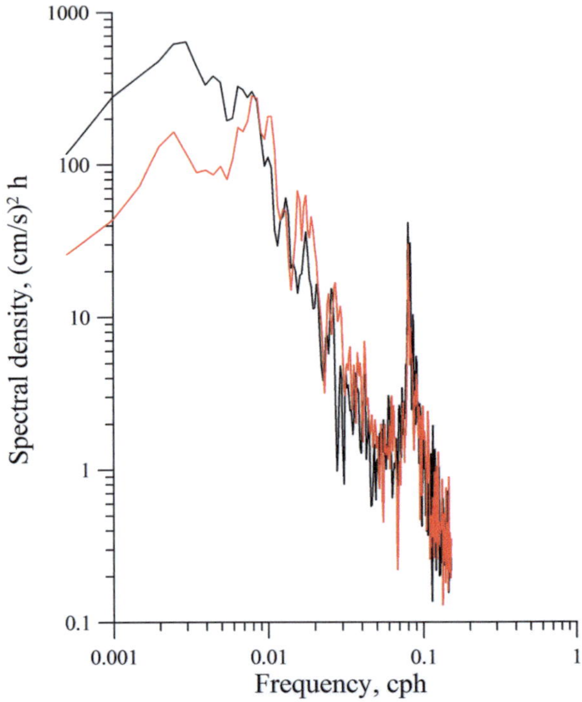

Fig. 1.24 Spectra of zonal (black) and meridional (red) components of velocity fluctuations at a depth of 1700 m over the ocean bottom at 1710 m at 73°42.6′ N, 13°06.0′ E. Adopted from Morozov et al. (2019a). Used with permission of the Russian Journal of Earth Sciences

there is no leakage of the radioactive substance. If unfortunately, such a leakage starts the deep location of the wreck will not produce much harm because the amount of water for dilution is enormous. The possible radioactive pollution would remain in the deep water and will be diluted slowly. It is not possible that polluted deep water would ascent to the surface from a depth of about 1700 m. The currents in the region are not uniform but fluctuating; hence polluted water would only slowly spread from this region.

1.4 Currents in the Barents Sea

The Barents Sea is of particular interest for several reasons. The Barents Sea is the only Arctic region which remains unfrozen throughout the year. During some years the ice boundary in winter passes along 75° N. The inflow of warm Atlantic waters makes the climate of the Kola Peninsula and adjacent regions relatively mild even in winter. The heat transported by the Barents Sea branch of the Norwegian Atlantic Current is effectively lost through intense ocean–atmosphere heat exchange (Häkkinen and Cavalieri 1989; Årthun and Schrum 2010; Årthun et al. 2011; Yashyaev and Seidov 2015), and according to the estimates based on atmospheric observations and oceanic heat budgets about half of the heat loss in the entire Nordic Seas takes place in the Barents Sea (Simonsen and Haugan 1996). The shallow shelf of the Barents Sea is one of the largest in the Arctic Ocean (1.4 × 10^6 km^2), and the deepest (230 m). The ice extent shows large seasonal variations (Kvingedal 2005), which provides favorable conditions for ice formation and subsequent brine release. Therefore, seasonal formation of dense water in the Barents Sea is high (Årthun et al. 2011).

Oceanographic conditions in the Barents Sea strongly depend on the variations in the inflow of Atlantic waters from the Norwegian Sea and on the balance of Arctic waters between the Barents and Kara seas. The outflow transport from the Barents Sea is mainly determined by the outflow of transformed Atlantic water to the Arctic Ocean. Recirculation back to the Norwegian Sea plays a minor role (Loeng et al. 1997).

The main inflow of Atlantic water to the Barents Sea takes place across the Barents Sea Opening (BSO) (Smedsrud et al. 2013), which is the southwestern boundary of the Barents Sea. The major portion of the Atlantic water passes through the Barents Sea and enters the Arctic Ocean through the northern boundary, which is the Northern Barents Sea Opening (NBSO) (Lind and Invaldsen 2012). This is a strait between Novaya Zemlya and Franz Josef Land (Loeng 1991). The eastern strait between Franz Josef Land and Novaya Zemlya is called the Barents Sea Exit (BSX) (Gammelsrød et al. 2009). A bathymetric chart of the Barents Sea is shown in Fig. 1.25.

Climatic Atlas of the Barents Sea (Atlas 1998) presents maps of temperature and salinity of the Barents Sea for each month of 1998. Examples of temperature and salinity in August are shown in Figs. 1.26 and 1.27. These maps correlate well with the chart of mean ice coverage of the sea and location of the Polar Front (Fig. 1.28). These

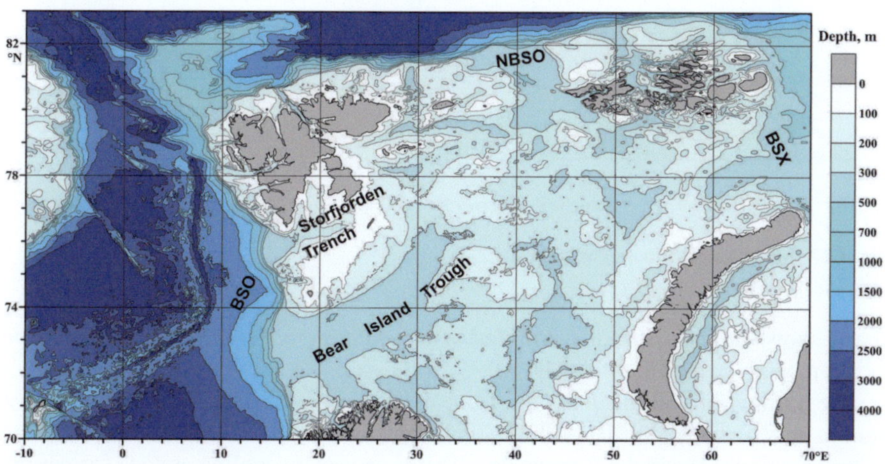

Fig. 1.25 Bathymetric chart of the Barents Sea. Bathymetry is based on GEBCO 2019 data. Abbreviations: Northern Barents Sea Opening (NBSO); Barents Sea Opening (BSO); Barents Sea Exit (BSX)

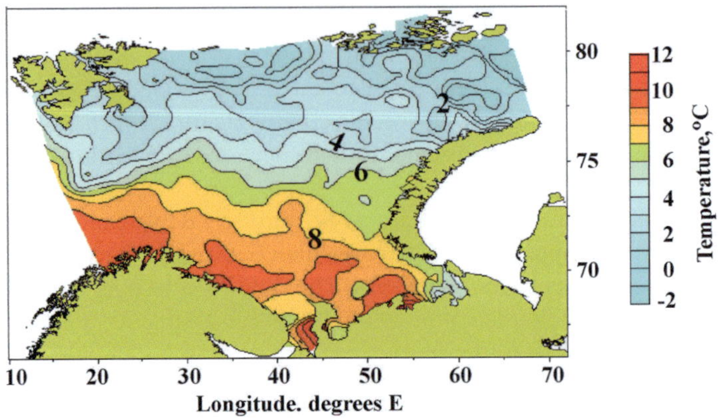

Fig. 1.26 Mean temperature in August (Atlas 1998)

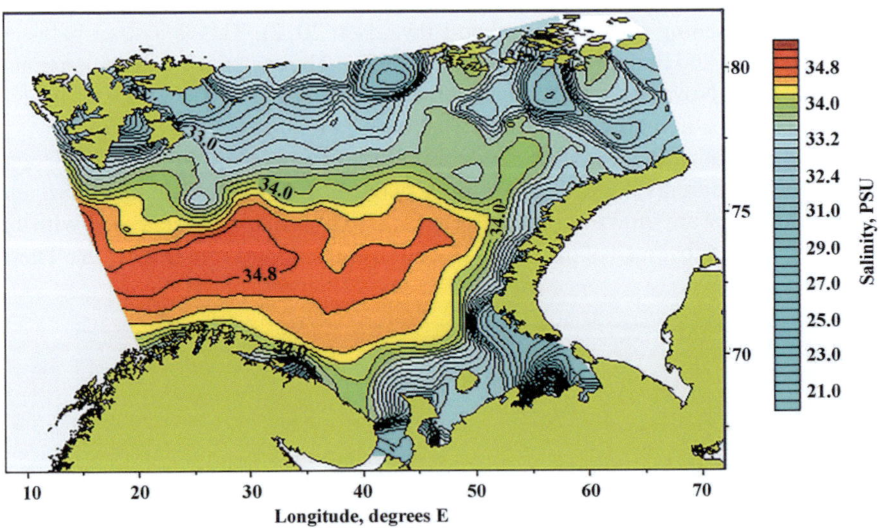

Fig. 1.27 Mean Salinity in August (Atlas 1998)

charts show the intrusion of warm and saline Atlantic waters and their permanent location southwest of the Polar Front.

The outflow of cold deep waters from the Barents Sea occurs through all three boundaries (Årthun et al. 2011). The Atlantic water changes its characteristics during its propagation through the Barents Sea. This is caused by cooling and ice formation, which is described in detail by Midttun (1985). Mixing also takes place while Atlantic waters pass through the Barents Sea. Loeng et al. (1997) presented a map of currents in the Barents Sea and estimated the transport of individual branches (Fig. 1.29).

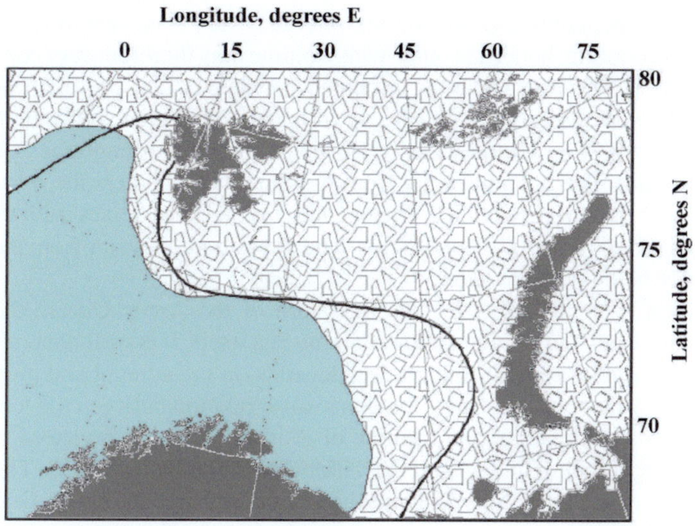

Fig. 1.28 Polar Front in April 1979 (Loeng et al. 1997). The front is shown with a thick black line

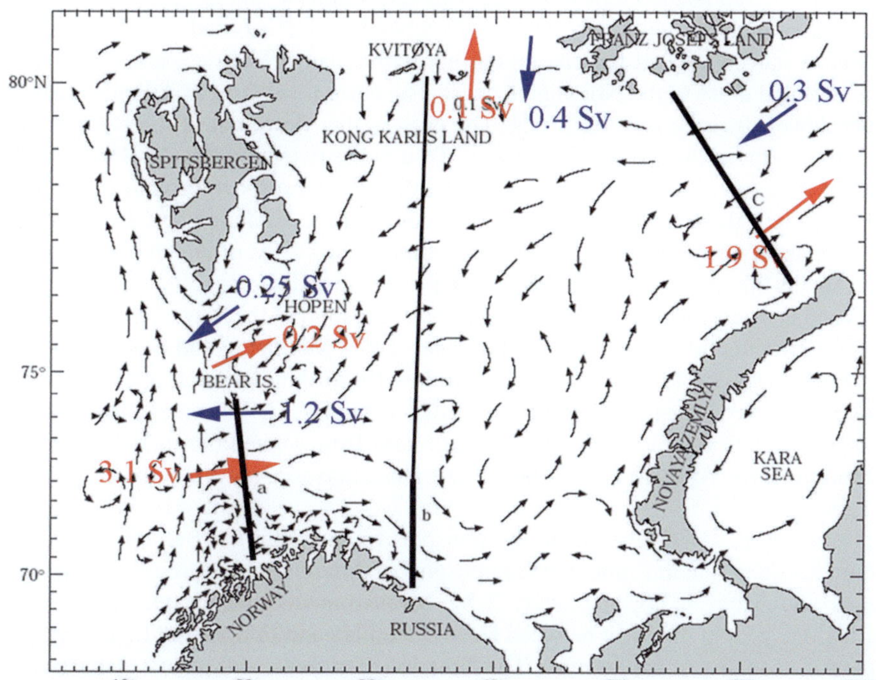

Fig. 1.29 Map of mean currents in the Barents Sea (arrow indicate velocity vectors). Three secular sections in the sea are shown: **a** Fugløya Island to Bear Island, **b** Kola Meridian; **c** Frantz Josef Land to Novaya Zemlya (Loeng et al. 1997)

The rate of production of dense bottom water may vary from year to year due to ice formation and cooling; hence the outflowing volume may therefore vary considerably (Loeng et al. 1997). The generally accepted concept is that the main portion of Atlantic water spreads to the Arctic Ocean from the Norwegian and Greenland Seas through the Fram Strait (Rudels 1987). However, the flow through the Barents Sea is also important. The Atlantic water from the Barents Sea enters the Arctic Ocean at the depth of the lower halocline or, depending on density, mixes with the deeper water layers (Anderson et al. 1994, 2013). The dense bottom water from the Barents Sea probably belongs to the latter.

In addition to contributing to the circulation of the Arctic Ocean, the flow of Atlantic water is also important for the Barents Sea itself. It is well documented that the climatic variability in the Barents Sea depends on the amount and properties of the inflowing Atlantic water. Since there is a close relation between the temperature variations in the Barents Sea and inflow of Atlantic water as reviewed by Loeng (1991), it is important to understand the variability of the Atlantic flow. The temperature conditions in the Barents Sea and the importance for the Arctic Ocean water balance is therefore the rationale for studying the Barents Sea through-flux.

Blindheim (1989) reported the analysis of mean velocities through the section from Norwegian Island Fugløya to Bear Island (section "a" in Fig. 2 in their publication) based on two months (September–October 1978) of moored current measurements. A narrow and strong coastal current with mean velocity exceeding 25 cm/s was revealed close to the Norwegian coast. The current velocity of Atlantic water is much lower; its core is at 72°30′ N. The current that flows out of the Barents Sea over the northern slope of the Bear Island Trough has two components. Based on the data of these measurements Blindheim calculated the transport through the section at 3.1 Sv into the Barents Sea, which includes both the Atlantic and Coastal water. The transport out of the Barents Sea was estimated at 1.2 Sv, of which bottom water contributes 0.8 Sv (Blindheim 1989; 1990).

Russian scientists estimated the geostrophic transport of the flow of Atlantic water into the Barents Sea (Uralov 1960; Timofeev 1963; Moretskiy and Stepanov 1974; Orlov and Poroshin 1988; Potanin and Korotkov 1988). Annual variations of transport based on different estimations do not differ strongly. All estimates clearly show seasonal variations (Fig. 1.30) with a maximum from September to January and a minimum from April to June. The geostrophic transport ranges between 1.2 and 2.1 Sv.

Loeng et al. (1997) estimated the transport to and from the Barents Sea. Their estimates show that the annual mean transport of Atlantic water through the Barents Sea to the Arctic is approximately 2 Sv. Measurements of currents on moorings during one year in the strait between Novaya Zemlya and Franz Josef Land revealed a clear seasonality of the currents. The direction of currents was almost stable, while the speed showed seasonal variability (Loeng et al. 1997). The maximum velocities were observed from November to March; a minimum was detected in autumn. The transport is shown in Fig. 1.31.

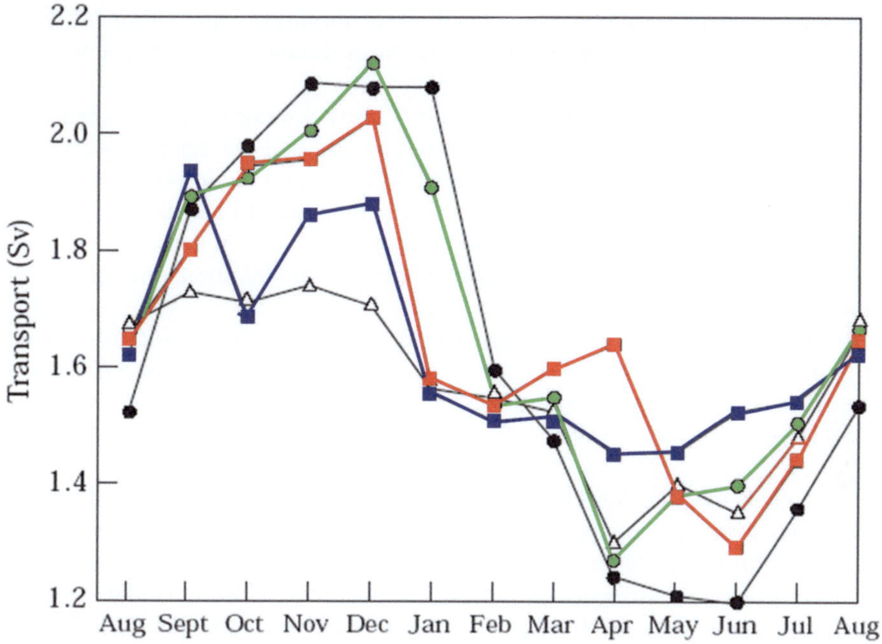

Fig. 1.30 Variations in the geostrophic transport in the Atlantic water inflow to the Barents Sea as estimated by Uralov (1960) (•), Timofeev (1963) (■), Moretskiy and Stepanov (1974) (Δ), Orlov and Poroshin (1988) (■), Potanin and Korotkov (1988) (●); adopted from (Loeng et al. 1997)

Fig. 1.31 Monthly mean volume transport (Sv) through the strait between Novaya Zemlya and Franz Josef Land estimated from the mean current velocities measured by current meters (Loeng et al. 1997)

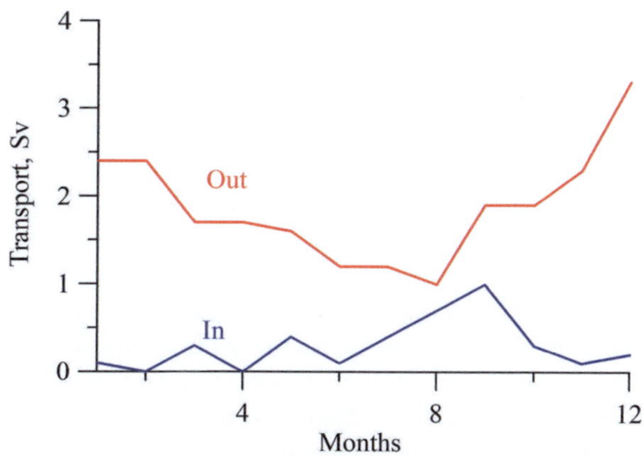

The period of high transport is in November–February with a maximum in December. The transport in December is twice as high as the minimum transport in summer. Geostrophic estimates of the Atlantic inflow to the Barents Sea (Fig. 1.30) support the seasonal variations observed by the current measurements. The geostrophic estimates vary from month to month but a seasonal signal is clearly seen; the ratio between the maximum and minimum inflow varies from 1.3 to 1.9 (Loeng et al. 1997). The balance of measured fluxes through the western and eastern borders of the Barents Sea seems adequate.

However, an increased inflow in the west may also be compensated by an increased outflow through the western boundary, which means that the net inflow does not change so much. It should also be mentioned that there was almost no outflow of dense bottom water between Novaya Zemlya and Franz Josef Land during the winter of 1992. In some years, such as 1983 and 1989, large volumes of dense bottom water from the eastern Barents Sea left the Barents Sea through the strait north of Novaya Zemlya (Loeng et al. 1997).

Estimates of the transport through the strait north of Novaya Zemlya based on geostrophic estimates showed much lower values than those based on current measurement. The outflow was estimated to vary between 0.3 and 0.7 Sv while the inflow varied between low values of 0.2 and 0.5 Sv (Uralov 1960; Pavlov and Pfirman 1995).

The authors of (Matishov et al. 1999, 2010) analyze the characteristics of modern climate trends in water temperature and ice cover in the Barents Sea. They indicate that the leading role in determining the climate of the Western Arctic seas belongs to heat advection by the Norwegian Atlantic Slope Current, which is a continuation of the North Atlantic Current. A stable frontal zone is formed in the Barents Sea due to the spreading of Atlantic waters. This frontal zone separates its ice-free southwestern part from the rest of the water basin with transformed water mass and seasonal ice cover. The authors of (Matishov et al. 2010) present a detailed chart of mean currents in the Barents Sea and the Polar Frontal Zone (Fig. 1.32). An example of the Polar Front position in April 1979 is shown in Fig. 1.28 (Loeng et al. 1997).

The structure of currents is based on the data of oceanographic sections crossing the Atlantic water flow. The position of the Polar Front is best determined by analyzing temperature and salinity measurement data on standard (secular) sections. Three of these sections are shown in Fig. 1.29. The main of them is the Kola meridian section approximately along 33°30′ E. This section is unique due to the duration of its observation period, which began in 1900 and was interrupted in 1907–1920 and 1942–1944, and continued from 1945 to the present. Regular measurements are made south of 74° N. Northern continuation of this section is made rarer.

The Kola Meridian is a good indicator of heat transport to the Barents Sea. Warm cycles were noted more than once in the secular dynamics of the Barents Sea climate. The warm climate phase was most pronounced in the 1920–1930s. It was associated with the climatic phenomenon of the first half of the twentieth century, known as "Arctic warming," which manifested itself in an increase in temperature on the coasts and islands, and a reduction in seasonal ice cover in the Arctic seas (Vinje 2001). Two warming waves were noted over the Kola section, with temperature anomalies up to +1.0 °C in the 1920–1930s. A similar period of warming was revealed, which began in the late 1980s. This indicates a clear increase in the advection of warm waters from the Atlantic to the Barents Sea. A short-term cooling in 1997–1999 was observed against the background of long-term warming (Fig. 1.33). Our data relate to the summer months, but they can rightfully be extended to all seasons of the year,

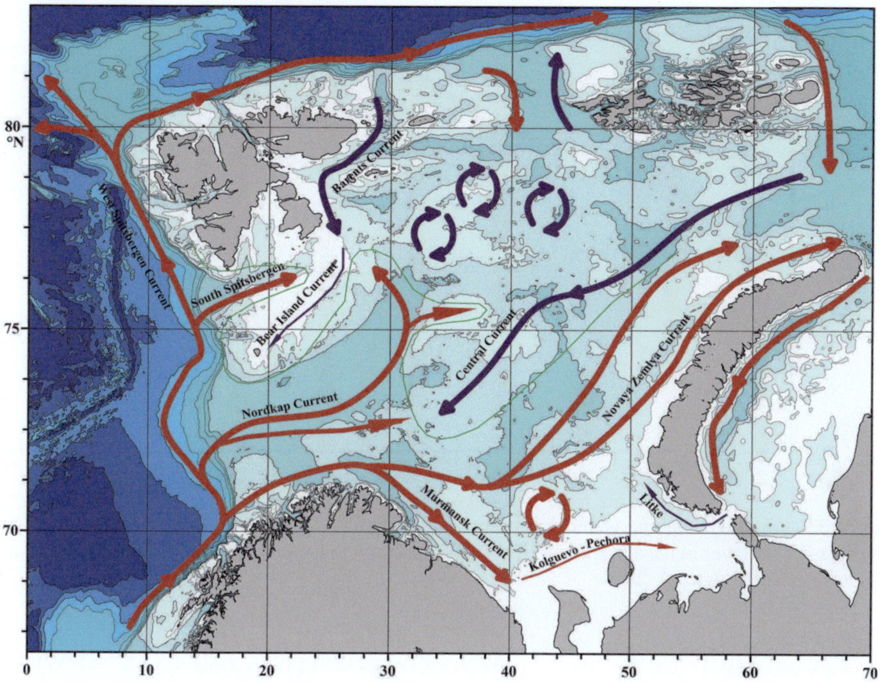

Fig. 1.32 Currents in the Barents Sea based on (Loeng et al. 1997; Matishov et al. 2010)

since the summer thermal regime is the most variable. The standard deviation of the monthly mean temperature of the surface layer in summer in different areas of the Barents Sea varies from 1 to 3 °C, while in winter the difference is within 0.5–1.0 °C (Matishov et al. 2010).

A paper by Harms (1992) describes results of numerical study of currents in the Barents Sea. Distinct evidence of topographic steering was revealed. His estimates of the barotropic transport only in the paper (Harms 1992) were as follows. The Norwegian Atlantic Current follows the topography on the shelf edge and transports 3.6 Sv of Atlantic water. According to his estimates, the major part of the Norwegian Current turns to the north as the West Spitsbergen Current (3 Sv) and only 0.24 Sv turns to the east. However more recent studies indicate that approximately 3 Sv of the flow continues to the east as the Nordkap Current.

McClimans and Nilsen (1993) modeled currents in the Barents Sea on a laboratory setup 5 m in diameter simulating topography and lows through the boundaries. Wind forcing was not simulated. Several physical scales were simulated. As a result, they obtained a realistic pattern of circulation in the sea and its seasonal variations.

Averkiev and Klevannyi (2010) characterize currents in the middle part of the Barents Sea near the Shtokman gas deposit based on observations and numerical model. The three-dimensional model considers all basic factors influencing the hydrodynamics of the sea. The Barents Sea model considered by Averkiev and

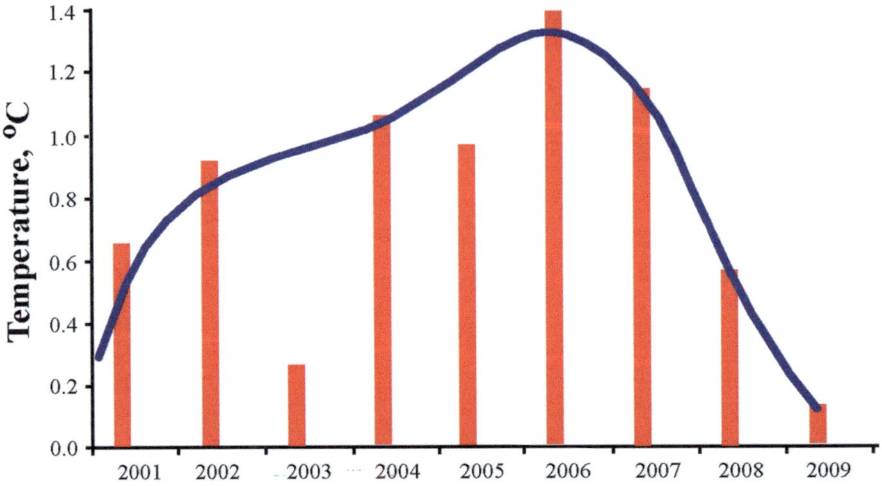

Fig. 1.33 Weighted average water temperature anomalies (°C) over the "Kola Meridian" section in the 0–200 m layer in the summer of 2001–2009 based on the expedition data on the R/V "*Dalnie Zelentsy*" (Matishov et al. 2010)

Klevannyi (2010) was developed with the software complex CARDINAL (Klevannyi and Shavykin 2008). Long-tern measurements at moorings are indicative of the current stability and general direction of the drift. Disturbances are associated with storms and are of an episodic character. Wind is the main factor only in the surface layer. The general drift in the middle part of the sea is directed to the northeast. Based on the data of moorings no northeastward transport occurred here in individual spring and summer months. In the middle region of the sea, annual mean current velocity at the surface was estimated at about 15 cm/s, while near the bottom it was 4–5 cm/s. Tidal currents in some regions of the sea exceed the mean currents by one order of magnitude. The maximum tidal current speed up to 100 cm/s was observed in the region of the Teriberskii Peninsula. The main inflow of Atlantic waters to the Barents Sea occurs across its southwestern boundary. Some water amount near the southern coast of Bear Island returns, but the most part of water passes the Barents Sea and enters the Arctic Ocean through the Makarov Strait between Novaya Zemlya and Franz Josef Land. Current measurements performed throughout the year at three stations in the Makarov Strait demonstrated that the direction of currents here is quite stable and their speed undergoes seasonal oscillations. The outflow of the "dense" Barents Sea waters through the Makarov Strait occurs mainly in the near-bottom layer; in the region of the maximum flow leaving the sea, the speed increases towards the bottom. Warm Nordkap Current is the strongest and most stable current of the Barents Sea. The speed in the surface layer here based on (Yakovlev 1999) reaches 25 cm/s. Two branches, the Northern branch (the speed is about 12 cm/s) and the Coastal branch separate from the Nordkap Current at 25° E; the Coastal branch divides at 30° E into the Murmansk Current flowing eastward with a speed at the surface of 35–45 cm/s and the Murmansk Coastal Current flowing along the Murmansk coast.

1.5 Drifting Ice and Surface Currents on Spitsbergen Bank

A shallow region of the Barents Sea with depths shallower than 100 m between Bear Island and Hopen Island is called the Spitsbergen Bank (Fig. 1.34, left panel). High biological productivity (Henrich et al. 1997) supports the significant fishery activities on the Spitsbergen Bank during the summertime (Gulliksen et al. 2009; ICES, 2019), which is performed mainly by fishing vessels with deadweights below 5000 gross tonnage (GT) (BarentsWatch 2020). The inflow of Arctic water into the shallow Spitsbergen Bank region causes a more extended ice season in this basin compared with the other regions of the Barents Sea at the same latitudes. During the winter, ice cover combined with polar darkness stops plankton production, resulting in deficient food supply for benthic organisms. Therefore, winter fishing is performed at the periphery of the Spitsbergen Bank around the Polar Front (BarentsWatch 2020). Tankers and research vessels perform other shipping activities in the region. Sea ice and icebergs influence the shipping routes in winter; they are also a potential danger for the offshore activity to the southwest of Bear Island.

Below, we review a paper (Marchenko et al. 2022). Cold water transported by the East Spitsbergen Current from the north interacts with warm Atlantic water in the Barents Sea (Loeng 1991). The boundary zone between these waters shown in Fig. 1.34 (left panel) by a double black line is called the Polar Front (Harris et al.

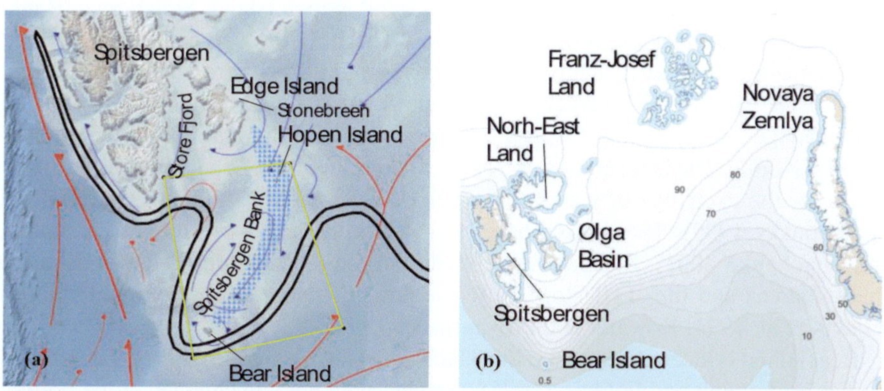

Fig. 1.34 Study region described in this section. The double black line shows the Polar Front between Arctic and Atlantic waters. **a** Left panel: red and blue lines and arrows show directions of the currents of Atlantic and Arctic waters, cloud of blue triangles shows the Tidal Front (Fer and Drinkwater 2014). The yellow square shows the area extending from 24° to 16.5° E, and 74° and 77° N. **b** Right panel: Gray lines show isolines of ice concentrations in April over 1998–2017 (https://kart.barentswatch.no/?epslanguage=no). Adopted from Marchenko et al. (2022)

1998). Salinity and density of Atlantic water are higher than salinity and density of the Arctic water transported into the Barents Sea from the Arctic Ocean by the East Spitsbergen Current through the strait between North-East Land and Franz Josef Land. The temperature of "ideal" Arctic water is close to the freezing point (–1.9 °C), and salinity is below 34.5 PSU. Salinity of "ideal" Atlantic water is around 35 PSU, and temperature is above 0 °C (Oziel et al. 2015). Therefore, Atlantic water tends to occupy deeper ocean layers when it meets with colder and fresher Arctic water. One branch of Atlantic water penetrates into the Barents Sea through the Barents Opening, and the other branch follows the west coast of Spitsbergen and penetrates into the Barents Sea from the north between Svalbard and Franz Josef Land (Lind and Ingvalsen 2012). Polar Front water is saltier than Arctic water but colder than Atlantic water (Fer and Drinkwater 2014; Kedra et al. 2013). The shallow part of the Svalbard Bank to the north of Bear Island is occupied by well-mixed Svalbard Bank water with lower salinity than that of Polar Front water. Mixing is caused by strong tidal currents (Gjevik et al. 1994; Kowalik and Proshutinsky 1995), wind forcing, and convection (Dobrovolsky 1977; Fer and Drinkwater 2014).

Gjevik et al. (1990) revealed a substantial increase of tidal velocities on Spitsbergen Bank with a dominant M_2 constituent and clockwise rotation. Principal axes of the M_2 tidal ellipses extend from east-southeast to west-northwest, and the lengths of their major axes reach 1 m/s on the shallow Spitsbergen Bank and near Bear Island. The semidiurnal components S_2 and N_2 have nearly the same dynamic features as M_2 with amphidromic points to the southeast from Hopen Island (Gjevik et al. 1994). Fer and Drinkwater (2014) specified the Tidal Front (Fig. 1.34, left panel) where the horizontal gradient of density, homogeneous in the water column, exceeded 0.01 kg/m^3 per km, dominated by negative salinity anomalies because of ice melt.

Numerical simulations of Gjevik et al. (1994) and Kowalik and Proshutinsky (1995) showed that clockwise eddies exist around Hopen Island, Spitsbergen Bank, and Bear Island under the influence of tides. The flow-through motion is realized around major bathymetric features and branches into flows interconnecting different regions. The speed of residual currents around Bear Island and shallow Spitsbergen Bank changes in time with the spring-neap period, and the maximal speed is estimated at 8 cm/s. The local eddies prevent residual flow-through and provide pathway routes for the exchange of water properties between various regions (Kowalik and Proshutinsky 1995). The diurnal tide resonates with shelf wave modes (Kasajima and Marchenko 2001). Residual currents due to the diurnal tides occur both in the shallow areas and the shelf slope in the regions of maximum topographic slopes (Gjevik et al. 1994; Kowalik and Proshutinsky 1995).

Bear Island is located near the critical latitude of the M_2 tide, 74.47° N. Theoretical studies by Furevik and Foldvik (1996) demonstrated notable changes in the bottom boundary layer of the M_2 tidal current at latitudes 72°–79° N which can lead to the Kelvin–Helmholtz instability of shear flow in stratified fluid. Analysis of CTD data near the critical latitude for the M_2 constituent in the eastern part of the Barents Sea indicates that the stability is closely linked to the velocity shear. Large vertical shear correlates with low stability and a lower shear with higher stability. Nevertheless, the observed density gradients and the M_2 tidal velocity shear yielded in the values of the Richardson numbers between 16 and 400 when fully developed turbulence does not exist. Furevik and Foldvik (1996) assumed that local values of the Richardson numbers could be higher comparing to the values estimated with vertically averaged and time-averaged density gradients.

We describe the oceanographic and ice conditions on Spitsbergen Bank during the ice season based on analysis of ice maps and results of field work in April 2018 performed from the research vessel *Polarsyssel*. The paper focuses on the investigation of the vertical structure and stability of the tidal current and on the influence of tidal current on sea ice drift in the shallow water region of the bank.

Sea ice was observed and investigated during the research cruises of the R/V *Polarsyssel* in the region of the Spitsbergen Bank on April 22–30, 2017; April 24–30, 2018; and April 23–29, 2019. The investigations were performed in ice covered regions of the bank. Since the Spitsbergen Bank is an open area of the Barents Sea, the ice conditions are very dynamic and weather-dependent. The extension of sea ice on the Spitsbergen Bank on April 27, 2018, is shown in the left panel of Fig. 1.35. The ice map was taken from the web-based resource https://cryo.met.no. The ice tongue extending from Hopen Island to Bear Island is a typical feature of the ice cover on the bank. Right panels in Fig. 1.35 show the development of ice conditions on the bank over five days. A cyclonic ice gyre formed near the underwater hill where minimal water depth reaches 17 m according to the map onboard the *Polarsyssel*. The underwater hill is also known as Kelp forest place (Henrich et al. 1997).

The areas covered by sea ice are associated with water depths shallower than 100 m. Depending on the weather conditions the ice edge may extend to Bear Island or be located north of the island. Winds from the north influence ice drift to the Spitsbergen Bank from the Olga Basin. Winds from the south might displace sea ice from the bank to the Store Fjord and Hopen Island. Winds from the west and east

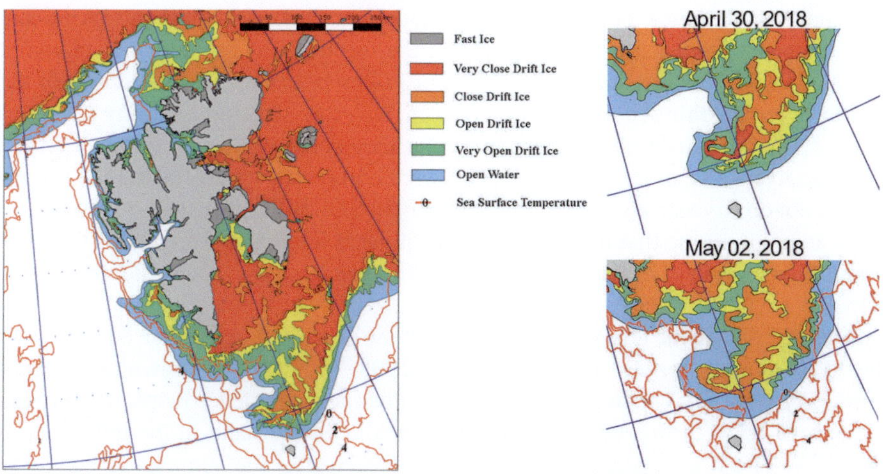

Fig. 1.35 Development of ice conditions on the Spitsbergen Bank from April 30 to May 02, 2018 based on the data of the Norwegian Meteorological Institute (Marchenko et al. 2022)

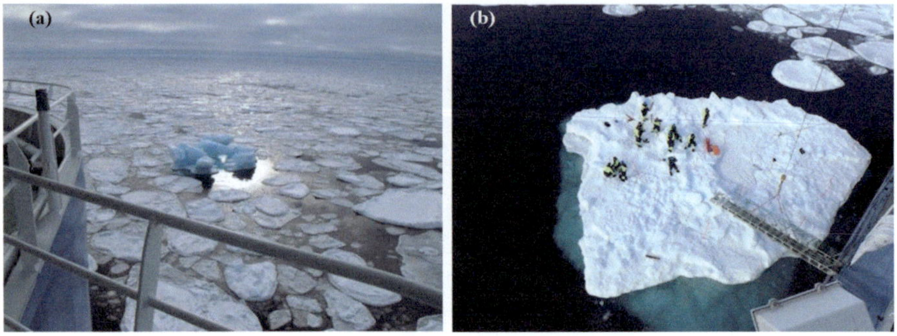

Fig. 1.36 Drift ice and bergy bit (**a**), ice floe with a maximum thickness of 6 m (**b**). Photographs taken from onboard the *Polarsyssel* on April 28, 2018 (Marchenko et al. 2022)

displace ice in the regions of warm Atlantic currents, where the oceanic heat flux to the ice bottom influences ice melt (Fig. 1.34, left panel). Latitude 75° N can be considered as a boundary characterizing the northern position of the ice edge on the Spitsbergen Bank.

Drift ice observed on the Spitsbergen Bank during the cruises consisted of consolidated sea ice floes with drafts up to 9 m and diameters of 20–50 m (Fig. 1.36, right panel); there are relatively large areas with slushy ice, sea ice floes with thicknesses of 30–40 cm and diameters 10–100 m, and bergy bits (Fig. 1.36, left panel) (Marchenko 2019). Thick consolidated floes were found on each cruise, and their occurrences were not rare events. Nevertheless, their number was smaller than the number of floes with 30–40 cm thicknesses. Large areas of slushy ice were observed when the *Polarsyssel* approached shallow areas with depths shallower than 50 m.

Laser scanning and drilling studies were performed to investigate the structure and shape of the thick floes (Marchenko and Marchenko 2019). Analysis of thin sections taken from ice cores showed that the top layer of the floes has almost a granular structure with small elongations of grains in the vertical direction. In the middle layer, the thin sections showed a more columnar structure. Salinity (2–7 ppt) of ice cores taken from the floes showed that they consist of first-year sea ice. The shape analysis suggested that these floes were completely consolidated ice ridges. We did not observe strong interaction between floes of drifting ice on the Spitsbergen Bank and we assume that the observed consolidated ice ridges were not of the local origin. The motion of drifting ice on the Spitsbergen Bank can be considered as free drift.

The blue dots in the left panel of Fig. 1.37 show the days when the ice edge crossed the 75° N latitude north of Bear Island since 1998. The data were downloaded from https://cryo.met.no. This web-based resource does not provide ice maps in 2011, and the lack of data for this year is shown by the solid white band in the left panel in Fig. 1.37. The first appearance of sea ice at 75° N after the summer seasons was recorded at the end of December, and the last observations of ice appearance at 75° N were recorded in June. Most frequently sea ice has been found at 75° N in

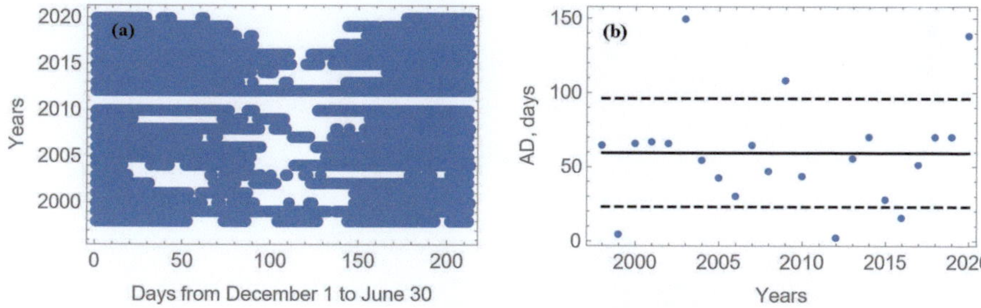

Fig. 1.37 Occurrence of sea ice at 75° N latitude north of Bear Island since 1998. **a** days when sea ice was recorded at 75° N. **b** Annual number of days (AD) when sea ice was recorded at 75° N. Data for 2011 are absent (Marchenko et al. 2022)

March and April. The right panel in Fig. 1.37 shows the annual number of days (AD) when the ice edge crossed the 75° N latitude north of Bear Island since 1998. The AD reached a maximum of 150 days in 2003 and it was equal to 138 in 2020. The minima of AD reached two days in 2012, and five days in 1999. The mean value of AD (60 days) is shown by the solid black line in Fig. 1.37. The standard deviation of the AD represented by the difference between the dotted black lines and the solid mean black line is 36.6 days.

The field measurements were performed on the drifting ice floe and onboard the *Polarsyssel* on April 27–28, 2018. The trajectory of the *Polarsyssel* (moored to the drifting ice floe) during the measurements is shown in the left panel of Fig. 1.38 by the blue line. The field work was performed in shallow water with depths shallower than 35 m near an underwater hill located north of Bear Island. The minimum depth in the region depicted on the navigational maps available on the bridge was 17 m. Seabed bathymetry in the area around the ship trajectory from 09:00 to 21:00 UTC, April 28, is shown in the right panel of Fig. 1.38. The numbers 1–7 show the ship's location at 09:30, 11:00, 13:00, 15:00, 17:00, 19:00, and 21:00 respectively.

An ice tracker (IT) (Oceanetic Measurements Ltd.) was deployed on April 28 in the middle of the floe shown in the right panel of Fig. 1.36. It transmitted Global Positioning System (GPS) coordinates with a 10-min sampling interval by Iridium modem via satellite. Seawater temperature and salinity were recorded by the Conductivity Temperature Depth (CTD) meter SBE-19P over the water column (CTD profiling). The Acoustic Doppler Current Profiler (ADCP) Nortek AWAC 400 kHz was used to record water velocities. The ADCP was fixed on a wire extending from the shipboard to a depth of around 5.5 m; the instrument recorded water velocities with a sampling interval of 10 min and a spatial resolution of 1 m (cell size) over the water column from 6.5 m depth to the bottom.

The trajectory of the IT recorded during April 28 to May 3 is shown by the black curve in the left panel of Fig. 1.38. The blue points in the left panel of Fig. 1.39 show the zonal and meridional components of the IT velocity calculated from the GPS data. The yellow points in the same figure show the components of depth averaged water velocity computed by the TPXO9 tide model (Egbert and Erofeeva 2002) for

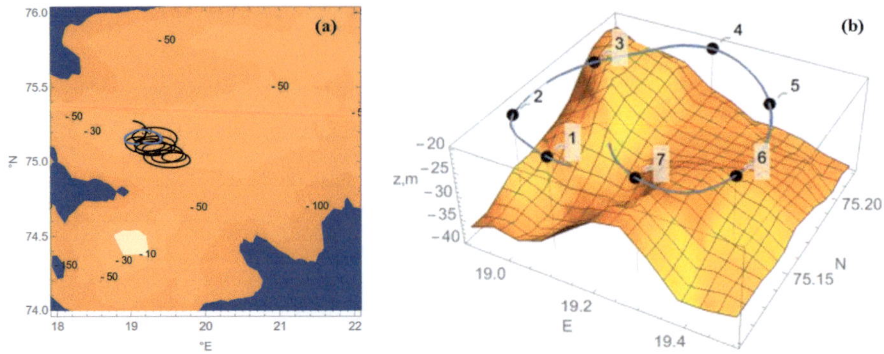

Fig. 1.38 Bathymetry map near Bear Island and trajectories of the *Polarsyssel* (blue line) and the ice tracker (black line) in April 2018 (**a**, left panel). Trajectory of the *Polarsyssel* drift on April 28, 2018 09:00–21:00 UTC, and seabed bathymetry (**b**, right panel) (Marchenko et al. 2022)

the region and the time under consideration. The right panel of Fig. 1.39 shows the IT speed (blue curve) and simulated tidal current speed (yellow curve) versus time. It is obvious that the IT speeds are greater than the simulated tidal current speeds. The mean values of the IT speeds and simulated tidal current speeds were 0.83 m/s and 0.62 m/s, respectively, as shown by the blue and yellow lines in the right panel of Fig. 1.39.

Figure 1.40 shows wind speed and direction measured by the meteorological station on Bear Island. Wind speeds were mainly lower than 6 m/s during April 28–May 01 and increased up to 14 m/s on May 03. The wind direction changed from the north to the west, south and southeast and east directions during April 28–May 01. Then, the wind turned back to the southeast direction during May 02–03. The right panel in Fig. 1.39 shows that fluctuations of IT speed increased during May 02–03 together with the increase in the wind speed. The wind forcing on ice was low in comparison to the water drag during April 28–May 01.

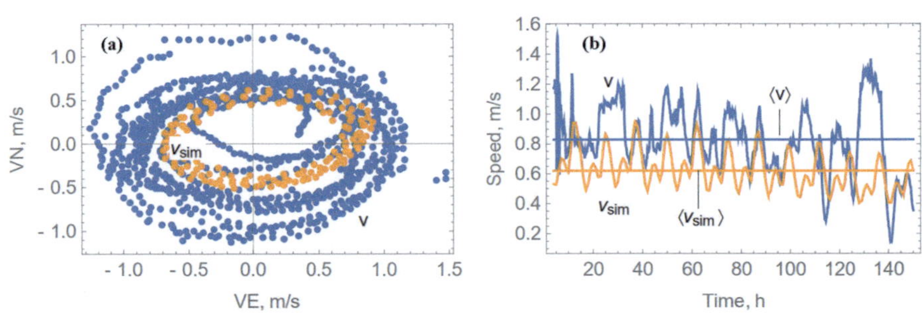

Fig. 1.39 Components of IT velocities (blue points) and simulated tidal current velocities (yellow points) computed for a 150-h time interval beginning from 15:00 UTC, April 28, 2018 (**a**, left panel). IT speeds (blue curve) and simulated tidal current speeds (yellow curve) versus time (**b**, right panel) (Marchenko et al. 2022)

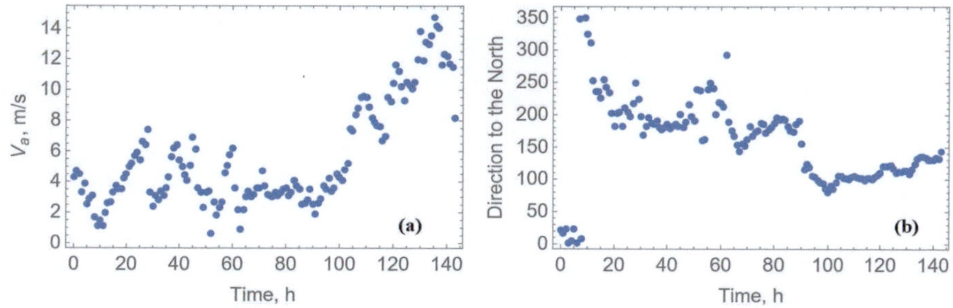

Fig. 1.40 Wind speed (**a**, left panel) and direction (**b**, right panel) measured at Bear Island beginning from 00:00 UTC, April 28, 2018 (Marchenko et al. 2022)

According to the data from CTD profiling, vertical profiles of the water density are well approximated by the formula $z = -5 \times (-200(\Delta\rho - 27.78))$, where z (m) is the vertical upward-directed coordinate, and $\Delta\rho = \rho_w - 1000$ (kg/m^3) is the difference between sea water density and 1000 kg/m^3. The vertical gradient of sea water densities is estimated by the formula $\partial\rho_w/\partial z = -5 \times 10^{-3}$ kg/m^4, and the squared Brunt-Väisälä frequency $N^2 = -g\rho^{-1}\partial\rho/\partial z$ equals to 4.7×10^{-5} s^{-1} over the entire water column.

Vertical profiles of ocean current velocities were measured by the ADCP Nortek AWAC (400 kHz) deployed from the vessel in a down-looking position. The bin size was 1 m, and the profile interval was 10 min. The ADCP was deployed at 5.5 m depth, and the first bin of the records corresponds to the depth interval from 6.5 to 7.5 m. The first measurement started at 09:07 UTC, and the last measurement was at 20:57 UTC on April 28, 2018. The ADCP recorded current velocities relative to the ship velocity. Therefore, the ship velocity vector was added to the ADCP velocity vector to calculate current velocities in the ground-based reference frame. The data were processed using the operation ListDensityPlot in Wolfram Mathematica software. The procedure Fit was used for the time-interpolation of the velocity components by third-order polynomials at each depth. This procedure reduced the noise level of the raw data.

Vertical profiles of current speed in the ground-based reference frame are shown in Fig. 1.41. The numbers show the time of measurement in hours since the first measurement at 09:07 UTC on April 28, 2018. One can see that the bottom boundary layer extended from the bottom to the first bin on the profiles recorded from $t = 1.0$–3.0 h. Maximal ocean current speeds in the range of 1.14–1.24 m/s were recorded at $t = 4.0$ h at the shallowest point along the ship trajectory with a water depth of 25 m. The vertical gradient of current speed is not significant along this profile. This can be explained by the influence of the Kelp forest underwater hill on the boundary layer. Boundary layers extending from the seafloor to 10 m are seen on the profiles recorded from $t = 7.5$–10.0 h. Several profiles show an increase of current speed near the seabed.

The dependencies of the meridional and zonal velocity components in the ship-fixed reference frame are shown in Fig. 1.42 versus time. Vertical gradients of the

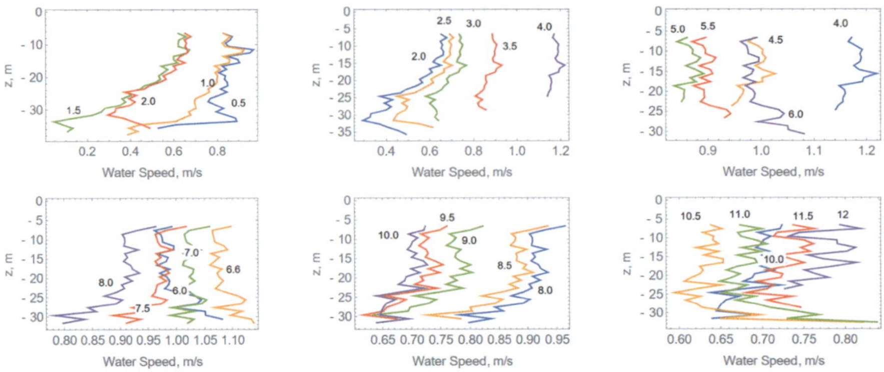

Fig. 1.41 Vertical profiles of the horizontal current speeds measured by the ADCP. The first measurement was performed at 09:07:37 UTC, April 28, 2017. Numbers show the time (hours) after the first measurement (Marchenko et al. 2022)

velocities are clearly seen in the figure and are consistent with the structure of the first mode of internal waves in the fluid with constant stratification (Gerkema and Zimmerman 2008). Since the velocities were measured in the moving reference frame their temporal variability can be related to topographic ocean currents. The magnitude of the vertical velocity gradient is found with the formula.

$$\Delta_z v = \surd(\partial v_E/\partial z)^2 + (\partial v_N/\partial z)^2,$$

where, v_E and v_N are the zonal and meridional components of current velocity. The Richardson number $Ri = N^2(\Delta_z v)^{-2}$ indicates dynamic instability of shear currents in stratified fluid when $Ri < 0.25$ leading to turbulence and water mixing (Taylor 1931). Figure 1.43 shows the magnitudes of the vertical velocity gradients (left panel) and the Richardson numbers (right panel) calculated with $N^2 = 4.7 \times 10^{-5}$ s^{-1} versus time and water depth. It is clear that $Ri < 0.25$ over almost the entire water column. This indicates dynamic shear instability of the water flow in the shallow region of the Spitsbergen Bank.

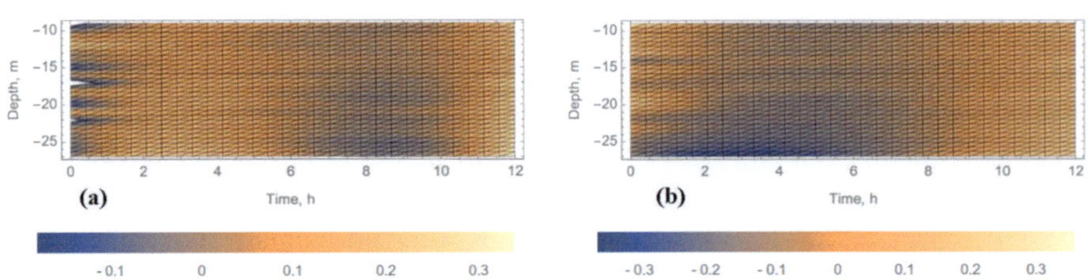

Fig. 1.42 The zonal (**a**) and meridional (**b**) ADCP-recorded current velocities (m/s) from April 28 from 09:07 to 20:57, UTC, 2018

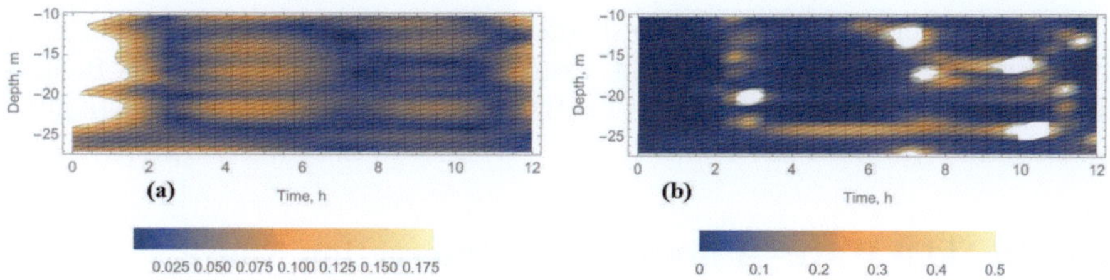

Fig. 1.43 Magnitudes of vertical gradients of the horizontal ADCP-recorded current velocities (1/s) from April 28 from 09:07 to 20:57 (UTC), 2018 (**a**). The Richardson numbers were calculated with the ADCP-recorded current velocities (**b**) (Marchenko et al. 2022)

Analysis of ice conditions on the Spitsbergen Bank showed that sea ice can be observed over a significant period of every year. A criterion for comparison is the annual number of days when ice was found at 75° N latitude. Most frequently, sea ice was observed at 75° N in March and April. Since 1998, the annual number of days (AD) when sea ice was present at 75° N reached a maximum of 150 days in 2003 and was equal to 138 days in 2020. The minimum numbers of AD reached two days in 2012 and five days in 1999. The mean value of AD is 60, and the standard deviation is 36.6 days.

Despite the relatively stable location of sea ice on the Spitsbergen Bank, the ice is generally transported by strong tidal currents. Measurements of the drifting ice floe velocity and current velocities were performed in the region with water depths of 25–35 m located north of Bear Island around 75.1° N, 19.3° E. The maximum speed of the ice tracker deployed on the floe reached 1.5 m/s, the minimum speed was 0.15 m/s, and the mean speed was 0.83 m/s. The ice tracker trajectories over one semidiurnal cycle were almost circular, and the ice tracker rotated clockwise. This motion corresponded well to the characteristics of tidal currents on the Spitsbergen Bank (Gjevik et al. 1994), but the mean speed of the tidal current calculated for the locations and times of the IT drift with the TPXO9 model was 0.62 m/s. Thus, the mean measured velocities exceeded the mean simulated tidal current velocities by 20%, and the maximum measured velocities exceeded the maximum simulated tidal current velocities by 50%.

We hypothesize that the boundary layer effect in the proximity of the M_2 critical latitude 74.47° N described by Furevik and Foldvik (1996) explains the difference. In shallow water the boundary layer can extend over the entire water column and cause significant differences between the mean current velocity averaged over the sea depth and the surface current velocity. The IT velocity was almost the same as the surface current velocity, while the tide model calculates current velocity averaged over the sea depth, and it leads to the differences. The Richardson numbers calculated from the measured current velocities were smaller than 0.25 over almost the entire water column in the shallow region with 25–35 m depths located north of Bear Island. This shows dynamic instability of the tidal currents on the Spitsbergen Bank whose existence was discussed by Furevik and Foldvik (1996) and Rippeth et al. (2017). The dynamic instability influences water mixing and reduction of ocean heat flux in shallow water regions of the Spitsbergen Bank. It causes fast formation of ice in cold weather, and partially explains extended ice season in Spitsbergen Bank in comparison with the other regions of the Barents Sea located at the same latitude.

1.6 In Tidal Wave—Elliptic Island Interaction Above the Critical Latitude

This section was written together with Z. Kowalik

The goal of this research is to examine analytically and explain specific features of the tidal variations around the elongated Hopen Island located at the Spitsbergen Bank in the Barents Sea. The trapping of the M_2 tide around certain islands leads to a dipole structure in the sea level that rotates clockwise. This sea level difference travels around the island while the primary tidal wave determines semidiurnal variations of the mean water level around the island, as explained by Kowalik and Marchenko (2002). In this section we review a paper (Marchenko and Kowalik 2023).

Proudman (1915) constructed the first analytical solutions of tidal wave scattering by a coastline or small circular and elliptical islands. Huthnance (1974) analytically investigated the tidal currents over the Rockall Bank in the North Atlantic Ocean, assuming non-divergent currents. The analysis was performed in polar and elliptic coordinates. Larsen (1977) and Reynolds (1978) analyzed the systematic variation in amplitude and phase around the Hawaiian Islands. They applied the scattering problem solution around the elliptical ridge and island to explain the observations. The analytical solutions for the tidal wave scattering around small islands were further developed by Lee and Kim (1999) by introducing bottom friction. The presence of bottom friction enabled them to extend the scattering theory to sub-inertial frequencies. Further analytical extension was achieved in (Kowalik and Marchenko 2002). The model of an island in the ocean of a constant depth was found to be insufficient to explain the behavior of tidal waves in the region of the Pribilof Islands, Bering Sea. However, the inclusion of a sill changed the situation. The "skirt" around the Pribilof Islands played an essential role in forming the observed tidal features.

Numerical models of the tides in the Barents Sea and measurements indicated enhanced currents over the Spitsbergen Bank due to the topographic amplification in the semidiurnal band of oscillations (Gjevik et al. 1990; Kowalik and Proshutinsky 1995; Padman and Erofeeva 2004; Chen et al. 2009). Over the entire region, the current vectors of M_2, S_2, and N_2 tides showed a clockwise rotation. At the same time, tidal charts demonstrated the counterclockwise motion of the tide phases around the amphidromic point southeast off Hopen Island. Furevik and Foldvik (1996) suggested that this pattern of tidal motion might be described by a superposition of axially symmetric solution formulated in terms of the Bessel functions and Sverdrup wave (south of the critical latitude) or complex wave (north of the critical latitude).

The inertial period equals the M_2 tidal period at the critical latitude (~74.5°N). At the Spitsbergen Bank, it crosses Bear Island, while Hopen Island is north of the critical latitude. From our computation (Kowalik and Marchenko 2023), we have concluded that since Spitsbergen Bank straddles over the critical latitude, the presence of the critical latitude does not influence the trapping of the semidiurnal wave. The difference in the island shapes between the near-circular Bear Island and the elongated Hopen Island also demonstrates that the trapping is not dependent on the island's geometry. To describe analytically the properties of trapped waves around an elongated island above the critical latitude we solve the tidal scattering problem for an elliptically shaped island invoking elliptical coordinates.

1.6.1 Observations of Ocean Currents Near Hopen Island

Hopen Island is located in the northwest region of the Barents Sea, 115 nautical miles east of Spitsbergen. The island's length is approximately 30 km, and its maximal width is less than 2 km. The Norwegian Pilot Guide (https://www.barentswatch.no/en/articles/the-norwegian-pilot-guide-guide-at-sea/) describes tides around the island as: "the tidal stream, which changes at high and low water, rushes around the island with unusual strength and in some places, especially around the south and north points, it sets up a violent sea which can be dangerous to smaller vessels."

Satellite images obtained using the Drift-Noise Polar Services application IcySea (https://icysea.app) show frequent formation of vortices near the southern and northern tips of Hopen Island (Fig. 1.44). Streamlines of drift ice following the surface current and wind drag are also seen in the island's vicinity. With respect to the zonal direction of the tidal current, the vortices form on the island's leeward side.

Drifting buoys (Oceanetic Measurements Ltd, Canada) were deployed on drift ice at the Spitsbergen Bank in April 2018 to observe the ice motion (Marchenko et al. 2022). One buoy survived the ice melt and kept transmitting GPS coordinates every 10 min until February 2019. In July 2018, the buoy drifted near Hopen Island (Fig. 1.45a). The buoy trajectory is subdivided into several fragments depicted by blue, black, yellow, and green lines. The blue line starts close to the island's southern tip on July 16, 18:50 UTC, and continues for four M_2 periods (4×12.42 h). The yellow line starts in the nearby location on July 20, 01:50 UTC, and continues for four M_2 periods as well. The green line starts at the yellow point near the island's north tip on July 22, 03:30 UTC, and finishes after thirty M_2 periods. Blue, yellow, and green dots mark the locations of the buoys after passing a single M_2 period. Red dots mark the maximum speed of the buoy. Figure 1.45b shows the speed of the buoy as a function of time at the western side of the island. The initial points on the graphs correspond to the buoy's speeds in the region 25.0–25.5° E, 76.4–76.5° N in Fig. 1.45a. The buoy circled the island two times. The starting yellow and green points almost coincide in Fig. 1.45a, but the speed of the buoy along the green line was lower than the speed along the blue and yellow lines. The green trajectory exhibits only partial trapping.

The buoy trajectory consists of many loops passed by the buoy in clockwise direction. Pease et al. (1995), Marchenko et al. (2011), Marchenko and Marchenko

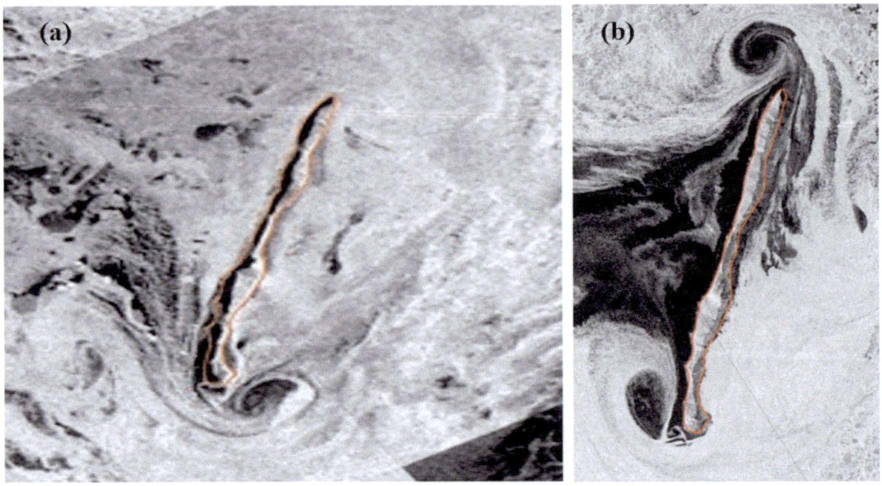

Fig. 1.44 Satellite images of Hopen Island on April 15 (**a**) and June 1 (**b**), 2021, describe through the pack ice concentration the eddy-like motion at the north and south ends of the island. Adopted from Marchenko and Kowalik (2023)

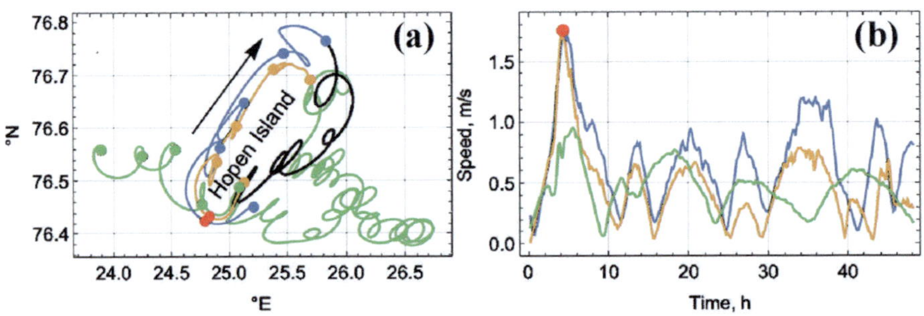

Fig. 1.45 Trajectory (**a**) and drift speeds (**b**) of the buoy deployed in the Barents Sea in 2018. The buoy trajectory is subdivided into several fragments depicted by blue, black, yellow, and green lines. Blue, yellow, and green dots mark the locations of the buoys after passing a single M_2 period. Speeds of the buoy are calculated at the western side of Hopen Island. Adopted from Marchenko and Kowalik (2023)

(2019), Turnbull and Marchenko (2022) also observed similar buoy drift patterns in different regions of the Spitsbergen Bank. Usually, buoys pass one round cycle over a time close to the semidiurnal cycle, as it is marked by blue, yellow, and green dots in Fig. 1.45a. In the open sea, when the wind drag is negligible, the buoys' trajectories are close to circles. Increasing wind speed influences the resulting drift of buoys in a specific direction and changes the diameter of the loops.

To identify the influence of wind on the trapped motion around the island, we constructed probability density functions of the wind direction and speed. The data from https://earth.nullschool.net at (25.2° E, 76.6° N) based on the data on July 16–24, 2018. At this time, the buoy drifted near the island, see, Fig. 1.45a. The colors of the Paired Histograms in Fig. 1.46 correspond to the colors of the buoy trajectory segments in Fig. 1.45. Because the green particle displays only a short trapping time and it is dragged both to the east and west of the island; this is a good case to observe the influence of the wind-driven motion. In July 24, strong southwestern, and western winds (30–40 km/h) displaced the buoy to the east, and until July 28 the buoy drifted in the region 76.4° N–76.5° N and 26.0° E– 26.8° E, see Fig. 1.45a. However, later due to the eastern wind (not shown in Fig. 1.46), the buoy drifted to the southern tip of the island, and proceeded to drift to the west after passing the southern tip of the island. This example shows that wind drag influences surface currents and the buoy drift, but it cannot explain high drift speeds and the trapped motion near the island tips as seen along the blue, yellow, and black trajectories.

Here, we describe the sea level and the buoy speed to be used later for comparison with the analytical computation of sea level and velocity of an incident wave. The sea level was estimated using Wolfram Mathematica TideData software, which calculates the data from the space altimetry and tide gauge stations. Figure 1.47a shows the sea level on July 16–28, 2018, at point 76.4°N, 26.3°E located to the east of Hopen Island. Semidiurnal period of about 12.42 h is well recognized. Due to the two unequal tides per 24.84 h tidal day, the amplitude of the water surface elevation shows substantial diurnal inequality between 6 and 16 cm. Daily high and low tides slightly change

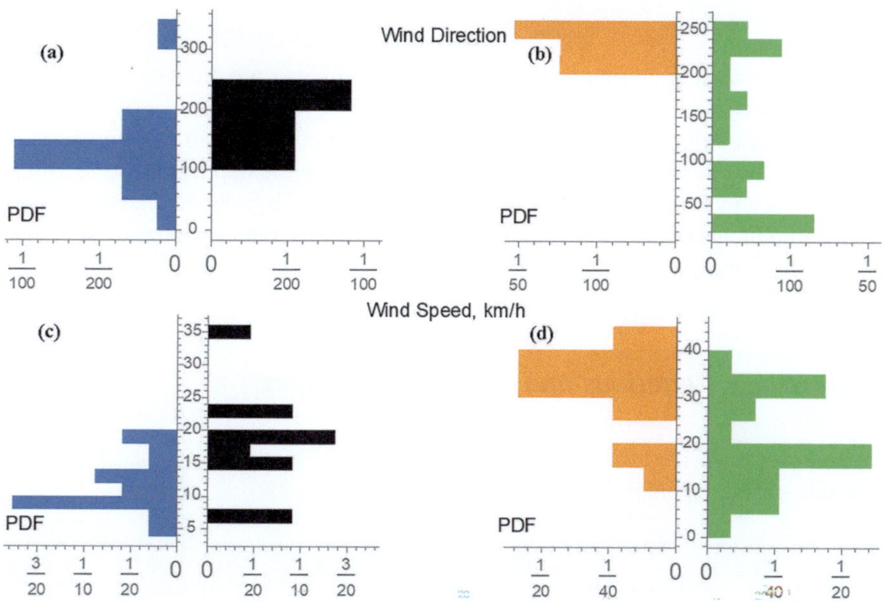

Fig. 1.46 Probability density functions of wind directions (**a**, **b**) and wind speeds (**c**, **d**), in the middle of Hopen Island, July 16–24, 2018. Colors correspond to the colors of the buoy trajectory segments shown in Fig. 1.45. Adopted from Marchenko and Kowalik (2023)

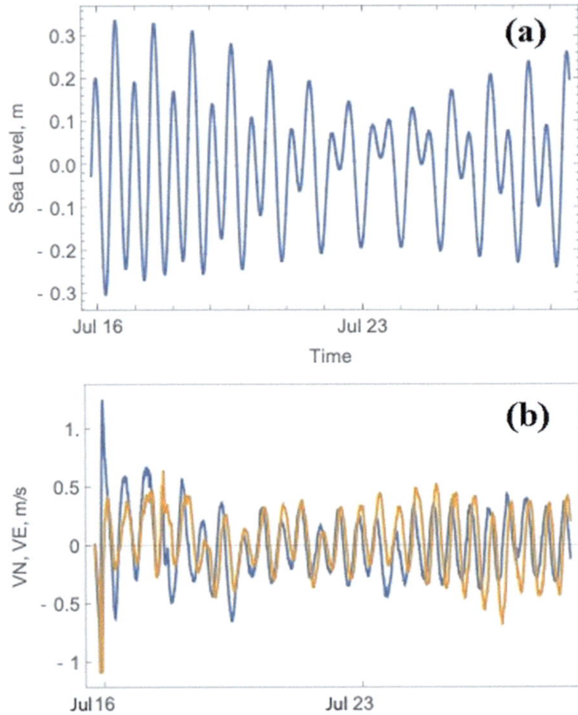

Fig. 1.47 Sea level in July 16–28, 2018, defined by Mathematica's TideData software at point 76.4° N, 26.3° E (**a**). Meridional (blue) and zonal (yellow) velocities of the drifting buoy over the same time in in the area 76.4° N–76.5° N and 26.0° E–26.8° E (**b**). Adopted from Marchenko and Kowalik (2023)

the amplitude at neighboring tidal days, and these changes are induced by the slight modulation caused by the fortnightly lunar tides.

The current velocities are calculated from the buoy drifting in the region 76.4–76.5° N and 26.0–26.8° E on July 24–28, 2018, where the depth is close to 100 m (Figs. 1.45a and 1.48). Blue and yellow lines in Fig. 1.47b show the meridional and zonal components of the buoy velocity, respectively. The main period is close to M_2, and velocity amplitudes reach 0.5 m/s.

1.6.2 Basic Equations and Coordinate Systems

We consider linearized shallow water equations in the rotating reference frame, see Kowalik and Marchenko (2002)

$$\frac{\partial u}{\partial t} - v = -\frac{\partial \zeta}{\partial x}, \quad \frac{\partial v}{\partial t} + u = -\frac{\partial \zeta}{\partial y}, \quad \frac{\partial \zeta}{\partial t} + d\left(\frac{\partial u}{\partial x} + \frac{\partial v}{\partial y}\right) = 0 \tag{1.1}$$

The set of Eqs. (1.1) is written in dimensionless form, where x and y are the Cartesian coordinates, t is time, u and v are components of water velocity, ζ is water

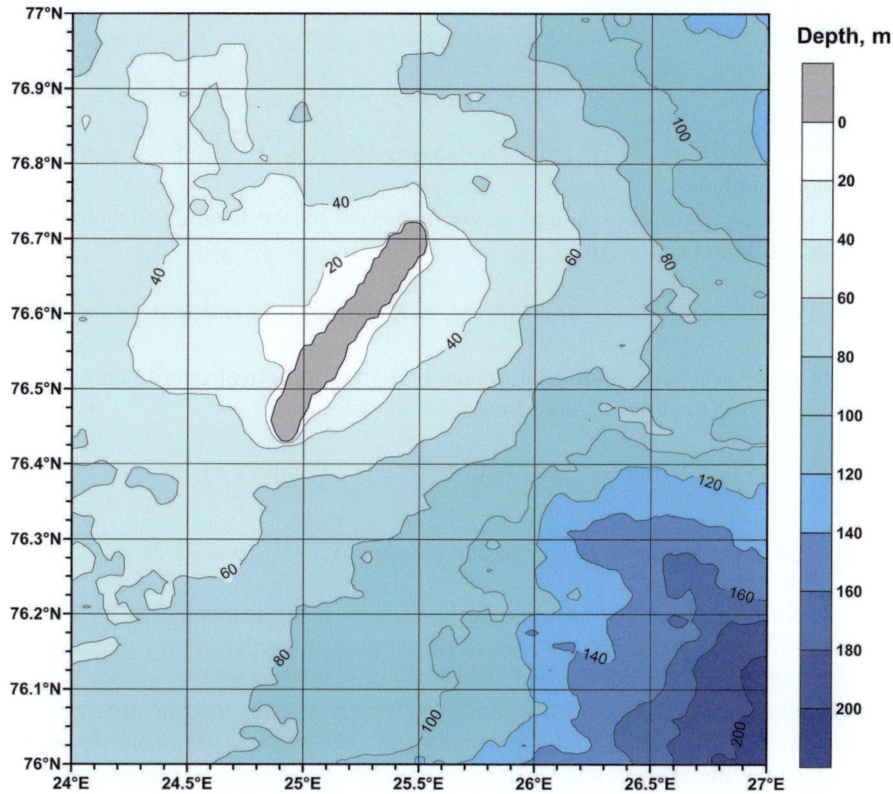

Fig. 1.48 Bathymetry (meters) around Hopen Island. The island extends between northern tip A_1 and southern tip A_2. Large letter A marks the location of the M_2 amphidromic point. Adopted from Marchenko and Kowalik (2023)

surface elevation, and d is water depth. The actual water depth equals $H = dH^*$, where H^* is the characteristic water depth (scale parameter). Dimensional variables denoted below by the subscript "*dim*" are related to the dimensionless variables as:

$$t_{dim} = t/f, \quad (x_{dim}, y_{dim}) = \alpha_l(x, y),$$
$$(u_{dim}, v_{dim}) = U^*(u, v), \quad \zeta_{dim} = \alpha_\zeta \zeta, \tag{1.2}$$

where, t_{dim} is time, x_{dim} and y_{dim} are horizontal coordinates, u_{dim} and v_{dim} are water velocities averaged over depth, ζ_{dim} is water surface elevation, U^* is characteristic amplitude of the velocity, f is the Coriolis parameter, and g is the gravity acceleration. Scale factors for the calculation of dimensional length and water surface elevation are $\alpha_l = \sqrt{gH^*}/f$ and $\alpha_\zeta = U^*\sqrt{H^*/g}$.

We consider solutions of Eqs. (1.1) in the form $(\zeta, u, v)=(\zeta_0, u_0, v_0)e^{i\omega t}$, where ω is the dimensionless angular frequency related to the dimensional frequency by formula$\omega_d = \omega f$. Introducing the above solution into (1.1) we get an equation for the sea level amplitude,

$$\left(d\Delta + K^2\right)\zeta_0 = 0, \tag{1.3}$$

where, $\Delta = \partial^2/\partial x^2 + \partial^2/\partial y^2$ is the Laplace operator, and $K^2 = \omega^2 - 1$ defines the dispersion equation.

The velocity amplitudes u_0 and v_0 are expressed by the formulas developed in Kowalik and Marchenko (2002):

$$u_0 = K^{-2}(i\omega\partial\zeta_0/\partial x + \partial\zeta_0/\partial y), \quad v_0 = K^{-2}(i\omega\partial\zeta_0/\partial y - \partial\zeta_0/\partial x). \tag{1.4}$$

Further, we consider polar coordinates (r, θ) and elliptical coordinates (ξ, η), which are related to cartesian coordinates (x, y) as follows:

$$x = r\cos\theta, \quad y = r\sin\theta; \quad r > 0, \quad \theta \in (0, 2\pi). \tag{1.5}$$

$$x = \mu\cosh\xi\cos\eta, \quad y = \mu\sinh\xi\sin\eta; \quad \xi > 0, \quad \eta \in (0, 2\pi). \tag{1.6}$$

The coordinate lines $\xi = const$ are ellipses with foci located at points $(\pm\mu, 0)$. The ratio of the major and minor axes of the ellipses along the x and y directions is equal to $\coth\xi$. The coordinate lines $\eta = const$ are hyperbolas.

The polar coordinates are further used to describe water motion generated by the semidiurnal tide near the amphidromic point. The water motion near the island surrounded by the elliptic sill (shelf) is studied in the elliptical coordinates. The island boundary coincides with coordinate line $\xi = \xi_0$, and the shelf boundary is defined by coordinate line $\xi = \xi_1$ $(\xi_1 > \xi_0)$. Water depth in the deep domain outside the shelf equals H^* at $\xi > \xi_1$, and at the shelf located at $\xi \in (\xi_0, \xi_1)$, the depth is $H_s < H^*$. The water depth in dimensionless units is $d = 1$ outside the shelf and $d < 1$ on the shelf.

The velocity amplitudes in polar coordinates $u_{0,r}$ and $u_{0,\theta}$, and elliptical coordinates $u_{0,\xi}$ and $u_{0,\xi}$ are expressed by the formulas:

$$u_{0,r} = K^{-2}\left(i\omega\partial\zeta_0/\partial r + r^{-1}\partial\zeta_0/\partial\theta\right), \quad u_{0,\theta} = K^{-2}\left(i\omega r^{-1}\partial\zeta_0/\partial\theta - \partial\zeta_0/\partial r\right), \tag{1.7}$$

$$u_{0,\xi} = K^{-2}J^{-0.5}(i\omega\partial\zeta_0/\partial\xi + \partial\zeta_0/\partial\eta), \quad u_{0,\eta} = K^{-2}J^{-0.5}(i\omega\partial\zeta_0/\partial\eta - \partial\zeta_0/\partial\xi), \tag{1.8}$$

where, $J = \mu^2\left(\sinh\xi^2 + \sin\xi^2\right)$ is the Jacobian of the coordinate transformation.

The condition

$$K = 0, \omega = \pm 1 \tag{1.9}$$

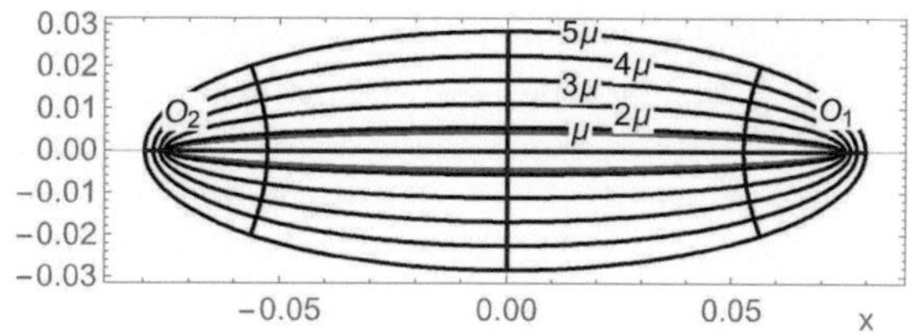

Fig. 1.49 Coordinate lines $\xi = const$ (ellipses) and $\eta = const$ (hyperbolas) expressed in the dimensionless variables. The frame of reference is related to Hopen Island. Points O_1 and O_2 correspond to the points A_1 and A_2 in Fig. 1.48. Red color ellipse describes the island boundary at $\xi = \xi_0$. Coordinates x and y are related to ξ and η by formulas (1.6). Adopted from Marchenko and Kowalik (2023)

corresponds to the definition of the critical latitude for the wave which frequency $\omega_{dim} = f$. The Coriolis parameter equals to $f = 2\Omega \sin\varnothing$, where $\Omega = 7.2921 \cdot 10^{-5}$ rad/s is the frequency of the Earth's rotation, and $\varnothing$ is latitude. The frequency of semidiurnal tide M_2 equals to $\omega_{M2} = 1.40519 \cdot 10^{-4}$ rad/s, therefore the critical latitude for the M_2 tide is $\varnothing_{M2} = 74.4717°$. Since in the proximity to the critical latitude $K \to 0$ it follows from (1.4), (1.7), and (1.8) that the current velocity tends to infinity when the sea surface slope is non-zero.

We investigate tidal currents near Hopen Island in the Barents Sea (Fig. 1.48). The island extends between (76.70° N, 25.46° E) and (76.45° N, 24.93° E), and is located north of the critical latitude $\varnothing_{M2}$. The island length is approximately $L_H \approx$ 31.7 km, and the width of the island is smaller than 2 km. In the middle part of Hopen Island, the Coriolis parameter equals to $f = 1.415 \cdot 10^{-4}$ rad/s, the dimensionless frequency of the M_2 tide equals to $\omega = \omega_{M2}/f = 0.9931$, and the squared value is $K^2 = -0.0138$.

Representative depth in the open sea at the amphidromic point is close to 100 m, and the representative M_2 tidal velocity is about 0.1 m/s (Gjevik et al. 1994). Assuming $H^* = 100$ m and $U^* = 0.1$ m/s we find the scale factor $\alpha_\zeta = 0.32$ m. The scale factor α_l at the middle point of Hopen Island equals to 221 km. Expressed by the dimensionless variables, the island length equals to $L_H/\alpha_l = 0.14$. Elliptic coordinate lines specified by formulas (1.6) with $\mu = 0.07$ are given in Fig. 1.49. The coordinate line $\xi = \xi_0 = 0.05$ corresponds to the contour of an elliptic island with dimensions similar to Hopen Island.

1.6.3 Semidiurnal Tidal Waves Near Hopen Island

To demonstrate the behavior of the semidiurnal M_2 tide in the proximity of Hopen Island, we use amplitudes and phases of sea surface elevation computed by Kowalik and Proshutinsky (1995), see Fig. 1.50. The main feature, the amphidromic point, is located to the south-southeast of Hopen Island (Gjevik et al. 1994; Kowalik and Proshutinsky 1995). A red line segment shows Hopen Island. Figure 1.50 depicts the approximate location of the amphidromic point at (75.7° N, 25.4° E), which sets this point north of the critical latitude. The M_2 tidal wave travels counter-clockwise (CCW) around the amphidromic point. Using a superposition of explicit analytical solutions of Eq. (1.3) we aim to reproduce the distribution of co-phase and amplitude lines of the M_2 tide. The analytical solution will also describe the tidal ellipses and orbital circulation of water particles in the vicinity of Hopen Island.

To study the direction of orbital circulation the sign of vertical coordinate of the following vector product is used

$$w = u|_{t=t_0} \times u|_{t=t_0+\Delta t}, \tag{1.10}$$

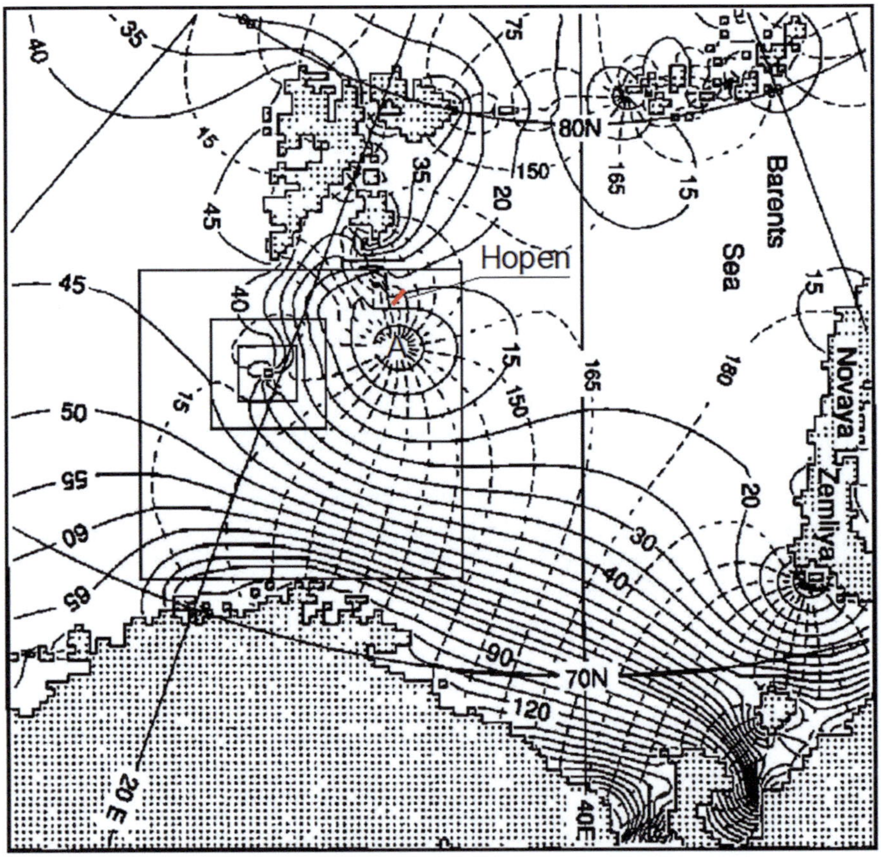

Fig. 1.50 Computed amplitude (solid lines, in cm) and phase (dashed lines, in deg) of sea surface elevation for the semidiurnal tide M_2 (from Kowalik and Proshutinsky 1995). Hopen Island is shown by a red line segment. Letter "A" marks the location of M_2 amphidromic point. Adopted from (Marchenko and Kowalik 2023)

where, u is the vector of the horizontal water velocity at fixed point $x = x_0$. In the right-handed reference frame (x, y), the positive and negative signs of w_z denote respectively the counterclockwise (CCW) and clockwise (CW) orbital motion of water particles located near point x_0 over the time interval $t \in (t_0, t_0 + \Delta t)$. The orbital motion of water particles located near point x_0 is expressed by the equations

$$\frac{dx}{dt} = \alpha_d Re\left[u_0 e^{i\omega t}\right], \quad \frac{dy}{dt} = \alpha_d Re\left[v_0 e^{i\omega t}\right], \tag{1.11}$$

where, $\alpha_d = U^*/(f\alpha_l)$, and u_0 and v_0 are calculated at $x = x_0$. The Lagrangian drift of water particles can be also described by Eqs. (1.11). In this case u_0 and v_0 are not any longer calculated at a fixed point but along a trajectory $x = x(t)$ and $y = y(t)$.

1.6.4 M_2 Complex Tidal Waves

We consider waves with the dimensionless M_2 tidal frequency $\omega = 0.9931$ transmitting energy in the direction of the x-axis in an unbounded ocean of constant depth $(d = 1)$. In this section, the direction of the x-axis is not associated with the axis of Hopen Island. The following expressions describe the periodic solution of Eq. (1.3)

$$\zeta_0 = \zeta_{C0} = Ae^{i\boldsymbol{k}\cdot\boldsymbol{x}}, \quad k_{y,\pm} = \pm i\sqrt{1 + k_x^2 - \omega^2}, \tag{1.12}$$

where, A is the amplitude of water surface elevation, $\boldsymbol{x}$ is the vector with (x, y) coordinates and $\boldsymbol{k}$ denotes the wave vector with coordinates (k_x, k_y). When $\omega < 1$ it is evident that $k_y = k_{y,\pm}$ is a pure imaginary quantity. In this case, the wave energy is transferred along the x-axis only. Solutions (1.12) are further referred to as M_2 complex tidal waves CTW+ and CTW-, where the symbols + and – correspond to the signs + and – in formula (1.12).

Velocity amplitudes are calculated from (1.4) as follows:

$$\begin{aligned} &u_0 = A_{u,\pm}e^{ik_x x \mp |k_y| y}, \quad v_0 = -iA_{v,\pm}e^{ik_x x \mp |k_y| y}, \\ &A_{u,\pm} = AK^{-2}\left(\mp|k_y| - \omega k_x\right), \quad A_{v,\pm} = AK^{-2}\left(\pm\omega|k_y| + k_x\right). \end{aligned} \tag{1.13}$$

The end of the velocity vector $\boldsymbol{u} = (Re[u_0], Re[v_0])$ traces an ellipse in the (u_0, v_0) coordinate system with the principal axes $|A_{u,\pm}|e^{\mp|k_y|y}$ and $|A_{v,\pm}|e^{\mp|k_y|y}$ at $x = x_0$. Simple algebra demonstrates that the endpoint of $\boldsymbol{u}$ travels the ellipse in CW direction when $A_u A_v < 0$, and in CCW direction when $A_u A_v > 0$(Kowalik and Luick 2019).

In Fig. 1.51, the CW and CCW orbital motion regions are defined for the water particles in the plane (k_x, ω). The orbital motion is induced by complex tidal waves CTW+ and CTW-. The regions are separated by the Kelvin waves' dispersion curves $\omega = \pm k_x$. These waves propagate in the positive and negative directions of the x-axis.

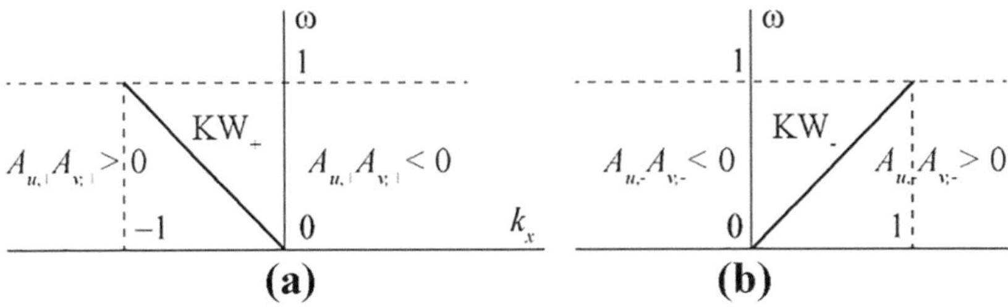

Fig. 1.51 Regions of CW ($A_uA_v < 0$) and CCW ($A_uA_v > 0$) motions plotted in the (k_x, ω) plane. The motion is induced by complex tidal waves CTW+ and CTW-. A_u and A_v are defined by Eq. (1.13). Lines KW_+ and KW_- represent the dispersion lines of the Kelvin waves propagating in the positive and negative directions of the k_x-axis. Adopted from Marchenko and Kowalik (2023)

In the Kelvin waves the velocity component $v = 0$, and the orbital motion is reduced to oscillations along the x-axis.

1.6.5 M_2 Tidal Wave in the Vicinity of an Amphidromic Point

Solutions of Eq. (1.3) in the vicinity of the amphidromic point are defined at $\omega^2 < 1$; they were developed by Martin and Dalrymple (1994)

$$\zeta_0 = \zeta_{A0} = AI_n(|K|r)e^{\pm ni\theta}, \qquad (1.14)$$

where, A is the sea surface elevation amplitude, $I_1(|K|r)$ is the Bessel function of the first kind, $r = \sqrt{x^2 + y^2}$ is the radius vector, and θ is the polar angle.

The Coriolis parameter at the amphidromic point equals to $f = 1.409 \cdot 10^{-4}$ rad/s, the dimensionless frequency of M_2 tide equals to $\omega=\omega_{M2}/f = 0.9973$, and $K^2 = -0.0054$. Solutions (1.14) are dubbed as M_2 amphidromic tidal waves ATW + and ATW-, where the symbols + and – correspond to the signs + and − in formulas (1.14). Further, we consider only the first modes of ATW corresponding to $n = 1$. To explore the rotational motion caused by ATW, we made an estimate based on the above parameters. The motion induced by ATW- is CCW ($w_z > 0$) inside the circle with a radius of 1.48 and CW ($w_z < 0$) outside the circle (Fig. 1.52a). Thus, the orbital motion is CCW inside the circle with a radius of 1.48 and is CW outside the circle (Fig. 1.52a). ATW + induces the orbital CCW circulation for the arbitrary values of r.

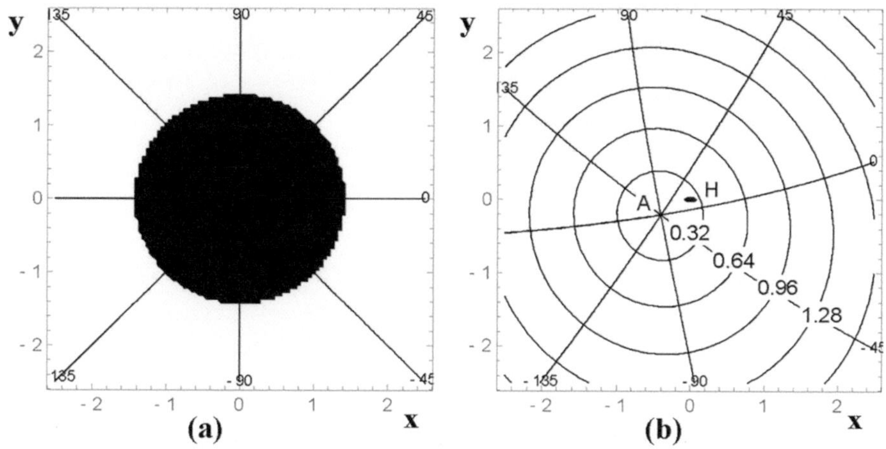

Fig. 1.52 Black and white regions show respectively regions of CCW and CW motions caused by ATW- (**a**) and superposition of ATW- and CTW +(**b**). Lines running from the center denote co-phase in degrees in (**a, b**), and concentric lines are amplitudes in (**b**). A denotes the amphidromic point, and H stands for the elliptic island. Adopted from Marchenko and Kowalik (2023)

1.6.6 Superposition of ATW- and CTW+ Waves

For the analytical description of tides observed in the Barents Sea north of the critical latitude, Furevik and Foldvik (1996) suggested the superposition of amphidromic waves ATW and complex waves CTW. This linear superposition is defined as

$$\zeta_0 = \zeta_{I0} = c\zeta_{A0} + (1-c)\zeta_{C0}, \tag{1.15}$$

where, ζ_{C0} and ζ_{A0} are specified by (1.12) and (1.14), respectively; and $c \in (0,1)$ is the weight coefficient. Subscript "I" means that (1.15) defines the incident tidal wave transmitting energy to Hopen Island.

Formulas (1.12) are rewritten to account for the arbitrary direction of the wavevector $\boldsymbol{k}$ of CTW + in the frame of reference (x, y) related to Hopen Island (Fig. 1.52). The direction of wavevector $\boldsymbol{k}$ and the weight coefficient c vary to adjust the location of the amphidromic point generated by the wave superposition defined by formula (1.15) to the observed amphidromic point A near Hopen Island (Figs. 1.48 and 1.50).

It is assumed that frequency $\omega = 0.9931$ is the dimensionless M_2 frequency at the middle of Hopen Island, and the origin of the Cartesian reference frame (x, y) is placed at that location. CTW+ transfers energy in the negative direction of the axis $\tilde{x}$ determined by the angle θ_k between the axes x and $\tilde{x}$. Positive direction of θ_k is counter-clockwise with respect to the axis x. The arrow in Fig. 1.53 shows the direction of CTW+ propagation coinciding with the direction of the vector $-\boldsymbol{k}$ (negative direction of the axis $\tilde{x}$). Accordingly, Cartesian coordinates (x, y) in formulas (1.12) are replaced by $(\tilde{x}, \tilde{y})$:

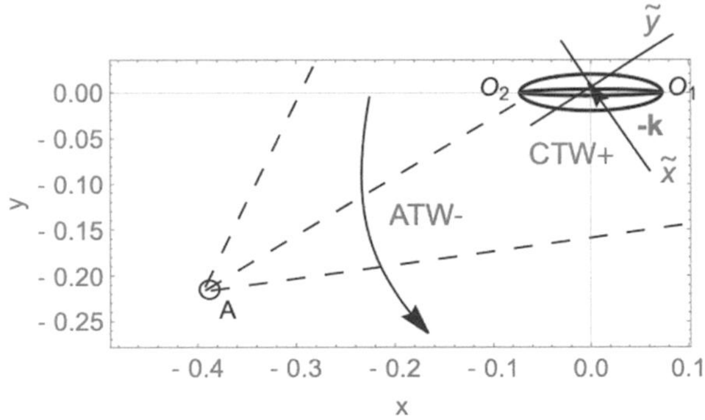

Fig. 1.53 Superposition of the M_2 tidal waves ATW- and CTW+ near amphidromic point A. Dashed lines depict lines of equal phase of the incident wave. Adopted from Marchenko and Kowalik (2023)

$$\tilde{x} = x\cos\theta_k + y\sin\theta_k, \quad \tilde{y} = -x\sin\theta_k + y\cos\theta_k. \tag{1.16}$$

Coordinates r and θ in (1.14) are modified as follows:

$$r^2 = (x - x_0)^2 + (y - y_0)^2, \quad \theta = \text{ArcTan}\left[x - x_0, y - y_0\right]. \tag{1.17}$$

The values of x_0 and y_0 are found from the condition that the point with zero amplitude (the amphidromic point due to the wave superposition) coincides with the location of amphidromic point A near Hopen Island (Fig. 1.51b).

Figure 1.52b shows co-phase lines and isolines of wave amplitudes constructed for

$$A = 1, \ c = 0.9, \ k_{\tilde{x}} = 0.17, \ \theta_k = -\pi/4, x_0 = 1.32, \ y_0 = -0.17. \tag{1.18}$$

The location of amphidromic point A in Fig. 1.52b is close to the location of amphidromic point A in Figs. 1.50 and 1.52. The wave superposition causes CW orbital motion in the vicinity of the elliptic island.

The superposition of ATW- and CTW+ qualitatively reproduces the incident tidal waves in the vicinity of Hopen Island. Figure 1.54 shows the real parts of the dimensional water surface elevation and water velocity as functions of time (ζ_I, u_I, v_I)=$(\zeta_{I0}, u_{I0}, v_{I0})e^{i\omega t}$. The dimensionless velocity amplitudes u_{I0} and v_{I0} are calculated by formulas (1.4). The computed amplitudes of the sea surface elevation and water velocities turned out to be in satisfactory agreement with those obtained from observation and presented in Fig. 1.47.

1.6.7 Semidiurnal Tide Diffraction by an Elliptic Island

Along with the general circulation around the amphidromic point described above, the analytical approach in this section will reproduce the trapped clockwise motion in proximity to Hopen Island. For this purpose, diffraction of an incident wave by

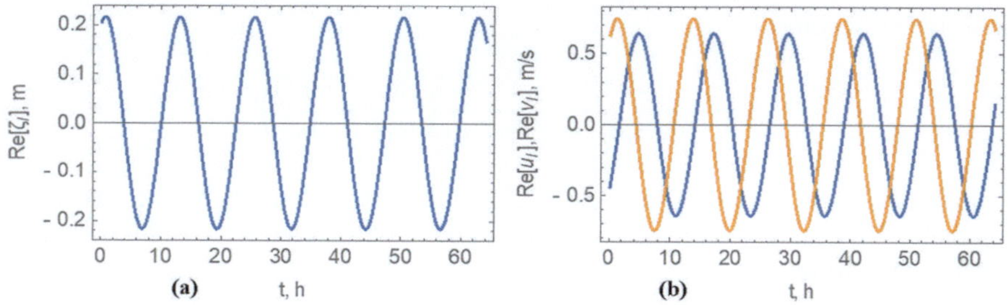

Fig. 1.54 Dimensional water surface elevation Re[ζ_I] (**a**) and water velocity components Re[u_I] (blue line) and Re[v_I] (yellow line) (**b**) as functions of time. These time series result from the superposition of ATW- and CTW+ waves. Adopted from Marchenko and Kowalik (2023)

an elliptic island with a shelf is considered. The Mathieu functions usually express the solution to the sea level oscillations in the elliptical coordinates. Fortunately, in proximity to the critical latitude, these functions can be developed into series of a small parameter, and solutions are expressed as cos and sine series for the angular coordinate (η) and as an exponential function for the radial coordinate (ξ).

The incident wave dimensionless frequency is equal to $\omega = 0.9931$. The island and the shelf boundaries coincide with coordinate lines $\xi_0 = 0.05$ and $\xi_1 = 4\mu$, here $\mu = 0.07$, see Fig. 1.49. The solution of Eq. (1.3) is constructed as a combination of an incident tide and a reflected wave:

$$\zeta_0 = \zeta_{I0} + \zeta_{R+}, \quad \xi > \xi_1; \quad \zeta_0 = \zeta_{R-}, \quad \xi_0 < \xi < \xi_1; \tag{1.19}$$

where, ζ_{I0} is given by (1.15) and (1.18). Functions $\zeta_{R\pm}$ describe waves reflected from the shelf break and the island. They satisfy Eq. (1.3) and the boundary conditions:

$$\lim_{\xi \to \xi_1 - 0} \zeta_0 = \lim_{\xi \to \xi_1 + 0} \zeta_0, \tag{1.20}$$

$$d \lim_{\xi \to \xi_1 - 0} u_{0,\xi} = \lim_{\xi \to \xi_1 + 0} u_{0,\xi}, \tag{1.21}$$

$$\lim_{\xi \to \xi_0} u_{0,\xi} = 0, \tag{1.22}$$

where, the velocity component $u_{0,\xi}$ is calculated from the first formula (1.8). Matching conditions (1.20) and (1.21) mean the continuity of the sea level and the conservation of the mass flux over the shelf break, respectively. Boundary condition (1.22) states the no-flow condition through the island. Conditions (1.20)–(1.22) are identical to the matching conditions used by Kowalik and Marchenko (2002) at the shelf break and at the contour of the circular island.

The amplitude of the reflected wave $\zeta_{R\pm}$ is constructed using Eq. (1.3) written in elliptic coordinates (1.6). In Eq. (1.3) the constant $d = 1$ at $\xi > \xi_1$ (open water), and $d < 1$ at $\xi_0 < \xi < \xi_1$ (shelf). The Laplace operator written in the elliptical coordinates becomes

$$\Delta = \frac{2}{\mu^2(\cosh 2\xi - \cos 2\eta)}\left(\frac{\partial^2}{\partial\xi^2} + \frac{\partial^2}{\partial\eta^2}\right). \quad (1.23)$$

First, we introduce (1.23) into Eq. (1.3), and afterward, applying the method of separation of variables, the sea level amplitude is written as $\zeta_0 = F(\xi)G(\eta)$. This results in the following equation

$$\frac{1}{\cosh 2\xi - \cos 2\eta}\left(\frac{1}{F}\frac{d^2F}{d\xi^2} + \frac{1}{G}\frac{d^2G}{d\eta^2}\right) = -2q, \quad (1.24)$$

where, $q = (\mu K/2)^2 d^{-1}$.

Equation (1.4) yields two ordinary differential equations, see Reynolds (1978)

$$\frac{d^2G}{d\eta^2} + (a - 2q\cos 2\eta)G = 0, \quad (1.25)$$

$$\frac{d^2F}{d\xi^2} - (a - 2q\cosh 2\xi)F = 0, \quad (1.26)$$

where, a is the constant of separation.

Equation (1.25) has two linearly independent solutions for arbitrary values of constants a and q. We consider only periodic solutions of Eq. (1.25) with arbitrary real q existing when $a = a_n(q)$ $(n = 0,1,\ldots)$, and $a = b_n(q)$, $(n = 1,2,\ldots)$ (McLachlan 1951). Such solutions are called Mathieu functions of the first kind $ce_n(\eta, q)$ and $se_n(\eta, q)$. Linearly independent solutions of Eq. (1.26) corresponding to $a = a_n(q)$ $(n = 0,1,\ldots)$, and $a = b_n(q)$, $(n = 1,2,\ldots)$ are expressed in terms of the modified Mathieu functions of the third kind $Me_m^{(1)}(\xi, q)$ and $Ne_m^{(1)}(\xi, q)$ (Mathieu-Hankel functions) to represent outgoing (reflected) waves (Bhatta 2005).

Even and odd Mathieu functions $ce_n(\eta, q)$ and $se_n(\eta, q)$ can be represented by Fourier series as follows:

$$ce_{2n}(\eta, q) = \sum_{r=0}^{\infty} A_{2r}^{(2n)}\cos(2r\eta), \quad ce_{2n+1}(\eta, q) = \sum_{r=0}^{\infty} A_{2r+1}^{(2n+1)}\cos((2r+1)\eta), \quad (1.27)$$

$$se_{2n+1}(\eta, q) = \sum_{r=0}^{\infty} B_{2r+1}^{(2n+1)}\sin((2r+1)\eta), \quad se_{2n+2}(\eta, q) = \sum_{r=0}^{\infty} B_{2r+2}^{(2n+2)}\sin((2r+1)\eta).$$

Since $K^2 = -0.0138$ and $\mu = 0.07$, we estimate $q \approx -1.7 \cdot 10^{-5}/d$. The depth at the shelf is greater than 10 m, and the open ocean depth is 100 m. It means that coefficient $|q| \approx 1.7 \cdot 10^{-5}$ at $\xi > \xi_1$, (open ocean) and $|q| < 1.7 \cdot 10^{-4}$ at $\xi \in (\xi_0, \xi_1)$ (shelf). Further, we consider solutions of Eq. (1.3) with small negative values of q, and set $d = 0.2$, i.e., the shelf depth is 20 m.

Constants $a_n(q)$ and $b_n(q)$ are calculated as power series of q, and the first terms in these expansions are equal to n^2 (McLachlan 1951). Functions sin$n\eta$ and cos$n\eta$ satisfy Eq. (1.24) with an accuracy of $O(q)$. Numerical estimates of the Fourier coefficients show that

$$A_n^{(n)} \approx 1,\ B_n^{(n)} \approx 1 \ and\ \left|A_n^{(m)}\right| \ll 1,\quad \left|B_n^{(m)}\right| \ll 1 (m \neq n),\quad when\ \ |q| \ll 1. \tag{1.28}$$

From (1.27) and (1.28) it follows that

$$ce_m(\eta, q) \approx \cos(m\eta),\quad m \geq 0;\quad se_m(\eta, q) \approx \sin(m\eta),\quad m > 0. \tag{1.29}$$

With the same order of accuracy, the linearly independent solutions of Eq. (1.26) expressed by the Mathieu functions of the third kind are approximated by hyperbolic functions cosh$n\xi$ and sinh$n\xi$ if $n > 0$. If $n = 0$ the linearly independent solutions of Eq. (1.26) are reduced to a linear function of ξ and a constant.

From (1.29), it follows that for small values of q the solution of Eq. (1.3) can be written as a series:

$$\zeta_R(\eta, \xi, q) = \alpha_{01}\xi + \alpha_{02} + \sum_{n=1}^{\infty} \cos n\eta[\alpha_{n1}e^{-n\xi} + \alpha_{n2}\cosh n\xi]$$
$$+ \sum_{n=1}^{\infty} \sin n\eta[\beta_{n1}e^{-n\xi} + \beta_{n2}\sinh n\xi], \tag{1.30}$$

where, constants α_0, α_{n1}, α_{n2}, β_{n1}, and β_{n2} are found from the boundary conditions.

The approximate solution (1.30) is valid when $2|q|\cosh 2\xi \ll 1$. For example, assuming $2|q|\cosh 2\xi = 0.1$ and $|q| = 1.7 \cdot 10^{-4}$ we arrive at an estimate for the coordinate $\xi < 3.7$. From (1.6) it follows that Eq. (1.30) approximates the solution of Eq. (1.3) inside a circle with coordinates $x, y < 1.4$. Figure 1.52 shows that this condition is satisfied in the vicinity of the elliptic island including amphidromic point A.

The values of q are different at the shelf and outside the shelf: $q = q^+ = (\mu K/2)^2$ at $\xi > \xi_1$ and $q = q^- = (\mu K/2)^2 d^{-1}$ at $\xi_0 < \xi < \xi_1$. Therefore, the constants α_{n1}, α_{n2}, β_{n1}, and β_{n2} are also different in these regions. Further we mark these values by superscripts "+" for ($\xi > \xi_1$) and "–" for ($\xi_0 < \xi < \xi_1$). Constants $\alpha_{n2}^+ = 0$ ($n \geq 0$) and $\beta_{n2}^+ = 0$ ($n > 0$) because the reflected waves decay as $\xi \to \infty$. From the boundary condition (1.22) integrated over the variable η from $-\pi$ to π and from the velocity in the elliptical coordinates (1.8) it follows that $\alpha_{01}^- = 0$.

Further, we represent the incident wave given by formulas (1.15) and (1.19) at $\xi = \xi_1$ as Fourier series

$$\zeta_{I0} = \alpha_{I,0} + \sum_{n=1}^{\infty}(\alpha_{I,n}\cos n\eta + \beta_{I,n}\sin n\eta). \tag{1.31}$$

Substituting of (1.30) and (1.31) into the boundary conditions (1.20)–(1.22) and equating the coefficients of $\cos n\eta$ $(n \geq 0)$ and $\sin n\eta$ $(n > 0)$ at both sides of the equations we find a system of nonhomogeneous linear algebraic equations for calculation constants $\alpha_{n1}^{\pm}$, $\alpha_{n2}^{\pm}$, $\beta_{n1}^{\pm}$, and $\beta_{n2}^{\pm}$, where the coefficients $\alpha_{I,n}$ and $\beta_{I,n}$ are the "source" terms associated with the incident wave.

Numerical estimates based on formulas (1.15) and (1.19) show that at the shelf break:

$$\left(\zeta_{I0} - \alpha_{I,0} - \alpha_{I,1}\cos\eta - \beta_{I,1}\sin\eta\right)\zeta_{I0}^{-1} \approx 10^{-3}, \xi = \xi_1. \tag{1.32}$$

It means that series (1.30) can be simplified using the same order $O(10^{-3})$ of accuracy

$$\zeta_R^+ = \alpha_{01}^+\xi + \alpha_{11}^+\cos\eta e^{-\xi} + \beta_{11}^+\sin\eta e^{-\xi}, \tag{1.33}$$

$$\zeta_R^- = \alpha_{02}^- + \cos\eta\left[\alpha_{11}^- e^{-\xi} + \alpha_{12}^-\cosh\xi\right] + \sin\eta\left[\beta_{11}^- e^{-\xi} + \beta_{12}^-\sinh\xi\right]. \tag{1.34}$$

Thus, the above solution is constructed by using trigonometric functions in the periodic (azimuthal) direction, and exponential functions along the radial coordinate. Eight constants α_{01}^+, α_{11}^+, β_{11}^+, α_{02}^-, α_{11}^-, α_{12}^-, β_{11}^-, and β_{12}^- are expressed through the coefficients $\alpha_{I,n}$ and $\beta_{I,n}$ using boundary conditions (1.20)–(1.22).

In Fig. 1.55, the calculated sea surface elevations, and in Fig. 1.56, velocity vectors are plotted at different phases of the semidiurnal tide. The sea-level dipole rotating CW is formed near the island. Maximal amplitudes of the sea level and the strongest currents are found near the island tips O_1 and O_2.

Figure 1.57a shows the magnitude of the velocity vector calculated by the formula

$$V = \sqrt{|u|^2 + |v|^2}, \tag{1.35}$$

where, u and v are the complex components of the water velocity, and V is strongly amplified at the island tips. The maximum water velocities are confined to boundary zones with the dimensionless radius of 0.05 near the tips. The dimensional radius of these zones is close to 10 km.

Figure 1.57b shows co-phase lines and isolines of the dimensional sea surface elevation. The highest amplitude occurs near O_1. The lowest amplitudes are near O_2. The co-phase lines highlight the CCW propagation of the M_2 tide at some distance from the island, while the dipole rotates CW around the island (Fig. 1.55).

An example of Lagrangian trajectory of a water particle trapped near the island, calculated for the initial position $\xi = \xi_1 - 0.5$, $\eta = 0.9\pi$ at $t = 0$, is given in Fig. 1.58a. The initial point of the trajectory is marked by a black dot. Simulations were performed from the initial time $t = 0$, to $t = 64$, i.e., for approximately 5.2×12.42 h of the dimensional time. Over this time interval the water particle travelled

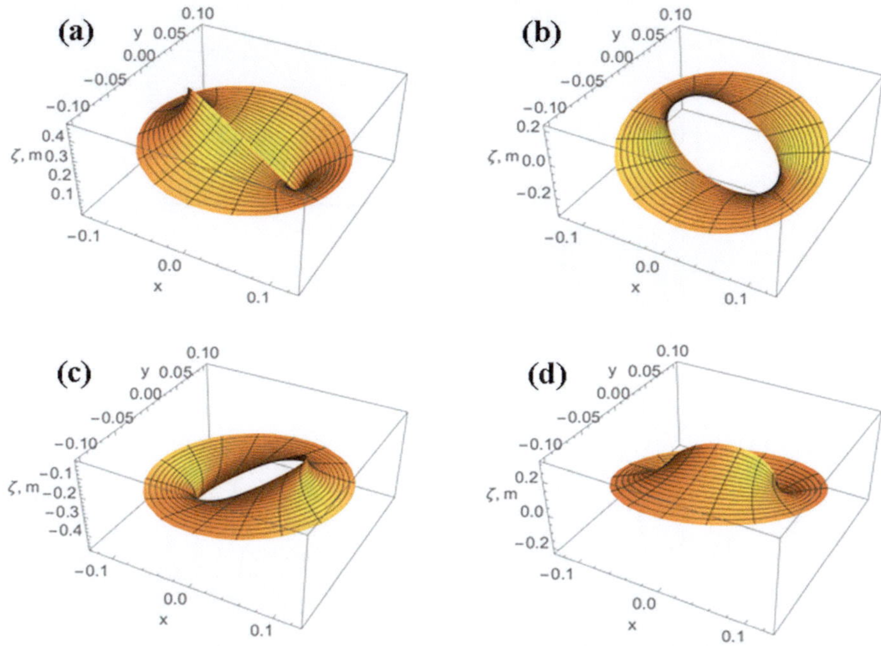

Fig. 1.55 Trapped dipole in the sea-level distribution around the elliptic island, revealed by sea level maximum and minimum traveling clockwise on the opposite sides of the island at $t = 0$ (**a**), $t = 0.25T$ (**b**), $t = 0.5T$ (**c**), and $t = 0.75T$ (**d**). T is the semidiurnal M_2 period. Adopted from Marchenko and Kowalik (2023)

CW around the island. The trajectory includes two CW loops on each side of the island. The dimensional speed of the particle as a function of time (Fig. 1.58b) reveals the highest speeds of approximately 3 m/s near points O_1 and O_2, while for the rest of the trajectory the speed ranges between zero and 1 m/s. The shape of the trajectory and the particle's speed are in good agreement with the observations shown in Fig. 1.45 a and b.

Figures 1.59a–d show respectively several Lagrangian trajectories calculated with the following initial conditions

$$\xi = 2\xi_1, \quad \eta = 0.5(m-1)\pi, \quad t = 0; \tag{1.36a}$$

$$\xi = 2\xi_1, \quad \eta = 0.5(m-1)\pi, \quad t = T/2; \tag{1.36b}$$

$$\xi = 4\xi_1, \quad \eta = 0.5(m-1)\pi, \quad t - 0; \tag{1.36c}$$

$$\xi = \xi_1 - 0.1, \quad \eta = 0.5(m-1)\pi, \quad t = 0. \tag{1.36d}$$

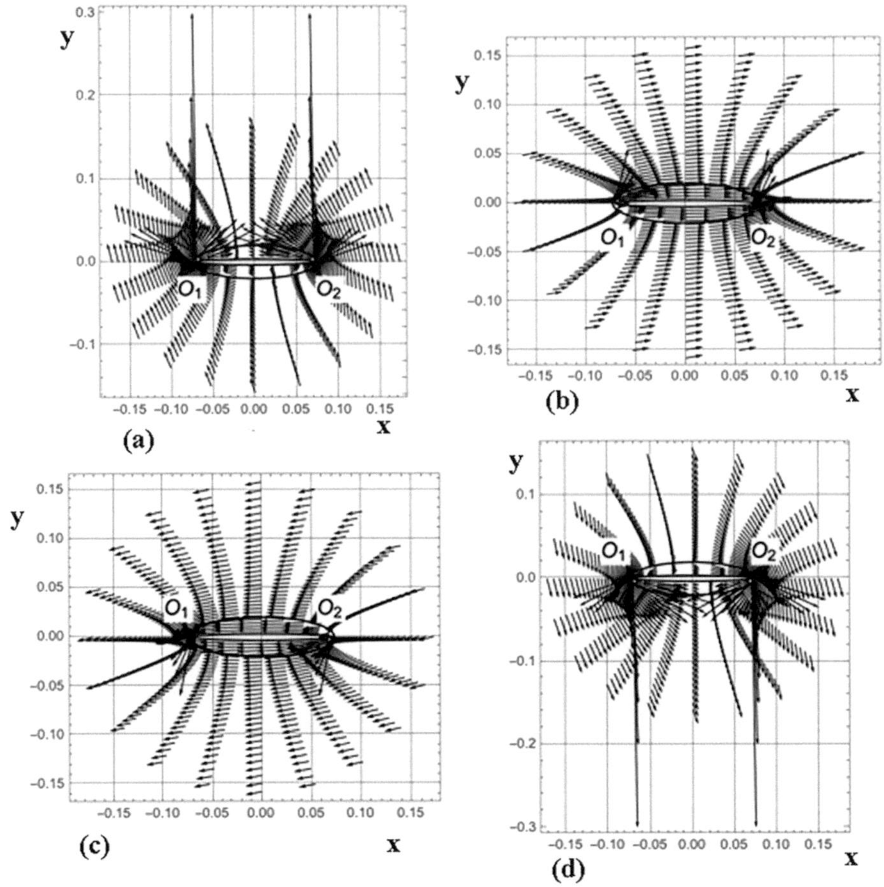

Fig. 1.56 Velocity vectors near the elliptic island, calculated at $t = 0$ (**a**), $t = 0.25T$ (**b**), $t = 0.5T$ (**c**), and $t = 0.75T$ (**d**). T is the semidiurnal M_2 period. Adopted from Marchenko and Kowalik (2023)

Blue, yellow, green, and red trajectories were constructed respectively for the values $m = 1, 2, 3, 4$. Simulations were performed for dimensionless time 80, starting from the initial time, or for approximately 12.6 tidal periods.

Figure 1.59a corresponds to the initial position (Eq. 1.36a), where the particles are placed outside of the shelf ($\xi = 2\xi_1$). The values of η in (Eq. 1.36a) specify the starting azimuthal coordinate of the particles. Figure 1.59a, b show how a half-period shift in the initial tidal phase affects the particle trapping. The particles initially located at $\eta = 0.5\pi$ (yellow) and $\eta = 1.5\pi$ (red) are trapped by the island for both values of the initial phase, while trapping of $\eta = 0$ (blue) and $\eta = \pi$ (green) particles depends on the initial phase. Figure 1.59c indicates that the trapping is not present when the particles start farther from the shelf ($\xi = 4\xi_1$). Figure 1.59d demonstrates

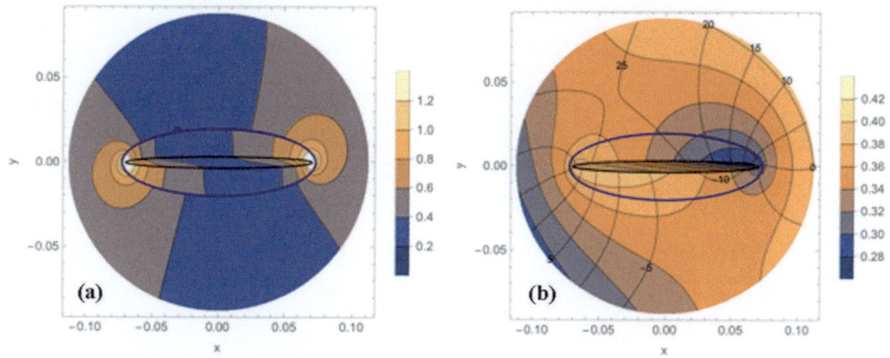

Fig. 1.57 Amplitude of the water speed (m/s) (**a**) and co-phase lines (degree) and isolines (colors) of dimensional amplitudes (m) of the sea surface elevation in the vicinity of the elliptic island (**b**). Adopted from Marchenko and Kowalik (2023)

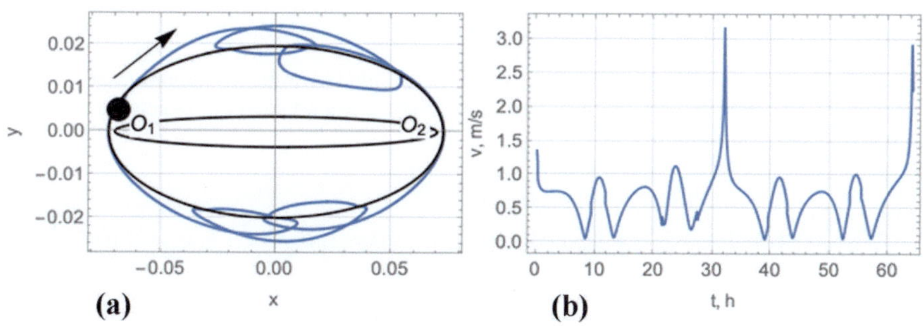

Fig. 1.58 Trajectory of a Lagrangian water particle trapped near the island (**a**) and dimensional speed of the water particle along the trajectory as a function of time (**b**). Adopted from Marchenko and Kowalik (2023)

that all trajectories initiated within the shelf ($\xi = \xi_1 - 0.1$) exhibit the island trapping, but its characteristics might substantially differ.

1.6.8 Elliptic Island with Shelf—Eigen Modes of Water Oscillations

We consider water motion resulting from waves trapped near an elliptic island with a shelf, assuming that incident wave is absent, and frequency ω of trapped waves is not specified. Amplitudes of sea surface elevation are described according to (1.33) and (1.34) by the formulas

$$\zeta_R^+ = \alpha_{n1}^+ \cos(n\eta) e^{-n\xi} + \beta_{n1}^+ \sin(n\eta) e^{-\xi n}, \tag{1.37}$$

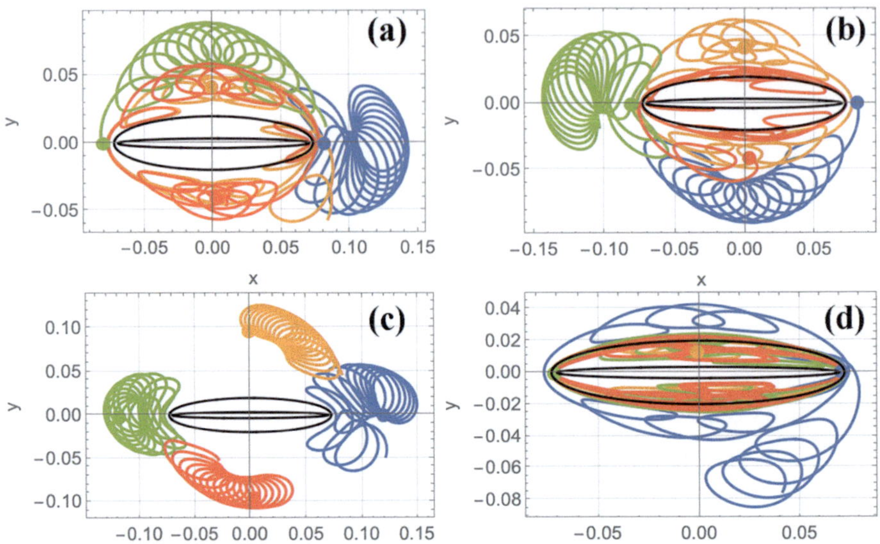

Fig. 1.59 Trajectories of Lagrangian water particles specified by initial conditions (1.36a, 1.36b, 1.36c and 1.36d). Blue, yellow, green, and red dots show the beginning of the trajectories. Adopted from Marchenko and Kowalik (2023)

$$\zeta_R^- = \cos(n\eta)\left[\alpha_{n1}^- e^{-n\xi} + \alpha_{n2}^- \cosh(n\xi)\right] + \sin(n\eta)\left[\beta_{n1}^- e^{-n\xi} + \beta_{n2}^- \sinh(n\xi)\right], \tag{1.38}$$

where, $n = 1, 2, 3, \ldots$ is the mode number representing azimuthal harmonics. Six constants α_{n1}^+, β_{n1}^+, α_{n1}^-, α_{n2}^-, β_{n1}^-, and β_{n2}^- are found from the boundary conditions (1.20)–(1.22). Substitution of formulas (1.37) and (1.38) into the boundary conditions and equating the coefficients at $\cos(n\eta)$ and $\sin(n\eta)$ lead to the homogeneous system of six linear algebraic equations to determine the six constants. The determinant of the system should be zero for the existence of a nontrivial solution. Solution of the determinant results in the dispersion relation between frequency ω and parameters ξ_0, ξ_1, and d. These parameters include gravity and Coriolis forces, the geometry of the island, and the surrounding shelf.

Eigenfrequencies of the first ($n = 1$), second ($n = 2$), third ($n = 3$), fourth ($n = 4$), and fifth ($n = 5$) harmonics of the eigenmodes were calculated versus water depth d. Eigenfrequencies of the first mode ($n = 1$) depicted in Fig. 1.60a explain dependence of the dispersion on the shelf width, specified by the value of ξ_1. Points N_1, N_2, N_3, N_4, and N_5 in Fig. 1.60b belong to the dispersion curves constructed with $\mu = 0.07$, $\xi_0 = 0.05$, $\xi_1 = 4\mu$, and $d = 0.2$. These values are chosen to model the geometry of Hopen Island. The non-dimensional frequencies at points N_n are $\omega_1 = 0.173$ ($n = 1$), $\omega_2 = 0.315$ ($n = 2$), $\omega_3 = 0.427$ ($n = 3$), $\omega_4 = 0.507$ ($n = 4$), $\omega_5 = 0.563$ ($n = 5$). The dimensional periods of the eigenmodes are respectively 3 days ($n = 1$), 1.64 day ($n = 2$), 1.21 day ($n = 3$), 1.02 day ($n = 4$), and 22.06 h

($n = 5$). The eigenfrequency of the fourth harmonic is close to the frequency of the diurnal tide, suggesting resonance in the diurnal band of oscillations, see Huthnance (1974).

As shown in Fig. 1.60a, the eigenfrequency can reach the M_2 frequency $\omega = 0.9931$ at a very small shelf depth d and an extremely wide shelf ($\xi_1 \gg \xi_0$). We conclude that it is almost impossible to find eigenfrequencies close to the semidiurnal band for realistic islands and shelves. Even if such oscillations could be generated, they would be quickly damped by bottom friction due to small depth.

To compare the dispersion curves for an elliptic island with the previous result for a circular island obtained in Kowalik and Marchenko (2002), we use the dispersion relation between frequency ω and the azimuthal wave number n of a free wave progressing CW around a cylindrical island with a surrounding shelf (see, formula (13) in Kowalik and Marchenko 2002).

Eigenfrequencies of the first mode ($n = 1$) versus depth d are plotted in Fig. 1.61a. The island radius is $a = 0.049$, and the shelf break is located at $r = A$, where $A = 0.05, 0.062, 0.082, 0.11$, and 0.15. The values of a and A are chosen to demonstrate similarity of the dispersion curves in Figs. 1.60a and 1.61a. Location of the islands and shelf breaks are shown in Fig. 1.61b by solid and dashed lines for the elliptic and circular cases respectively. The contour and the shelf break in the elliptic case are constructed for the following parameters, $\mu = 0.07$, $\xi_0 = 0.05$ (island) and $\xi_1 = \mu$, 4μ, 8μ, 12μ, and 16μ (shelf).

Thus, we examined semidiurnal tide dynamics around elongated Hopen Island. Observations and numerical computations (Kowalik and Marchenko 2023) suggest that the island traps M_2 tidal wave as a dipole moving clockwise. Concurrently, the primary tidal wave is progressing counterclockwise around the amphidromic

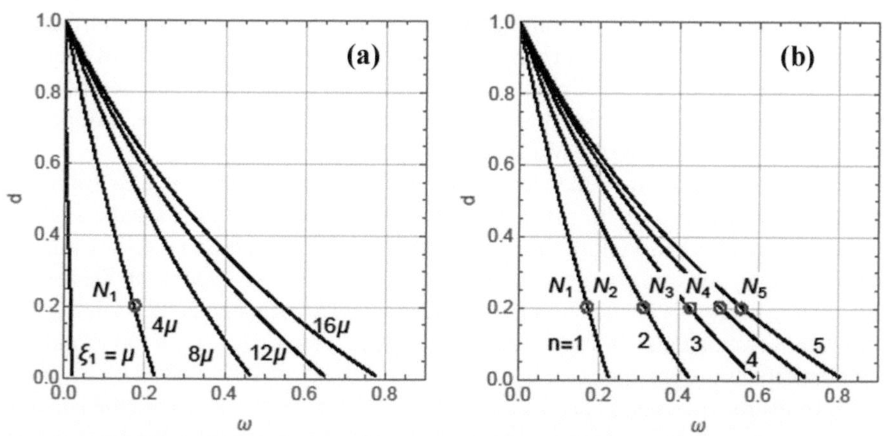

Fig. 1.60 Dispersion curves for an elliptic island: eigenfrequency ω as function of depth d. The first natural mode ($n = 1$) constructed for different shelf widths is described by ξ_1, (**a**). First five modes of the natural oscillations are determined by numbers n and the shelf width $\xi_1 = 4\mu$, (**b**). Points N_1, N_2, N_3, N_4, and N_5 belong to the dispersion curves at $d = 0.2$. Adopted from Marchenko and Kowalik (2023)

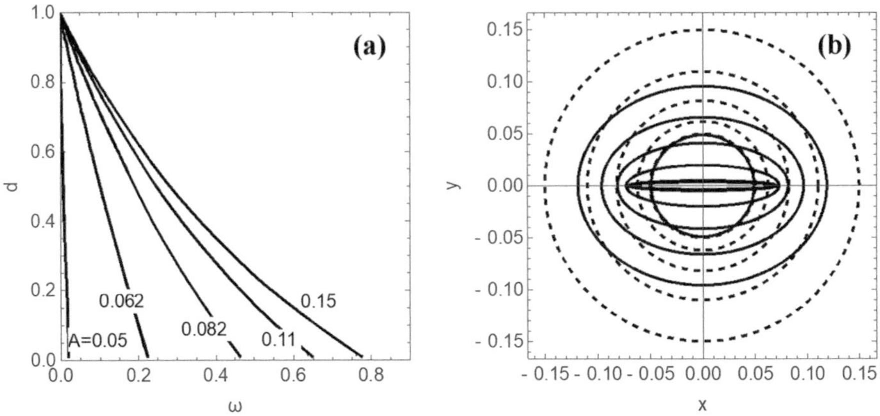

Fig. 1.61 Dispersion curves for a circular island: eigenfrequency ω as function of depth d. The island radius is $a = 0.049$, the shelf break is located at $r = A$. The first natural mode ($n = 1$) is constructed for varying A (**a**). Contours of elliptic and circular islands are given by solid lines. Shelf breaks: solid lines around the elliptic island, dashed lines around the circular island (**b**). Adopted from Marchenko and Kowalik (2023)

point at some distance from the island. To explain the trapping phenomenon, an analytical model includes both the elliptic island and the surrounding shelf. The analytical solution is subdivided into two steps. In the section where we discuss semidiurnal tidal waves near Hopen Island, the primary wave pattern is described by the incident wave as a superposition of axially symmetric solution expressed by the Bessel functions and complex wave (Furevik and Foldvik 1996). The sea level and currents generated by the incident wave agree well with the observed sea level and speed measured by the drifting buoy. During the discussion on semidiurnal tide diffraction by an elliptic island we solved a diffraction problem for both incident and reflected waves in the elliptic coordinates corresponding to the geometry of Hopen Island. In the constructed analytical solution, the reflected wave's amplitude is expressed by a series of the Mathieu functions.

The incident tidal wave generates mainly the first angular (azimuthal) harmonic at the shelf break, therefore, the diffraction problem was also simplified to the analysis of the first angular harmonics of the reflected wave. Because this is the lowest mode of oscillations, one would expect that it displays the strongest response to the forcing and a stable behavior; see discussion in (Kowalik and Marchenko 2002). The derived analytical solution detailed the interaction of the incident waves with the island and trapping effects around the island.

The constructed solution shows that the sea surface elevation around the island is organized into a dipole structure rotating CW, and the maximum amplitude occurs near the south and north points of the island. Narrow regions of the amplified amplitude behave like boundary zones with the radiuses of about 10 km near the island tips;

they are associated with the largest currents as well. The boundary regions correspond to observed locations of significant buoy drift. These currents also explain frequent formation of vortices clearly seen in satellite images.

Simulated Lagrangian drift of near-island water particles demonstrated the existence of trajectories confined to the island's shelf. Shape and number of loops in these trajectories depend on the initial location of water particles and the tidal phase at the initial time. We managed to construct trapped trajectories with four loops circled by the water particles while travelling around the island. In this case observed trajectories and speed of a buoy are similar to the simulated trajectories and speeds of the Lagrangian particles.

Finally, we conclude that the interaction of the semidiurnal tide with Hopen Island causes high tidal currents near the southern and northern tips of the island and induces a CW circulation around the island. These tidal features would govern drift of the open ice. Fast moving ice floes could be dangerous to navigation. Strong currents may also force intense interaction of ice keels with the seabed.

Theoretical investigations presented here cannot explain all physical effects controlling drift of real floaters (buoys, floes) near Hopen Island. Many factors influence barotropic tidal currents calculated in the paper. The bathymetry near Hopen Island does not display elliptic symmetry (Fig. 1.48). The isobaths reproduce the island shape on the eastern side of Hopen Island, while on the western side, they deviate further to the west from the island. This shallow bathymetry influences the M_2 tides around the island. The tidal wave along the western side propagates slower when compared with the eastern side. This introduces a phase difference into symmetrical dipole structures found via the elliptic island approximation, see Kowalik and Marchenko (2023).

Winds may destroy the trapped motion of floats and force them to drift away from the island due to the wind drag and waves (Turnbull and Marchenko 2022). Hence, the barotropic tide can establish the trapped motion in shallow waters near Hopen Island only in calm weather. Interaction of the tides with the continental slope in the Barents Sea generates internal waves with amplitudes up to 10 m at a depth of 60–80 m (Marchenko et al. 2021). Currents caused by the internal waves modify the barotropic tidal currents predominantly in the deeper water and could be omitted in the considered problem.

Interaction of waves with islands could cause resonance when incident wave frequencies coincide with eigenfrequencies of island trapped waves. Our analysis showed the existence of eigen modes propagating in the CW direction near an elliptic island with a shelf. The frequencies of the eigen modes are consistently below the semidiurnal frequency but can be close or even equal to the diurnal frequency (Huthnance 1974). Therefore, the general picture of the M_2 tide interaction with an island would not change due to the resonance; such phenomenon may only influence diurnal tide interaction with an island.

Field observations, analytical and numerical computations show that most frequent pattern of tidally trapped sea surface around an island with a shelf is a dipole structure rotating clockwise. We related this pattern to the first azimuthal harmonic of waves trapped around the island. Our previous solution (Kowalik and

Marchenko 2002) for a wave trapped around a cylindrical island surrounded by a sill and the present solution for the elliptical case demonstrate that the trapping and the dipole pattern occur both above and below the critical latitude, while the eigen modes of the trapped waves exist only in sub-inertial frequency range. This analytical validation of the trapping for a wide variety of the island geometry, from circular to elliptical, supports the assumption made by Dyke (2005) that the semidiurnal dipole pattern for an island is relatively immune to variations in geometry.

1.7 Arctic Water Current in the Bear Island Trough

This section was written together with N. A. Diansky and V. V. Fomin

Interaction between dense and cold Arctic waters and relatively warmer Atlantic waters is an important property of the entire North Atlantic region and is a key mechanism of climate formation in whole Europe and Russia. The Barents Sea is an important region of strong interaction between these waters. Together with the Fram Strait it contributes to the water mass exchange between the Arctic and Atlantic oceans (Giraudeau et al. 2016). The Barents Sea is located in a relatively shallow basin north of Europe (Fig. 1.62).

The terms for geographical names in the region are the following: the western boundary of the sea is the Barents Sea Opening (BSO) (Smedsrud et al. 2013); the northern boundary is the Northern Barents Sea Opening (NBSO) (Lind and Ingvaldsen 2012). The eastern strait between Franz Josef Land and Novaya Zemlya

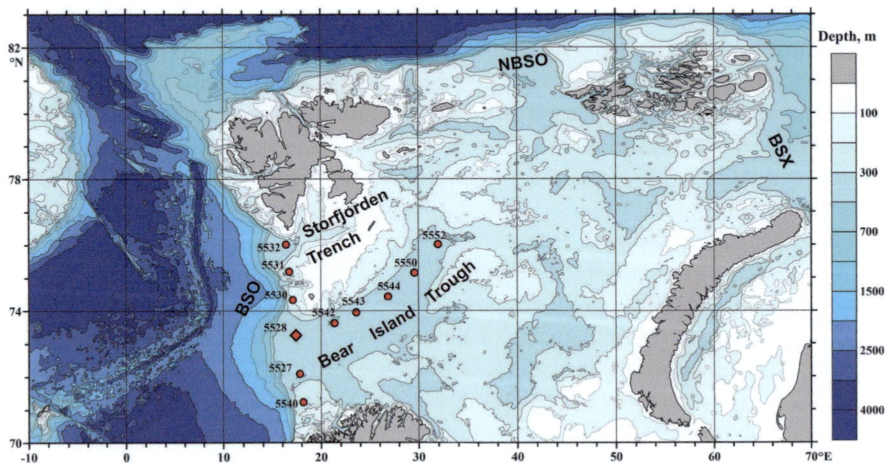

Fig. 1.62 Locations of measurements in the western Barents Sea in July–August 2017. Red circles show our CTD stations; red diamond is the site of mooring deployment in the Bear Island Trough. Bathymetry is based on GEBCO 2019 data. Abbreviations: Northern Barents Sea Opening (NBSO); Barents Sea Opening (BSO); Barents Sea Exit (BSX)

is the Barents Sea Exit (BSX) (Gammelsrød et al. 2009). Water exchange occurs through all three boundaries (Årthun et al. 2011). The first hydrographic measurements in this region were performed in the beginning of previous century (Nansen 1906).

The Bear Island Trough is the deepest channel for the water exchange between the Norwegian and Barents seas. Its depths reach 500 m; the background depths of the upper parts of its walls are 100–200 m shallower. The length of the trough exceeds 500 km and its width is approximately 100 km. The slow bottom current of Deep Barents Sea water in the Bear Island Trough is directed to the Norwegian Sea (Lukashin and Scherbinin 2007b).

The flow of cold and dense Barents Sea waters through the Bear Island Tough is one of the partial sources of Iceland-Scotland Overflow Water, which is generally formed in the Norwegian Sea. In winter, cold dense water is formed here due to intense heat transfer to the atmosphere. The water mass that is formed when flowing over the sill in the Faroe-Shetland Channel is called Iceland-Scotland Overflow Water (ISOW). This water mass mixes with the warmer and saltier waters of the Northeast Atlantic. As a result, ISOW turns out to be significantly saltier and warmer than the deep water that penetrates into the Atlantic through the Denmark Strait between Greenland and Iceland (Denmark Strait Overflow Water, DSOW).

In the Atlantic Ocean, North Atlantic Deep Water occupies a layer between 1200 and 4000 m. According to Koltermann et al. (1999), the densest waters are located in the bottom layers of the Arctic Ocean, but they cannot propagate to the Atlantic owing to the shallow thresholds east of Iceland: Iceland-Faroe and Faroe-Shetland submarine ridges with depths of 480 and 840 m, respectively (Dickson and Brown 1994; van Aken and de Boer 1995; Lankhorst and Zenk 2006). A deep (650 m) pathway is located between Greenland and Iceland in the Denmark Strait.

Waters located above the limiting ridges flow to the Atlantic through the Denmark Strait (Denmark Strait Overflow Water, DSOW) (Jochumsen et al. 2012) and Iceland-Scotland thresholds (Iceland-Scotland Overflow Water, ISOW) (Mauritzen et al. 2005; Olsen et al. 2008). Iceland-Scotland Overflow Water is formed when water from the Norwegian Sea overflows the thresholds east of Iceland. It is generally accepted that North Atlantic Deep Water is formed as a result of mixing between Labrador Sea Water, Iceland-Scotland Overflow Water, Denmark Strait Overflow Water, and Mediterranean Sea Water (Talley 2011). Oceanographers divide the flow of NADW into three layers: Upper (UNADW), Middle (MNADW), and Lower (LNADW). There is no commonly accepted opinion about the origin of North Atlantic Deep Water components.

In this section we review a paper (Morozov et al. 2024). On July 23–August 3, 2017, an expedition onboard the R/V *Akademik Ioffe* occupied two CTD (Conductivity-Temperature-Depth) sections across and along the Bear Island Trough in the western Barents Sea. Each section included six stations. A mooring was deployed with an Acoustic Current Doppler Profiler (ADCP) close to the bottom at the point of intersection (station 5528 at 73°15.6° N, 17°24.6° E) between these two sections in the deepest part of the Bear Island Trough. The bottom topography along the sections was measured by the Kongsberg EA600 12 kHz single beam echo sounder. The locations of the CTD stations and mooring are shown in Fig. 1.62.

The CTD profiles were measured from the surface to the bottom (3–4 m above the seafloor) using a Sea-Bird SBE 911plus profiler. This CTD system was equipped with two parallel temperature and conductivity sensors; the mean temperature difference between them did not exceed 0.001ºC, while that of salinity was not greater than 0.001 PSU. The raw CTD data were processed by SBE Data Processing software version 7.23.2 with standard parameters described in (Sea-Bird Electronics 2014).

Mooring observations of the velocity time series were carried out at 73°15.6' N, 17º24.6' E in 25 m above the seafloor using downward looking Teledyne RD Instruments Doppler Volume Sampler (DVS) 2400 kHz ADCP. It was deployed for 4 days (July 24–28, 2017) with a 10 min sampling interval. Each velocity measurement consisted of 120 pings. The data were processed by standard DVS software (Teledyne RD Instruments 2012). We also used the satellite altimetry data and the TPXO9 model (Egbert and Erofeeva 2002) to plot tidal ellipses and analyze tides from the mooring measurements and altimetry.

Surface currents in the region of the Bear Island Trough are well studied (Beszczynska-Möller et al. 2012). Geostrophic currents based on satellite altimetry averaged over 1993–2020 are shown in Fig. 1.62. Atlantic water transported by the North Atlantic Current overflows the Iceland–Scotland Ridge and continues as the Norwegian Atlantic Current. This current maintains a two-branch structure in the Nordic Seas towards the Fram Strait. Both branches follow topography, the eastern branch (the Norwegian Atlantic Slope Current) is a barotropic current along the continental slope west of Norway. Part of this flow separates and turns to the north. The western branch of the Norwegian Atlantic Current or Norwegian Atlantic Front Current flows as a topographically guided stream to the Fram Strait. Closer to Spitsbergen, this current is called the West Spitsbergen Current. However, only a part of the West Spitsbergen Current continues into the Arctic Ocean, while a significant amount recirculates in the Fram Strait and returns south to the Nordic Seas (Schauer et al. 2004). North of Cape Nordkap, the eastern branch of the Norwegian Atlantic Slope Current enters the Barents Sea. Part of the West Spitsbergen Current near Bear Island separates from the main current and turns east. Geostrophic currents based on long term satellite data are shown in Fig. 1.63 and confirm this scheme.

A similar scheme of currents but based on the altimetry only on the data on July 26, 2017 (the day of our mooring operation) is shown in Fig. 1.64. A distinguishing difference is a stronger southern branch separated from the West Spitsbergen Current west of Bear Island, which turns to the east at 74° N following bottom topography. In addition, the current north of Cape Nordkap is poorly pronounced.

To study the bottom circulation in the region, we applied the Institute of Numerical Mathematics Ocean Model (INMOM) to simulate the circulation in the bottom layer of the Norwegian Sea (Diansky et al. 2002; Zalesny et al. 2012). This model is based on the system of the primitive ocean hydrodynamic equations in spherical horizontal coordinates. We applied the hydrostatic and Boussinesq approximations and the vertical σ-coordinate system. The results of model simulations are shown in Fig. 1.65, which presents the vectors of currents in the bottom layer in July (Morozov

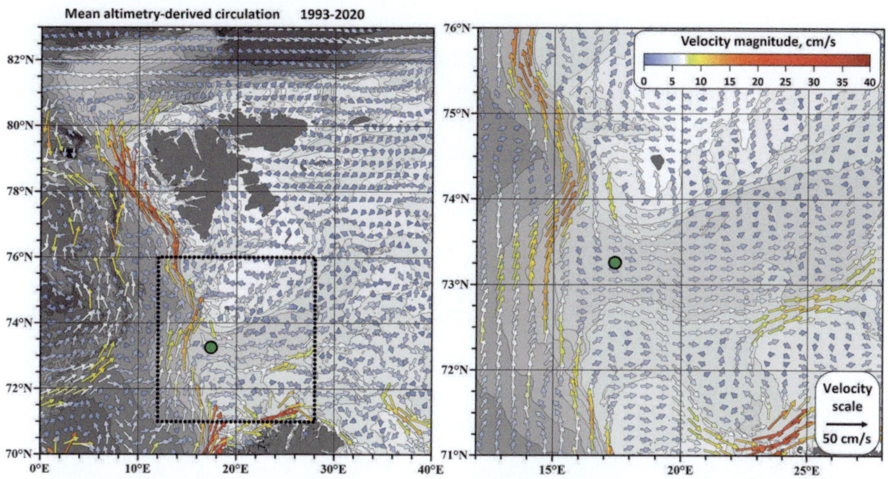

Fig. 1.63 Surface geostrophic currents in the Norwegian Sea based on satellite altimetry averaged over 1993–2020. Part of this map confined by black square is shown in the right panel. The green dot shows location of our mooring in the Bear Island Trough (station 5528). Adopted from Morozov et al. (2024). Used with permission of the Russian Journal of Earth Sciences

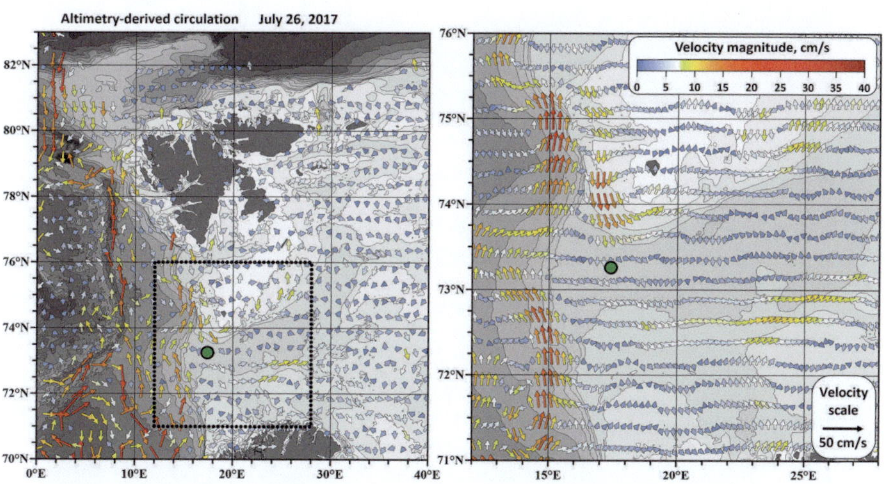

Fig. 1.64 Same as in Fig. 1.63, but on the day of July 26, 2017. Adopted from (Morozov et al. 2024). Used with permission of the Russian Journal of Earth Sciences

et al. 2019a. Bottom currents are quite weak and generally there are no high-velocity jets in the entire sea except for the bottom stream of the bottom water from the Barents Sea flowing through the Bear Island Trough to the southwest. This bottom stream turns to the north around Spitsbergen. These conclusions are consistent with the conclusions based on the direct measurements in the Bear Island Trough in 2017 (Frey et al. 2017). The model results for July 2017 are shown in Fig. 1.65.

This descending flow of cold water in the Bear Island is also illustrated by the water structure over the temperature and salinity sections along the trough. Sections of potential temperature, salinity, and density along the trough are shown in Fig. 1.66a–c. Potential temperature and salinity sections across the through in its deepest part are shown in Fig. 1.67 a and b.

One can see from Fig. 1.66a and b that a tongue of cold and low saline water descends down the trough. Based on the numerical simulations the velocity of the flow is low. It is of the order of 2–4 cm/s, which gives a time of half a year for a water

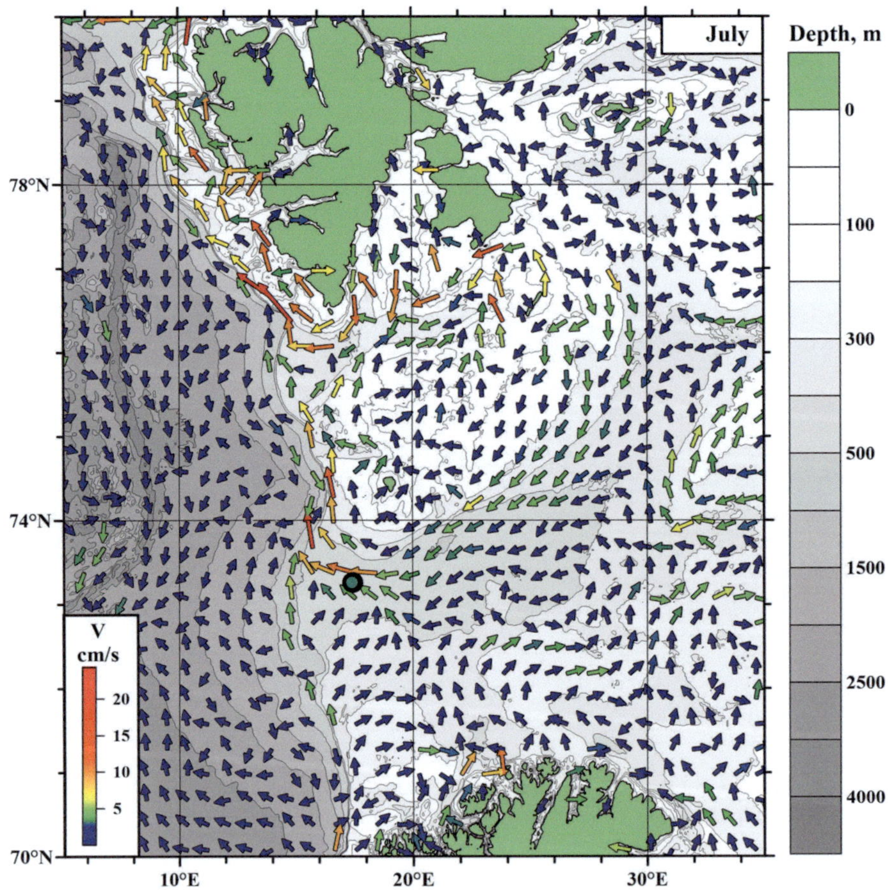

Fig. 1.65 Bottom model currents in the Norwegian Sea based on simulations using the INMOM model for July 2017. The green dot shows the location of our mooring in the Bear Island Trough (station 5528). Adopted from Morozov et al. (2024). Used with permission of the Russian Journal of Earth Sciences

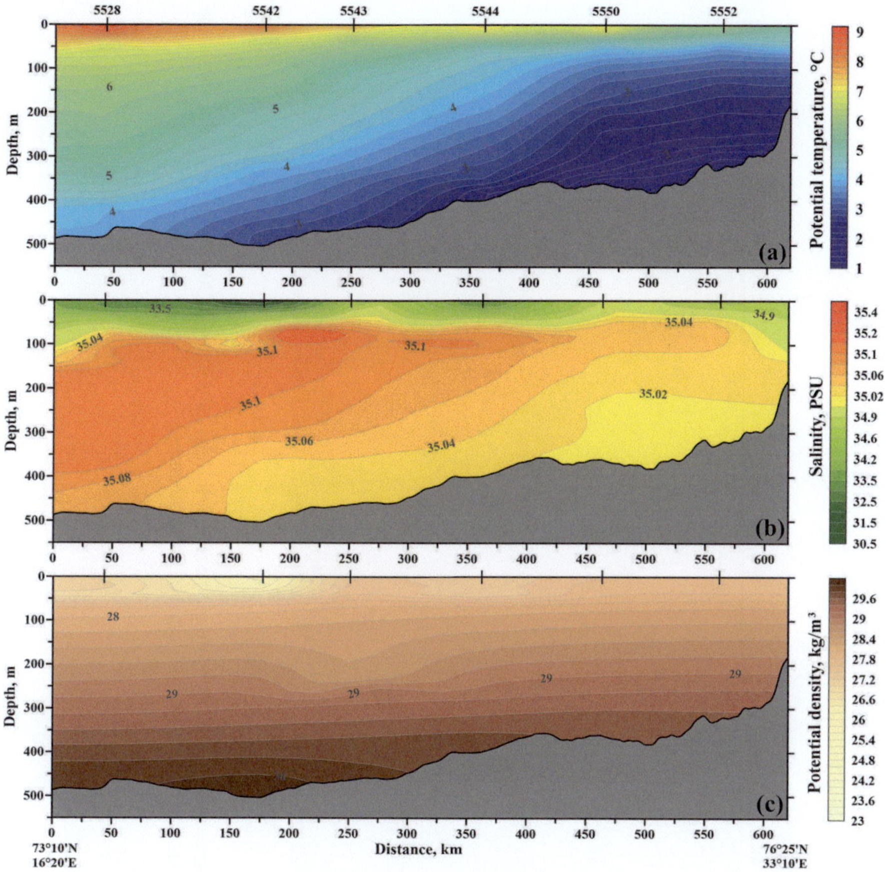

Fig. 1.66 Sections of potential temperature (**a**), salinity, (**b**), and sigma-t density (**c**) along the Bear Island Trough. Adopted from Morozov et al. (2024). Used with permission of the Russian Journal of Earth Sciences

particle to flow along the trough. In the lower part of the flow the current accelerates, but arrives to the mooring with a lower speed.

The surface temperature was approximately 7–8ºC along the entire quasi-meridional section from Norway to 76º N near Svalbard slowly decreasing to the north. Unlike the surface temperatures, the bottom temperatures decreased significantly to the north forming a strong density gradient that drives the geostrophic currents. Even though the bottom water temperature in the deepest part of the Bear Island Trough was quite low (3.52 ºC), it was not the coldest water over this section. The Storfjorden Trench located immediately south of Svalbard is 150 m shallower but the bottom temperature there was much lower (1.05 ºC), which is approximately 2.5 ºC lower than in the Bear Island Trough.

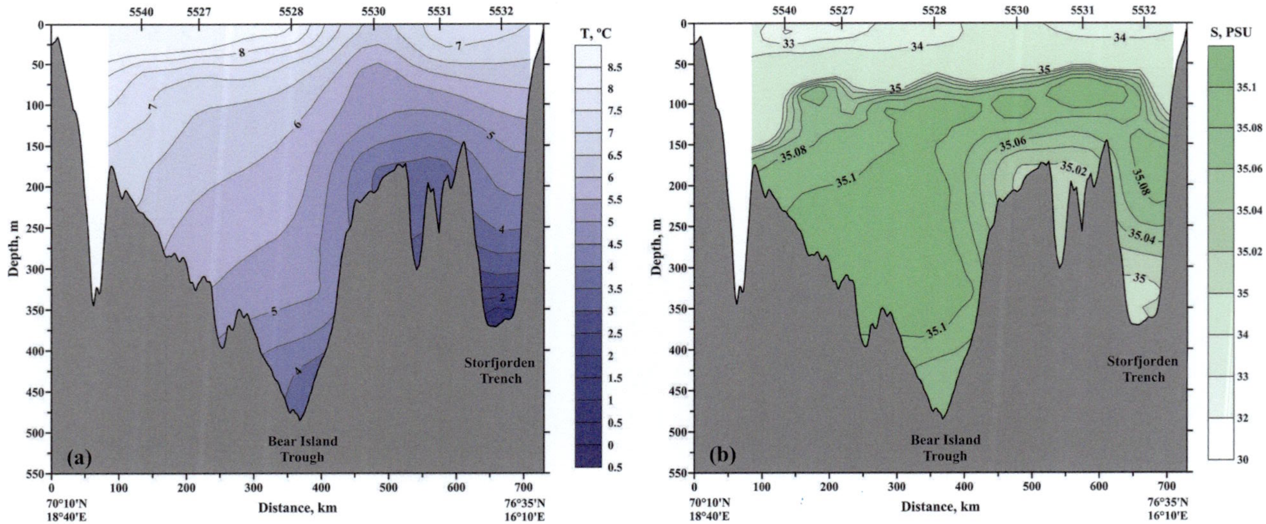

Fig. 1.67 Potential temperature (**a**) and salinity (**b**) sections across the Bear Island Trough measured on July 23–August 3, 2017. The grey color shows the ocean bottom. Adopted from Frey et al. (2017). Used with permission of the Russian Journal of Earth Sciences

The section along the trough (Fig. 1.66a) shows significant increase of bottom temperature in the southwestern direction along this trough. The minimum temperature here was 1.53°C at a depth of 400 m in the northeastern part; the temperature near the exit to the Bear Island Trough (station 5528) was 3.52 °C at a depth of 500 m. The temperature increase can be caused by the influence of warm Atlantic waters in the western part of the Barents Sea and by mixing of cold bottom water with warmer waters above the bottom layer during its southwestward propagation.

The salinity section along the trough reveals a tongue of low saline waters descending down the slope from northeast to southwest (Fig. 1.66). Salinity increases during the propagation of this water downslope the Bear Island Trough. The θ,S-diagrams at stations along the trough, show that the coldest water was found at the bottom of the shallowest station in the northeastern part of the trough. As the water flows down the trough, it becomes warmer and the cold "tails" on the θ,S-diagram are sequentially truncated (stations 5552, 5544, 5528) (Fig. 1.68). The curves differ only above the density level 27.85, which corresponds to 200 m. The θ,S-diagram for station 5516 in the western part of the Norwegian Sea is absolutely different from the diagrams in the Bear Island Trough, which indicates that the water structure in the ocean and in the trough are different. The bottom water in the Bear Island Trough is formed predominantly by descending of cold low saline water from the slopes that becomes warmer and more saline as it flows. The bottom water in the region west of the trough is generally of the oceanic origin. The difference between them drives the geostrophic current to the north.

The flow in the Bear Island Trough enters the Norwegian Sea at the depths of ~450 m. Thus, the inflowing water appears in the upper part of the water column. The water in the columns up to 850 m deep overflows the Faroe Threshold and becomes part of Iceland-Scotland Overflow Water. Figure 1.69 shows comparison between the water density in the deep open part of the Norwegian Sea and that in the Bear Island Trough. The water density in the Bear Island Trough is lower than in the open sea because of the excess of freshwater from land, while the water in the open sea is in direct contact with the Arctic Ocean.

Temperature and salinity profiles at station 5528 are shown in Fig. 1.70. Salinity decreases significantly (from 35.11 to 35.03 PSU) from 200 m to the bottom. The bottom layer about 30 m thick between depths of 450 and 480 m was well mixed. The bottom potential temperature at the westernmost station in the trough was 3.52°C, and salinity was 35.04 PSU. The ADCP velocity measurements were carried out near the upper boundary of this layer. The θ,S-diagrams for four stations along the Bear Island Trough are shown in Fig. 1.68.

No water descent is seen in the density section because the process is density compensated by the process of double diffusion developing over a time period of approximately half a year. Let us estimate whether conditions allow the existence of double diffusion. Double diffusion occurs when cold and low saline water is located above warmer and more saline water. In our case cold, low saline Arctic water descends into the layers of warmer Atlantic water with greater salinity. Thus, the density flux occurs from top to bottom, and the effective density diffusion is positive, which means that the colder water warms (Kantha and Clayson 2000).

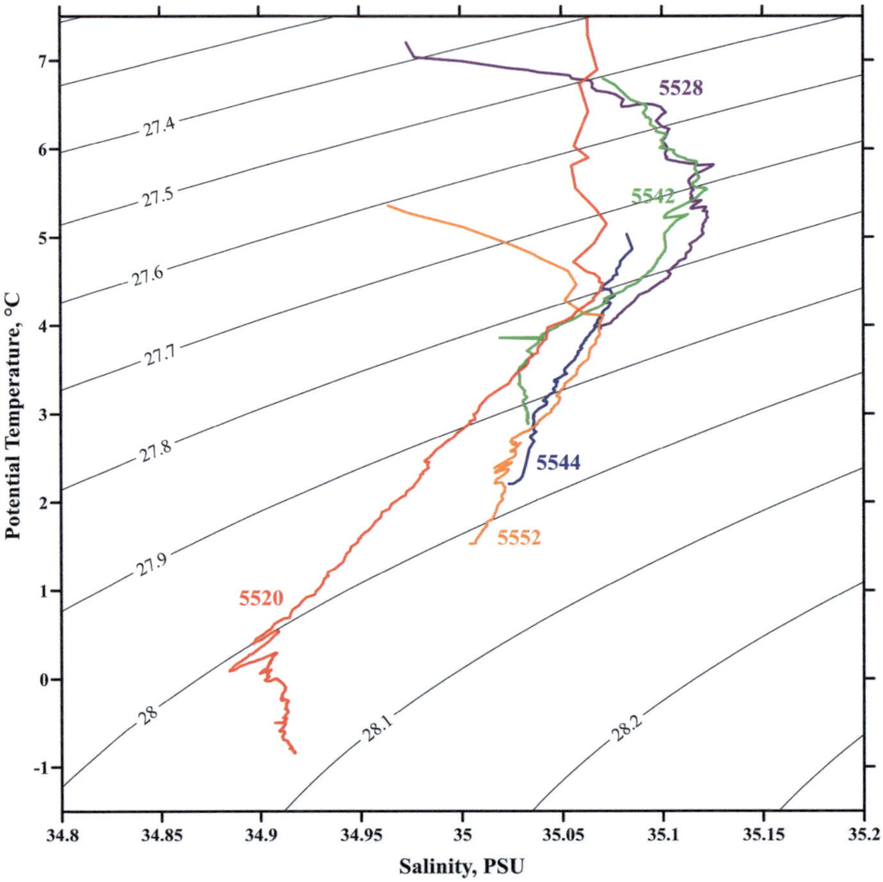

Fig. 1.68 The θ,S-diagrams for four stations along the trough. Stations 5528, 5544, and 5552 are in the Bear Island Trough, station 5520 is southwest of the trough near Jan Mayen Island. Adopted from Frey et al. (2017). Used with permission of the Russian Journal of Earth Sciences

A horizontal flow of colder and less saline water descends in the trough and colder water appears in warmer water of higher salinity. Mechanical mixing occurs, which is accompanied by a faster heat exchange than that with salt and other ions. Warm water of higher salinity in this case ascends. This is a slow process but we estimated the time of the particle motion along the trough as six months, which is enough for this mechanism to be in force.

It was proposed in (Turner 1973) to use the gradient ratio $R\rho = \alpha Tz/\beta Sz$ to estimate the relative intensity of double diffusion. Here, $\alpha = -\rho^{-1}\, \partial\rho/\partial T$ is the coefficient of thermal expansion, and $\beta = \rho^{-1}\, \partial\rho/\partial S$ is the coefficient of salinity compression, ρ is density, and Tz, Sz are vertical gradients of temperature and salinity. The change in density with a change in salinity is approximately 10 times greater than with a change in temperature if they change by one unit. In the ocean, where cold and

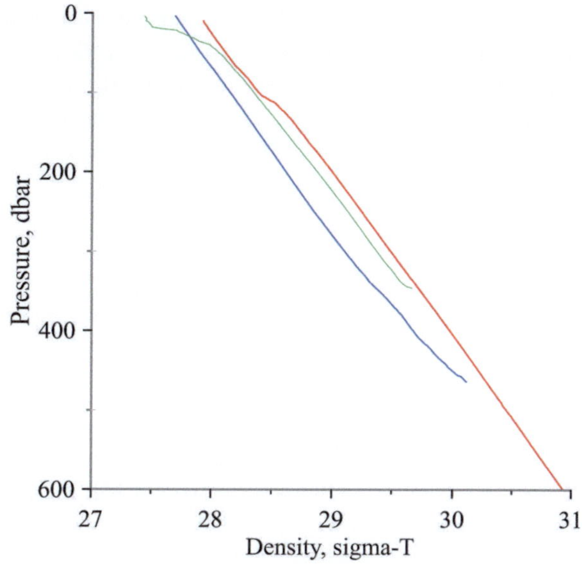

Fig. 1.69 Vertical variation of density at the station in the deep part of the sea at a depth of 3475 m in March 1982 (74.5° N, 4.6° E) (red line) and at stations 5532 (green) and 5550 (blue). Adopted from Morozov et al. (2024). Used with permission of the Russian Journal of Earth Sciences

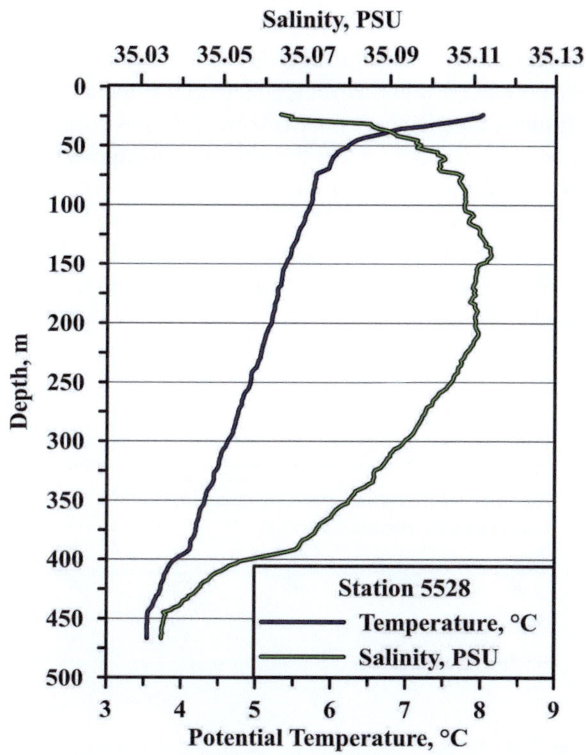

Fig. 1.70 Vertical profiles of potential temperature and salinity at the location of mooring in the Bear Island Trough (station 5528). Adopted from Frey et al. (2017). Used with permission of the Russian Journal of Earth Sciences

low saline waters border with warm and salty waters, coefficient Rρ is close to unity since temperature and salinity have opposite effects on density. Both forms of double diffusion are most intense when coefficient Rρ is close to unity (Shi and Wei 2007).

Physical parameters are determined from the data of measurements, which makes it possible to study double diffusion structures. Turner angles (Tu) are used to study and classify double diffusion (Ruddick 1983), which are determined by the density coefficient (Rρ). In our study area in the Bear Island Trough at depths of 300–350 m we have:

$$R_\rho = \frac{\alpha}{\beta}\frac{T_z}{S_z} = 2.6,$$

where, α is the coefficient of thermal expansion, $\alpha = 150 \times 10^{-6}$ 1/°C for temperatures close to 0 °C; β is salinity compression coefficient is $\beta = -7 \times 10^{-4}$ or 0.7 kg/g. The vertical temperature gradient in the study area at depths of 300–350 m is $T_z = 0.01$ °C/m, the salinity gradient is $S_z = 0.0008$ 1/m. Coefficient $R_\rho = 2.6$ is of the order of unity, which gives reason to believe that double diffusion is possible. Double diffusion occurs at positive values of the density ratio R_ρ, that is, in other words, when the vertical gradients of temperature and salinity have the same sign.

The velocity measurements reveal a westward bottom flow in the Bear Island Trough (Fig. 1.71). Strong tides in this region are responsible for the variations in the bottom currents. We calculated tidal velocities by TPXO7.1 model using satellite altimetry data (Egbert and Erofeeva 2002). A comparison of measured and calculated tidal velocities shows that the observed change of the bottom flow direction is caused by the semidiurnal tides. Nevertheless, the mean current is directed from the Barents to the Norwegian Sea. The four-day mean westward component was 4.1 cm/s; the northward component was 6.6 cm/s.

We compared our recent velocity data with the mooring bottom current measurements in July 1995 (Aleynik et al. 2002). Velocities near the bottom were measured

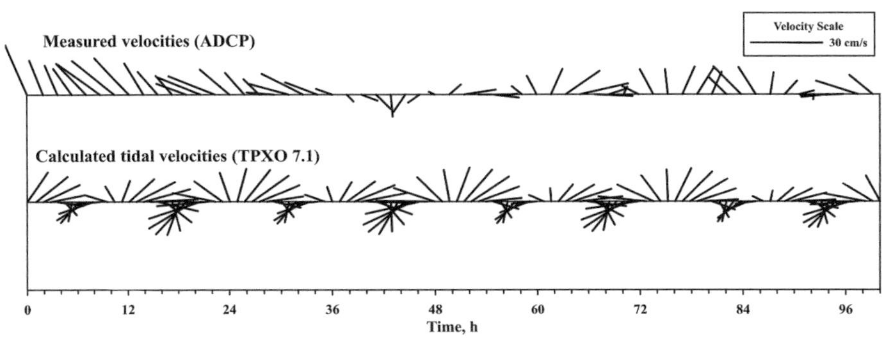

Fig. 1.71 Velocity vectors based on four-day mooring measurements (station 5528) of bottom flow in the Bear Island Trough on July 24–28, 2017 at 73°15.6' N, 17°24.6' E; and calculated tidal velocities for the same period. Adopted from Frey et al. (2017). Used with permission of the Russian Journal of Earth Sciences

in summer of 1995 in the cruise of the R/V *Akademik Mstislav Keldysh*. The mooring was deployed at 73°42' N, 13°06' E. The instrument was located at a depth of 1674 m over the ocean depth at 1687 m. The sampling interval was 1 h. The bottom flow was generally directed to the north (Fig. 1.72). In the period of measurements from July 28 – August 7, the mean northward velocity was 13.2 cm/s, the mean eastward component was 5.2 cm/s. The influence of the semidiurnal tides is also observed over this time series.

The mean values of velocity components based on the mooring measurements were U = –0.04 m/s, V = 0.06 m/s. The study region is characterized by strong tides, which influence the current in the trough and generate strong tidal internal waves, which in turn facilitate mixing (Morozov et al. 2024). The spectrum of velocity fluctuations of the zonal and meridional components reveals peaks at the semidiurnal tidal frequency but we cannot separate the M_2 and S_2 constituents because the time series is very short. The spectra of velocity components are shown in Fig. 1.73.

Let us estimate how strong is the tide using the data of measurements and the TPXO9 tidal model (Egbert and Erofeeva 2002) of the Oregon University. We filtered the velocities of the zonal and meridional components so that all other tidal frequencies except the M_2 and S_2 constituents were filtered out. The time period of the ellipses is from July 25, 18:30 to July 27, 06:50. The tidal ellipses are shown in Fig. 1.74 together with the model ellipses for both the M_2 and S_2 constituents. The measured and model ellipses are quite close, which indicates that the TPXO model is adequate.

After we subtracted the velocities of the tidal ellipse from the initial data, the mean currents changed from U = –0.04 m/s, V = 0.06 m/s to U = –0.03 m/s, V = 0.02 m/s. We consider that these values are more reliable than the initial ones that include averaging with tidal velocities. However, the time series is very short but it shows that an opposite mean flow can be present in the trough. Figure 1.75 illustrates this.

Figure 1.76 shows a comparison of the tidal ellipses for four tidal constituents, which shows that all other constituents they are significantly weaker than the M_2 constituent.

Thermohaline structure and velocity data in the Bear Island Trough showed the bottom water spreading along the trough from the Barents to the Norwegian Sea. The four-day mean velocity at the exit from the trough was 7.8 cm/s directed to the northwest. This flow of warm Norwegian Atlantic Current waters transported heat to the northwest and western part of the Svalbard Archipelago, which does not freeze

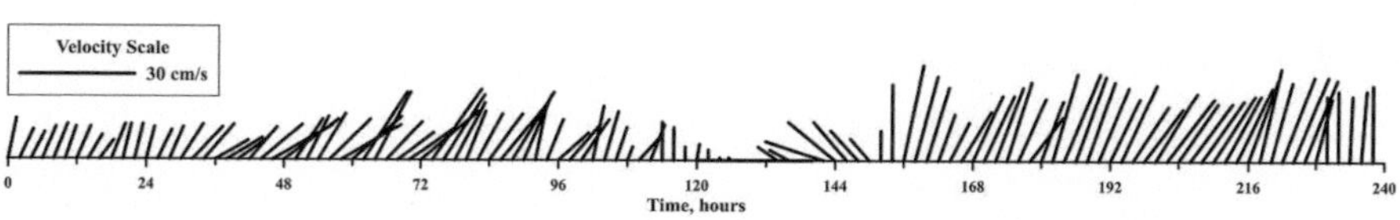

Fig. 1.72 Vectors of bottom currents measured at 73°42' N, 13°06' E in July–August 1995. Adopted from Frey et al. (2017). Used with permission of the Russian Journal of Earth Sciences

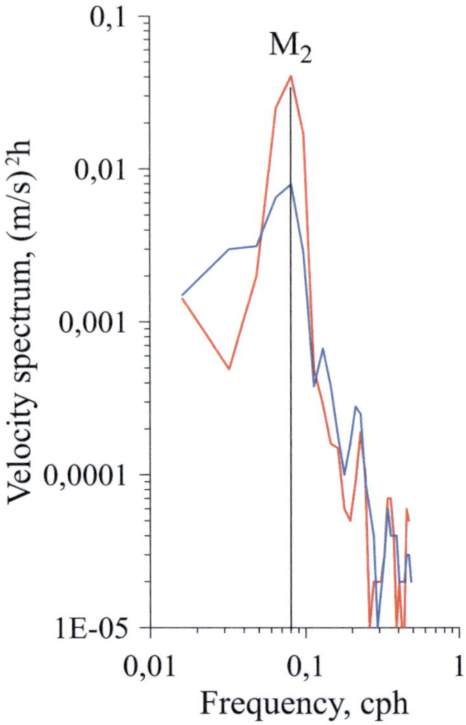

Fig. 1.73 Spectra of velocity fluctuations of the zonal (red) and meridional components (blue) from the mooring. Adopted from Morozov et al. (2024). Used with permission of the Russian Journal of Earth Sciences

in winter. The bottom temperature in the Bear Island Trough was 3.52 ºC, while it was lower (1.05 ºC) in the Storfjorden Trench, which is 150 m shallower. The bottom temperatures along the trough increased from 1.53 ºC in its eastern part to 3.52 ºC in the western boundary of the Barents Sea. Strong semidiurnal tides in this region influence bottom currents significantly, but the mean flow is directed from the Barents Sea to the Norwegian Sea.

Satellite altimetry was used to map surface geostrophic currents averaged over a long-term period and on the day of measurements. Numerical modeling reveals a bottom current in the trough, which is supported by ADCP measurements on a mooring. Water structure (temperature and salinity sections along the trough) also supports the existence of the bottom flow. Mixing occurs in the trough caused by different processes. Tidal ellipses were calculated from the data of mooring and using the TPXO9 tidal model based on satellite altimetry. The M_2 tide dominates in the region.

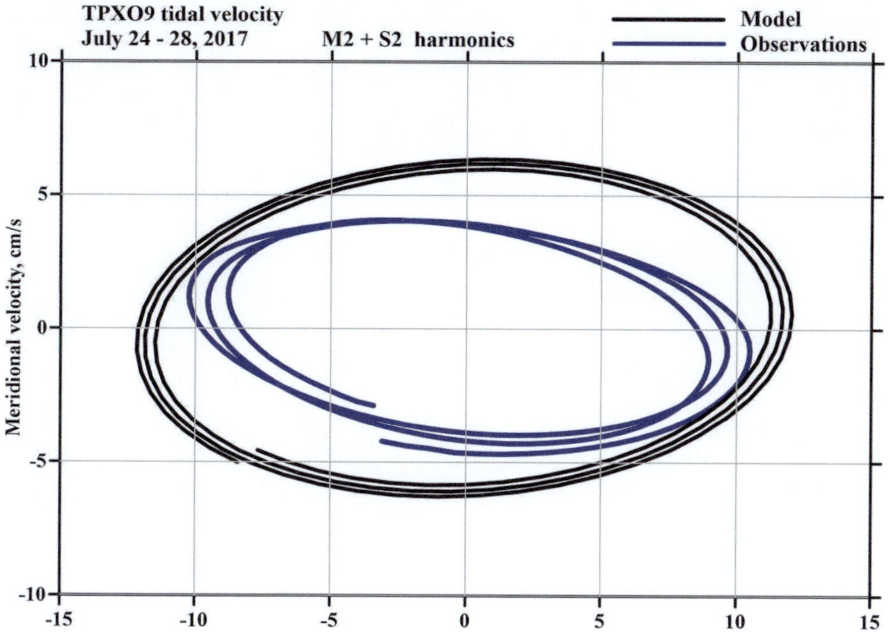

Fig. 1.74 Model (black) and experimental (blue) tidal ellipses. The model ellipses cover a period from July 24 to July 28. Adopted from Morozov et al. (2024). Used with permission of the Russian Journal of Earth Sciences

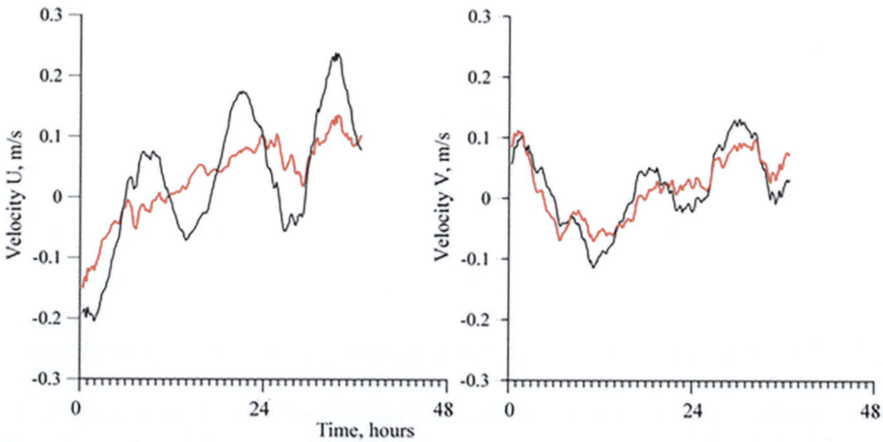

Fig. 1.75 Time series of velocity components (U, left; V, right) before subtracting tidal currents (black) and after subtracting (red). Adopted from Morozov et al. (2024). Used with permission of the Russian Journal of Earth Sciences

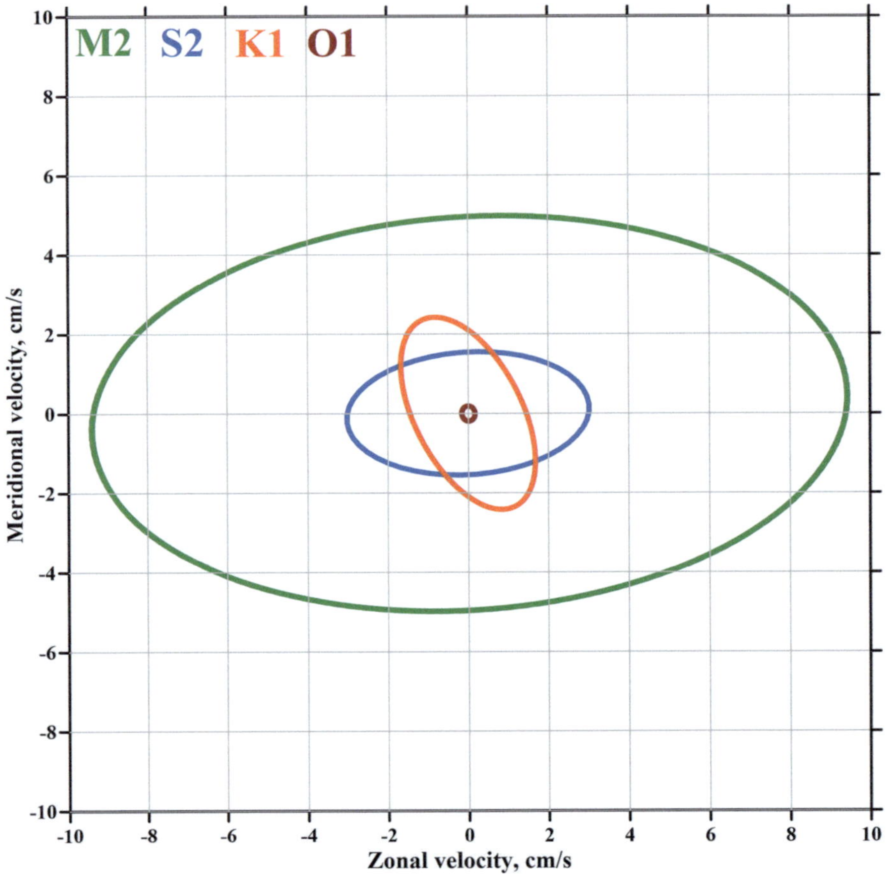

Fig. 1.76 Tidal ellipses for the M_2, S_2, K_1, and O_1 tidal constituents. Adopted from Morozov et al. (2024). Used with permission of the Russian Journal of Earth Sciences

1.8 Currents in the Greenland Sea and in the Fram Strait

Research on water dynamics in polar and subpolar regions is extremely important for studying the variability of the Earth's climate. The Greenland Sea is one of the most important regions of the World Ocean where processes of heat exchange between the atmosphere and the ocean occur. These interactions in the subarctic zone of the North Atlantic influence the climate of Europe and Russia, and this has been repeatedly noted in the literature (Seager et al. 2002). Currents in the upper layer of the ocean have the greatest influence on heat transport in the ocean and interaction with the atmosphere. In this section we review (Morozov and Frey 2024).

Water exchange between the North Atlantic and the Arctic Ocean is one of the most important processes that lead to the existence of meridional circulation and variability of Atlantic water masses. Warm and salty Atlantic waters flowing into the Arctic

Ocean are cooled in the northern seas. Freezing and thawing processes occur, and as a result, fresh water and ice appear in the shallow areas, and salt water sinks to greater depths. The southward flow of Arctic waters gives rise to the deep limb of the global thermohaline circulation cell in the Atlantic (Meridional Overturning Circulation). These processes play an important role in climate variability on interannual and decadal time scales (Daniault et al. 2016).

Northwest of Norway, the Norwegian Atlantic Slope Current merges with the Norwegian Coastal Current branches into the warm jet (the Nordkap Current) flowing into the Barents Sea and the West Spitsbergen Current transporting warm and salty Atlantic waters further north to Spitsbergen. There are two main currents in the Greenland Sea: the West Spitsbergen Current, transporting Atlantic waters to the Arctic, and the East Greenland Current transporting cold Arctic waters to the Atlantic. As the warm waters move north, they are cooled (by 0.25 °C per 100 km of distance traveled) (Cokelet et al. 2008). Water transport by the West Spitsbergen Current is estimated at 6–9 Sv (Beszczynska-Möller et al. 2012; Fahrbach et al. 2001; Schauer et al. 2004). In contrast to the West Spitsbergen Current, water transport by the East Greenland Current remains almost constant and equal to 4 Sv (Fahrbach et al. 2001; Schauer et al. 2004), which exceeds the estimate of 2 Sv in Foldvik et al. (1988). About 2 Sv of recirculation of the West Spitsbergen Current is added to the southward transport (Fahrbach et al. 2001; Hattermann et al. 2016; de Steur et al. 2014). This current makes the main contribution to the removal of sea ice from the Arctic (Woodgate et al. 2020) and determines a significant transport of fresh water from the Arctic to the Atlantic. A scheme of currents in the northern seas is shown in Fig. 1.77.

The basin has been intensely studied using deployments of moorings, CTD sections (Beszczynska-Möller et al. 2012; Fahrbach et al. 2001; Schauer et al. 2004), and observations of drifting ice and icebergs (Yulmetov et al. 2016). We analyzed current measurement data at moorings at 13 points in the Fram Strait in 1997–1998 and 2016–2018, carried out under the ASOF international project (https://asof.awi.de/science/projects/13-monitoring-of-oceanic-fluxes-across-fram-strait/) by scientists from the German Alfred Wegener Institute (AWI, Germany) and the Norwegian Polar Institute (NPI, Norway). Moorings were deployed along a transect across the Fram Strait at approximately 79° N (Fig. 1.77). Coordinates of moorings in 1997–1998 are given in Table 1.1.

Comparative analysis of volume, heat, and salt fluxes estimated from instrumental observations at autonomous moorings in the Fram Strait is given in (Smirnov et al. 2024). It was carried out with similar fluxes calculated from the GLORYS2v4, ORAS5, GloSea5, and C-GLORSv7 reanalyses. We also analyze CTD sections in this region. The data are in a free access. From these data, we selected two CTD sections at 79° and 75° N, from which we can form a concept about hydrology of the basin.

The RCM and Aanderaa current meters were used in 1987–1998; in later deployments of moorings, Aquadopp and ADCP upward-looking current meters were used. In this case, the buoyancy was deepened by approximately 50 m, and the upward radiation of the ADCP current meter was sufficient to record the entire upper layer

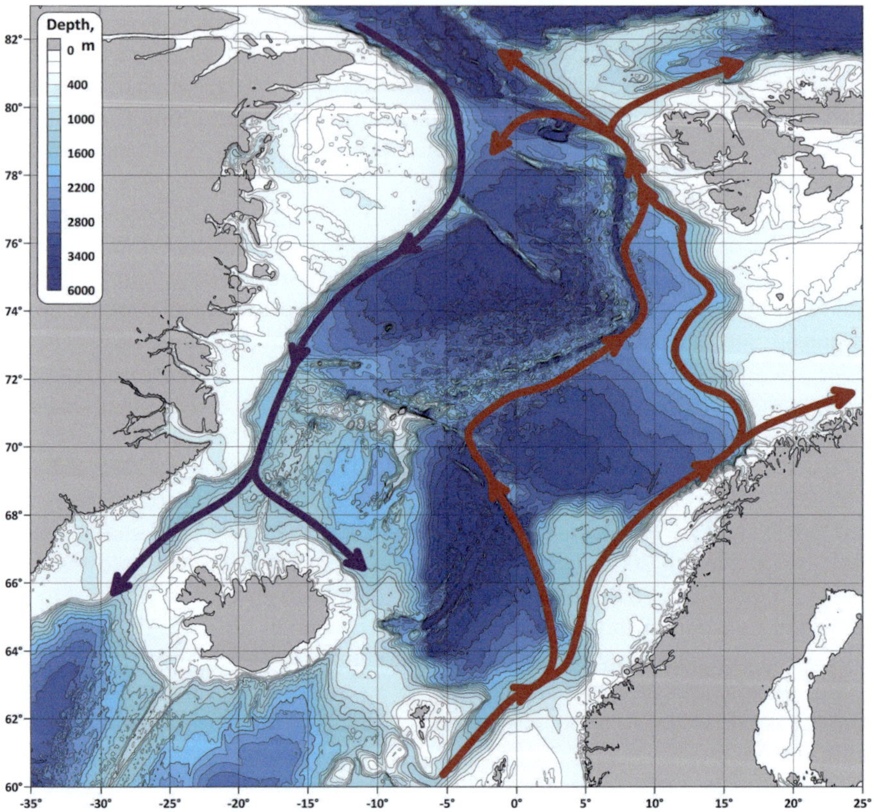

Fig. 1.77 Scheme of currents in the Nordic Seas (Beszczynska-Möller et al. 2012; Fieg et al. 2010). The red line shows the Norwegian Atlantic Current and its continuation, the West Spitsbergen Current. Blue line shows the East Greenland Current

of the sea (above the buoyancy of the mooring), which could contain icebergs and ice.

In addition, we used the data from 12 moorings of the Greenland Sea experiment, which were deployed on the eastern slope of Greenland and in the open sea in 1987–1989. The experiment was carried out by the NOAA/PMEL research laboratory (USA). Data in the Fram Strait are freely available on the website (Fahrbach et al. 2012). Experimental data in 1987–1989 was distributed on disk with archived data from the WOCE experiment.

Oceanographic conditions in the Greenland Sea were studied along two sections in July–August 1998 at 75° and 79° N latitudes (Figs. 1.78 and 1.79) and other stations in this region. Temperature and salinity sections show that cold, low saline Arctic waters occupy the shelf and continental slope of Greenland; the East Greenland Current transports them to the south. The warm West Spitsbergen Current transports warm (up to 7 °C on the surface in summer and +4 °C on the surface in winter)

Table 1.1 Coordinates of moorings in the Fram Strait in 1997–1998

Mooring index	Northern latitude	Eastern longitude	Depth, m
F1	78.830	8.636	305
F2	78.860	8.335	751
F3	78.846	7.961	1045
F4	78.828	6.933	1490
F5	78.823	6.450	1989
F6	78.828	4.995	2629
F7	78.811	4.045	2299
F8	78.831	2.611	2454
F9	78.990	0.261	2449
F10	79.008	–2.041	2564
F12	78.997	–4.227	1826
F13	78.972	–5.319	990
F14	79.029	–6.852	264

waters of continuation branches of the North Atlantic Current north along the coast of Spitsbergen. These waters west of Spitsbergen and the waters in the fjords west of Spitsbergen do not freeze in winter.

The deep layers are occupied by cold and dense Arctic waters. Due to this distribution of water masses an induced maximum of water temperature (of the tropical origin), exists even in summer at depths of 200–500 m in the region of latitude 80° N in the western Greenland Sea. This is caused by the presence of cold waters of the East Greenland Current in the upper layer and cold dense Arctic waters in the deep layers. This warm layer of saltier Atlantic waters thins towards the west (Figs. 1.78 and 1.79). South of 79–80° N, this subsurface layer outcrops to the surface as cold Arctic waters penetrate south, mainly only along the coast of Greenland.

Warm and salty Atlantic water over a section along 79° N is more displaced to the west relative to the displacement over the section along 75° N. This is due to the recirculation of the flow of these waters, which occupy more space in the Fram Strait than in the southern Greenland Sea.

To judge about the constant currents in the Fram Strait, a map of geostrophic currents based on the absolute dynamic topography over the period 1993–2020 and a similar map for March 1998 (the period of joint operation of the moorings, Fig. 1.80) were constructed. Both maps show the East Greenland Current generally flowing to the south, and the West Spitsbergen Current, which transports Atlantic waters to the north. Both maps show the recirculation of the West Spitsbergen Current.

Mean meridional currents were estimated from the meridional component of currents at moorings in the strait. The maximum velocities of the West Spitsbergen Current exceed 50 cm/s during short periods of time. The mean velocities of the East Greenland Current are not high: 6–10 cm/s. The maximum velocities are observed on the surface and reach 60 cm/s. The current occupies the depths up to 500 m on the

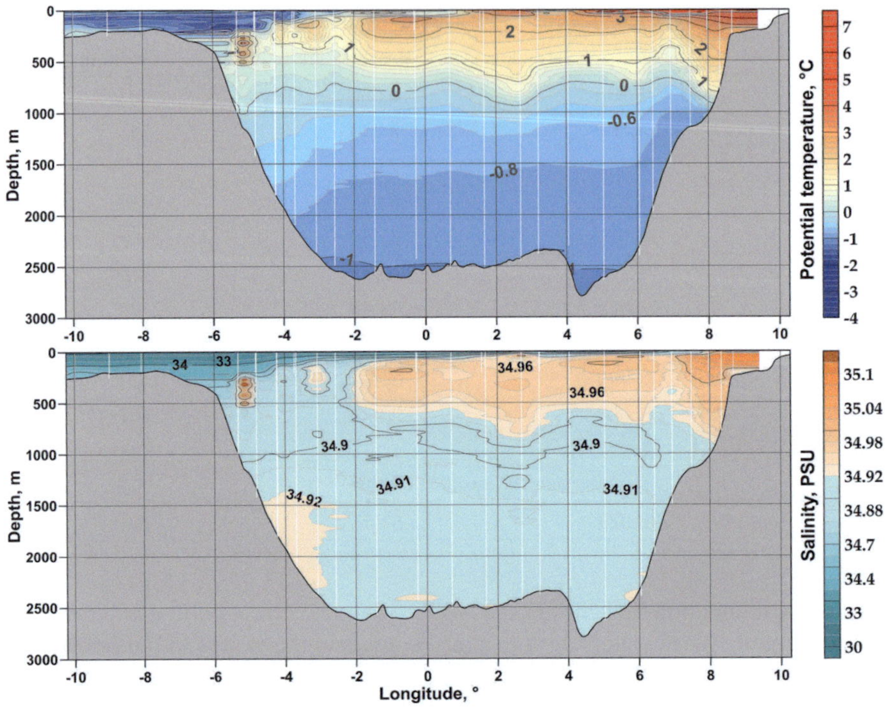

Fig. 1.78 Sections of potential temperature and salinity at 79° N based on the section data in the summer of 1998. White vertical lines show the positions of CTD casts. Adopted from Morozov and Frey (2024). Used with permission of the Pleiades Publishing, Ltd.

Greenland shelf (Fig. 1.81). Tidal currents are characterized by higher speeds than mean background ones. A cross-section of meridional current velocities on the shelf and continental slope of Greenland in the middle part of the Greenland Sea (75° N) is shown in Fig. 1.82.

The East Greenland Current transports cold polar water with low salinity in the upper 150 m. Low salinity is largely due to freshwater runoff from melting sea ice. Typical characteristics of polar water are temperatures below 0 °C to near freezing temperatures of around –1.7 °C. Salinity varies greatly from 30 PSU (at the surface) to 34 PSU at a depth of 150 m. The layer below the polar waters is known as the Atlantic water layer. It extends to about 1000 m. This layer of relatively warm water is more saline. The temperatures typically exceed 0 °C, and salinity increases from 34 PSU at a depth of 150 m to approximately 35 PSU at a depth of 1000 m. Atlantic water in the East Greenland Current flows from two different sources. The first source of Atlantic water originates from the recirculation of Atlantic water from the West Spitsbergen Current in the Fram Strait. Because these waters are denser than surface polar waters, they sink to intermediate depths. The second source of Atlantic waters is the recirculation of Atlantic waters in the Arctic.

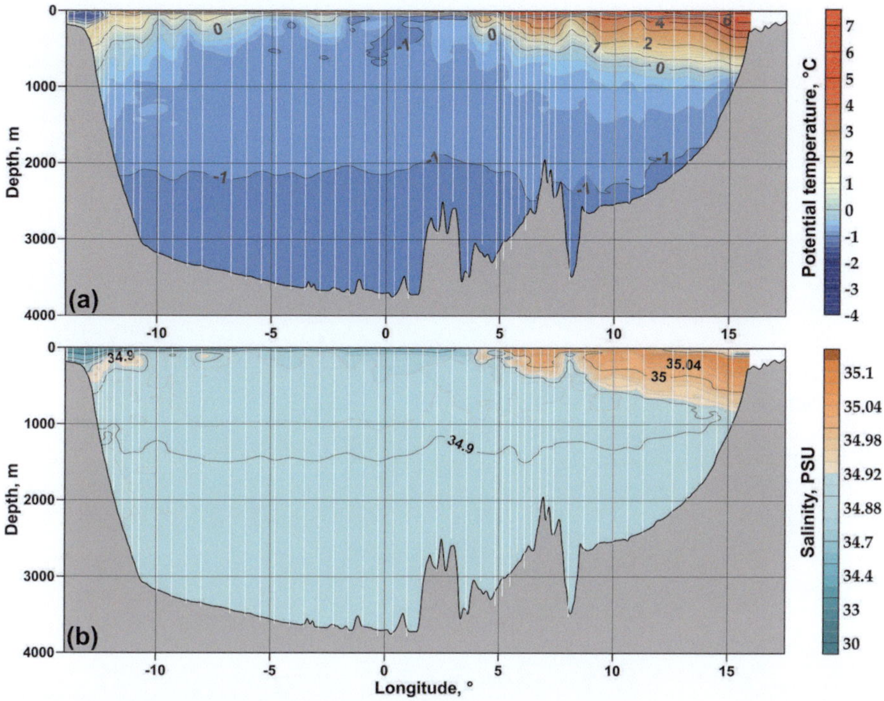

Fig. 1.79 Sections of potential temperature and salinity at 75° N based on the section data made in the summer of 1998. White lines show positions of CTD casts. Adopted from Morozov and Frey (2024). Used with permission of the Pleiades Publishing, Ltd.

We constructed current vectors in the surface layer (0–100 m) based on the measurement data from 1997–1998. Annual mean temperatures differed from the summer and winter values by no more than a few hundredths of a degree. Therefore, the current vectors are colored in a palette of colors corresponding to the annual mean water temperature as measured by the temperature sensor of the current meter. A map of the annual mean current vectors in the surface layer and the corresponding temperatures of these waters are shown in Fig. 1.83. The map clearly shows the recirculation of warm waters of the West Spitsbergen Current. Part of the current jet with lower velocities on the outer side of the shelf turns first to the northwest, then to the west and forms a cyclonic recirculation jet.

Seasonal variability of currents was studied in von Appen et al. (2016), Woodgate et al. (1999), and it is low. To illustrate this, we give an example of daily mean and monthly mean values of the meridional component of the West Spitsbergen Current on mooring F3, depth 1040 m (Fig. 1.84). As waters are transported south, they flow through the Denmark Strait and become part of North Atlantic Deep Water (Mauritzen 1996; Morozov et al. 2020).

The amplitude of the velocity of tidal currents is much greater than the velocity of the mean transport. Tidal currents are stronger in the upper 1500 m. Bottom tidal

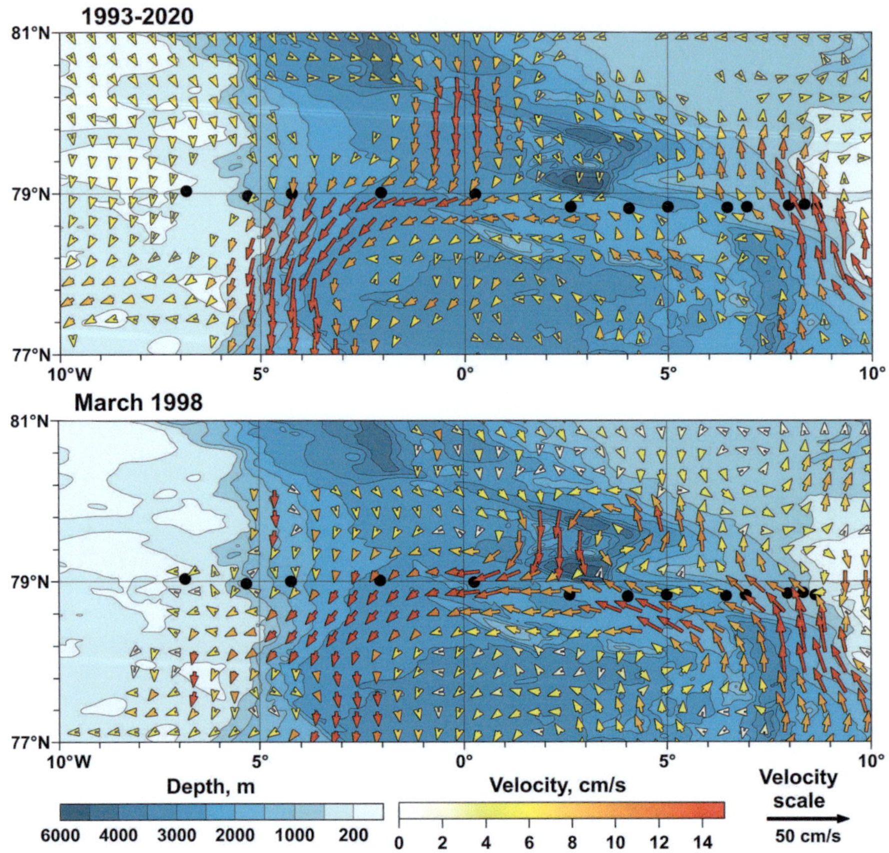

Fig. 1.80 Geostrophic currents based on the absolute dynamic topography of the Fram Strait region using satellite altimetry over the period 1993–2020 and in March 1998. Black dots indicate locations of moorings. Adopted from Morozov and Frey (2024). Used with permission of the Pleiades Publishing, Ltd.

currents in the deep part of the strait are weaker. This is clearly seen in the spectra of tidal currents (Fig. 1.85). Usually, the M_2 tide is the strongest in the spectra. The diurnal component of the tide and the S_2 tide are weaker, but on individual moorings at the upper depths the S_2 semidiurnal tide prevails over the M_2 tide. As to the moorings deployed on the shelf, bottom currents are somewhat stronger than in the upper layer. Figure 1.86 shows an example of the spectrum of tidal currents in the upper and lower layers at mooring F1 on the Spitsbergen shelf.

Tidal currents in the deepest parts of the strait are weaker than throughout the entire water column. On the shelves of Spitsbergen and Greenland, tidal currents are stronger than in the central part of the strait. This is probably because a large mass of water accelerates during the tide as the depth decreases. For example, Fig. 1.87 shows the spectra of the meridional component of currents at depths of 260–280 m

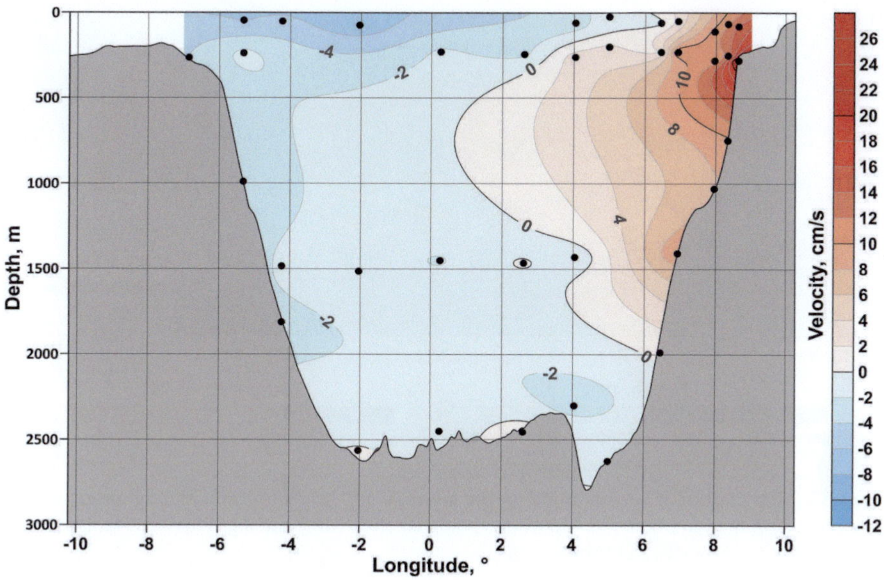

Fig. 1.81 Section of mean velocities of meridional currents over 1997–1998 at 78°50′ N. Used with permission of the Pleiades Publishing, Ltd.

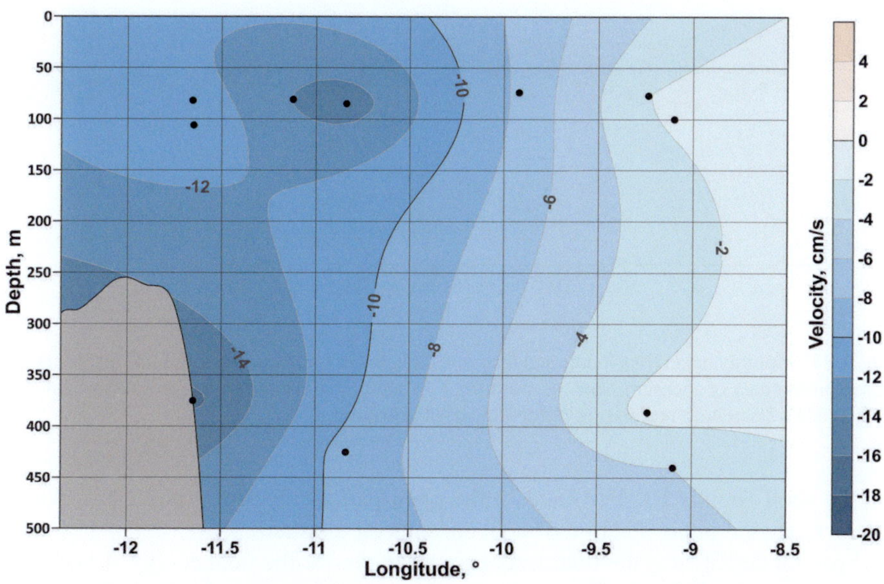

Fig. 1.82 Section of mean velocities of meridional currents over 1997–1998 at 75°30′ N. Adopted from Morozov and Frey (2024). Used with permission of the Pleiades Publishing, Ltd.

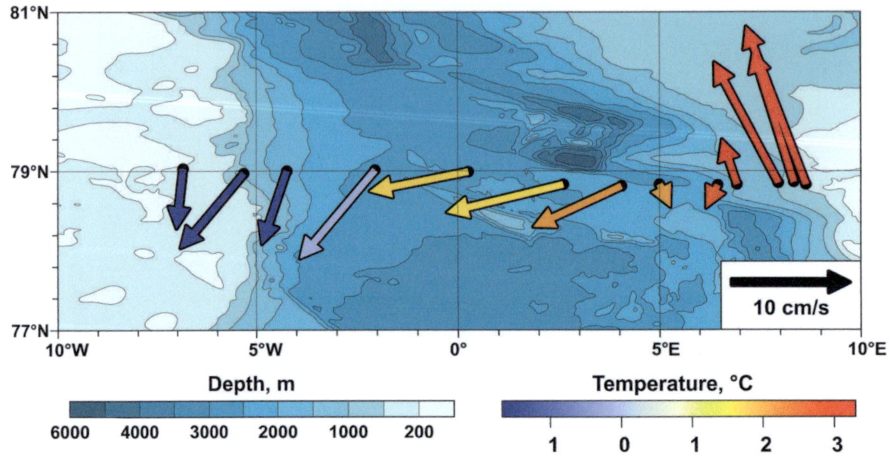

Fig. 1.83 Map of current vectors in the upper layer (0–300 m) and corresponding water temperatures based on measurements on moorings. Adopted from Morozov and Frey (2024). Used with permission of the Pleiades Publishing, Ltd.

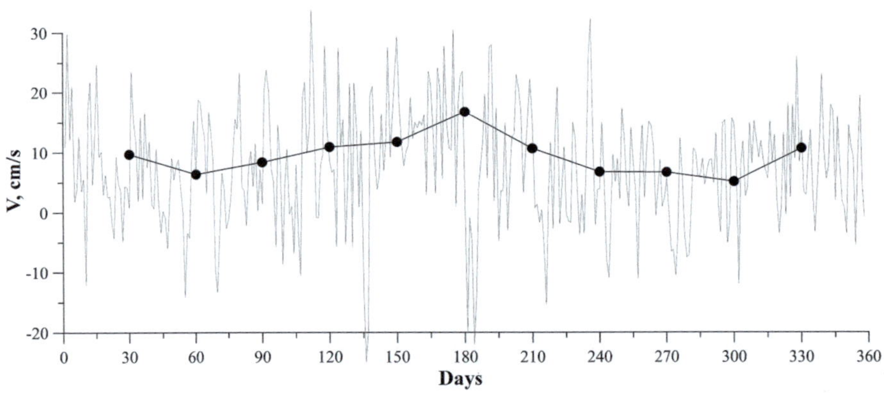

Fig. 1.84 Monthly mean (black dots and line) and daily mean (gray line) meridional current speeds based on the data of measurements on mooring F3 (depth 1040 m). Adopted from Morozov and Frey (2024). Used with permission of the Pleiades Publishing, Ltd.

on the shelf (points F1, F13) and in the open part of the strait (point F7), which characterize the intensification of tidal currents on the shelf.

Spectral densities at the tidal frequency characterize the tidal energy. To demonstrate the variability of tidal energy along a section across the Fram Strait, we constructed a section of the spectral densities of the velocity components U and V at depths of 200–300 m at frequency M_2. Graphs in Fig. 1.88 clearly show an increase in tidal energy in the shelf areas and on slopes. It is notable that tidal currents over the eastern slope are greater than those over the western slope.

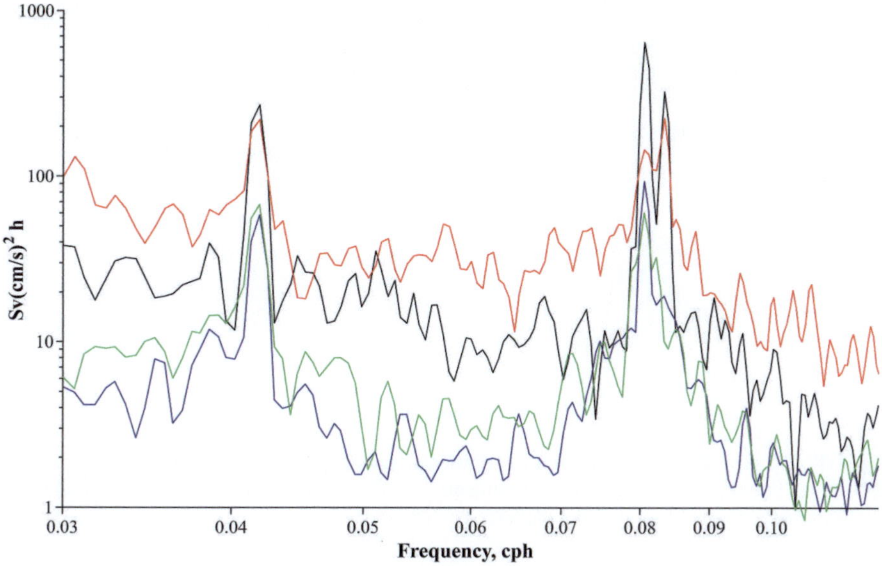

Fig. 1.85 Spectra of tidal currents (meridional component) at point F7 at depths 60 (black), 280 (red), 1430 (blue) and 2299 m (green) based on the measurements in 1997–1998. Logarithmic scales on the axes were selected for the graphs not to merge. Adopted from Morozov and Frey (2024). Used with permission of the Pleiades Publishing, Ltd.

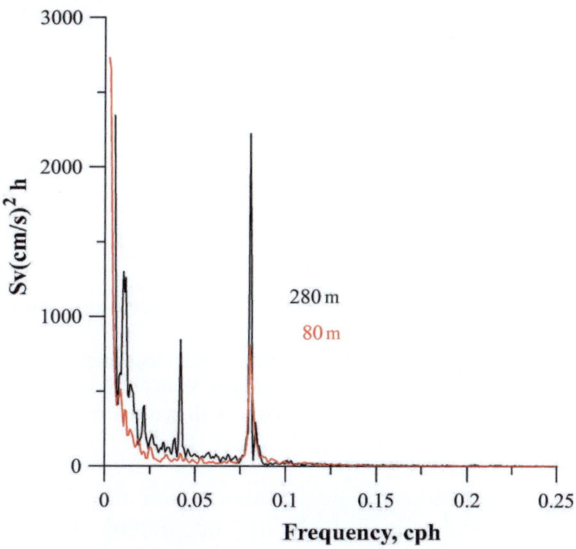

Fig. 1.86 Spectra of tidal currents (meridional component) at point F1 at depths of 80 and 280 m based on a time series of measurements in 1997–1998. Adopted from Morozov and Frey (2024). Used with permission of the Pleiades Publishing, Ltd.

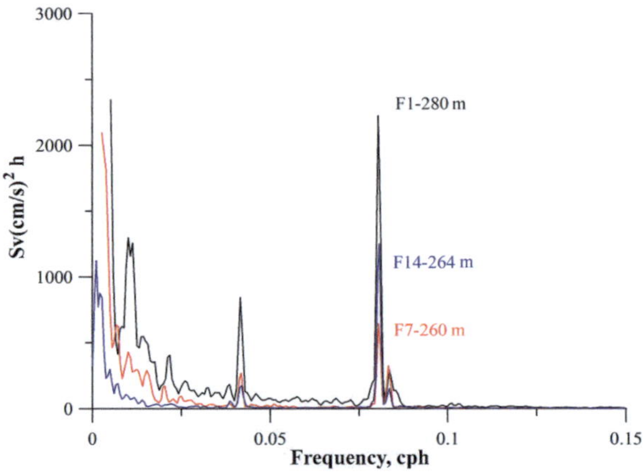

Fig. 1.87 Spectra of current velocities (meridional component) at points F1, F7, F13 at depths of 260–280 m based on a tile series of measurements in 1997–1998. Adopted from Morozov and Frey (2024). Used with permission of the Pleiades Publishing, Ltd.

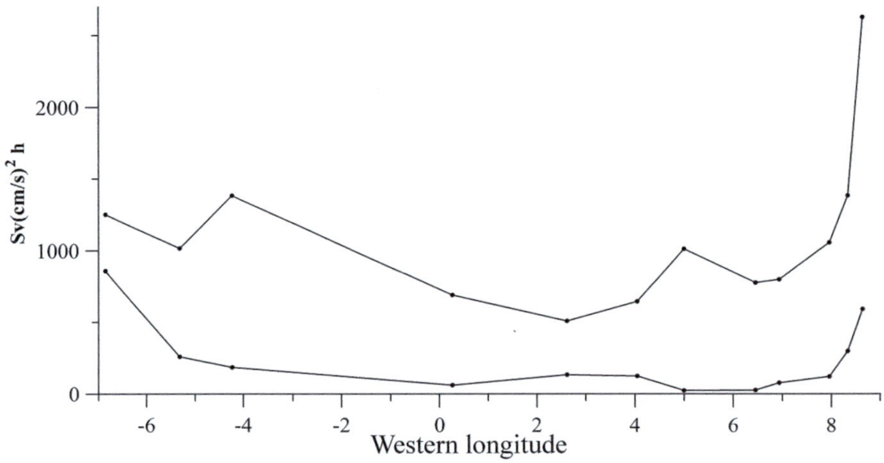

Fig. 1.88 Horizontal variations in the spectral densities of the velocity components U (lower curve) and V (upper curve) along the section at depths of 200–300 m at the M_2 tidal frequency. Adopted from Morozov and Frey (2024). Used with permission of the Pleiades Publishing, Ltd.

Intensification of tidal currents over the shelf is also confirmed by comparison of tidal ellipses (velocity hodographs). The series of current measurements were subjected to band-pass filtering, which passes semi-diurnal fluctuations in the frequency range corresponding to periods of 11.5–13.0 h. We have chosen two dates: intensification of the tides at the full moon and their weakening after 7 days. In 1998, the full moon was observed on March 13, which corresponded to spring tides, and the first quarter of the Moon was on March 20 (neap tides). Tidal ellipses were

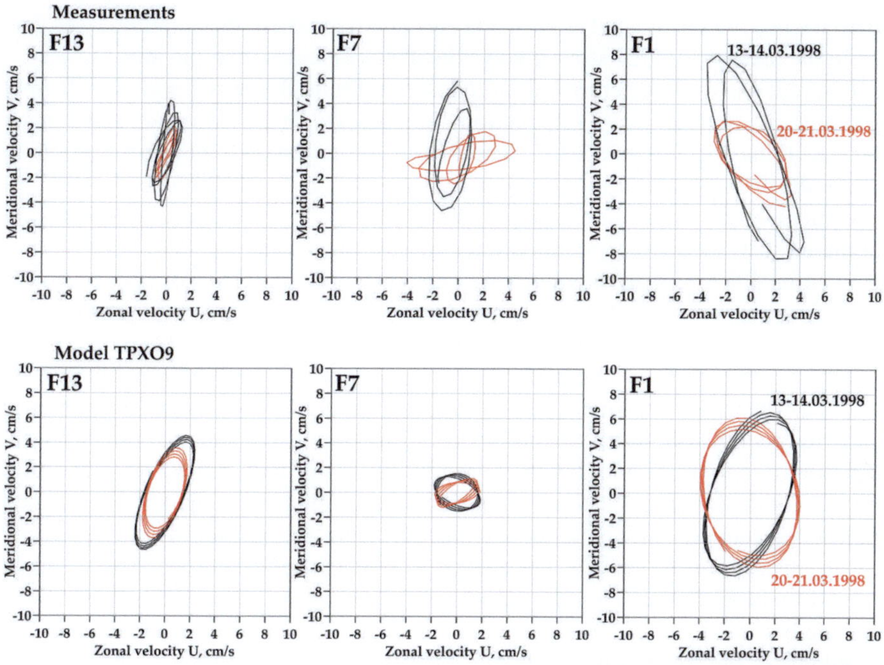

Fig. 1.89 Tidal ellipses (velocity hodographs) based on the measurements on moorings F7 (260 m) and F1 (280 m) on March 13–14 and March 20–21, 1998 (top row). The full moon was on March 13, and March 20 was the first quarter of the moon. The bottom row of the figure shows ellipses based on the TPXO9 model data (Egbert and Erofeeva 2002). Adopted from Morozov and Frey (2024). Used with permission of the Pleiades Publishing, Ltd.

constructed based on the series of measured currents after a bandpass filter, which showed that both spring and neap tides are stronger on the shelf. For illustration, we present tidal ellipses for measurements on mooring F7 in the middle of the strait and mooring F1 on the Spitsbergen shelf (Fig. 1.89).

A similar procedure was performed for temperature series based on mooring measurements. Temperature fluctuations reflect fluctuations caused by internal waves, although the fluctuations themselves are very noisy by fine temperature structure and other small-scale processes. The spectra calculated from temperature fluctuations are also very noisy, although all spectra have semidiurnal peaks. Determining the amplitudes of oscillations associated with internal waves requires multiple vertical temperature sensors, but no such sensors were available in this experiment. Temperature fluctuations with a semi-diurnal period after band-pass filtering were processed to extract a time series of semi-diurnal fluctuation amplitudes at measurement depths of 230–260 m. After this, the oscillation amplitudes were divided by the vertical temperature gradient from the data of CTD casts over the section across the strait. The noise in the time series was quite strong. However, reasonable results were obtained. On the Spitsbergen shelf, the annual mean amplitude of semidiurnal internal waves was estimated as 12 m. On the Greenland shelf, the amplitude was 8 m, and in the deep part of the strait it was 7 m. This coincides with the idea of internal semidiurnal waves at supercritical latitudes. Semidiurnal internal waves exist only on submarine slopes. Above a flat bottom they quickly decay (Morozov and Pisarev 2002; Kurkina and Talipova 2011). There is no flat bottom in the Fram Strait, but in the middle of the strait the slopes are not as steep as on the shelf break.

Vertical variations in tidal energy can be studied from the available data and compared with the barotropic tidal model (Egbert and Erofeeva 2002). To do this, we examined the vertical variability of tidal currents. The series of measurements on the moorings were subjected to bandpass filtering, which leaves semi-diurnal fluctuations in the frequency range corresponding to periods of 11.5–13.0 h. As a result, oscillating time series of current velocities with varying amplitude were obtained. The mean amplitudes of oscillations during the observation period at each of the four depths of measurements on the F7 mooring over an annual period were as follows: at 80 m depth: 3.3 cm/s, at 260 m: 3.0 cm/s, at 1430 m: 3.15 cm/s, and at 2229 m: 1.2 cm/s. On the F1 mooring on the Spitsbergen shelf, the oscillation amplitudes were greater: at a depth of 80 m, they were 4.8 cm/s and at 280 m, 5.6 cm/s. Thus, measurements show that in the entire layer, except for the near-bottom depth in the deep part of the strait, tidal velocities are almost the same, but become lower near the bottom. The velocities recorded on the shelf mooring were higher than in the middle of the strait, and the velocities are slightly higher near the bottom.

Based on the results of such a comparison using one example, it can be concluded that the barotropic tidal model describes the tides in the strait well and reflects the differences between the tides on the shelf and in the deep part of the strait. However, real instrumental measurements differ from model calculations, which should always be the case.

We obtained a year-long time-series of amplitudes with a time step of about 12.4 h as a result of bandpass filtering and isolating a number of changing amplitudes of oscillations. A frequency spectrum was constructed from this series. The spectrum contains a reliable peak at a frequency corresponding to the period of spring-neap variability (14.77 days) (Fig. 1.90), arising from the beating of two sinusoidal oscillations with periods of 12.4 h (M_2) and 12.0 h (S_2).

Long-term measurements on moorings and CTD sections in the Greenland Sea and Fram Strait have revealed several important oceanographic features of the water structure, mean currents, and tidal fluctuations in this basin. The West Spitsbergen Current transports warm Atlantic waters to the Arctic. Its velocity is maximum on the shelf and continental slope. Waters with temperatures above 0 °C occupy a layer up to 500 m deep. Cold Arctic waters are in the deeper layers. Mean current speeds are about 10–12 cm/s, but the maximum during short periods of time exceeds 50 cm/s. Water transport varies between 6 and 9 Sv, and part of the flow to the north recirculates in the Fram Strait to the south. These waters merge with the East Greenland Current. Mean velocities of the East Greenland Current are not high: 6–10 cm/s. The current transports the coldest waters with temperature below zero within the depths of 500 m at longitudes west of 4° W. South of the Fram Strait, the current accelerates and transports cold Arctic waters into the Atlantic through the Denmark Strait. The

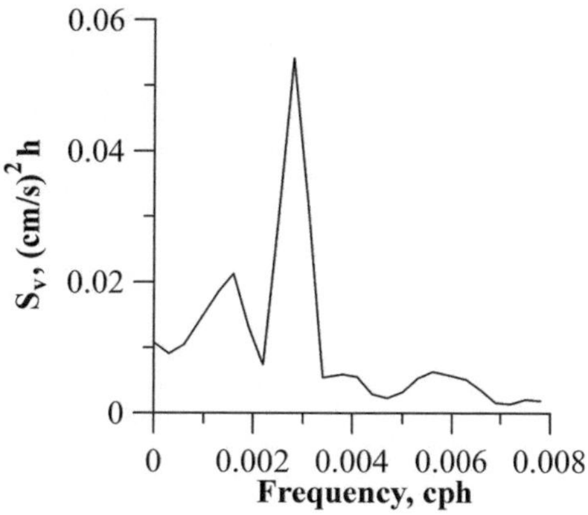

Fig. 1.90 Spectrum over a time series of amplitudes of tidal currents of the meridional component on mooring F7, depth 280 m. Adopted from Morozov and Frey (2024). Used with permission of the Pleiades Publishing, Ltd.

southward transport of this current is about 4 Sv, but the recirculation waters of the West Spitsbergen Current (about 2 Sv) are added to it; hence, the slow southward flow expands from the slope of Greenland to 4° E. Recirculation of warm, salty Atlantic water causes water motion further west in the northern Greenland Sea than in its central part.

Tidal currents in the strait are much stronger than the mean currents. Tidal currents are almost constant in depth, but decrease near the greatest depths in the middle of the strait. Tidal ellipses (tidal current velocities) decrease and increase with the period of tidal spring-neap variations (14.77 days). They are reasonably described by the TPXO9 barotropic tidal model. In the middle of the strait, tidal currents are lower than on the shelves of Greenland and Spitsbergen.

1.9 Bottom Water Flow from the Arctic to the Atlantic Through the Strait of Denmark

Cold and dense water forms in the Greenland Sea. This water gives rise to North Atlantic Deep Water after it overflows the threshold of the Strait of Denmark into the Atlantic (according to the modern concepts, the densest water in the Strait of Denmark is formed in the Iceland Sea). As a result, the densest lower part of this water mass is formed. This water is called Denmark Strait Overflow Water (DSOW). Further, this stream flows along the continental slope of America to the south in the form of the Deep Western Boundary Current. Several water masses with different properties are the sources of this water in the Greenland Sea (Mauritzen 1996; Köhl 2010; Mastropole et al. 2017). A map of the Strait of Denmark with a scheme of the stations in 2018 along sections is shown in Fig. 1.91.

The flow of water from the Greenland Sea to the Atlantic is known as a powerful underwater spillway. The depth of the threshold in the strait is about 650 m. Then, the water flows down the slope to the Irminger Sea to a depth exceeding 3000 m.

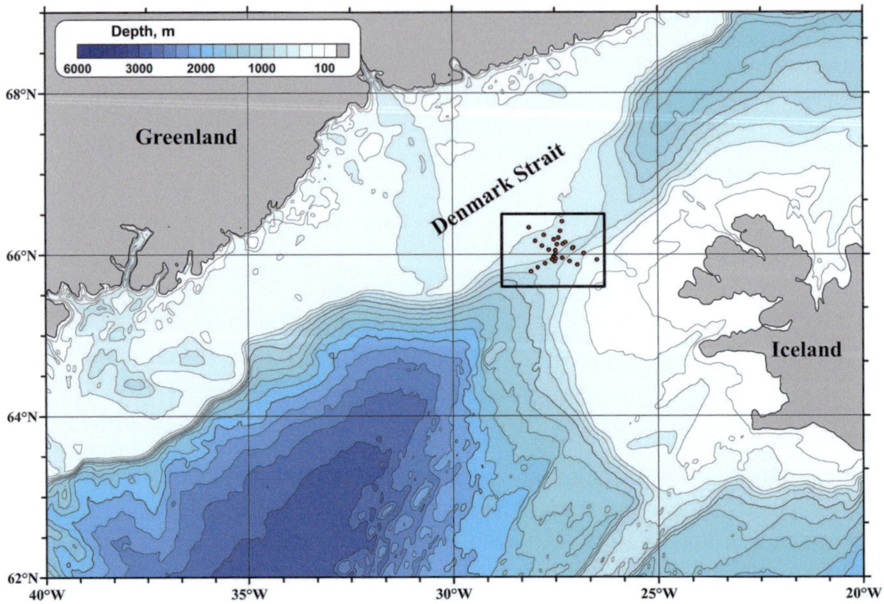

Fig. 1.91 Bathymetry of the Strait of Denmark and scheme of the stations along the sections

Based on a 20-year series of measurements and model estimates, the water transport with a potential density of $\sigma_0 > 27.80$ in the Strait of Denmark is estimated at 3.18 ± 0.56 Sv (Köhl et al. 2007; Macrander et al. 2007; Haine 2010; Filyushkin et al. 2013; Jochumsen et al. 2012, 2017). Instrumental observations show that water transport exceeding 6 Sv is often observed in the strait. In this section we review our publication (Morozov et al. 2020).

The flow of Arctic water over the threshold of the Strait of Denmark has been investigated many times. It was shown that the interannual or intra-annual variability is low (Jochumsen et al. 2012), but strong fluctuations of the flow and its structure were detected on a time scale of several days. It was found that throughout the year large volumes of cold water in the form of strongly vertically mixed lenses overflow the threshold once in a few (3–6) days. Such lenses were called boluses (Cooper 1955). These are the lenses with cyclonic rotation and low stratification. During the passage of the bolus, the depth of the boundary (isopycnal 28.0) with the surrounding water shifts upward by a distance of 50 m or more, and the temperature at the bottom drops by 0.5 °C. The appearance of lenses is associated with the variability of meteorological forcing (Haine et al. 2009) or the influence of internal waves of large amplitude incident on the underwater slope (Bourgault et al. 2008; Moore et al. 2016; Venayagamoorthy and Fringer 2007).

The density of water in boluses is on average by 0.049 ± 0.003 kg/m^3 higher than in a flow without a bolus due to the entrainment of denser Arctic water. The average cross-sectional area of the bolus is 4 km^2. The proportion of different waters that make up a bolus is given in (Mastropole et al. 2017): 80 ± 9% of water of the Arctic origin (not surface); 14 ± 9% of the Atlantic origin water; 4 ± 3% of Polar Surface Water; and 2 ± 2% of the Irminger Water. In a normal flow of water without a bolus, the proportion is as follows: 63 ± 9% of the Arctic origin water; 23 ± 9% of the Atlantic origin water; 8 ± 3% of the Polar Surface Water; and 6 ± 1% of the Irminger Water. Thus, when flowing in the form of a bolus, the proportion of Arctic water is much larger than with a normal stream, in which more water is returned from the Atlantic.

Let us consider the typical vertical distribution of temperature and salinity for various flow patterns. Figure 1.92 shows several characteristic temperature distributions at the threshold of the Strait of Denmark. The curves were plotted from the measurements on cruise 71 of the R/V *Akademik Mstislav Keldysh* in July 2018 with the addition of similar historical data. When flowing in the bolus mode, the lower layer with a thickness up to 300 m is almost completely mixed. As a rule, the thickness of the mixed layer is 150–250 m. The temperature of the flowing water is slightly below zero from –0.10 °C to –0.23 °C. Above the bolus there is a sharp temperature gradient of up to 1 °C per 50 m. In the overflow regime without a bolus, the whole stream consists of a warmer water mass, and the thickness of the cold lower mixed layer does not exceed 50 m. The water temperature in this layer is about 0 °C. Above this layer a layer of much warmer water is located with a temperature exceeding 6 °C. The temperature gradient is one degree per 7–10 m. At the same time, vertical salinity gradients are small.

The bolus occupies the lower layer about 250 m thick from a depth of 400 to 650 m near the bottom. The entire upper layer above 400 m with a stream without a bolus is on average warmer by 4 °C up to the surface than in a stream with a bolus, which indicates that these are waters from different regions of the Greenland Sea. The colder water in the region of the eastern coast of Greenland forms a lens and then a bolus, and the warmer water from the region of northern Iceland flows in an ordinary stream. Such a mechanism was proposed in Mauritzen (1996), Köhl (2010). The source of cold water is the shelf of eastern Greenland, and warm waters flowing over the threshold are formed in the region of the formation of the North Icelandic Jet on a slope at a depth of about 650 m. This is confirmed by temperature-salinity relations for these regions.

Various modes of flow over the Strait of Denmark threshold are clearly seen in the sections of the temperature distribution through the strait constructed based on the data of CTD profiling using the SBE 911 probe. The casts were made together with a pair of LADCP velocity up-looking and down-looking profilers (Fig. 1.93). The cold and dense Denmark Strait Overflow Water (DSOW) is formed northwest of our section across the strait. The warm and saltier water of the Irminger Sea, which enters this region from the subtropical zone, occupies the shelf of Iceland. These waters enter the strait and give significant differences in the temperature distribution in the section. A sharp temperature and salinity front separates these waters.

The existence of a flow of bottom water in the bolus regime was noted in many publications (Spall and Price 1998; Girton et al. 2000; Käse et al. 2003; Koszalka

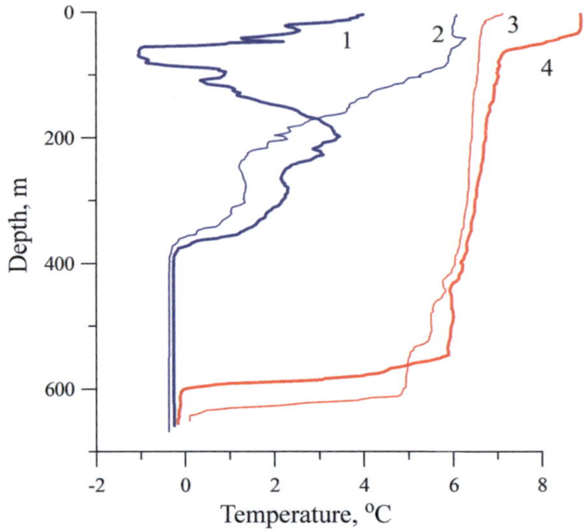

Fig. 1.92 Vertical distribution of potential temperature in various waters of the Strait of Denmark. Curves 1, 2 correspond to the flow regime with a bolus; curves 3, 4 are related to the overflow without bolus: Station 5894 of the cruise of the R/V *Akademik Mstislav Keldysh* (AMK) at point 66.076° N, 27.079° E on July 29, 2018 at a depth of 658 m (curve 1); station from the WODB archive at point 66.08° N, 05.27° on May 18, 2003 at a depth of 657 m (curve 2); station from the WODB archive at point 66.08° N, 27.05° E on May 17, 2002 at a depth of 646 m (curve 3), station 5837 of the R/V *AMK* at point 66.076° N, 27.080° E on July 25, 2018 at a depth of 654 m (curve 4). Adopted from Morozov et al. (2020) with permission. Used with permission of the Pleiades Publishing, Ltd.

et al. 2013). The velocity spectra have a pronounced peak at a period of 4–5 days. This peak is much stronger than the semidiurnal tidal peak M_2 (Morozov et al. 2019b).

A possible cause of the five-day variations in the bottom flow regime may be internal seiches. The oscillation period of seiches is determined by the Merian formula (Merian 1828):

$$T_{\text{seich}} = 2L \Big/ n\sqrt{gH_{\text{mean}}},$$

here, T is the seiche period, L is the characteristic size of the basin, H_{mean} is the average depth, g is the acceleration due to gravity, n is the mode number. Seiches are natural oscillations of the basin. The low frequency mode ($n = 1$) is the fundamental mode. In the basin north of the Strait of Denmark, the seiches of surface oscillations have a frequency of several hours. In our case, internal seiches are important, that is, seiche oscillations of the lower layer. A typical change in the height of the upper boundary of the bottom flow in the usual flow regime and in the regime with the presence of a bolus is 150–300 m, which is possibly due to internal seiche oscillations. We assume that the bolus formation region is the basin north of the strait at a distance of 100–150 km from a threshold with a characteristic transverse dimension of 250 km

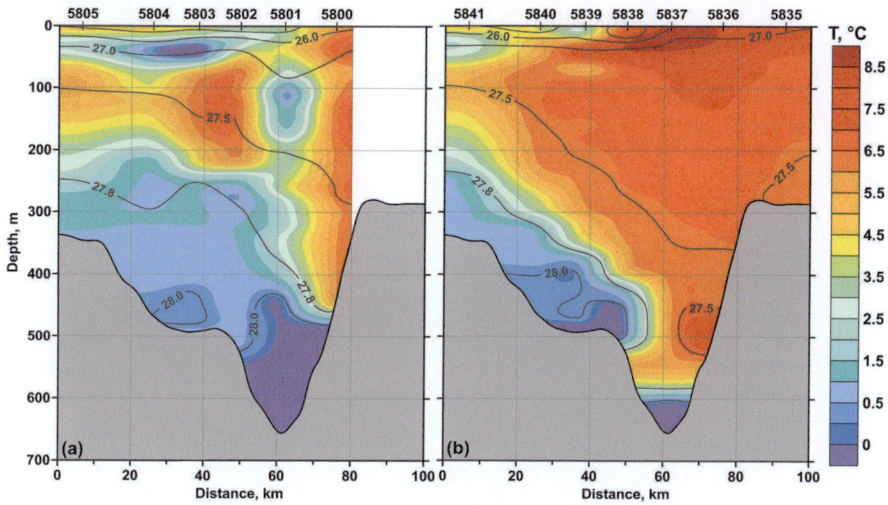

Fig. 1.93 Temperature sections across the strait during the overflow mode with a bolus on July 23, 2018 (**a**) and without a bolus on July 25, 2018 (**b**). The section shows the 28.0 kg/m^3 isopycnal limiting the bolus, and isopycnal 27.8 kg/m^3, restricting the flow of the bottom water (other isopycnals are also shown). Adopted from Morozov et al. (2020). Used with permission of the Pleiades Publishing, Ltd.

along the shelf isobath 200 m. For internal oscillations, acceleration of gravity g must be replaced by reduced acceleration due to gravity

$$g' = g\frac{\Delta\rho}{\rho},$$

where, $\Delta\rho$ is the density difference between the upper and lower layers equal to 2×10^{-3} g/cm^3. The average depth between the 200 m isobaths in the section across the strait is estimated as 600 m. Then, the period of internal seiche oscillations of the lower layer is estimated as 5.3 days, which is close to the periodicity of changes in the flow modes. Since all initial values for estimating this period should be considered as approximate, the result is reasonable.

Let us consider the vertical temperature distributions over several stations on the Greenland slope north of the threshold in the strait, performed in September in various years in a small square: from 68.13° to 68.24° N and from 25.08° to 25.25° W. The plots of vertical temperature distributions are shown in Fig. 1.94.

Approximately the same law of temperature decrease to the bottom is noted in the lower part of each curve, however, the curves are shifted vertically from each other by 100–300 m, which is caused by internal seiches of the lower layer. The mixing and vertical uniformity in the bolus is apparently determined by the cascading of cold waters on the shelf of Greenland down the slope. This process of water descent cools the water in the bolus and makes it homogeneous. There is no similar process on the shelf of Iceland, where the shelf water is slightly warmer.

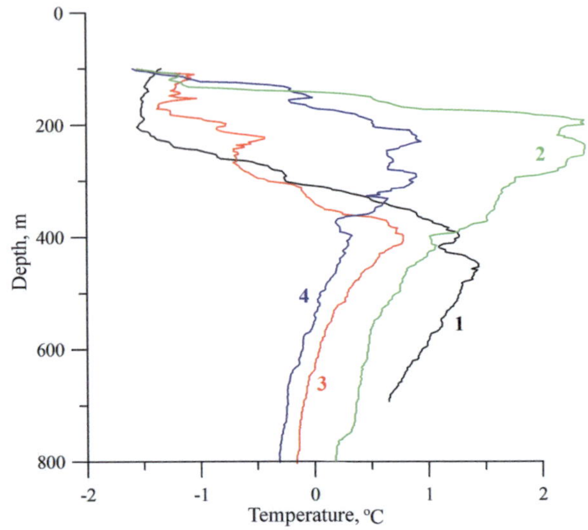

Fig. 1.94 Vertical temperature distributions at stations in September in various years in the region north of the threshold. Station 1 on September 3, 1997 (black curve); station 2 on September 26, 2007 (green); station 3 on September 7, 1998 (red); and station 4 on September 11, 1995 (blue). Adopted from Morozov et al. (2020). Used with permission of the Pleiades Publishing, Ltd.

Let us consider the spatial structure of the bolus. To study the spatial structure of the bolus, two sections across this structure with CTD/LADCP stations from south to north, and then from west to east, were performed. This made it possible to construct a map of isothermal and isopycnic lines from the data near the bottom to outline a bolus. A map of the position of the bolus on July 21–23, 2018 is shown in Fig. 1.95. The temperature section along the bolus is shown in Fig. 1.96.

The bolus is an almost circular structure with a diameter of about 70 km, slightly elongated forward along its motion in the southwestern direction. It is displaced from the center of the channel to its northwest slope. After overflowing the threshold, the bolus rolls down under the action of gravity and shifts to the right to the Greenland slope in the direction of propagation under the influence of the Coriolis force (Girton et al. 2000). The lower part of bottom water up to 50 m thick flows forward at a higher speed and underflows a layer of less dense water. This motion of the bolus transports this water in the direction to the southwest, as can be seen from Fig. 1.97. A thin layer of the leading front of the bolus is ahead of its main body, whose thickness is about 250 m. This is also seen in the section of temperature across the bolus from south to north (Fig. 1.96).

The bolus we crossed is characterized by the presence of high horizontal velocities exceeding 50 cm s^{-1}, which are measured at the leading edge of the bolus. This is seen on the section of velocities along the bolus from south to north (Fig. 1.97). The cross-section of the meridional velocity across the bolus is shown in Fig. 1.98. The velocities in the bolus core exceed 60 cm/s. In the overflow mode without a bolus, the characteristic velocities do not exceed 40 cm/s. Two cores of the flow are seen in the section: one core propagates along the deep part of the strait, and the second is displaced to the northwest (to the right along flow stream) due to the Coriolis force.

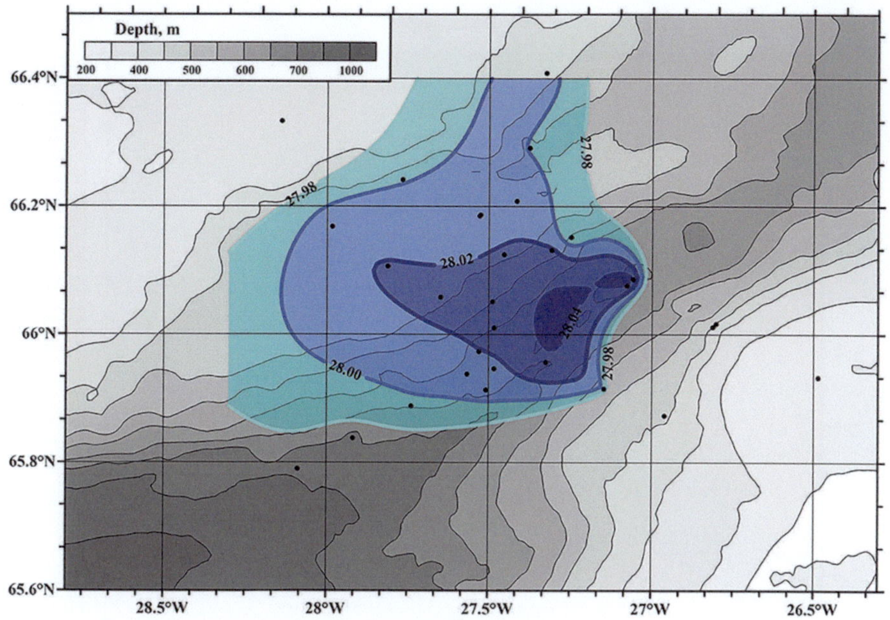

Fig. 1.95 Map of isopycnals that outline a bolus in the Strait of Denmark based on the data on July 21–23, 2018. Station points and four isopycnals are shown: 1.02798, 1.02800, 1.02802, 1.02804 g/ cm^3. Adopted from Morozov et al. (2020). Used with permission of the Pleiades Publishing, Ltd.

The data of velocity measurements at two sections across the bolus using a lowered velocity profiler made it possible to construct a velocity map in the bolus (Fig. 1.99). The maximum speeds were found in the front of the bolus where the bolus rolls down the slope and its forward part is ahead of the central one (tidal currents are removed from the LADCP data). The minimum of velocity was found in the center of the bolus. In the rear part of the bolus, which is in the narrowest part of the strait above the threshold, the velocities increase again.

The leading part of the bolus separates from its center, while velocities are high in the rear part because it is in the narrowest part of the strait. In the central part of the bolus, the speed is lower. After passing through the narrowness of the strait, its width increases slightly, which leads to a decrease in speed.

Numerical simulation of the bottom flow of arctic waters through the Strait of Denmark has been carried out using the three-dimensional numerical model INMOM developed at the Institute of Computational Mathematics, Russian Academy of Sciences (Diansky et al. 2013). The model is based on a complete system of equations of hydrodynamics in spherical coordinates with the Boussinesq and hydrostatics approximations. The INMOM model has been repeatedly tested to simulate bottom currents in the underwater channels of the Atlantic (Frey et al. 2019).

A rectangular region was selected for modeling in the strait between 65° N and 67° N and between 26° W and 29° W. The topography of the ocean floor with a resolution

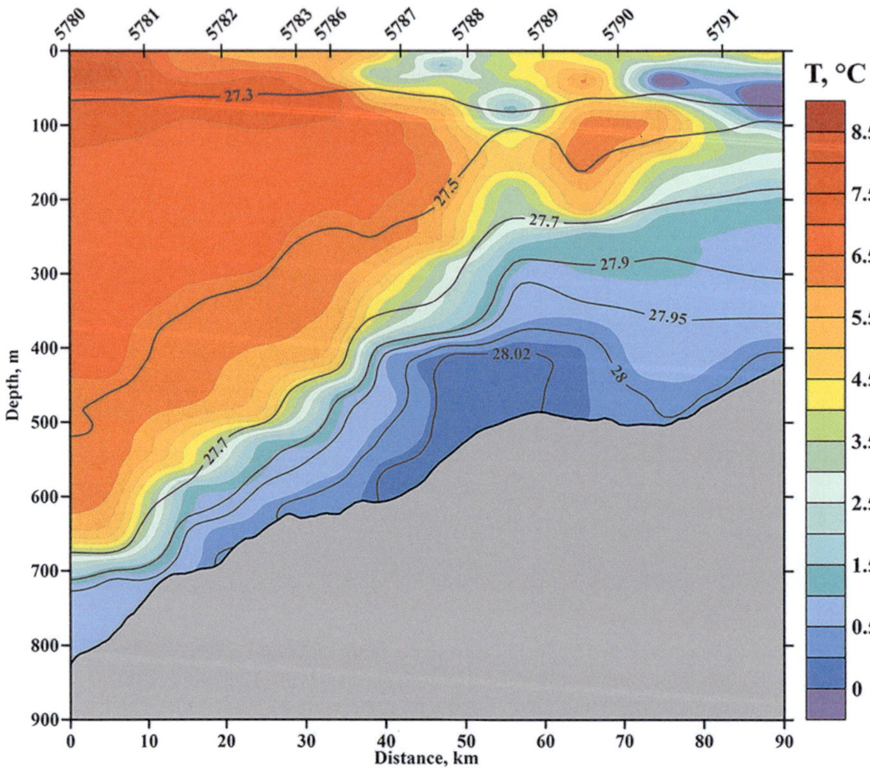

Fig. 1.96 Temperature section along the bolus from south to north based on the data on July 21–22, 2018. Adopted from Morozov et al. (2020). Used with permission of the Pleiades Publishing, Ltd.

of 30' was taken from the GEBCO 2024 electronic database (https://www.gebco.net/data-products-gridded-bathymetry-data/gebco2024-grid). The temperature and salinity fields as initial conditions were taken from the monthly average data of the atlas (Antonov et al. 2010); initial velocity fields were assumed zero.

Fields of horizontal velocities, potential temperature, and salinity at the sigma levels and fields of sea surface height deviation from the unperturbed state are the predicted model parameters. The horizontal resolution of the model was 0.02° by latitude and longitude. The vertical sigma levels were not evenly distributed. The best resolution was set near the bottom. The total number of sigma levels was 33. The total calculation time was 30 days.

The density distribution in the bolus was specified based on our CTD casts. According to the simulation data, the flow velocities north and south of the threshold exceed 100 cm s^{-1}. A map of simulated velocities is shown in Fig. 1.100. Based on the measurements, it was determined that the core of the maximum speeds was displaced closer to the slope of Greenland. A similar displacement was found during the simulation. In addition, it was shown above that according to the CTD/LADCP

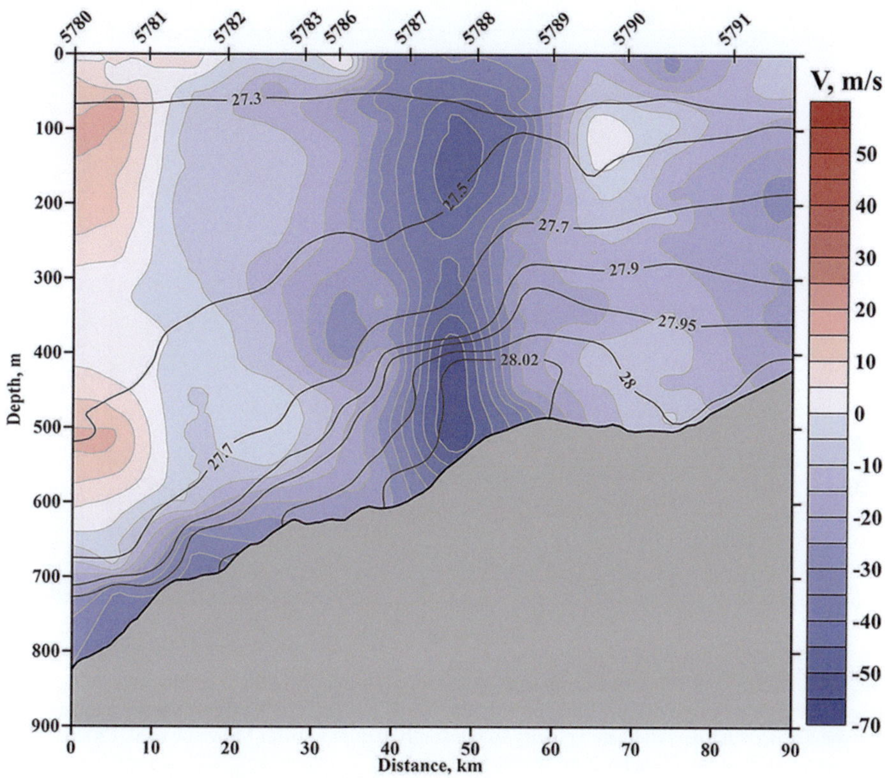

Fig. 1.97 Section of meridional velocity along the bolus from south to north based on the data on July 23, 2018. Adopted from Morozov et al. (2020). Used with permission of the Pleiades Publishing, Ltd.

casts, the minimum velocity was found at the center of the bolus. A similar minimum of velocities at the center of the bolus appeared in the numerical simulations of the flow in the strait. A decrease in the velocity of the flow and a slight rotation of the flow driven by the bottom topography was noted in (Käse and Oschlies 2000). According to their modeling data, the maximum speeds were found in the front of the bolus, which causes an extension of its shape in the direction of flow propagation to the southwest.

We studied the spatial structure of the bottom water overflowing the threshold in the Strait of Denmark in the presence of a bolus. The intermittent regime of bottom water overflow from two sources in the Arctic north of the strait has been confirmed. A unique physical and hydrodynamic effect has been noted when bottom waters overflow the threshold. A large volume of cold mixed water flows in the form of a lens (bolus). The frequency of occurrence of boluses is about 5 days. It is shown that the transverse size of the bolus limited to the 1.028 g/cm^3 isopycnal is 70 km. Its thickness reaches 300 m. This form of overflowing through a threshold in the form

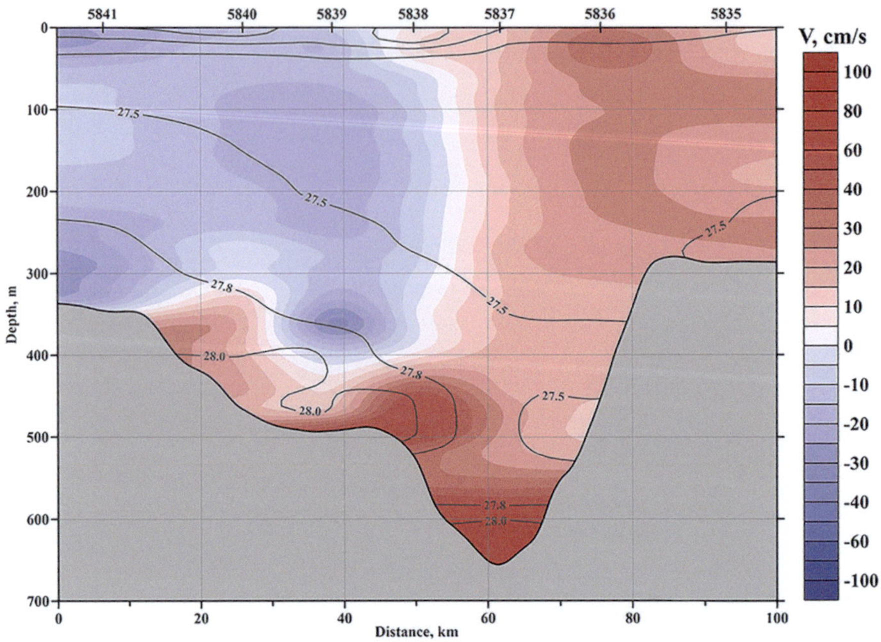

Fig. 1.98 Meridional velocity section across the bolus (from northwest to southeast, see Fig. 1.91) based on the data of LADCP profiling on July 25, 2018. The lines in the figure show the position of isopycnals. Brown tones indicate velocity to the south. Adopted from Morozov et al. (2020). Used with permission of the Pleiades Publishing, Ltd.

of a bolus alternates with ordinary overflowing. When flowing in the bolus mode, the maximum speeds are detected on the leading edge, there is a minimum of speeds at the center of the bolus, and in the rear part of the bolus the speeds increase again.

1.10 Sea Currents in the Navigational Strait Akselsundet in Spitsbergen

The Akselsundet Strait is located between Bellsund Bay and Van Mijen Fjord of the West Spitsbergen. Coal boats up to 70,000 tons deadweight pass the strait on the way to and from the Kapp Amsterdam coal quay near former coal mining settlement Svea, which is close to the tip of the Van Mijen Fjord (Fig. 1.101). Navigational activity of coal boats in the Van Mijen Fjord is performed only in ice free season 5 months long from the end of June to November. During this time the coal produced over the winter and summer in Svea mines is transported from the coal quay. One passage of a coal boat through the Akselsundet Strait occurs in three-five days. In this section we review (Marchenko et al. 2015).

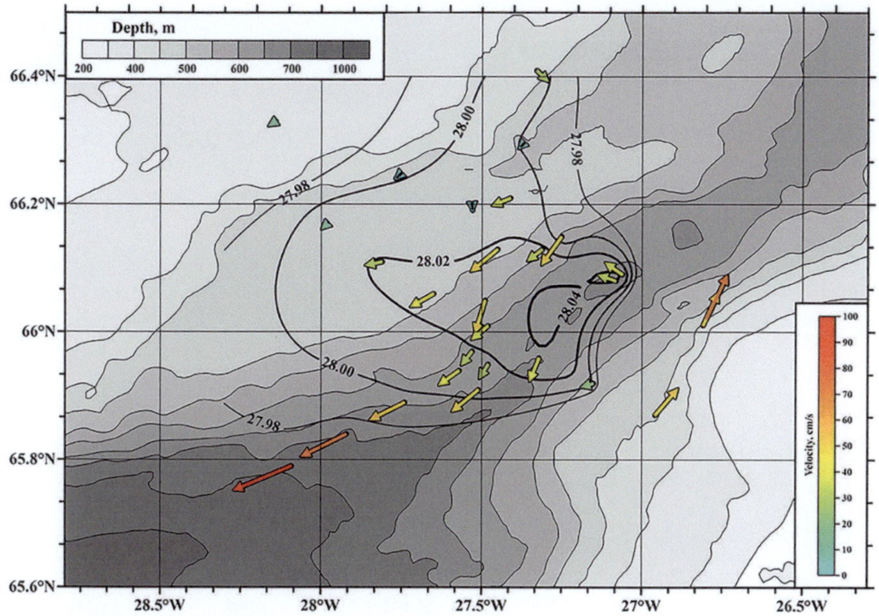

Fig. 1.99 Map of horizontal speeds in the bolus. Isopycnals of the bolus structure and velocity vectors are shown. Adopted from Morozov et al. (2020). Used with permission of the Pleiades Publishing, Ltd.

The Norwegian Pilot (2014) characterizes navigational conditions in the strait by tidal streams with eddies and speeds up to 5–6 knots (2.5–3.0 m/s). Coal boats pass the strait with assistance of tugboats in navigable window when the current speed is low. It happens two times per one period of the semidiurnal tide 12.42 h. Navigable window duration is based on the measurements of sea current speed from tugboats. Here, we estimate the navigable window and describe characteristics of tides around Akseløya Island.

Effect of jet flow formation in the constriction at the entrance to the fjord was described by Stigebrandt (1980); the local conditions in the strait between Svea Bay and Lake Vallunden in the Van Mijen Fjord were investigated by Marchenko and Morozov (2013). We revealed similar phenomena in the Akselsundet Strait at the entrance to the Van Mijen Fjord.

A 3D view of seabed topography around Akseløya Island is shown in Fig. 1.102a. Coordinate axes x and y are directed to the east and north. The width of the Akselsundet Strait between the northern tip of Akseløya Island and the northern shore of the Van Mijen Fjord is about 1 km at the water level. Further to the east the channel extends so that the angle between Akseløya Island and the coastal slope of the Van Mijen Fjord is about 53° (Fig. 1.102b). A 3D view of the channel near the entrance to the Van Mijen Fjord is shown in Fig. 1.103a, where plane $z = 0$ shows the water-level. Vertical cross-sections of the channel constructed at $x = -3000$ m and $x = -2000$ m

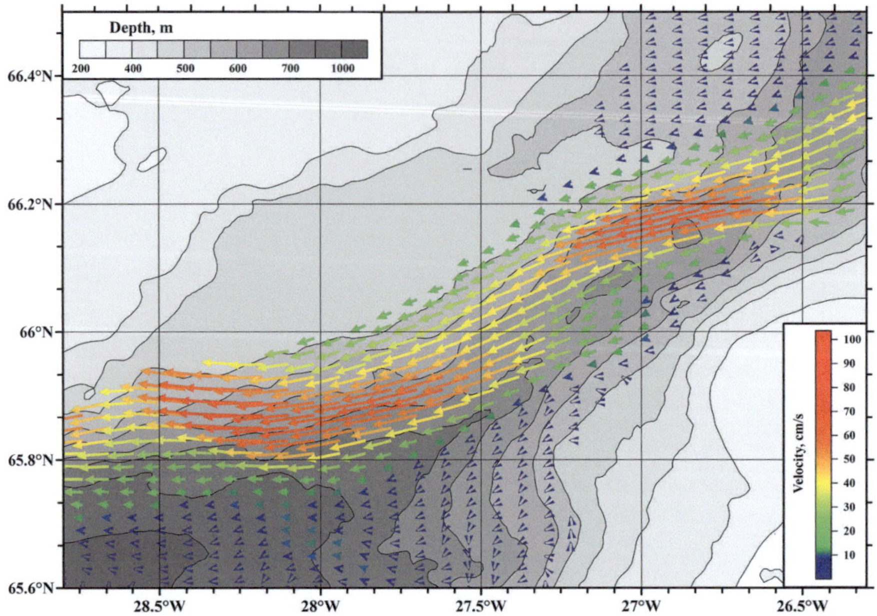

Fig. 1.100 Map of the estimated model horizontal velocities. Adopted from Morozov et al. (2020). Used with permission of the Pleiades Publishing, Ltd.

Fig. 1.101 Location of the Akselsundet Strait in Spitsbergen (a), a coal boat in the strait (b). Red arrow and rectangle show the Akselsundet Strait. Adopted from Marchenko et al. (2015)

are shown in Fig. 1.103b by blue and purple lines respectively. The deepest part of the Akselsundet Strait is associated with a channel bounded by vertical cross-sections S_0 and S_1 shown in Fig. 1.103b. The mean water depth in these cross-sections reaches 40 and 60 m respectively. The width of the channel is estimated as 300 and 1000 m in the cross-sections S_0 and S_1 respectively.

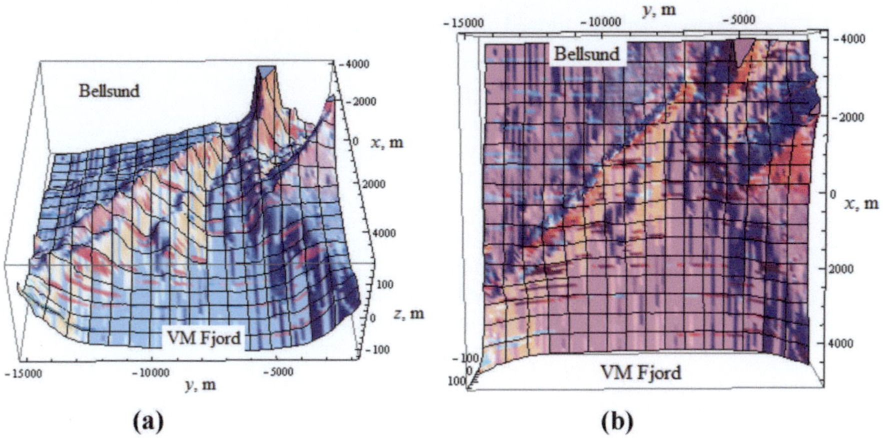

Fig. 1.102 Bathymetry around Akseløya Island: 3D view (**a**) and view from the top (**b**). Bellsund is the offshore part of the Van Mijen Fjord west of Akseløya Island. Adopted from Marchenko et al. (2015)

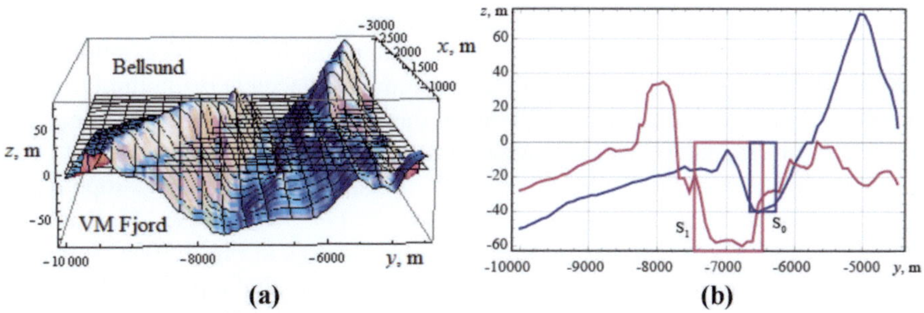

Fig. 1.103 3D bathymetry of the navigational channel Akselsundet (**a**). Vertical cross-sections of the Akselsundet navigational channel are shown at $x = -3000$ m (blue line) and $x = -2000$ m (purple line) (**b**). Adopted from Marchenko et al. (2015)

Field works around Akseløya Island performed in 2010–2013 are specified in Table 1.2. The main goals of the field works were to measure tide induced variations of the water level near Bellsund and Van Mijen shores of Akseløya Island and in the Akselsundet Strait, and to measure the velocity of sea currents in the Akselsundet Strait. It is assumed that the tidal current in the strait is driven by the local gradient of the water pressure determined by the water level gradient in the strait. Synchronous measurements of the water pressure from different sides of Akseløya Island and sea current velocity in the strait can be used for the calculation of the transfer function showing the dependence of the sea current velocity in the strait from the water level gradient over Akseløya Island measured at two specific points.

Field works were organized from the R/V *Lance* in 2010 and R/V *Viking Explorer* in 2011 and 2013. In May 2010, the Van Mijen Fjord was covered by sea ice. The

Table 1.2 Field measurements performed near Akseløya Island

Time	Water level measurements	Water velocity measurements
May 5–6, 2010	Records of SBE 39 deployed on the sea bottom near Bellsund and Van Mijen shores of Akseløya Island (red dots in Fig. 1.105a)	
August 30–31, 2011	Records of SBE 37 deployed on the sea bottom near Akseløya Island (blue dot in Fig. 1.105a), pressure records of ADCP Aquadopp (green dot in Fig. 1.105a)	Records of ADCP, Aquadopp (green dot in Fig. 1.105a)
September 24–October 24, 2013	Records of SBE 39 deployed on the sea bottom near Bellsund and Van Mijen shores of the Akseløya (red dots in Fig. 1.105a)	Surface drifters equipped with GPS antennas AstroDog-320 were deployed in the Akselsundet

R/V *Lance* was moored to the ice and pressure sensors were delivered from the ship to the island by snow scooters. Records of the water pressure at the sea bottom were performed using the SBE 39 and SBE 37 instruments placed at the bottom near the shoreline of Akseløya Island. In 2010 one of the SBE 39 sensors deployed near Van Mijen shore of the island was mounted on a rope fixed on the ice, while the other sensor deployed near the Bellsund shore of the island was fixed by a rope to the shoreline (Fig. 1.105a). Locations of these sensors (77° 42.556' N, 14° 41.846' E and 77° 43.451' N, 14° 39.397' E) are shown by red dots in Fig. 1.105a. Sampling interval of both sensors was 1 s. The measurements revealed a phase delay of the tide near Van Mijen shore of about 30 min in comparison with the tide near Bellsund shore of Akseløya Island (Marchenko et al. 2011).

In August 2011, the SBE 37 sensor (sampling frequency 6 s) was deployed near the northern tip of Akseløya Island (blue dot in Fig. 1.105a). It was connected by a rope to a floating buoy. At the same time Acoustic Doppler Current Profiler equipped with a pressure sensor (Aquadopp, operating frequency 2 MHz) was deployed in the Akselsundet Strait (green dot in Fig. 1.105a at 77° 45.367' N and 14° 36.835' E) for the measurements of sea current velocity and water pressure (Fig. 1.104). The ADCP was mounted to the anchor and deployed at a bottom slope with pitch angle $\alpha_p \approx -30°$ at a depth of 20 m. The anchor was connected to a floating buoy by a rope (Fig. 1.104a). Small inclination of the ADCP axis to the bottom surface provided records of sea current velocities within a boundary layer of 1 m thick near the bottom (Fig. 1.104a). The ADCP recorded the data with a sampling interval of 60 s in 20 cells with a bin size of 0.5 m. The time variations of water level measured at the points marked with blue and green dots in Fig. 1.105a are shown in Fig. 1.106a by blue and green lines. The time axis in Fig. 1.106a is local time on August 30 starting from 12:00. The records in Fig. 1.106a show the difference of tidal phases in the blue and green dots about 1 h, i.e., the maximum flood and ebb tides occurred at the green dot 1 h later than at the blue dot.

Figure 1.106a shows the absolute values of horizontal velocity measured by the ADCP at different distances from the sea bottom laid off along the z-axis shown in Fig. 1.106a. Figure 1.108b shows vertical velocities measured at the same points. Nonzero vertical velocities near the bottom are explained by 30° bottom slope as measured according to the pitch angle α_p of the ADCP (Fig. 1.104b). The maximal horizontal velocities exceeded 1.5 m/s at 1 m distance from the bottom. The maximal horizontal velocities were below 0.5 m/s near the bottom. The maximal absolute vertical velocities are below 0.5 m/s near the bottom and reach 1 m/s at 1 m distance from the bottom. The heading angle of the ADCP changed its value during 2–3 h from 100° to 240° during the high-water tidal phase around $t = 15$ h and $t = 27$ h. The values of the pitch angle α_p changed during this time by 10°. The mean values of the pitch angle were $\alpha_p \approx 30°$. Figure 1.108a shows a hodograph of horizontal velocities measured at different depths with the ADCP within 1 m boundary layer

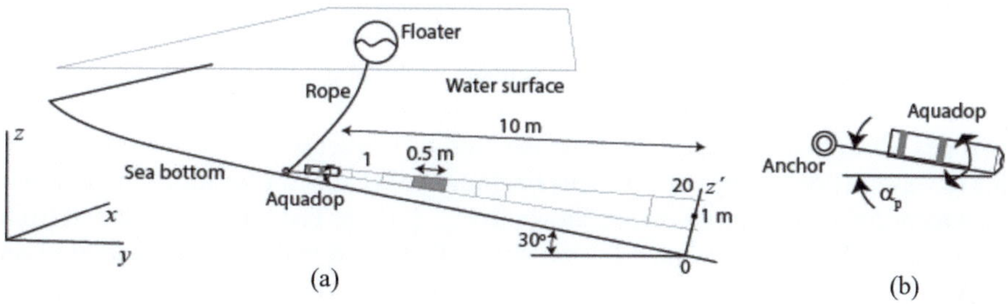

Fig. 1.104 Deployment of ADCP Aquadopp (**a**) and its pitch angle α_p (**b**). Adopted from Marchenko et al. (2015)

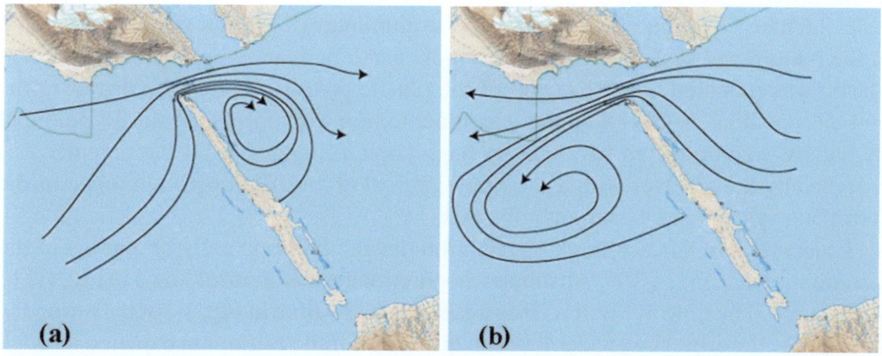

Fig. 1.105 Akseløya Island at the entrance to the Van Mijen Fjord. Qualitative scheme of flow lines in the Akselsundet Strait during the flood (**a**) and ebb (**b**) phases of the tide. Locations of field measurements are shown with red, blue, and green dots in panel (**b**). Adopted from Marchenko et al. (2015)

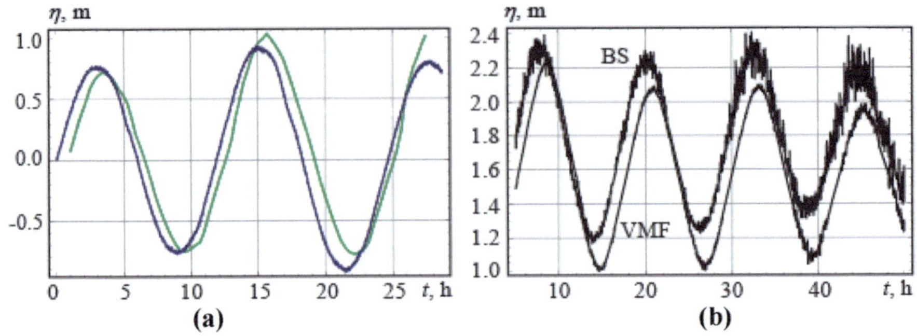

Fig. 1.106 Water level elevations measured at the locations marked with blue and green dots (Fig. 1.105a) are shown with blue and green lines (**a**), and near Bellsund (BS line) and Van Mijen Fjord (VMF line) shores of Akseløya Island (**b**). Adopted from Marchenko et al. (2015)

near the bottom. The maximal absolute values of velocities measured in the east–west directions reach 1 m/s, while the maximal absolute values of the south-north velocities are within 0.6 m/s.

On September 24, 2013, two SBE 39 sensors were deployed with sampling interval 2 min at the locations shown by red dots in Fig. 1.105a and fixed on the shoreline by ropes. The sensors operated during one month. The sensors were pushed onshore due to wave action during storms in the beginning of October. Elevations of the water level recorded before the storm are shown in Fig. 1.106b. The time is accounted from 09:00 of September 24. Figure 1.106b shows that flood and ebb tides occurred on the Van Mijen shore of Akseløya Island one hour later than on the Bellsund shore of the island.

Trajectory of surface water particles and their velocities were investigated in September 2013 by the surface drifters equipped with GPS antennas AstroDog 320. Each drifter was constructed from four aluminum poles fixed on a vertical plastic pole. Submerged drogue was mounted on the aluminum poles. A weight was mounted at the bottom end of the vertical plastic pole and a buoyancy ring made from foam plastic was mounted at the top end of the plastic pole. Both kept the drifter in the vertical position. A GPS antenna was mounted on the top of the plastic pole and its battery was protected by a plastic bag. The GPS signal from the antenna was recorded by the receiver with a sampling interval of 5 s. The operation range of the AstroDog system was 10 km on a flat area.

Trajectories of surface trackers recorded during the flood and ebb phases of the tide are shown in Fig. 1.109. Absolute velocities of the drifters are shown in Fig. 1.110 versus time. The time in Fig. 1.110 corresponds to the time in Fig. 1.106b. During the flood phase the drifters were deployed in different initial positions and drifted through the Akselsundet Strait almost along the same trajectory showing the existence of water jet with strong current in the middle part of the strait. The water jet coincided with the deepest area of the strait indicated by vertical rectangles S_0 and S_1 in Fig. 1.103b. The maximal speed of the drifters reached 2.5 m/s during the flood phase of the tide. It was recorded inside the red square in Fig. 1.109a around $t = 6$ h

(Fig. 1.110a). During the ebb phase the maximal drift speed was recorded inside the red rectangle in Fig. 1.109b around $t = 24$ h (Fig. 1.110b). Other local maxima of the drift velocities are related to the rotation of the drifters captured by small eddies in the water jet. Figure 1.108b shows a hodograph of the velocities of drifters measured when the drifters passed the Akselsundet Strait and their locations were between the red rectangles shown in Fig. 1.109. It shows that the maximal speeds of the drifters in the east–west direction exceed 2 m/s, South projections of the drifter velocities reached 1 m/s and North projections of the drifter velocities were as high as 0.5 m/s.

Satellite image provided by the Norwegian Polar Institute (Fig. 1.111a) shows the formation of eddies on the eastern side of Akseløya Island during the flood phase of the tide. Similar eddies are seen in Fig. 1.109a. An aerial image made by Trond Jensen at the ebb phase of the tide demonstrate the existence of a large eddy on the western side of the island (Fig. 1.111b). A qualitative scheme of the stream lines reconstructed by the analysis of trajectories of the surface drifters and satellite and aerial images is shown in Fig. 1.105a for flood phase of the tide and in Fig. 1.105b for the ebb phase of the tide. The influence of the sea current in the strait south of Akseløya Island is not taken into account in the scheme. The schemes shown in Fig. 1.105 show that a part of water around the shore of Akseløya Island is not involved in the water jet in the middle part of the Akselsundet Strait and participates only in the formations of eddies near the northern tip of the island. It is explained by the joint action of the sea bottom topography and friction between the water and seabed, the Coriolis force, and nonlinear hydrodynamics effects.

We performed numerical simulations using a high-resolution tide model (Kowalik and Proshutinsky 1994). A numerical model of the barotropic tides was used to investigate strong tidal currents and fine scale eddies observed in the vicinity of Akseløya Island. The computation domain in the western fjords of Spitsbergen (see Fig. 1.101a) is based on the high spatial resolution with a grid size close to 90 m. The model is forced through the open boundary by five tidal constituents, namely: M_2, S_2, N_2, K_1, and O_1. These data were adopted from the large-scale computation made by Kowalik and Proshutinsky (1994). Figure 1.107 shows high resolution of the tidal currents in the Akselsundet Strait. The scheme in the left panel is related to the flood phase and in the right panel to the ebb phase. Strong currents are marked by the red and brown colors. The jet-like current of up to 250 cm/s is generated in the passage. The ebb and flood currents are directed by extremely variable different geometry of the fjord's shore and Akseløya Island.

On the western side of the passage Akseløya Island orientation relative to the shore generated an obtuse angle, as the flood waters (Fig. 1.112a) are not strongly constrained by this geometry, therefore the flow through the passage does not entrain all the inflowing waters. The part of the flow is deflected to the northern shore and backward to the west resulting in the large sluggish eddy. Concurrently at the eastern side of the passage the compact eddy is generated by the inflowing flood current (Fig. 1.112a). The eddy's shape and strength are defined by the influence of Akseløya Island orientation relative to the shore. The ebb flow phase (right panel) on the eastern side of the passage due to acute angle between Akseløya and the fjord's shore shows the strong amplification of current. This jet-like current impinges on the

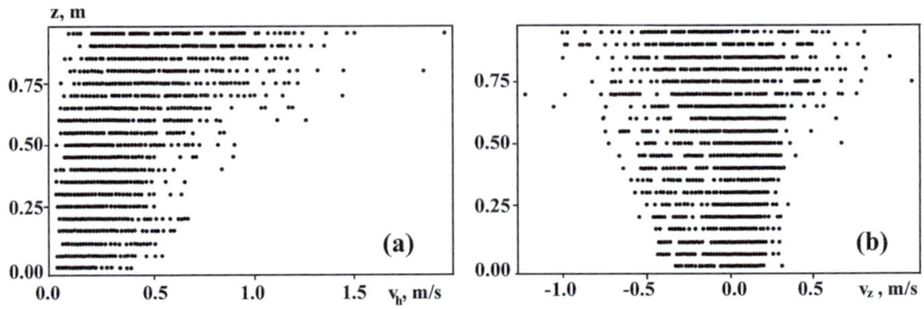

Fig. 1.107 Vertical profiles of the horizontal (**a**) and vertical (**b**) velocities. Adopted from Marchenko et al. (2015)

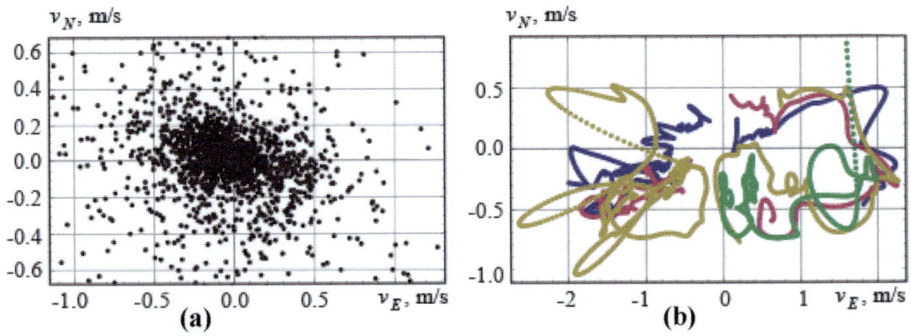

Fig. 1.108 Hodographs of horizontal velocities in the Akselsundet Strait measured with ADCP (**a**) and surface drifters (**b**). Adopted from Marchenko et al. (2015)

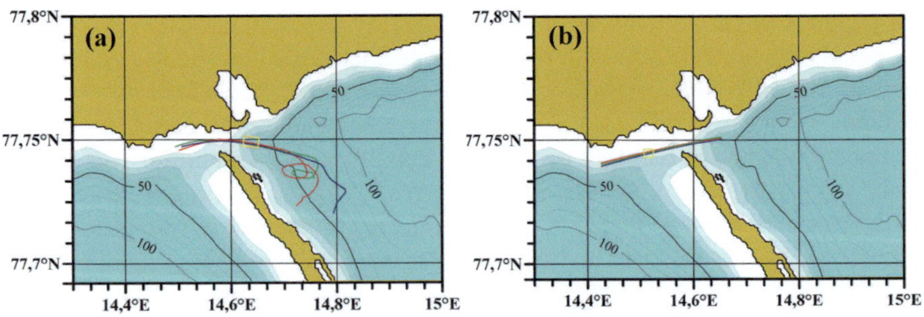

Fig. 1.109 Trajectories of ice drifters deployed in the Akselsundet Strait during the flood (**a**) and ebb (**b**) tidal phases. Adopted from Marchenko et al. (2015). Velocities were measured in the region of yellow rectangles

water at the western side of the passage resulting in the large domain of the enhanced currents and the large eddy south of the main flow. The regular jet is associated with two eddies but the eddy at the northern side is damped by the shore. The right panel

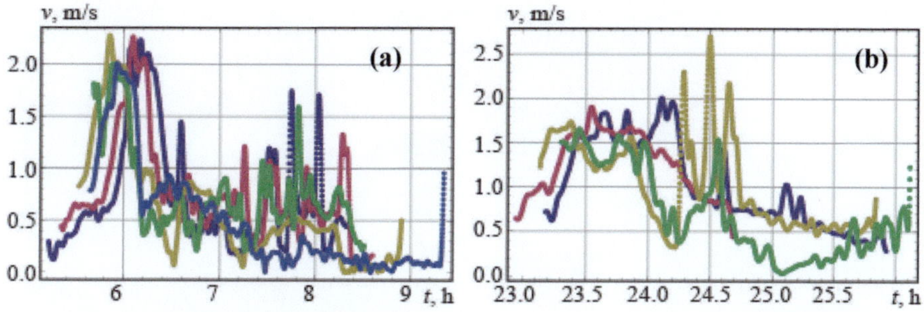

Fig. 1.110 Absolute velocities measured by surface drifters during the flood (**a**) and ebb (**b**) tide. Adopted from Marchenko et al. (2015)

Fig. 1.111 Aerial images of the Akselsundet Strait when the water flows into the Van Mijen Fjord from Bellsund Bay (**a**) and when the current flows from the Van Mijen Fjord into Bellsund Bay (**b**). Adopted from Marchenko et al. (2015)

shows two aspects of the jet-eddy interaction in Bellsund Bay, namely that on the one hand this large eddy is spawned by the jet but on the other hand the eddy is feeding the main jet as well. The influence of Akseløya Island orientation relative to the shore is seen well when the areas of the enhanced current are compared with the flood/ebb patterns in Fig. 1.112.

A simplified box model of the tide was developed for the description of tidal current and water level elevation in the strait connecting the Vallunden Lake and Svea Bay in the end of the Van Mijen Fjord (Marchenko and Morozov 2013). The model equations are derived by the integration over the strait volume of mass balance and momentum balance equations written in shallow water approximation. Geometrical parameterization of seabed topography in the strait is necessary to derive the model equations. Here, we consider the parameterization of the deepest part of the Akselsundet Strait where the jet flow is observed.

The law of mass balance in the fjord is written as follows

$$S_{VM}\frac{d\eta_1}{dt} = S_1 u_1, \tag{1.39}$$

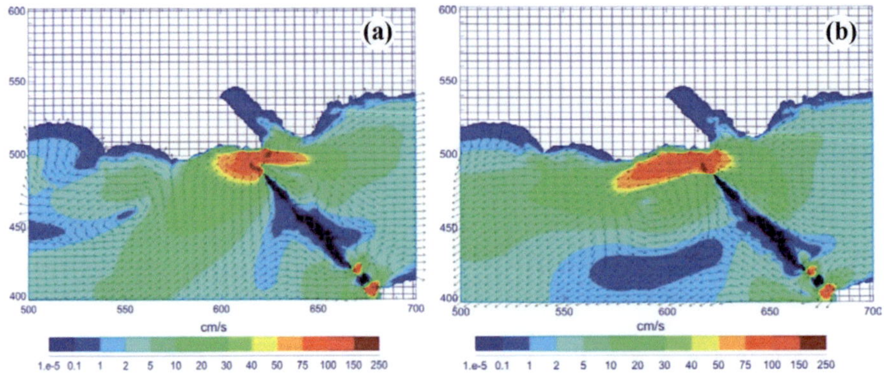

Fig. 1.112 Flood (**a**) and ebb (**b**) tidal currents computed using the numerical model by Kowalik and Proshutinsky (1994). Vectors define direction and the colors correspond to the velocity magnitude in cm/s. Adopted from Marchenko et al. (2015)

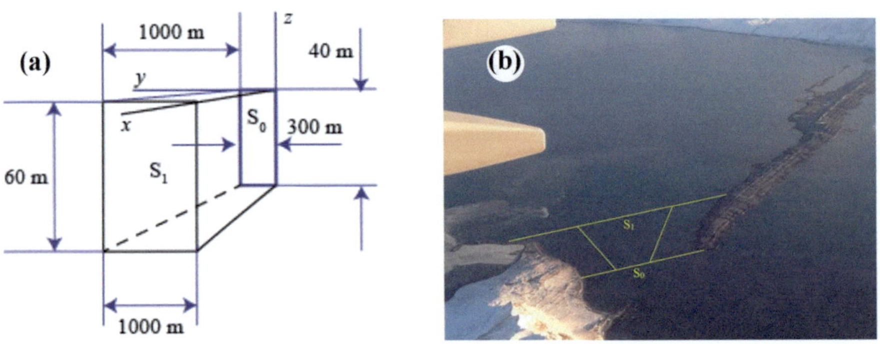

Fig. 1.113 Size of the water jet in the Akselsundet Strait (**a**). Location of the water jet on the aerial photo (**b**). Adopted from Marchenko et al. (2015)

where, $\eta_1(t)$ is equal to the elevation of the water level in the fjord, S_1 is the area of vertical cross-section of the water jet penetrating in the fjord through the channel (Fig. 1.103b), and u_1 is the water speed averaged over area S_1. The area of water surface S_{VM} in the Van Mijen Fjord is about 600 km^2.

Spatial configuration and size of the water jet in the Akselsundet Strait used in the numerical simulations are shown in Fig. 1.112. The length of the channel in the x-direction is $L = 1$ km. The sea depth changes from $h_0 = 40$ m to $h_1 = 60$ m along the channel. The vertical cross-sections of the jet are rectangular. The width of the jet is equal to $B_0 = 300$ m from Bellsund side and $B_1 = 1000$ m from the fjord side. Figure 1.113b shows the location of the jet on the aerial photo made on October 10, 2013 at 13:50 of the local time. The size of the water jet were selected after comparison of the phase shifts of the semidiurnal tide measured near Akseløya Island with the results of numerical simulations.

Total water depth is denoted as $H0$ at Bellsund and $H1$ at Van Mijen boundaries of the channel, and undisturbed water depth is equal to $h0 = H0$-η0 at Bellsund and $h1 = H1$-η1 at Van Mijen boundaries of the channel. Numerical simulations were performed with the drag coefficient C = 0.003 and κ=1. Water level variations at the fjord side of the channel (forcing) are determined by function

$$\eta_0 = A_{M2}\cos(\omega_{M2}t - G_{M2}) + A_{S2}\cos(\omega_{S2}t - G_{S2}) + A_{K1}\cos(\omega_{K1}t - G_{K1}) + A_{N2}\cos(\omega_{N2}t - G_{N2}) + A_{O1}\cos(\omega_{O1}t - G_{O1}),$$

where, numerical values of harmonic constants A, ω, and G are given in Table 1.3.

The results of the numerical simulations during 30 days are illustrated in Fig. 1.114. Figure 1.114a shows that the variations in the water level at the left (L) and right (R) boundaries of the channel are similar, but tidal amplitude on the left boundary of the channel is slightly higher. There is a phase shift about 0.5 h between tides at the left and right boundaries of the channel (Fig. 1.114b). Figure 1.114c shows that the maximal speed of sea current at the left boundary of the channel reaches 3 m/s during the maximal tide, while the maximal water velocities at the right boundary of the channel are close to 0.5 m/s. This effect is explained by the extension of the water channel in the eastern direction (Fig. 1.114b). Figure 1.115 shows water speeds at the boundaries of the channel versus the difference of the water levels at the boundaries of the channel. It is seen that the water speed at the left boundary of the channel is smaller than 0.5 m/s when $|\eta_0 - \eta_1| < 1$ cm. The navigable window is calculated using the following formula

$$T_N = \sum_{i=1}^{30\times24\times3600} \text{UnitStep}[0.01\ \text{m} - |\eta_0 - \eta_0|_{t=i}](2 \cdot 3600 \cdot N)^{-1}, \qquad (1.40)$$

where, function UnitStep[z] is equal to 1 when z > 0 and 0 when z < 0, and N = 58 is the amount of semidiurnal tide cycles in 30 days. Numerical estimates show that $T_N = 13.66$ min.

Thus, field investigations around Akseløya Island confirmed that the phase shift between the semidiurnal tide in the Van Mijen Fjord and Bellsund Bay is about 0.5–1.0 h. This phase shift creates water level gradient over the Akselsundet Strait and influences the generation of tidal currents with the maximal speeds up to 3 m/s. The jet flow is formed inside the Akselsundet Strait within its deepest part approximately 300 m wide. The mean duration of the time window when the water speed in the

Table 1.3 Characteristics of tidal constituents

	M2	N2	S2	K1	O1
A, m	0.485	0.113	0.184	0.06	0.024
ω, day^{-1}	0.0805114	0.07899925	0.08333334	0.04178075	0.03873065
G, rad	6.2	5.8	0.8	4.4	1.4

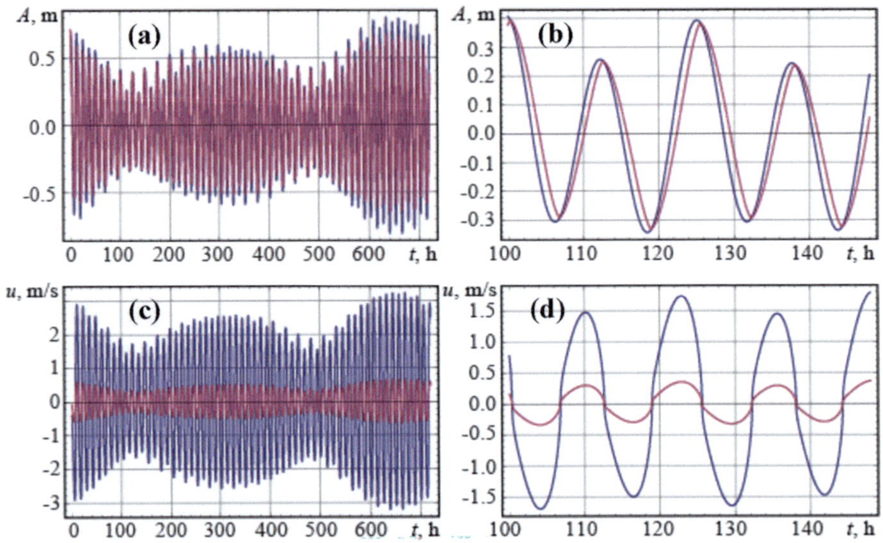

Fig. 1.114 Model simulations of the water level (**a**, **b**) and sea current speed (**c**, **d**) during one month (**a**, **c**) and 48 h (**b**, **d**). Blue and purple lines are related to the Bellsund and Van Mijen Fjord sides of the Akselsundet Strait. Adopted from Marchenko et al. (2015)

Fig. 1.115 Water speeds at the boundaries of the channel versus the difference of the water levels at the boundaries of the channel. Adopted from Marchenko et al. (2015)

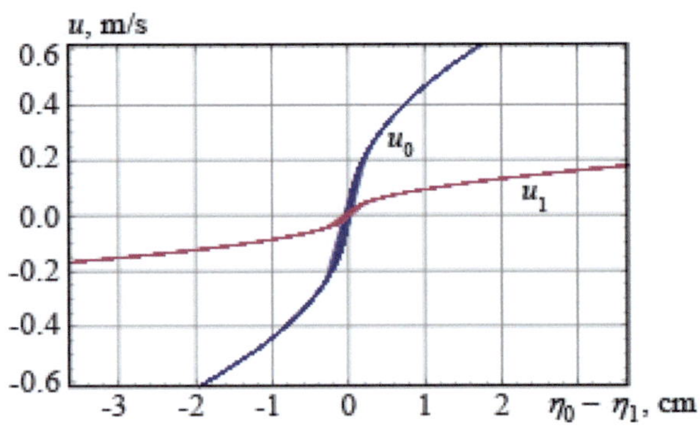

strait is lower than 0.5 m/s is about 15 min, while the actual size of the window is varying from neap to spring tide.

References

Aleinik DL, Byshev VI, Shcherbinin AD (1999) Measured currents in the near-bottom layer of the Norwegian Sea within the region of catastrophe of the atomic submarine Komsomolets. Doklady Earth Sci 369 (9):1393–1397

Aleinik DL, Byshev VI, Shcherbinin AD (2002) Water dynamics in the Norwegian Sea at the site of the accident of the *Komsomolets* nuclear submarine. Oceanology 42(1):7–16

Alekseev PA, Istoshin BV (1956) Scheme of constant currents in the Norwegian and Greenland seas. Proc Knipovich Polar Sci Inst (PINRO) (9): 62–68. [in Russian]. (English translation: Fish and wildlife service special scientific report–fisheries, no. 327: 69–76)

Alexeev GV, Bagryantsev MV, Bogorodsky PV, Vasin VB, Shirokov PE (1991) Structure and circulation of water masses in the area of an anticyclonic vortex in the north-eastern part of the Norwegian Sea. Probl. Arctic Antarct. 65:14–23 [in Russian]

Anderson LG, Björk G, Holby O., Kattner G, Koltermann PK et al (1994) Water masses and circulation in the Eurasian Basin: Results from the Oden91 North Pole Expedition. J Geophys Res Oceans 99:3273–3283

Anderson LG, Andersson PS, Björk G et al (2013) Source and formation of the upper halocline of the Arctic Ocean. J Geophys Res Oceans 118:410–421. https://doi.org/10.1029/2012JC008291

Antonov JI, Seidov D, Boyer TP et al (2010) World Ocean Atlas 2009. Levitus S (ed) V. 2: Salinity. U.S. Government Print. Office, Washington, DC, p 184

Årthun M, Schrum C (2010) Ocean surface heat flux variability in the Barents Sea. J Mar Syst 83:88–98

Årthun M, Ingvaldsen RB, Smedsrud LH, Schrum C (2011) Dense water formation and circulation in the Barents Sea. Deep-Sea Res I 58:801–817

Atlas (1998) In: Matishov G, Zuev V, Golubev V, Adrov N, Slobodin V, Levitus S, Smolyar I (eds) Climatic atlas of the Barents Sea. https://www.nodc.noaa.gov/OC5/barsea/barindex1.html

Averkiev AS, Klevannyi KA (2010) Model simulation of the currents in the Barents Sea in connection with the development of the Shtokman gas condensate field. Russ Meteorol Hydrol 35(11):750–759

BarentsWatch (2020) Map service. https://kart.barentswatch.no/?epslanguage=no. (DNV GL (2020) The Arctic risk map https://maps.dnvgl.com/arcticriskmap/)

Bashmachnikov I, Belonenko T, Kuibin P et al (2018) Pattern of vertical velocity in the Lofoten Vortex (the Norwegian Sea). Ocean Dyn 68(12):1711–1725. https://doi.org/10.1007/s10236-018-1213-1

Belonenko TV, Travkin VS, Koldunov AV, Volkov DL (2021a) Topographic experiments over dynamical processes in the Norwegian Sea. Russ J Earth Sci 21: ES1006. https://doi.org/10.2205/2020ES000747

Belonenko TV, Zinchenko VA, Fedorov AM, Budyansky MV, Prants SV, Uleysky MY (2021b) Interaction of the Lofoten Vortex with a satellite cyclone. Pure Appl Geophys 178:287–300. https://doi.org/10.1007/s00024-020-02647-1

Belonenko TV, Volkov DL, Norden YE, Ozhigin VK (2014) Water circulation in the Lofoten Basin of the Norwegian Sea. Vestnik St Petersburg State Univ Ser 7(2):108–121

Beszczynska-Möller A, Fahrbach E, Schauer U, Hansen E (2012) Variability in Atlantic water temperature and transport at the entrance to the Arctic Ocean, 1997–2010. ICES J Mar Sci 69(5):852–863. https://doi.org/10.1093/icesjms/fss056

Bhatta DD (2005) Wave diffraction by circular and elliptical cylinders in finite depth water. Int. J Pure Appl Math 19(1):69–87

Blindheim J (1989) Arctic intermediate water in the Norwegian Sea. Deep-Sea Res 37:1475–1489

Blindheim J (1990) Cascading of Barents Sea bottom water into the Norwegian Sea. Rapports Et Procès-Verbaux des Reunions Du Conseil International Par L'exploration De La Mer 188:49–58

Blindheim J, Føyn L, Martinsen EA, Svendsen E, Sætre R (1994) The sunken nuclear submarine in the Norwegian Sea-A potential environmental problem? Research report 1994, Institute of Marine Research, Bergen, https://brage.bibsys.no/xmlui/handle/11250/112899

Blindheim J, Østerhus S, (2005) The Nordic Seas, Main oceanographic features. In: Drange H, Dokken T, Furevik T, Gerdes R, Berger W (eds) The Nordic Seas: an integrated perspective, vol. 158. https://doi.org/10.1029/158GM03

Borenäs KM, Lake IL, Lundberg PA (2001) On the intermediate water masses of the Faroe-bank channel overflow. J Phys Oceanogr 31:1904–1914

Bosse A, Fer I, Lilly JM, Søiland H (2019) Dynamical controls on the longevity of a non-linear vortex: the case of the Lofoten Basin eddy. Sci Rep 9:13448. https://doi.org/10.1038/s41598-019-49599-8

Bourgault D, Kelley DE, Galbraith PS (2008) Turbulence and boluses on an internal beach. J Marine Res 66(5):563–588

Brydon D, San S, Bleck R (1999) A new approximation of the equation of state for seawater, suitable for numerical ocean models. J Geophys Res Oceans 104(C1):1537–1540

Chen C, Gao G, Qi J, Proshutinsky A, Beardsley RC, Kowalik Z et al (2009) A new high-resolution unstructured grid finite volume Arctic Ocean model (AO-FVCOM): an application for tidal studies. J Geophys Res Oceans 114(C8):C08017. https://doi.org/10.1029/2008JC004941

Cokelet ED, Tervalon N, Bellingham JG (2008) Hydrography of the West Spitsbergen Current, Svalbard branch: autumn 2001. J Geophys Res Oceans 113:C01006. https://doi.org/10.1029/2007JC004150

Cooper LHN (1955) Deep water movements in the North Atlantic as a link between climatic changes around Iceland and biological productivity of the English Channel and Celtic Sea. J Mar Res 14:347–362

Daniault N, Mercier H, Ferron B, Lherminier P et al (2016) The northern North Atlantic Ocean, mean circulation in the early 21st century. Prog Oceanography 146:142–158

de Steur L, Hansen E, Mauritzen C, Beszczynska-Möller A, Fahrbach E (2014) Impact of recirculation on the East Greenland Current in Fram Strait: Results from moored current meter measurements between 1997 and 2009. Deep-Sea Res I 92:26–40. https://doi.org/10.1016/j.dsr.2014.05.018

Diansky NA, Bagno AV, Zalesny VB (2002) Sigma model of global ocean circulation and its sensitivity to variations in wind stress. Izv Atmos Ocean Phys 38(4):537–556

Diansky NA (2013) Modeling of ocean circulation and investigation of its response to short- and long-period atmospheric forcing. Fizmatlit, Moscow [in Russian]

Diansky NA, Fomin VV, Zhokhova NV, Korshenko AN (2013) Simulations of currents and pollution transport in the coastal waters of Big Sochi. Izv Atmos Ocean Phys 49(6):611–621

Diansky NA, Morozov EG, Fomin VV, Frey DI (2021) Spread of pollution from a bottom source in the Norwegian Sea. Izv Atmosph Ocean Phys 57(2):197–207. https://doi.org/10.1134/S0001433821020043

Dickson RR, Brown J (1994) The production of North Atlantic Deep Water: sources, rates, and pathways. J Geophys Res 99:12319

Dobrovolsky AD (ed) (1977) Convective mixing in the Sea. University Press, Moscow, pp 239

Dyke P (2005) Wave trapping and flow around an irregular near circular island in a stratified Sea. Ocean Dyn 55(3–4):238–247. https://doi.org/10.1007/s10236-005-0002-9

Egbert GD, Erofeeva S (2002) Efficient inverse modeling of barotropic ocean tides. J Atmosph Ocean Tech 19(2):183–204. https://doi.org/10.1175/1520-0426(2002)019%3c0183:EIMOBO%3e2.0.CO;2

Fahrbach E, Meincke J, Østerhus S et al (2001) Direct measurements of volume transports through Fram Strait. Polar Res 20(2):217–224. https://doi.org/10.1111/j.1751-8369.2001.tb00059.x

Fahrbach E, Rohardt G, Schauer U (2012) Physical oceanography and current meter data from mooring F1-1–F1-14. AWI, Helmholtz Centre for Polar and Marine Research, Bremerhaven, PANGAEA, https://doi.org/10.1594/PANGAEA.800280, https://doi.org/10.1594/PANGAEA.800324, https://doi.org/10.1594/PANGAEA.800338, https://doi.org/10.1594/PANGAEA.800349, https://doi.org/10.1594/PANGAEA.800362, https://doi.org/10.1594/PANGAEA.800375, https://doi.org/10.1594/PANGAEA.800388, https://doi.org/10.1594/PANGAEA.800398, https://doi.org/10.1594/PANGAEA.800408, https://doi.org/10.1594/PANGAEA.800290, https://doi.org/10.1594/PANGAEA.800424, https://doi.org/10.1594/PANGAEA.800304, https://doi.org/10.1594/PANGAEA.800307

Falina AS, Sarafanov AA, Gladyshev SV, Sokov AV et al (2015) Water exchange between the Atlantic and Arctic Oceans: Russian research expeditions in 2011–2013. Oceanology 55:441–444. https://doi.org/10.1134/S0001437015030054

Fedorov AM, Bashmachnikov IL, Belonenko TV (2019). Winter convection in the Lofoten Basin according to ARGO buoys and hydrodynamic modeling. Vestnik St. Petersburg. Univ Earth Sci 4(3):491–511. https://doi.org/10.21638/spbu07.2019.308. [in Russian]

Fer I, Voet G, Seim KS, Rudels B, Latarius K (2010) Intense mixing of the Faroe Bank Channel overflow. Geophys Res Lett 37:L02604. https://doi.org/10.1029/2009GL041924

Fer I, Bosse A, Dugstad J (2020) Norwegian Atlantic slope current along the Lofoten Escarpment. Ocean Sci 16(3):685–701. https://doi.org/10.5194/os-2020-15

Fer I, Drinkwater K (2014) Mixing in the Barents Sea polar front near Hopen in spring. J Mar Systems 130:206–218. https://doi.org/10.1016/j.jmarsys.2012.01.005

Fieg K, Gerdes R, Fahrbach E, Beszczynska-Möller A, Schauer U (2010) Simulation of oceanic volume transports through Fram Strait 1995–2005. Ocean Dyn 60(3):491–502

Filyushkin BN, Moshonkin SN, Myslenkov SA, Zalesnyi VB, Kozhelupova NG (2013) Simulation of the interannual and seasonal variability of the overflow transport through the Denmark Strait. Oceanology 53(6):643–654

Filyushkin BN, Sokolovskiy MA, Lebedev KV (2018) Evolution of an intra-thermocline lens over the Lofoten Basin. In: Velarde MG, Tarakanov RY, Marchenko AV (eds) The ocean in motion. Springer Oceanography, pp 333–347

Foldvik A, Aagaard K, Tørresen T (1988) On the velocity field of the East Greenland Current. Deep-Sea Res 35(8):1335–1354

Fomin LM, Shcherbinin AD, Vinogradova KG et al (1996) Currents in the northeastern part of the Norwegian Sea from the results of direct measurements. In: Vinogradov ME, Sagalevich AM, Khetagurov SV (eds) Oceanological studies and underwater and engineering operations at the site of the accident of the Komsomolets nuclear submarine. Nauka, Moscow, pp 263–287. [in Russian]

Frey DI, Novigatsky AN, Kravchishina MD, Morozov EG (2017) Water structure and currents in the Bear Island Trough in July–August 2017. Russ J Earth Sci 17: ES3003

Frey DI, Morozov EG, Fomin VV et al (2019) Regional modeling of Antarctic Bottom Water flows in the key passages of the Atlantic. J Geophys Res Oceans 124(11):8414. https://doi.org/10.1029/2019JC015315

Furevik T, Foldvik A (1996) Stability at M2 critical latitude in the Barents Sea. J Geophys Res 101(C4):8823–8837. https://doi.org/10.1029/96JC00081

Gammelsrød T, Leikvin Ø, Lien V et al (2009) Mass and heat transports in the NE Barents Sea: observations and models. J Mar Syst 75(1):56–69

Gascard J-C, Raisbeck G, Sequeira S et al (2004) The Norwegian Atlantic Current in the Lofoten Basin from hydrological and tracer data (^{129}I) and its interaction with the Norwegian Coastal Current. Geophys Res Lett 31:L01308. https://doi.org/10.1029/2003GL018303

Gerkema T, Zimmerman JTF (2008) An introduction to internal waves. Lecture notes. Royal Netherland Institute for Sea Research, Den Burg, pp 207

Giraudeau J, Hulot V, Hanquiez V et al (2016) A survey of the summer coccolithophore community in the western Barents Sea. J Mar Syst 158:93–105

Girton JB, Sanford TB, Käse R (2000) Synoptic sections of the Denmark Strait Overflow. Geophys Res Let 28(8):1619–1622

Gjevik B, Nost E, Straume T (1990) Atlas of tides on the shelves of the Norwegian and the Barents Seas. Report F&U-ST 90012. Department of Mathematics at the University of Oslo. Report to Statoil, Stavanger

Gjevik B, Nost E, Straume T (1994) Model simulations of the tides in the Barents Sea. J Geoph Res 99(C2):3337–3350. https://doi.org/10.1029/93JC02743

Gjevik B, Moe H, Ommundsen A (1997) Sources of the Maelstrom. Nature 388(28):837–838

Gordeeva S, Zinchenko V, Koldunov A, Roshin Raj P, Belonenko T (2021) Statistical analysis of long-lived mesoscale eddies in the Lofoten basin from satellite altimetry. Adv Space Res 68(2):364–377. https://doi.org/10.1016/j.asr.2020.05
Gulliksen B, Hop H, Nielssen M (2009) Benthic life. Ch 15. In: Sakshaug E, Johnsen G, Kovacs K (eds) Ecosystem Barents Sea. Tapir Academic Press, Trondheim
Haine TWN, Zhang S, Moore GWK, Renfrew IA (2009) On the impact of high-resolution, high-frequency meteorological forcing on Denmark Strait ocean circulation. Quart J Roy Meteorol Soc 135:2067–2085. https://doi.org/10.1002/qj.505
Haine TWN (2010) High-frequency fluctuations in Denmark Strait transport. Geophys Res Let 37:L14601. https://doi.org/10.1029/2010GL043272
Häkkinen S, Cavalier DJ (1989) A study of oceanic surface heat fluxes in the Greenland, Norwegian and Barents Sea. J Geophys Res 94:6145–6157
Hansen B, Østerhus S (2007) Faroe bank channel overflow 1995–2005. Prog Oceanogr 75:817–856
Harms IH (1992) A numerical study of the barotropic circulation in the Barents and Kara Seas. Cont Shelf Res 12(9):1043–1058. https://doi.org/10.1016/0278-4343(92)90015-C
Harris C, Plueddemann A, Gawarkiewicz G (1998) Water mass distribution and polar front structure in the Western Barents Sea. J Geophys Res 103(C2):2905–2917. https://doi.org/10.1029/97JC02790
Hattermann T, Isachsen PE, von Appen W-J, Albretsen J, Sundfjord A (2016) Eddy-driven recirculation of Atlantic Water in Fram Strait. Geophys Res Let 43(7):3406–3414. https://doi.org/10.1002/2016GL068323
Hátún H, Chafik L, Larsen KMH (2021) The Norwegian Sea Gyre, a regulator of Iceland-Scotland Ridge exchanges. Front Mar Sci 8:694614. https://doi.org/10.3389/fmars.2021.694614
Henrich R, Freiwald A, Bickert T, Schäfer P (1997). Evolution of an Arctic open-shelf carbonate platform, Spitsbergen Bank (Barents Sea). In: James NP, Clarke JAD (eds) Cool-water carbonates, vol 56. SEPM, Special. Pub No 56, pp 163–181
Huthnance JM (1974) On the diurnal tidal currents over Rockall Bank. Deep-Sea Res 21:23–35. https://doi.org/10.1016/0011-7471(74)90016-3
ICES (2019) Barents Sea Ecoregion–Ecosystem overview. In Report of the ICES advisory committee. ICES Advice 2019, Section 5.1, pp 12
Ivanov V, Korablev A (1995a) Formation and regeneration of the pycnocline lens in the Norwegian Sea. Russ Meteor Hydrol 9:62–69
Ivanov V, Korablev A (1995b) Interpycnocline lens dynamics in the Norwegian Sea. Russ Meteor Hydrol 10:32–37
Jakobsen PK, Ribergaad MH, Quadfasel D et al (2003) Near-surface circulation in the northern North Atlantic as inferred from Lagrangian drifters: Variability from the mesoscale to interannual. J Geophys Res 108(C8):3251–3264
Jochumsen K, Quadfasel D, Valdimarsson H, Jónsson S (2012) Variability of the Denmark Strait overflow: moored time series from 1996–2011. J Geophys Res 117:C12003. https://doi.org/10.1029/2012JC008244
Jochumsen K, Moritz M, Nunes N, Quadfasel D et al (2017) Revised transport estimates of the Denmark Strait overflow. J Geophys Res Oceans 2017 122: 3434–3450. https://doi.org/10.1002/2017JC012803
Kantha LH, Clayson CA (2000) Small scale processes in geophysical fluid flows, vol 67. Elsevier. Academic Press, Hardback. ISBN: 9780124340701
Kasajima Y, Marchenko A (2001) On the excitation of resonant double Kelvin waves in the Barents Sea Opening. Polar Res 20(2):241–248
Käse RH, Oschlies A (2000) Flow through Denmark Strait. J Geophys Res 105(C12):28527–28546. https://doi.org/10.1029/2000JC900111
Käse R, Girton JB, Sanford TB (2003) Structure and variability of the Denmark Strait overflow: model and observations. J Geophys Res Oceans 108(C6):3181. https://doi.org/10.1029/2002JC001548

Kedra M, Renau PE, Andrad H et al (2013) Benthic community structure, diversity, and productivity in the shallow Barents Sea bank (Svalbard Bank). Mar Biol 160:805–819

Klevannyi KA, Shavykin AA (2008) Modeling transport and sedimentation of suspended substances when laying subwater gas pipeline of the Shtokman Field in the Barents Sea. In: Protection of the environment in the oil and gas complex, vol 6. [in Russian]

Köhl A (2007) Generation and stability of a quasi-permanent vortex in the Lofoten Basin. J Phys Oceanogr 37:2637–2751

Köhl A, Käse RH, Stammer D (2007) Causes of changes in the Denmark Strait overflow. J Phys Oceanogr 37(6):1678–1696

Köhl A (2010) Variable source regions of Denmark Strait and Faroe Bank Channel overflow waters. Tellus 62A:551–568. https://doi.org/10.1111/j.1600-0870.2010.00454.x

Koldunov AV, Belonenko TV (2020) Hydrodynamic modeling of vertical velocities in the Lofoten Vortex. Izv Atmosph Oceanic Phys 56(5):502–511. https://doi.org/10.1134/s0001433820040040

Koltermann KP, Sokov AV, Tereschenkov VP, Dobroliubov SA, Lorbacher K, Sy A (1999) Decadal changes in the thermohaline circulation of the north Atlantic. Deep-Sea Res II 46:109–138

Koszalka I, LaCasce JH, Andersson M, Orvik KA, Mauritzen C (2011) Surface circulation in the Nordic Seas from clustered drifters. Deep-Sea Res I 58:468–485

Koszalka IM, Haine T, Magaldi M (2013) Fates and travel times of Denmark Strait Overflow Water in the Irminger Basin. J Phys Oceanogr 43(12):2611–2628. https://doi.org/10.1175/jpo-d-13-023.1

Kowalik Z, Proshutinsky A (1994) The Arctic Ocean tides. In: Johannessen OM (ed) The Polar Oceans and their role in shaping the global environment. Geophysics monographic series, vol 85. AGU, pp 137–158. https://doi.org/10.1029/GM085p0137

Kowalik Z, Proshutinsky AY (1995) Topographic enhancement of tidal motion in the western Barents Sea. J Geoph Res Oceans 100(C2):2613–2637. https://doi.org/10.1029/94JC02838

Kowalik Z, Luick J (2019). Modern theory and practice of tidal analysis and tidal power. Austides Consulting. Eden Hills. South Australia, 5050 Australia. www.austides.com. https://uaf.edu/cfos/files/research-projects/people/kowalik/Book2019_tides.pdf

Kowalik Z, Marchenko AV (2002) Tidal motion enhancement around Islands. J Marine Res 60(4):551–581. https://doi.org/10.1357/002224002762324211

Kowalik Z, Marchenko A (2023) Tidal motion enhancement on Spitsbergen Bank, Barents Sea. J Geophys Res Oceans 128:e2022JC018539. https://doi.org/10.1029/2022JC018539

Kurkina OE, Talipova TG (2011) Huge internal waves in the vicinity of the Spitsbergen Island (Barents Sea). Nat Haz Earth Syst Stud 11:981–986. https://doi.org/10.5194/nhess-11-981-2011

Kvingedal B (2005) Sea-ice extent and variability in the Nordic Seas, 1967–2002. In: The Nordic Seas: an integrated perspective. Geophysical monograph, vol 158. American Geophysical Union, pp 137–156

Lankhorst M, Zenk W (2006) Lagrangian observations of the mid-depth and deep velocity fields of the northeastern Atlantic Ocean. J Phys Oceanogr 36:43–63

Larsen JC (1977) Cotidal charts for the Pacific Ocean near Hawaii using f-plane solutions. J Phys Oceanogr 7:100–109. https://doi.org/10.1175/1520-0485(1977)007%3c0100:CCFTPO%3e2.0.CO;2

Lee SH, Kim K (1999) Scattering of tidal frequency waves around a cylindrical island. J Phys Oceanogr 29:436–448. https://doi.org/10.1175/1520-0485(1999)029%3c0436:SOTFWA%3e2.0.CO;2

Lind S, Ingvaldsen RB (2012) Variability and impacts of Atlantic Water entering the Barents Sea from the north. Deep-Sea Res I 62:70–88. https://doi.org/10.1016/j.dsr.2011.12.007

Locarnini RA, Mishonov AV, Antonov JI, et al. (2010) World Ocean Atlas 2009, vol 1. Temperature, U.S. Government Printing Office, Washington, D.C

Loeng H (1991) Features of the physical oceanographic conditions of the Barents Sea. Polar Res 10(1):5–18

Loeng H, Ozhigin V, Ådlandsvik B (1997) Water fluxes through the Barents Sea. ICES J Mar Sci 54:310–317

Lukashin VN, Shcherbinin AD (2007a) Hydrological properties, suspended matter, and particulate fluxes in the water column of the Bear Island Trough. Oceanology 47(1):68–79

Lukashin VN, Shcherbinin AD (2007b) Nepheloid layer and horizontal flux of the sedimentary material in the Norwegian Sea. Oceanology 47(6):894–908

Lundrigan S, Demirov EK (2019) Mean and eddy-driven heat advection in the ocean region adjacent to the Greenland-Scotland Ridge derived from satellite altimetry. J Geophys Res Oceans 124:2239–2260. https://doi.org/10.1029/2018JC014854

Lynge BK, Berntsen J, Gjevik B (2010) Numerical studies of dispersion due to tidal flow through Moskstraumen, northern Norway. Ocean Dyn 60:907–920. https://doi.org/10.1007/s10236-010-0309-z

Lynge BK (2011) High resolution tidal models for the Norwegian coast. PhD-dissertation Department of Mathematics, University of Oslo

Macrander A, Käse RH, Send U et al (2007) Spatial and temporal structure of the Denmark Strait overflow revealed by acoustic observations. Ocean Dyn 57:75–89. https://doi.org/10.1007/s10236-007-0101-x

Marchenko A, Kvamstad B, Fjørtoft K et al (2011) Characteristics of ice drift in the Western Barents Sea reconstructed by the data of the trackers deployed on drifting ice in 2008 and 2010. In: Proceedings of international conference on port ocean engineering under arctic conditions, POAC11-147, Montreal, Canada

Marchenko AV, Morozov EG (2013) Asymmetric tide in Lake Vallunden (Spitsbergen). Nonlinear Process Geophys 20:935–944

Marchenko A, Kowalik Z, Brazhnikov D et al (2015) Characteristics of sea currents in navigational strait Akselsundet in Spitsbergen. In: Proceedings of 23rd international conference on port and ocean engineering under arctic conditions June 14–18, 2015 Trondheim, Norway

Marchenko N (2019) Ice at its southern limit in the Barents Sea: field investigation near Bear Island in April 2017–2018. In: Proceedings of international conference on port and ocean engineering under arctic conditions POAC19–58, Delft, Netherlands

Marchenko A, Marchenko N (2019). Characteristics of ice drift and waves on Spitsbergenbanken. In: Proceedings of international conference on port and ocean engineering under arctic conditions, POAC25-33, Delft, Netherlands

Marchenko AV, Morozov EG, Kozlov IE, Frey DI (2021) High-amplitude internal waves southeast of Spitsbergen. Cont Shelf Res 227:104523. https://doi.org/10.1016/j.csr.2021.104523

Marchenko A, Turnbull I, Kowalik Z et al (2022). Properties of drift ice and sea currents on Spitsbergen Bank. In: Proceedings of 26th international conference on port and ocean engineering under arctic conditions POAC2021, Moscow, Russia

Marchenko A, Kowalik Z (2023) Tidal wave—Elliptic island interaction above the critical latitude. J Phys Oceanogr 53(3):683–698. https://doi.org/10.1175/JPO-D-22-0018.1

Marchuk GI, Rusakov AS, Zalesny VB, Diansky NA (2005) Splitting numerical technique with application to the high resolution simulation of the Indian ocean circulation. Pure Appl Geoph 162:1407–1429

Marshall J, Adcroft A, Hill C et al (1997) A finite-volume, incompressible Navier Stokes model for studies of the ocean on parallel computers. J Geophys Res Oceans 102:5753–5766

Martin PA, Dalrymple RA (1994) On amphidromic points. Proc Roy Soc Lond A 444: 91–104. http://www.jstor.org/stable/52571

Mastropole D, Pickart RS, Valdimarsson H et al (2017) On the hydrography of Denmark Strait. J Geophys Res Oceans 122:306–321. https://doi.org/10.1002/2016JC012007

Matishov GG, Denisov VV, Zuev AN et al (1999) Climate atlas of the Barents Sea. Doklady Russ Acad Sciences 366(5):692–694

Matishov GG, Dzhenyuk SL, Moiseev DV, Zhichkin AP (2010) Climatic changes in marine ecosystems of the European Arctic. Probl Arctic Antarctic 3:86–93

Mauritzen C, Price J, Sanford T, Torres D (2005) Circulation and mixing in the Faroese Channels. Deep-Sea Res I 52:883–913

Mauritzen C (1996) Production of dense overflow waters feeding the North Atlantic across the Greenland-Scotland Ridge. Part 1: Evidence for a revised circulation scheme. Deep-Sea Res I 43(6):769–806. https://doi.org/10.1016/0967-0637(96)00037-4

McClimans TA, Nilsen JH (1993) Laboratory simulation of the ocean currents in the Barents Sea. Dyn Atmosph Ocean 19:3–25

McKenna C, Berx B, Austin WEN (2016) The decomposition of the Faroe-Shetland Channel water masses using Parametric Optimum Multi-Parameter analysis. Deep-Sea Res I 107:9–21

McLachlan NW (1951) Theory and applications of Mathieu functions. Oxford University Press. (Reprinted in Great Britain at the University Press, Oxford, 1951 from corrected sheets of the (1947) first edition)

Merian JR (1828) Ueber die Bewegung tropfbarer Flüssigkeiten in Gefässen [On the motion of drippable liquids in containers] (thesis), Schweighauser, Basel. [in German]

Midttun L (1985) Formation of dense bottom water in the Barents Sea. Deep-Sea Res I 32(10):1233–1241

Moe H, Ommundsen A, Gjevik B (2002) A high resolution tidal model for the area around the Lofoten Islands, northern Norway. Contin Shelf Res 22:485–504

Moore CD, Koseff JR, Hult EL (2016) Characteristics of bolus formation and propagation from breaking internal waves on shelf slopes. J Fluid Mech 791:260–283. https://doi.org/10.1017/jfm.2016.58

Moretskiy VN, Stepanov SI (1974) Atmospheric pressure over the Arctic Ocean and the North Atlantic related to water transport in the North Cape Current. Trudy Arctic and Antarctic Res Inst 325:92–95 [in Russian]

Mork KA, Blindheim J (2000) Variations in the Atlantic inflow to the Nordic Seas, 1955–1996. Deep-Sea Res I 47:1035–1057

Mork M (1981) Circulation and fronts in continental shelf seas. Philos Trans R Soc Lond Ser A, Math Phys Sci 302(1472):635–647. http://www.jstor.org/stable/37043

Morozov EG, Pisarev SV (2002) Internal tides at the Arctic latitudes (numerical experiments). Oceanology 42(2):153–161

Morozov EG, Frey DI, Diansky NA, Fomin VV (2019a) Bottom circulation in the Norwegian Sea. Russ J Earth Sci 19(2):ES2004. https://doi.org/10.2205/2019ES000655

Morozov EG, Frey DI, Gladyshev SV, Klyuvitkin AA, Novigatsky AN (2019b) Internal tides in the Denmark Strait. Izv Atmos Oceanic Phys 55:295–302. https://doi.org/10.1134/S0001433819030034

Morozov EG, Frey DI, Gladyshev SV, Gladyshev VS (2020) Hydrodynamics of the bottom-water flow from the Arctic to the Atlantic through the Strait of Denmark. Izv Atmos Oceanic Phys 56(5):479–487. https://doi.org/10.1134/S0001433820050096

Morozov EG, Frey DI, Novigatsky AN (2024) Arctic water current in the Bear Island Trough. Russ J Earth Sci 24: ES4001. https://doi.org/10.2205/2024es000925

Morozov EG, Frey DI (2024) Currents and tides in the Fram Strait and Greenland Sea. Oceanology 64(5):658–669. https://doi.org/10.1134/S0001437024700346

Nansen F (1906) Northernwaters: captain Roald Amundsen's oceanographic observations in the Arctic Seas in 1901. With a discussion of the origin of the Bottom-Waters of the Northern Seas. No. 3 (in Commission by Jacob Dybwad)

Olsen SM, Hansen B, Quadfasel D, Østerhus S (2008) Observed and modelled stability of overflow across the Greenland-Scotland ridge. Nature 455:519–522. https://doi.org/10.1038/nature07302

Orlov NF, Poroshin VV (1988) Water and heat transport in the North Cape Current in 1961–1980. Nature and Economy of the North. Murmansk 16:31–34 [in Russian]

Orvik KA, Skagseth Ø, Mork M (2001) Atlantic inflow to the Nordic Seas: current structure and volume #uxes from moored current meters, VM-ADCP and SeaSoar-CTD observations, 1995–1999. Deep-Sea Res I 48:937–957

Oziel L, Sirven J, Gascard J-C (2015) The Barents Sea polar front and water masses variability (1980–2011). Ocean Sci Discuss 12:449–492
Padman L, Erofeeva S (2004) A barotropic inverse tidal model for the Arctic Ocean. Geophys Res Let 31(2):L02303. https://doi.org/10.1029/2003GL019003
Pavlov VK, Pfirman SL (1995) Hydrographic structure and variability of the Kara Sea: implications for pollutant distribution. Deep-Sea Res II 42:1369–1390
Pease CH, Turet P, Pritchard RS (1995) Barents Sea tidal and inertial motions from Argos ice buoys during the Coordinated Eastern Arctic Experiment. J Geophys Res Oceans 100(C12):24705–24718. https://doi.org/10.1029/95JC03014
Potanin VA, Korotkov SV (1988) Seasonal variability of the main currents in the southern Barents Sea and water exchange with the adjacent areas. In: Geological and geographical problems of natural resources exploitation in the Northern Seas. Murmansk, pp 81–90 [in Russian]
Poulain P-M, Warn-Varnas A, Niiler PP (1996) Near-surface circulation of the Nordic seas as measured by Lagrangian drifters. J Geophys Res Oceans 101:18237–18258
Proudman J (1915) Diffraction of tidal waves on flat rotating sheets of water. Proc Lond Math Soc 14:89–102. https://doi.org/10.1112/plms/s2_14.1.89
Raj RP, Chafik L, Nilsen JEØ, Eldevik T, Halo I (2015) The Lofoten Vortex of the Nordic Seas. Deep-Sea Res I 96:1–14
Raj RP, Johannessen JA, Eldevik T, Nilsen JEØ Halo I (2016) Quantifying mesoscale eddies in the Lofoten Basin. J Geophys Res Oceans 121:4503–4521. https://doi.org/10.1002/2016JC011637
Raj RP, Halo I, Chatterjee S, Belonenko T, Bakhoday-Paskyabi M, Bashmachnikov I et al. (2020) Interaction between mesoscale eddies and the gyre circulation in the Lofoten Basin. J Geoph Res Oceans 125:e2020JC016102. https://doi.org/10.1029/2020JC016102
Reynolds RW (1978) Some effects of an elliptic ridge on wave of tidal frequency. J Phys Oceanogr 8:38–46. https://doi.org/10.1175/1520-0485(1978)008%3c0038:SEOAER%3e2.0.CO;2
Rippeth TP, Vlasenko V, Stashchuk N et al (2017) Tidal conversion and mixing poleward of the critical latitude (an Arctic case study). Geophys Res Let 44:12349–12357
Rossby T, Ozhigin V, Ivshin V, Bacon S (2009a) An isopycnal view of the Nordic seas hydrography with focus on properties of the Lofoten Basin. Deep-Sea Res I 56:1955–1971
Rossby T, Prater MD, Søiland H (2009b) Pathways of inflow and dispersion of warm waters in the Nordic seas. J Geophys Res Oceans 114:C04011. https://doi.org/10.1029/2008JC005073
Ruddick B (1983) A practical indicator of the water column to double-diffusive activity. Deep-Sea Res 30:1105–1107
Rudels B (1987) On the mass balance of the Polar Ocean, with special emphasis on the Fram Strait. Nor Polarinst Skr 188:1–53
Sandalyuk NV, Bosse A, Belonenko TV (2020) The 3-D structure of mesoscale eddies in the Lofoten Basin of the Norwegian Sea: a composite analysis from altimetry and in situ data. J Geophys Res Oceans 125:e2020JC016331. https://doi.org/10.1029/2020JC016331
Santeva EK, Bashmachnikov IL Sokolovskiy MA (2021) On the stability of the Lofoten Vortex in the Norwegian Sea. Oceanology 61(3):308–318
Saunders PM (1990) Cold outflow from the Faroe Bank Channel. J Phys Oceanogr 20:28–43
Saunders PM (2001) The dense northern overflows. In: Siedler G, Church J, Gould J (eds) Ocean circulation and climate. Academic Press, San Diego, pp 401–417
Schauer U, Fahrbach E, Osterhus S, Rohardt G (2004) Arctic warming through the Fram Strait: Oceanic heat transport from 3 years of measurements. J Geophys Res Oceans 109:C06026. https://doi.org/10.1029/2003JC001823
Sea-Bird Electronics Inc (2014) Seasoft V2: SBE data processing. http://www.seabird.com/software/SBEDataProcforWindows.htm
Seager R, Battisti DS, Yin J et al (2002) Is the Gulf Stream responsible for Europe's mild winters? Quart J Roy Met Soc 128(586):2563–2586. https://doi.org/10.1256/qj.01.128
Seim KS, Fer I (2011) Mixing in the stratified interface of the Faroe Bank Channel overflow: the role of transverse circulation and internal waves. J Geophys Res Oceans 116:C07022. https://doi.org/10.1029/2010JC006805

Sherwin TJ, Griffiths CR, Inall ME, Turrell WR (2008) Quantifying the overflow across the Wyville Thomson Ridge into the Rockall Trough. Deep-Sea Res I 55(4):396–404

Shi J, Wei H (2007) Evidence of double diffusion in the East China Sea. J Mar Sys 67:272–281. https://doi.org/10.1016/j.jmarsys.2006.04.017

Sidorova AN, Shcherbinin AD (2004) Intra-annual variability of the thermohaline structure and circulation in the Barents Sea based on the results of model calculations. Oceanology 44(5):629–637

Simonsen K, Haugan PM (1996) Heat budgets of the Arctic Mediterranean and sea surface heat flux parameterizations for the Nordic Seas. J Geophys Res Oceans 101:6553–6576

Smedsrud LH, Esau I, Ingvaldsen RB, Eldevik T et al (2013) The role of the Barents Sea in the Arctic climate system. Rev Geophys 51(3):415–449

Smirnov AV, Ivanov VV, Sokolov AA (2024) Comparison analysis of heat and mass transport through Fram Strait calculated using the mooring and ocean reanalysis data. Phys Oceanogr 31(3):364–386

Søiland H, Rossby T (2013) On the structure of the Lofoten Basin Eddy. J Geophys Res Oceans 118:4201–4212. https://doi.org/10.1002/jgrc.20301

Spall M, Price J (1998) Mesoscale variability in Denmark Strait: the PV outflow hypothesis. J Phys Oceanogr 28(8):1598–1623

Stigebrandt A (1980) Some aspects of tidal interactions with fjord constrictions. Estuarine Coastal Marine Sci 2:151–166

Talley LD (2011) Descriptive physical oceanography. Academic Press, Elsevier. https://doi.org/10.1016/C2009-0-24322-4

Taylor GI (1931) Effect of variation in density on the stability of superposed streams of fluid. Proc R Soc Lond A 132(4):99–523

Teledyne RD Instruments (2012) Doppler Volume Sampler (DVS) Operation Manual

The Norwegian Pilot (2014) Sailing Directions. Svalbard and Van Mijen. The Norwegian Hydrographic Service and Norwegian Polar Institute

Timofeev VT (1963) Water interaction between the Arctic, Atlantic and Pacific oceans. Oceanology 3:569–578 [in Russian]

Travkin VS, Belonenko TV (2019) Seasonal variability of mesoscale eddies of the Lofoten Basin using satellite and model data. Russ J Earth Sci 19: ES5004. https://doi.org/10.2205/2019ES000676

Trodahl M, Isachsen PE, Lilly JM et al (2020) The regeneration of the Lofoten Vortex through vertical alignment. J Phys Oceanogr 50:2689–2711

Turnbull I, Marchenko A (2022) Deformation of an ice pack influenced by waves and topographic enhancement of tidal motion near Hopen Island in the Barents Sea. Cold Regions Sci Tech 194:103463. https://doi.org/10.1016/j.coldregions.2021.103463

Turner JS (1973) Buoyancy effects in fluids. Cambridge University Press, New York, p 368

Uralov NS (1960) On the advective component of the heat balance in the southern Barents Sea. Trudy Gos Okeanograf Inst 55:3–20 [in Russian]

van Aken HM (1995) Hydrographic variability in the Iceland Basin. J Phys Oceanogr 25:1716–1722

van Aken HM, de Boer CJ (1955) On the synoptic hydrography of intermediate and deep-water masses in the Iceland Basin. Deep-Sea Res I 42(2):165–189

Venayagamoorthy SK, Fringer OB (2007) On the formation and propagation of nonlinear internal boluses across a shelf break. J Fluid Mech 577:137–159. https://doi.org/10.1017/S0022112007004624

Vinje T (2001) Anomalies and trends of sea-ice extent and atmospheric circulation in the Nordic Seas during the period 1864–1998. J Climate 14(3):255–267

Vinogradov ME, Sagalevich AM, Khetagurov SV (eds) (1996) Oceanological studies and underwater and engineering operations at the site of the accident of the komsomolets nuclear submarine. Nauka, Moscow

Volkov DL, Belonenko TV, Foux VR (2013) Puzzling over the dynamics of the Lofoten Basin-a sub-Arctic hot spot of ocean variability. Geophys Res Lett 40(4):738–743. https://doi.org/10.1002/grl.50126

Volkov DL, Kubryakov AA, Lumpkin R (2015) Formation and variability of the Lofoten basin vortex in a high-resolution ocean model. Deep-Sea Res I 105:142–157

von Appen W-J, Schauer U, Hattermann T, Beszczynska-Möller A (2016) Seasonal cycle of mesoscale instability of the West Spitsbergen Current. J Phys Oceanogr 46(4):1231–1254. https://doi.org/10.1175/JPO-D-15-0184.1

Woodgate RA, Fahrbach E, Rohardt G (1999) Structure and transports of the East Greenland Current at 75° N from moored current meters. J Geophys Res Oceans 104(C8):18059–18072. https://doi.org/10.1029/1999JC900146

Yakovlev NG (1999) Reconstruction of monthly mean circulation of the Barents Sea and analysis of its sensitivity to water exchange at open boundaries. Izv Atmos Ocean Phys 35(6):767–777

Yashayaev I, Seidov D (2015) The role of the Atlantic Water in multidecadal ocean variability in the Nordic and Barents Seas. Prog Oceanogr 132:68–127

Yulmetov R, Marchenko A, Loset S (2016) Iceberg and sea ice drift tracking and analysis off north-east Greenland. Ocean Eng 123:223–237

Yu L-S, Bosse A, Fer I, Orvik KA et al (2017) The Lofoten Basin eddy: Three years of evolution as observed by Seagliders. J Geophys Res Oceans 122:6814–6834. https://doi.org/10.1002/2017JC012982

Zalesny VB, Diansky NA, Fomin VV, Moshonkin SN, Demyshev SG (2012) Numerical model of the circulation of the Black Sea and the Sea of Azov. Russ J Numer Analysis Math Model 27(1):95–111. https://doi.org/10.1515/rnam-2012-0006

Zhmur VV, Novoselova EV, Belonenko TV (2021) Peculiarities of formation of the density field in mesoscale eddies of the Lofoten Basin: Part 1. Oceanology 61(6):830–838. https://doi.org/10.1134/S0001437021060333

Zhmur VV, Novoselova EV, Belonenko TV (2022) Peculiarities of formation of the density field in mesoscale eddies of the Lofoten Basin: Part 2. Oceanology 62(3):289–302

Zinchenko VA, Gordeeva SM, Sobko YV, Belonenko TV (2019) Analysis of mesoscale eddies in the Lofoten Basin based on satellite altimetry. Fund Prikl Gidrofiz 12(3). https://doi.org/10.7868/S207366731903006

Chapter 2
Ice and Icebergs Drift in the Svalbard Region

2.1 General Notes. Historical Overview of Studies

The ice regime in the Svalbard region (especially in the Barents Sea) has been studied since the middle of the last century, and the main features (variability of the ice edge, ice formation, concentration, drift), and driving forces (wind, currents, wave and tidal processes) have been investigated (Vinje and Kvambekk 1991; Mironov 2004; Frolov et al. 2009; Zubakin 2006; Divine et al. 2025). A great volume of research and literature is devoted to the ice field and iceberg drift (Keghouche et al. 2009; Zubakin et al. 2007; Dmitriev and Nesterov 2007; Marchenko et al. 2010).

The first idea about ice drift came from the famous ships *Jeannette, Fram,* and *St. Anna* trapped in the ice (Nansen and Sverdrup 1897; Albanov 1917). Driftwood found in Svalbard and Greenland also suggested that ocean currents transport driftwood beneath the Arctic ice sheet from east to west.

Features of ice drift in the Barents Sea via the strait between Franz Josef Land (FJL) and Spitsbergen were first described by Albanov (1917), Barr (1975). In June 1914, he measured ice drift velocity equal to 8.5 nm/day (approximately 0.16 m/s) at the point where he saw FJL for the first time on the way from schooner *Svyataya Anna* (Saint Anne) frozen in the ice. Albanov estimated the coordinates as 80°522 N and 40°202 E. In his diary, Albanov noted that the drift velocity was too large to be caused only by the wind influence. Later, the current which transports water from the strait between Spitsbergen and FJL, was named the East Svalbard Current. In 1988, the M/V *Polarbjørn* drifted from the Arctic Ocean into the Barents Sea between Kvitøya and Nordaustlandet and further south passing east of Hopen Island before emerging at the ice edge (Pease et al. 1995). Driftwood found in Spitsbergen (Petermann 1865) also suggested that an ocean current transported it beneath the Arctic ice sheet from east to west (Fig. 2.1).

However, ships can provide little data about ice motion over large spaces, comparatively to the more common now two methods of remote monitoring: drifting buoys and Sea Ice Products. Drifting buoys send signals containing coordinates and time

A. Marchenko et al., *Geophysical Hydrodynamics of Currents, Waves, and Ice in the Atlantic Arctic*, https://doi.org/10.1007/978-3-032-14122-4_2

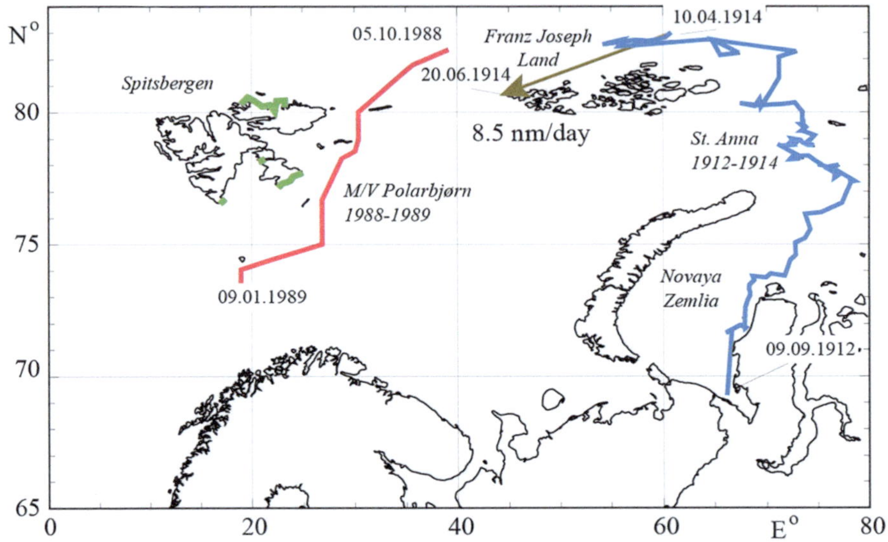

Fig. 2.1 Drift trajectories of the *St. Anna* (1912–1914) (blue line) and the M/V *Polarbjørn* (1988–1989) (red line). The brown arrow indicates Albanov's route to Franz Joseph Land. Green marks are the locations of driftwood on the beach. Adopted from Marchenko et al. (2013), Marchenko and Marchenko (2015)

with a determined frequency and can provide massive data, some of them with additional sensors, measuring temperature, wind speed, or other parameters.

Below, we provide an overview of scientific programs and campaigns revealing ice drift by means of installing satellite radio beacons or drifting buoys on ice.

2.1.1 Norwegian Polar Institute in Cooperation with NASA. 1975–1978

The use of satellite radio beacons for studying the ice drift in the Eurasian Arctic can be traced back to the implementation of the experimental program "Ice Drift Experiment in the Svalbard-Greenland Area", carried out by the Norwegian Polar Institute in cooperation with NASA in the period 1975–1978. The program implied the use of radio beacons in test mode transmitting signals to the American satellite Nimbus-6 (Vinje 1978). The simplest versions of radio beacons were equipped only with a radio transmitter; some buoys were also equipped with temperature and battery charge sensors. The weight of these devices was about 55 kg. The satellite received radio signals during each flight over the area of the radio beacon operation, the signal transmission resolution was 62 s, for a relatively accurate determination of the buoy location it was necessary to receive 4–5 signals per one flight of the satellite. The accuracy of determining the location of the buoy radio transmitter was from 3 to

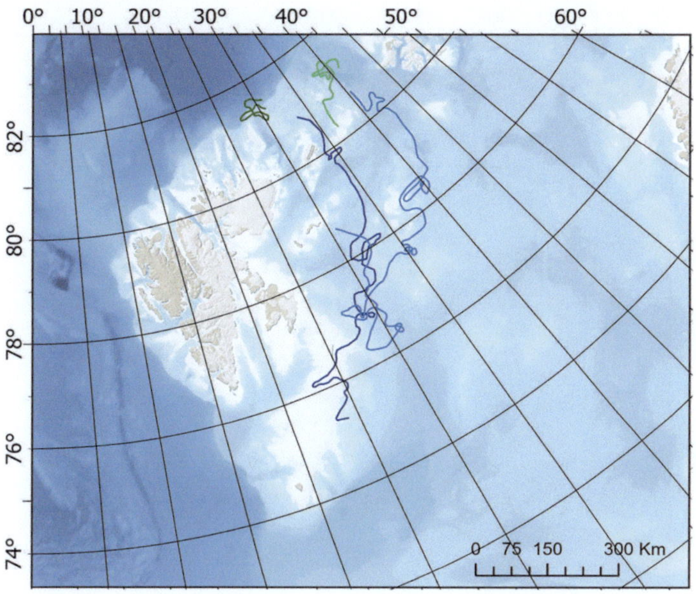

Fig. 2.2 Drift trajectories of Nimbus-6 beacons placed on ice floes in April 1977 (green lines) and in April 1978 (blue lines). The map is based on the data from Vinje (1978, 1979)

5 km. The buoys were deployed on the ice cover in spring and summer. Four radio buoys were drifting in the western Barents Sea (two in 1977 and two in 1978). Given the accuracy of the radio buoy locations, the ice cover drift could be determined with a large error. The ice field trajectories constructed from these data provide an idea of the nature of the ice cover motion in the Spitsbergen region and show its predominant removal from the northern parts of the sea in summer (Fig. 2.2). The map of drift vectors calculated from these data shows that southeast of Spitsbergen the ice cover drift velocity generally did not exceed 0.2 m/s (Vinje 1985).

Overall, the tested method has proven its reliability and success, and the use of this type of equipment has opened great prospects for obtaining information about ice drift and ambient air temperature in hard-to-reach areas of the Arctic.

2.1.2 ICEXAIR Buoys Program. 1986–1987

During the ICEXAIR buoy program, three buoys were parachuted and deployed on ice floes at the same time and longitude (along 40° E) but at different latitudes (76.8° N, 78.5° N, and 80.0° N) on December 16, 1986 (Fig. 2.3). They showed a great variety in the drift directions. The results, summarized in (Vinje and Kvambekk 1991) are as follows:

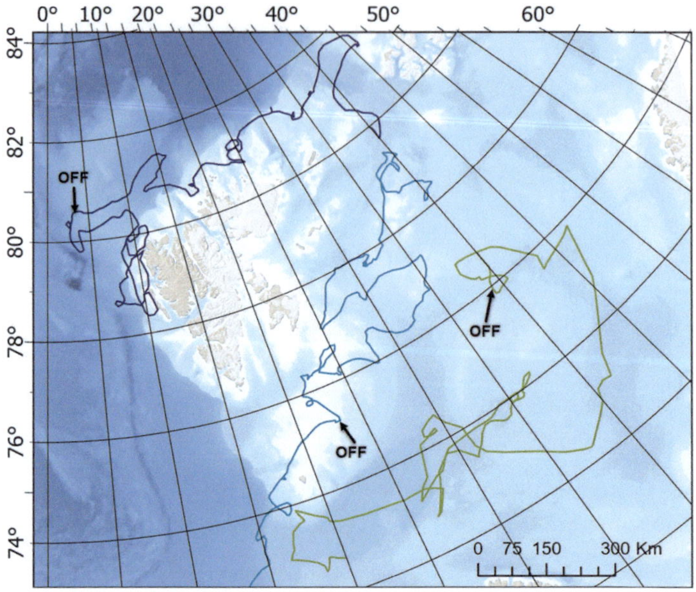

Fig. 2.3 Drift of ICEXAIR-buoys in 1986–1987. OFF indicates when the buoy became afloat. Based on the data from Vinje and Kvambekk (1991)

- the average simultaneous ice velocity in the outer margin is about 0.21 m/s with a mean net displacement of 0.06 m/s towards southwest;
- the ice velocity amounts to 1–2% of the geostrophic wind speed. This agrees with wind-induced ice edge displacements observed on satellite images;
- the wind effect on the ice increases towards the margin, probably because of the increased form drag due to wave effects (Stokes drift) as well as to the scattering of ice floes.

2.1.3 IDAP (Ice Data Acquisition Program). 1988–1994

The measurements within the IDAP were conducted by order of a consortium of Norwegian and international companies in 1988–1994 to insure the safe extraction of hydrocarbon raw materials in the Norwegian sector of the Barents Sea. Significant field data and archival materials were collected on the ice cover and icebergs (Spring 1994). During the work program, 53 buoys equipped with radio transmitters of the ARGOS satellite system were installed on icebergs (15 buoys in the FJL area), and their drift was tracked. The coordinates of the radio beacon were determined by the Doppler shift of the radio signal frequency during the flight of the ARGOS satellite over the area of the beacons. This method allowed determining the location of moving objects with an accuracy of 0–250, 250–500, and 500–1500 m (i.e., there are three accuracy classes) depending on the conditions for receiving radio signals during the flight of the satellite; hence, further analysis required unification. The devices were buried in the ice and equipped with air and ice temperature sensors.

Analysis of iceberg motion with ARGOS radio beacons showed that icebergs from the FJL region drift predominantly in the SW direction, while in the Kvitøya (White Island) region they can drift both south and north, depending on the prevailing wind currents. Average individual iceberg speeds were in the range of 0.02–0.29 m/s, the average drift speed for the entire sample was 0.19 ± 0.1 m/s. The maximum individual speeds (8 different icebergs) reached 1.0–1.1 m/s. Trajectory analysis showed that the main factors regulating iceberg drift in the Spitsbergen Bank region are tidal currents, in the FJL region, quasi-permanent currents, and in the region between Hopen Island and Kong Karls Land (King Charles Land) iceberg drift depends on the combined action of tides and quasi-permanent currents. Two areas were revealed where icebergs run aground and disintegrate: at the northern end of Hopen Island and on the Spitsbergen Bank.

Overall, the mass instrumental observations carried out under the IDAP program were the first of this kind and provided valuable new information on the ice cover and icebergs, significantly supplemented the existing archival data, and served as the basis for the conceptual design of hydrocarbon development facilities in the Norwegian sector of the Barents Sea. Icebergs were recognized as the main factor influencing the choice of the type of production platform and the concept of exploitation of shelf deposits north of 74° N (Spring 1994). Examples of drift of Argos buoys are presented in Fig. 2.4.

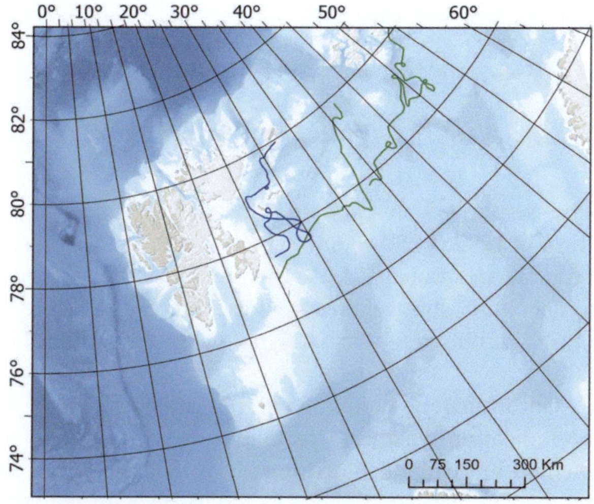

Fig. 2.4 IDAP Argos buoy trajectories. The map is based on the data from Spring (1994)

2.1.4 Coordinated Eastern Arctic Experiment (CEAREX). 1988–1989

The Coordinated Eastern Arctic Experiment (CEAREX) was a broadly interdisciplinary research program, sponsored by the U.S. Office of Naval Research Arctic Sciences Program, and conducted from autumn 1988 through spring 1989. The results have been summarised in (Pease et al. 1995). Two ships, the Norwegian vessel *Polarbjørn* and the U.S. Coast Guard icebreaker *Northwind,* entered the residual pack ice and proceeded to approximately to 82°40′ N, 32°30′ E, northeast of Svalbard where the *Polarbjørn* was intentionally frozen in the ice on September 16, 1988. See the drifting track in Fig. 2.1 (red line). A large variety of geophysical and biological measurements were made either from the ship or from the ice near the ship. After escorting the *Polarbjørn* into the ice, the *Northwind* maneuvered to support the helicopter deployment of six ARGOS buoys on ice floes in a 60-km radius around the *Polarbjørn* on September 16–17, 1988. The CEAREX participants expected that the *Polarbjørn* with the satellite buoys would drift westward along the north side of Svalbard towards the Fram Strait and exit the Arctic Ocean with the East Greenland Current. However, sustained northerly winds pushed the ice with the instruments toward the south. One of the buoys (7100) was lost at this time and presumably crushed during deformation related to the pack ice compression against the Svalbard Archipelago.

All other buoys drifted until they apparently melted out when the floes entered the relatively warm Norwegian Sea or during the Spring 1989 melt season. The general trajectory was southward along the east side of the Svalbard Archipelago along the Barents Sea shelf (Fig. 2.5).

2.1.5 NOAA Global Drifter Program (GDP). 1991–2016

The largest NOAA Global Drifter Program began in 1979 and reached its current form in 1992: 1250 satellite-tracked surface drifting buoys were deployed to make accurate and globally dense in-situ observations of mixed layer currents, sea surface temperature, atmospheric pressure, wind, and salinity. The GDP resulted in massive data with buoys sending a signal every 6 h from all over the world ocean (NOAA 2025).

Unfortunately, there are not very many buoys in the Arctic Region and there is no indication whether a particular buoy is floating or drifting with ice. The only method to distinguish drifting buoys is to compare the location with sea ice maps on the sending signal days. Comparison with the ice maps showed that the buoys were mostly floating in water and only several buoys to the north of Svalbard were drifting. An evaluation of data from NOAA floating buoys and those fixed on the ice to study the sea currents and ice drift pattern in the western part of Barents Sea has been presented in (Marchenko and Marchenko 2015) and analyzed below.

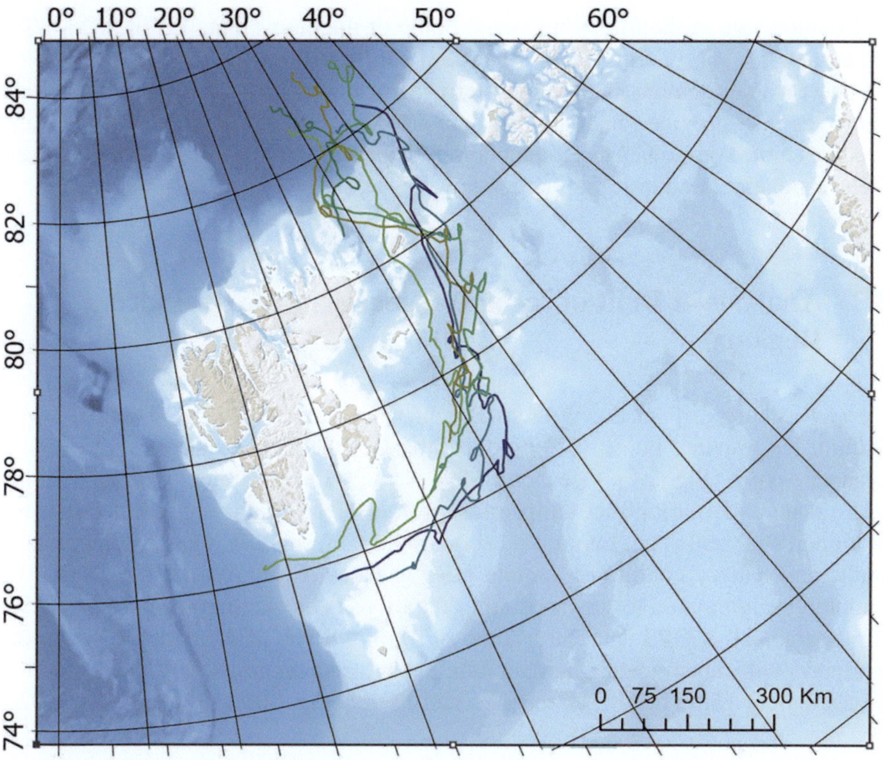

Fig. 2.5 ARGOS buoy tracks during (CEAREX). The map is based on the data from Pease et al. (1995)

2.1.6 Research in the Russian Arctic. 2001–2015

Extensive investigations were performed by the Arctic and Antarctic Research Institute (AARI) to support the development of the field in 2000–2010, supported by oil and gas producing companies. As part of these works on Shtokman gas condensate, eight winter expeditions were performed in the northeastern part of the Barents Sea in 2001–2009. The main results are reflected in (Buzin et al. 2020). Totally, 15 INMARSAT-C and 40 ARGOS buoys were deployed near FJL and drifted in the northeastern part of the Barents Sea; hence, they are not completely related to the Svalbard region.

The collected data formed the basis for the Temporary Local Technical Conditions (TLTC) documents used for the planning of the Shtokman field development. These TLTC described about 100 parameters of the ice regime of the field area and that in the northeastern part of the Barents Sea, of which about 10 parameters characterized the drift of the ice cover and icebergs. The investigation was continued in the Kara and Laptev seas in the following years. Some of the buoys installed in 2013–2015 penetrated into the Barents Sea northeastern corner.

These studies demonstrate the long-term interest in ice drift research using drifting buoys, as they provide real, valuable data and results. The research continues and becomes more comprehensive as technology improves, both in the buoys themselves, their installation, and in signal processing.

2.2 Analysis of Drift of Ice and Icebergs in the Svalbard Region

Ice and iceberg drift in the Western Barents Sea was studied in 2008–2019 with Iridium Ice tracking buoys (IT) produced by Oceanetic Measurement Ltd (www.oceanetic.com) (Fig. 2.6). Oceanetic Measurement Ltd. manufactures ice-tracking instruments and a through-ice temperature profiler, all of which report ice position and motion in near real-time. Internal sensors provide information regarding instrument status and battery condition. Sample rates and device reporting intervals can be remotely programmed to vary operating durations. The standard sampling interval for the transfer of GPS data via Iridium is 10 min. Some of the deployed ITs were equipped with anemometer, temperature string, or a VectorNav VN-100 Inertial Measurement Unit (IMU) (Marchenko and Marchenko 2019; Rabault et al. 2020, 2023).

All Oceanetic instruments are designed to endure long deployments in severe Arctic and Antarctic conditions on sea ice surface above water. Drop of IT into the water may lead to a leak and end of operating. If a leak does not happen, IT operates as a floater providing information on surface currents. The length and width of the 703 IT model is about 70 cm and 15 cm, respectively. The ITs were placed into a

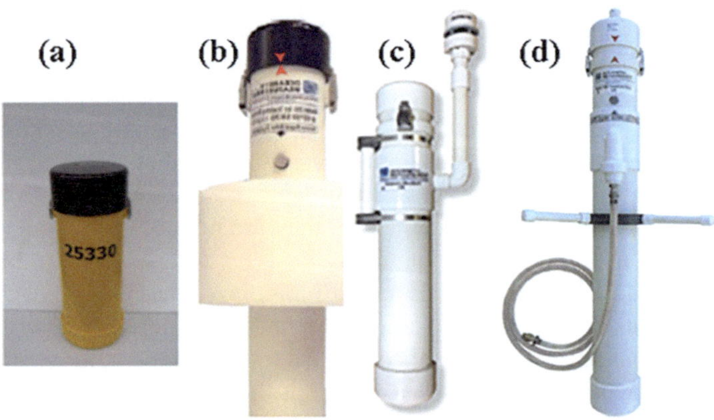

Fig. 2.6 Ice tracking buoys Oceanetic Model 703 transmitting only GPS data before 2010 (**a**), after 2010 (**b**), equipped with anemometer or IMU (**c**), equipped with temperature string (**d**)

hole drilled in an ice floe or an iceberg to a depth of about 30 cm. For measuring ice temperature, a well of 2-inch diameter is drilled through the ice near IT, and temperature string is placed in the well. The length of the temperature string is about 10 m, and the distance between neighbor thermistors is 20–30 cm.

The ITs were deployed on sea ice and icebergs during cruises of the R/V *Lance* (2008–2016) and *Polarsyssel* (2017–2019) organized within the study program of the University Centre in Svalbard with the support of the SFI Project SAMCoT. Based on bathymetry and the structure of sea currents Svalbard region is subdivided into four sub-regions: the western part of the Stoor Fjord (SFW), the eastern part of the Stoor Fjord (SFE), and the Barents Sea (BS) (Fig. 2.7a), the Spitsbergen Bank is the shallowest part of the Svalbard region with strong tidal currents. The Stoor Fjord is bounded by Spitsbergen from the west, Edge and Barents islands from the east, and the Storfjordrenna (trough) from the south. The Sea depth in the SFW is deeper than in the SFE. The region to the south of Edge Island called Tusenoyane including many small islands is included in the SFE. The Barents Sea subregion is extended between Edge and Barents islands in the west, Kong Karls Land in the east, and Hopen Island in the south. The IT data were analyzed separately in each region and compared. Figure 2.7b shows selected trajectories of ice trackers deployed on drift ice from 2008 to 2017.

Raw data transmitted via Iridium satellites were processed with the Mathematica Wolfram Software (MW). Despiking procedure was used to filter the raw data. Functions GeoDistance and GeoDirection of MW were used for the calculation of distances between consequent GPS locations of IT, and for the calculation of distances and azimuthal angles between different ITs operating within the same time window. Function Interpolation of MW was used to construct continuous dependences of IT coordinates from time. Numerical differentiation of interpolated data was used to calculate IT velocities depending on time. Fourier spectra of ice drift speeds were calculated using the Fourier function of MW. Function Histogram was used to show the distribution of the drift velocities over the entire range, divided into 50 bins. The dependence of the ice drift velocity from local sea depth was calculated using function GeoElevationData of MW along each trajectory of ITs.

Radii of curvature of IT trajectories were calculated using the formula:

$$R = \frac{V^3}{\left|V_E \dot{V}_N - V_N \dot{V}_E\right|}, \tag{2.1}$$

where V is the drift velocity, V_E and V_N are the east and north components of the drift velocity, and $\dot{V}_E$ and $\dot{V}_N$ are the east and north components of the acceleration.

The numerical procedure is based on constructing the second order approximations of the trajectory intervals passed by IT in 3 h. Each interval includes 30 points passed by the IT every 0.1 h. The second order approximations were constructed using function Fit of MW applied to the IT displacements in the eastern and northern directions. Then, the constructed approximations were substituted into Formula (2.1) to calculate R in the middle of the 30-point intervals, i.e., at the 15th point of each

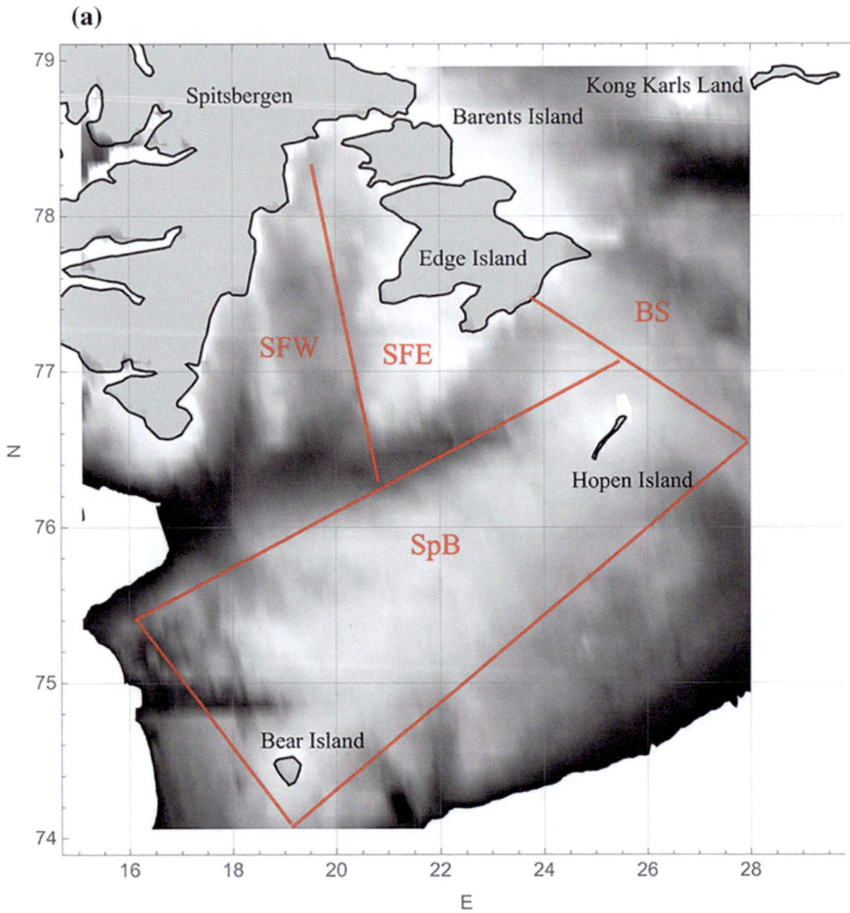

Fig. 2.7 **a** Svalbard region. Red lines show boundaries of the Spitsbergen Bank (SpB), the western part of the Stoor Fjord (SFW), the eastern part of the Stoor Fjord (SFE), and the Barents Sea (BS). **b** Selected trajectories of ice trackers deployed on drift ice from 2008 to 2017

interval. The $R(V)$ dependence characterizes the rate of change of the ice drift velocity depending on the drift speed. Further, in this chapter we consider characteristics of the ice drift reconstructed from collected data.

Trajectories, drift velocities depending on time, Fourier spectra and histograms, dependences of ice drift velocities on the local depth, and radius of curvature of the trajectory are given in the Appendix for each IT. The main results of the IT data processing are considered in the next section.

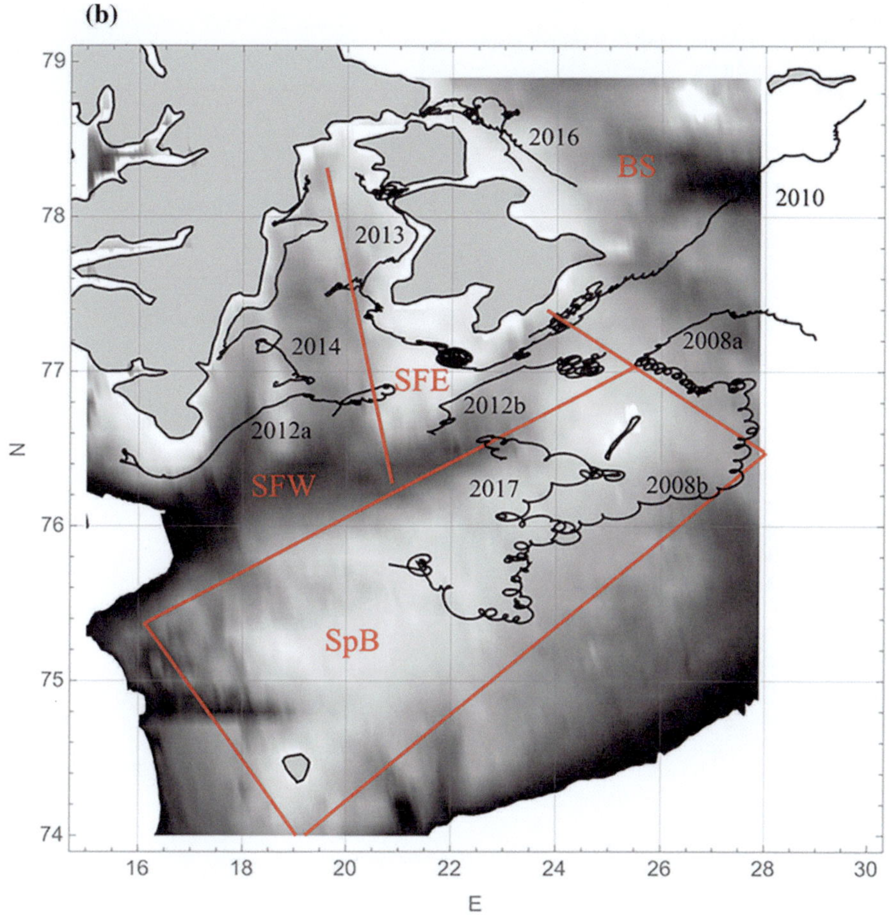

Fig. 2.7 (continued)

2.3 Results of the IT Data Processing

2.3.1 Measurements in 2008

The IT was deployed in the marginal ice zone (MIZ) of the Barents Sea during the expedition of the R/V *Lance* on May 9, 2008 (Fig. 2.8) (Marchenko et al. 2011). The sampling interval of transmitted data on the IT GPS position was 20 min. The IT trajectory is shown in Fig. 2.9 by a blue line from May 9 to May 31, and by a yellow line from June 1 to June 20. The ice maps in Fig. 2.10 show that the IT was in the region covered by continuous ice on May 9 and in the region covered by sea ice with 70%-concentration or lower on June 20. The IT drifted mostly in the BS in May and in SpB in June.

Fig. 2.8 Aleksey Marchenko after deployment of IT on drift ice in the Barents Sea on May 9, 2008

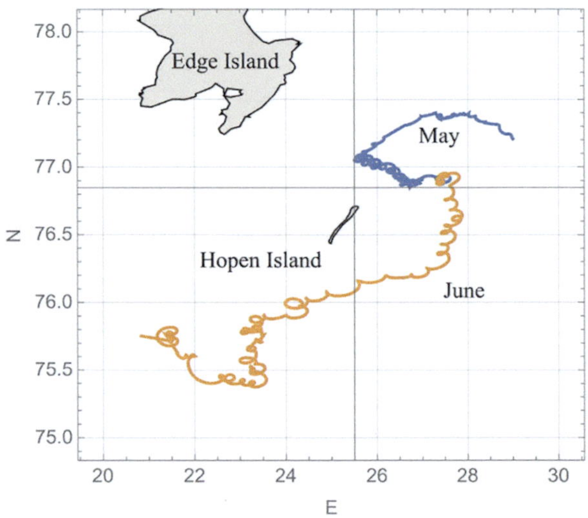

Fig. 2.9 Trajectory of the IT is shown by blue line on May 9–31, and yellow line on June 1–20, 2008

Wind speed and wind direction provided by the meteorological station of Hopen Island are shown in Fig. 2.11 from May 9 to June 20, 2008. Winds of dominating northern directions influenced the general drift of sea ice to the southwest. The mean wind speed was around 5 m/s. Land fast ice between Hopen Island and Edge Island restricted ice drift between them. As a result, the IT drifted around Hopen Island from the eastern side. The trajectory of the IT drift includes many loops reflecting combined action of wind and water drags caused by semidiurnal and diurnal tidal currents. The IT passed every loop in the clockwise direction. Figures 2.88 and 2.89b in the Appendix show the drift speed depending on time. Fourier spectrum of the drift velocity has local maxima at the frequencies $f \approx 0.04$, 0.08, and 0.16 cyc/h (Figs. 2.88 and 2.89c). It corresponds to the diurnal and semidiurnal periods of

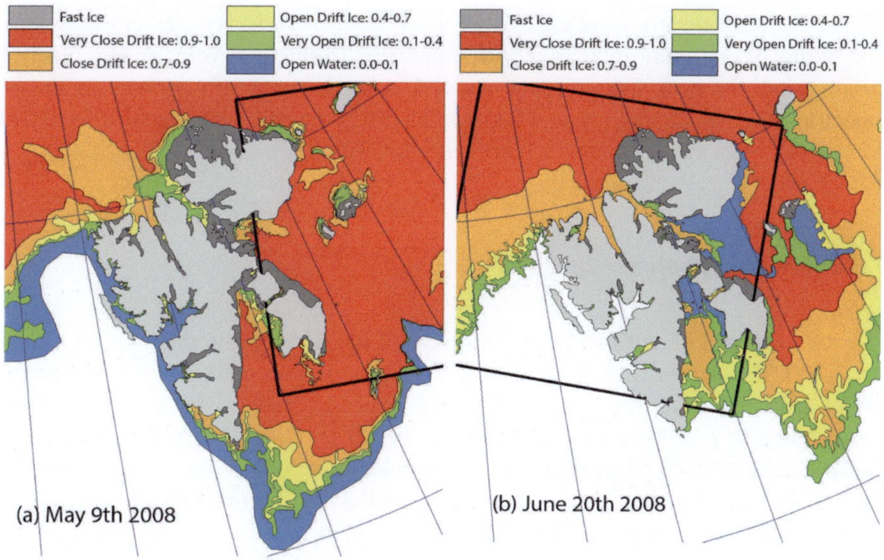

Fig. 2.10 Ice conditions on May 9 and June 20, 2008

around 25 and 12.5 h, and the second harmonic of the semidiurnal frequency with a period of around 6.25 h. Dominant oscillations occur with the semidiurnal frequency.

Histograms of the IT drift velocities in May and June show increasing drift velocity as the IT proceeded to the south (Figs. 2.88 and 2.89d). The mean drift velocities ($\langle V \rangle$) in May and June were respectively 0.23 m/s and 0.41 m/s. The maximum drift velocities $V_{\max}$ were 0.55 m/s in May and 1.36 m/s in June. Figures 2.88 and 2.89e show an increase in the maximum drift velocity with decreasing local sea depth D. Points in Figs. 2.88 and 2.89f show the radius of curvature of the IT trajectory (R) depending on the drift velocity V. The point cloud has well defined bottom envelope on the plane (V, R). The envelope shows how quickly the drift velocity may change over time depending on the drift velocity. The minimum radii of curvature in the

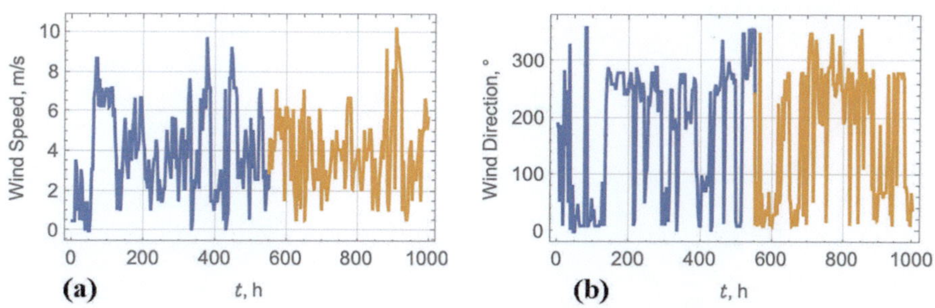

Fig. 2.11 Wind speed and direction depending on time measured on meteorological station of Hopen Island on May 9–31 (blue), and June 1–20 (yellow), 2008

range of several meters were calculated when the drift velocity was close to zero. The radius of curvature increased with increasing of the drift velocity and reached 10 km when the drift speed was of about 1 m/s. The radii of curvature of the IT trajectory were of about 1 km and 2 km when the IT drifted with a mean velocity of 0.23 m/s in May and 0.41 m/s in June, respectively.

2.3.2 Measurements in 2009

The ice trackers ITa and ITb were deployed on the iceberg and drift ice on May 2 at 02:30 UTC and on May 5 at 15:20 UTC in the region marked by blue squares in Fig. 2.12 and drifted in the BS (Marchenko et al. 2010, 2019). The R/V *Lance* was moored to a drifting floe on May 2 at 06:00 UTC, and drifted together with the floe until May 5. Measurements of sea current velocities and wind velocities were performed from the *Lance* board and from the floe. Figure 2.12 shows changes of ice conditions from April 30 to May 5. The wind speed and wind direction are shown in Fig. 2.13 depending on time during this period of operation. Winds of the southern directions influenced ice drift to the north against the East Spitsbergen Current supporting regular transport of drift ice along the eastern coast of Spitsbergen in the southwestern direction to the Spitsbergen Bank.

Ice maps in Fig. 2.12 show that the iceberg drifted in the marginal ice zone consisting of different ice types changing from very open drift ice to the ice with a concentration of 70–90%. Visual observations and analysis of echo-sounder records show that the drift ice consisted of floes with diameters of 2–3 km (Marchenko et al.

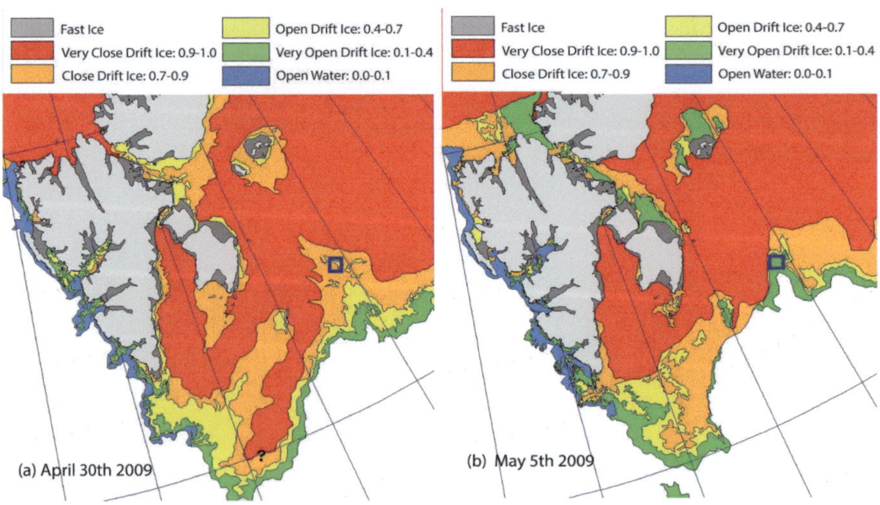

Fig. 2.12 Ice conditions on April 30 (**a**) and May 5 (**b**), 2009. Open square shows the location of an iceberg

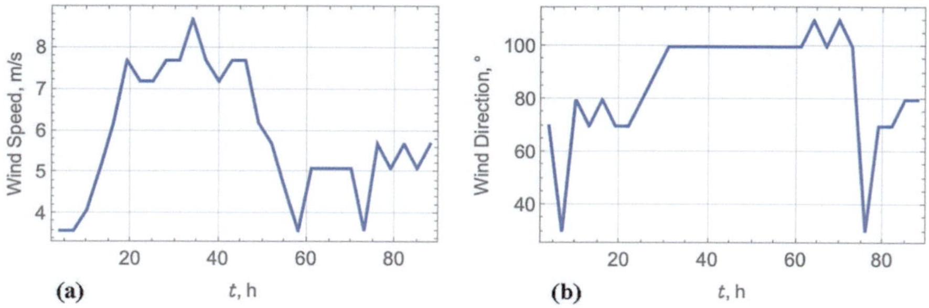

Fig. 2.13 Wind speed and direction depending on time measured on the meteorological station of Hopen Island from May 2, 01:00 UTC, to May 5, 13:00, UTC, 2009

2010). Ice thickness varied from 0.25 to 1 m. The IT1 was deployed on a tabular iceberg on May 2, 02:30 UTC (Fig. 2.14). Similar ice tracking buoy IT2 was deployed on the floe near the *Lance* on May 5, 15:20 UTC. The GPS positions of the ITs were determined at an interval of 20 min, and the reports were transmitted to satellite by tracking modem at an interval of 120 min. The IT1 and IT2 communicated via Iridium for 94.5 h and 27.6 h, respectively, and then the communications ended.

Figure 2.15a shows the trajectory of the *Lance* (thick blue line) before and after the deployment of the IT1 and IT2. The red segment of the *Lance* trajectory BC and the thick blue segment of the IT1 trajectory correspond to the time interval when the *Lance* was moored along the iceberg face. This face is shown on the photograph made after the deployment of the IT1 on May 2 (Fig. 2.16a). Figure 2.15b shows the IT1 and IT2 trajectories on May 5. Photo in Fig. 2.16b shows the iceberg during the time when the IT2 was deployed. The iceberg faces BC and CD were clearly seen from the *Lance*. Estimated spatial orientations of the iceberg on May 2, 02:30 UTC and May 5, 15:20 UTC are shown in Fig. 2.15b by a dashed square. It is clear that the iceberg has rotated over about 180° during this time.

Fig. 2.14 Ice tracking buoy (IT1) deployed on the drift iceberg. Barents Sea, May 2, 2009

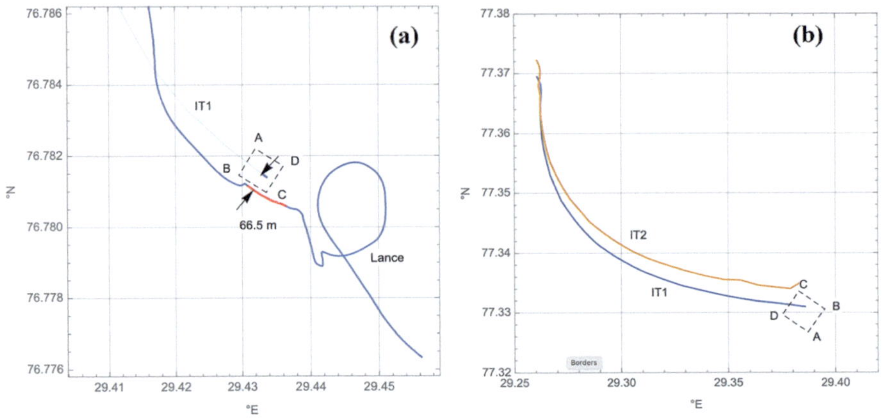

Fig. 2.15 Trajectory of the *Lance* before and after the deployment of IT1 and trajectory of IT1 deployed on the iceberg on May 2, 2009 (**a**). Trajectories of IT1 and IT2 after the deployment of IT2 on May 5, 2009 (**b**)

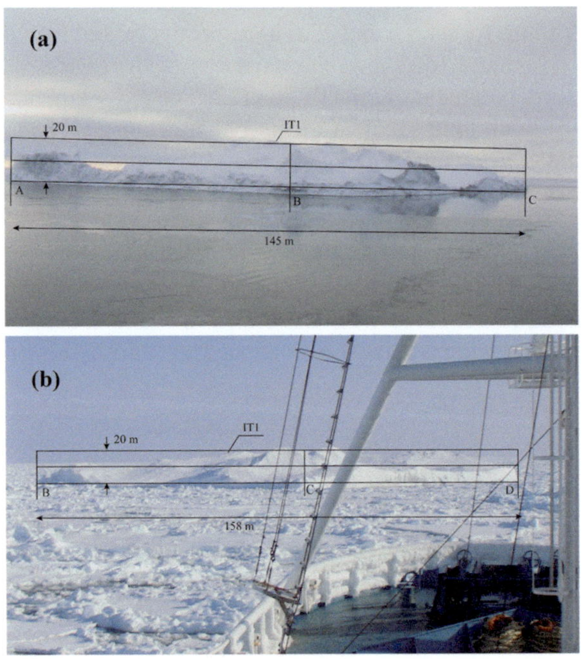

Fig. 2.16 Iceberg photographs made on 04.30, May 2 (**a**), and 16.30, May 5 (**b**) (UTC). Iceberg sizes are estimated using zoomed view of the ice tracking buoy on the iceberg

The iceberg size was estimated using photographs in Fig. 2.16 from the comparison of iceberg size with the sizes of zoomed IT1 deployed on the iceberg. The IT1 diameter is equal to 25 cm. The mean free side of the iceberg was estimated at 10 m. Iceberg diagonals AC and BD were estimated at 145 m and 158 m, respectively. It is assumed that the iceberg had square shape in the horizontal plane with a side of about 100 m. The mean iceberg draft is estimated from the hydrostatic equilibrium as 90 m. The iceberg mass was estimated approximately at 0.9 Mt.

The *Lance* was moored to a drifting floe at about 10 km from the iceberg on May 2, 06:00 UTC, and drifted with sea ice for 37 h. The navigational system of the *Lance* provided records of the geographical position, heading and other characteristics at a sampling interval of one second. Meteorological station on board the *Lance* provided the data on wind velocity. The trajectories of the IT1, IT2, and the *Lance* are shown in Fig. 2.17 by the lines marked as IT1, IT2, and Lance. Thin line L is the *Lance* trajectory whose initial point was moved to the location of the IT1 on May 2, 06:00 UTC. Points on line IT1 show locations of the IT1 over each 2 h. Points on line L correspond to locations of the *Lance* every 2 h. Arrows show wind velocities measured by the *Lance* meteorological station.

The ice and iceberg trajectories are similar, but not the same. Figure 2.18 shows the zonal and meridional components of the drift velocities of the IT1 and the *Lance*. The zonal component of the *Lance* drift velocity exceeded the zonal component of the IT1 velocity by about 5 cm/s during 15 h after $t = 20$ h. It can be explained by the iceberg interaction with compressed ice from the western side of the iceberg. Figure 2.19 shows counter clockwise rotation of the *Lance* moored to a drift floe together with the floe.

Figure 2.18 shows fluctuations of the IT1 velocity appearing on the trajectory patterns with smaller drift velocity, i.e., in the locations with higher density of points on line IT1 in Fig. 2.17. These oscillations may correspond to rotational oscillations of the iceberg or events of the iceberg interaction with large floes. The dependencies of the *Lance* trajectory and the IT1 drift velocity on time are smooth.

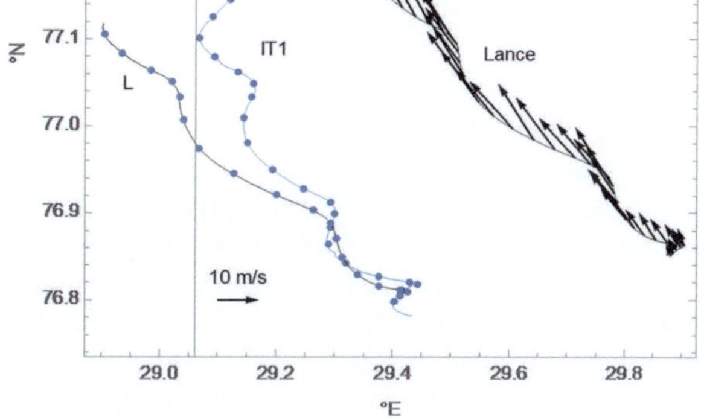

Fig. 2.17 Trajectories of the IT1, IT2 and the *Lance* when she was moored to the drifting ice. Line L is the *Lance* trajectory moved to the initial point of the IT1 trajectory. Points on the lines IT1 and L correspond to locations of the IT1 and the *Lance* at an interval of 2 h. Arrows show wind velocity measured by the *Lance* meteorological station

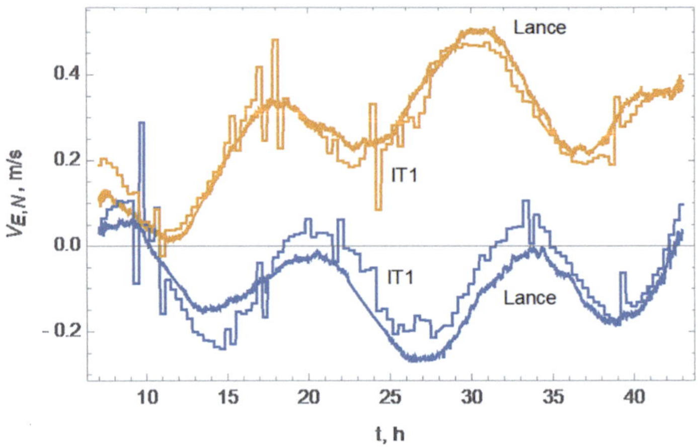

Fig. 2.18 East (blue lines) and North (yellow lines) components of velocities of the IT1 and the *Lance*. The time is accounted from 00:00 UTC of May 2, 2009

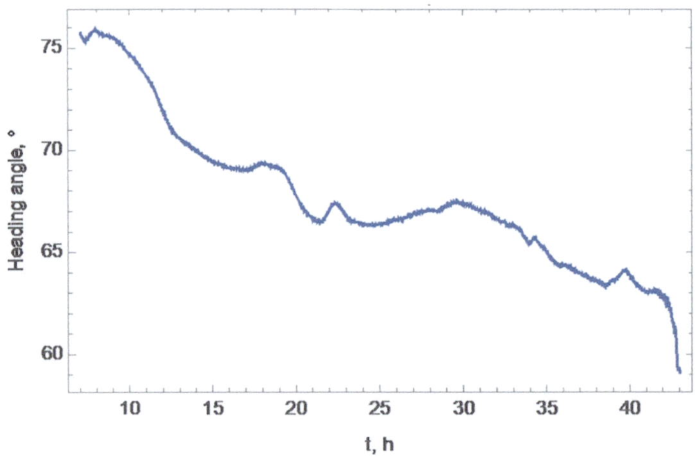

Fig. 2.19 Heading angle of the *Lance* depending on time. The time is accounted from 00:00 UTC of May 2, 2009

Figure 2.90b shows oscillations in the IT1 drift velocity. Spectrum in Fig. 2.90c shows semidiurnal period of the oscillations. The maximum drift velocity was 0.5 m/s, and the mean drift velocity was approximately 0.2 m/s. Histogram of the IT1 drift velocity is shown in Fig. 2.90d. Figure 2.90e does not show any visible dependence of IT drift velocity depending on local sea depth in the range from 250 to 180 m. The minimal values of the radius of curvature of the IT1 trajectory *R* are associated with the minimal drift speeds in Fig. 2.90f. The radius of curvature of the IT1 trajectory is about 1 km when the IT1 drifted with the mean velocity of 0.2 m/s.

2.3.3 Measurements in 2010

Five ITs were deployed on drift sea ice in the Barents Sea to the east of Kong Karls Land by helicopter LIMSAR SA 365 N2 on March 15, 2010 (Fig. 2.20) (Marchenko et al. 2011). The helicopter track from Longyearbyen to the place of deployment is shown in Fig. 2.21. The initial locations of the ITs on four neighboring ice floes are shown schematically in Fig. 2.22. The ITd and ITe were deployed on the same floe; the other ITs were deployed on different floes. Ice thickness varied from 60 to 70 cm within the floes. All five ITs transmitted their GPS data during approximately one month with a sampling interval of 10 min. Ice maps in Fig. 2.23 show that during this month the ITs were in the region covered by very compact drift ice with concentration 90–100%. The ITs trajectories in the BS are shown in Fig. 2.24. General direction of ice drift to the southwest was disturbed by the southeastern winds at 25 h $< t <$ 50 h, and 500 h $< t <$ 700 h (Fig. 2.25). The event of the southeastern wind was more persistent in April. It influenced the ice drift to the northwest near Edge Island with numerous loops in the IT trajectories. The ITs passed all loops in the clockwise direction.

Figures 2.91, 2.92, 2.93, 2.94 and 2.95b shows the drift velocities of the ITs depending on time. The minimum amplitudes of the oscillations of the drift velocities were measured at around t = 200 h and t = 600 h. It correlates well with the sea level amplitude at the point with coordinates 78° N, 27° E constructed with using function TideData of MW (Fig. 2.26). Fourier spectra of the drift velocities in Figs. 2.91, 2.92, 2.93, 2.94 and 2.95c show the presence of the oscillations with the semidiurnal frequency and the second and the third harmonics of the semidiurnal frequency. The amplitude of the third harmonic is higher in the spectra of the drift velocities in April shown by yellow lines in Figs. 2.91, 2.92, 2.93, 2.94 and 2.95c.

Histograms in Figs. 2.91, 2.92, 2.93, 2.94 and 2.95d and Table 2.1 show the characteristics of the ITs drift in March and April 2010. The minimum water depths along the ITs trajectories were approximately two times shallower in April than in

Fig. 2.20 Aleksey Marchenko after deployment of IT on a floe on March 15, 2010

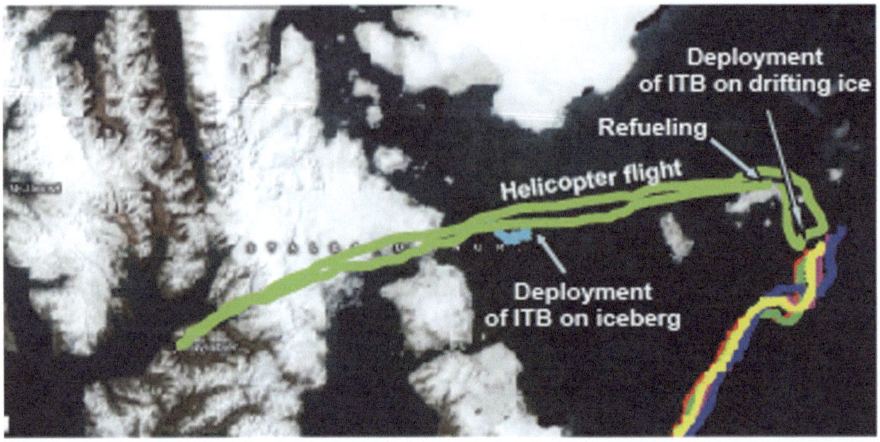

Fig. 2.21 Helicopter track for deployment of ITs on drift sea ice of the Barents Sea on March 15, 2010

Fig. 2.22 Schematic of IT deployment drift floes schematic on March 15, 2010

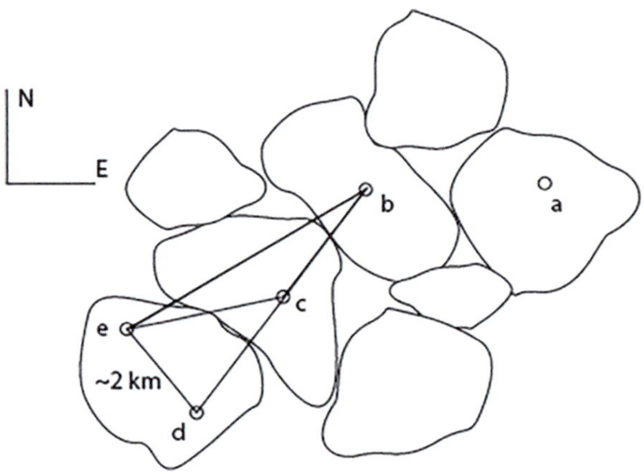

March. The mean drift velocity was approximately 0.15 m/s in March, and 0.21 m/s in April. The maximum drift velocity was approximately 0.48 m/s in March, and 0.57 m/s in April. Figures 2.91, 2.92, 2.93, 2.94 and 2.95e confirm an increase of the ITs drift velocities in April in comparison with March, which can be explained by decreasing sea depth.

Figures 2.91, 2.92, 2.93, 2.94 and 2.95f show that the minimum radii of curvature of the ITs trajectories were approximately 3 km in March and approximately 1 km in April at a mean ITs drift velocity of 0.15 m/s in March, and 0.21 m/s in April. Thus, the minimum radius of curvature decreases with decreasing drift velocity. This property seems natural when the dominant period of oscillations of the drift velocity is semidiurnal: the radius of curvature must be larger to travel around the loop at a higher speed in the same period of time.

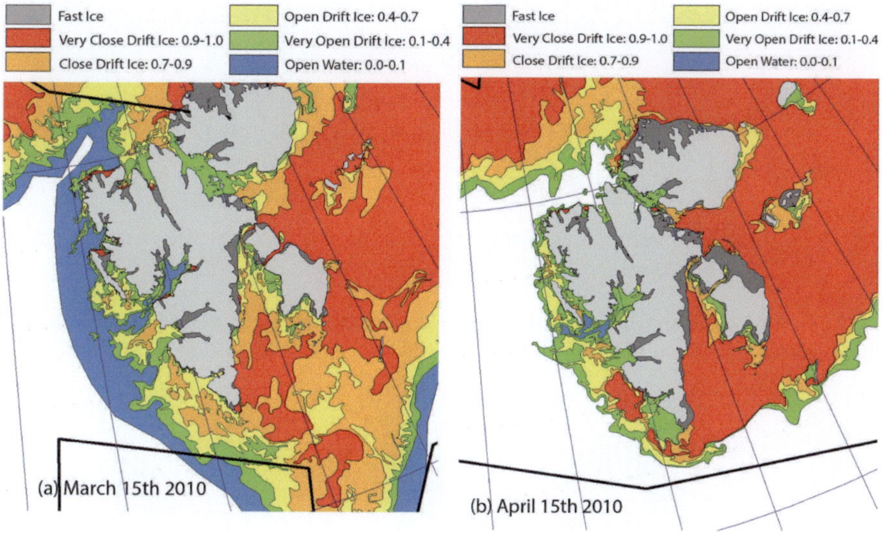

Fig. 2.23 Ice conditions on March 15 and April 15, 2010

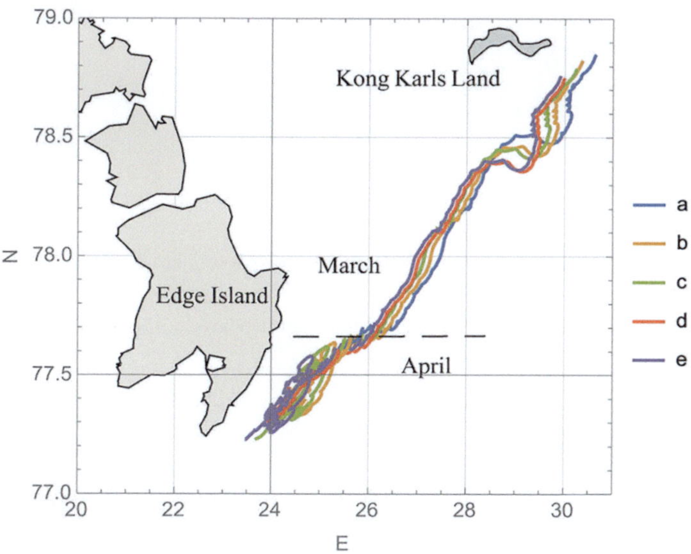

Fig. 2.24 Trajectories of five ITs on March 15–31 and April 1–15, 2010. Numbers correspond to different ITs

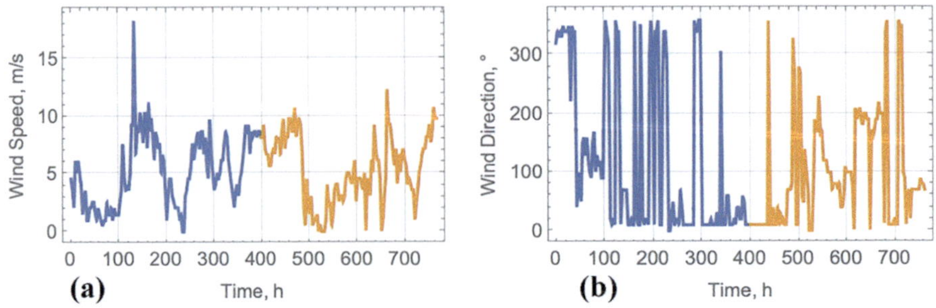

Fig. 2.25 Wind speed and direction depending on time measured on the meteorological station of Hopen Island from March 15 to March 31 (blue), and from April 1 to April 15 (yellow), 2010

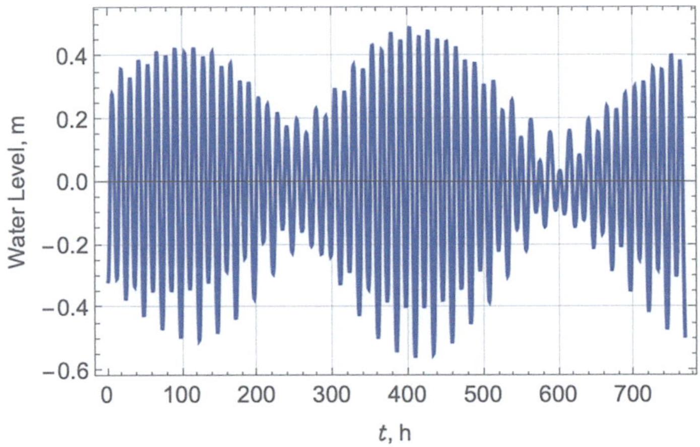

Fig. 2.26 Water level depending on time at the point with coordinates 78° N, 27° E

Table 2.1 Duration of the ITs drift Δt, minimum water depth $D_{\min}$, mean drift velocity $\langle V \rangle$, and maximum drift velocity of the ITs in March and April 2010

		Δt, h	$D_{\min}$, m	$\langle V \rangle$, m/s	$V_{\max}$, m/s
a	March	396	100	0.16	0.45
	April	85	100	0.2	0.47
b	March	396	100	0.16	0.49
	April	362	50	0.2	0.6
c	March	396	100	0.16	0.59
	April	413	<50	0.23	0.6
d	March	396	100	0.15	0.43
	April	211	50	0.2	0.58
e	March	396	100	0.16	0.43
	April	365	50	0.23	0.6

Figure 2.27 shows the changes in distances between the ITs. The ITd and ITe were deployed on the same floe (Fig. 2.22), and the distance between them was stable. Distances between the ITs deployed on different floes changed in time within 5 km, excluding the ITa deployed initially at the most northern point. The rates of distances were greater during the event of the southeastern wind in April. The distance between the ITa and ITb changed by 15 km during this event. Semidiurnal oscillations in the distances between the ITs during the event of the southeastern wind in April are clearly seen in Fig. 2.27.

Semidiurnal oscillations associated with the floe rotation are seen in the azimuthal angle of the line connecting the ITd and ITe deployed on the same floe (Fig. 2.28). Figure 2.29a, b shows the dependences of volumetric deformations of drift ice on time calculated from the formula:

$$\varepsilon_V = \frac{\Delta S}{S}, \tag{2.2}$$

where S is the area of a triangle, which vortices have coordinates of three ITs. Figure 2.29a shows the deformations of a triangle with coordinates of ITb, ITc, and ITe. Figure 2.29b shows the deformations of a triangle with coordinates of ITa, ITd, and ITe. Both triangles are shown in Fig. 2.22. Positive and negative deformations are associated with tension and compression. The largest deformations occurred during the event of the southeastern wind in April. Semidiurnal oscillations are clearly seen in the deformations. Figure 2.29c, d show semidiurnal oscillations in the deformations calculated as a difference between total deformations shown in Fig. 2.29a, b and smoothed deformations calculated using function MovingAverage of WM over 120 h. Amplitudes of semidiurnal oscillations of ε_V calculated using the data of ITb, ITc, and ITe deployed on different floes vary from 0.1 to 0.2 (Fig. 2.29c).

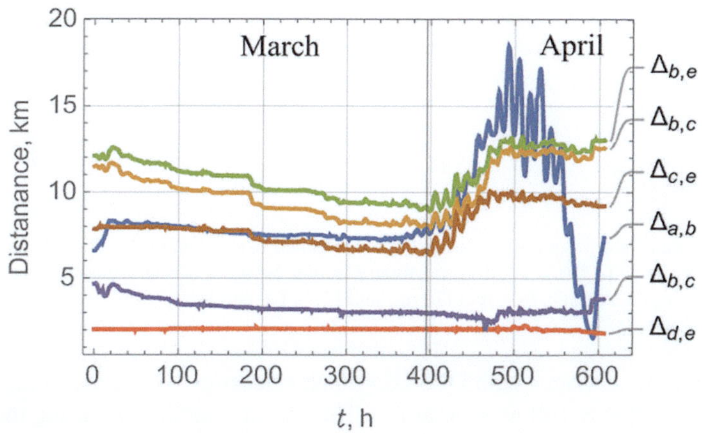

Fig. 2.27 Distances between the ITs depending on time in March and April

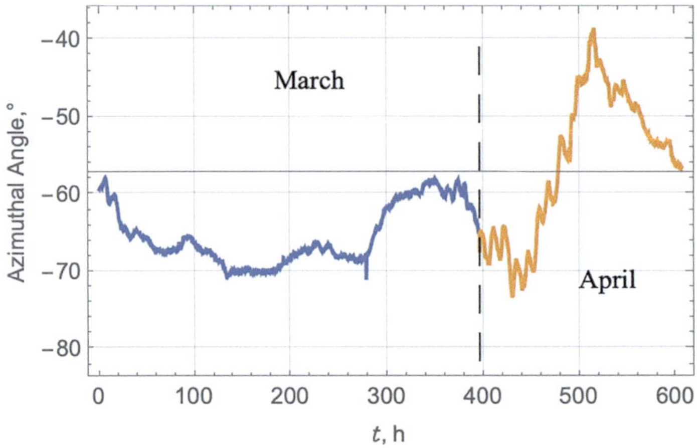

Fig. 2.28 Azimuthal angle of the line connecting ITd and ITe depending on time

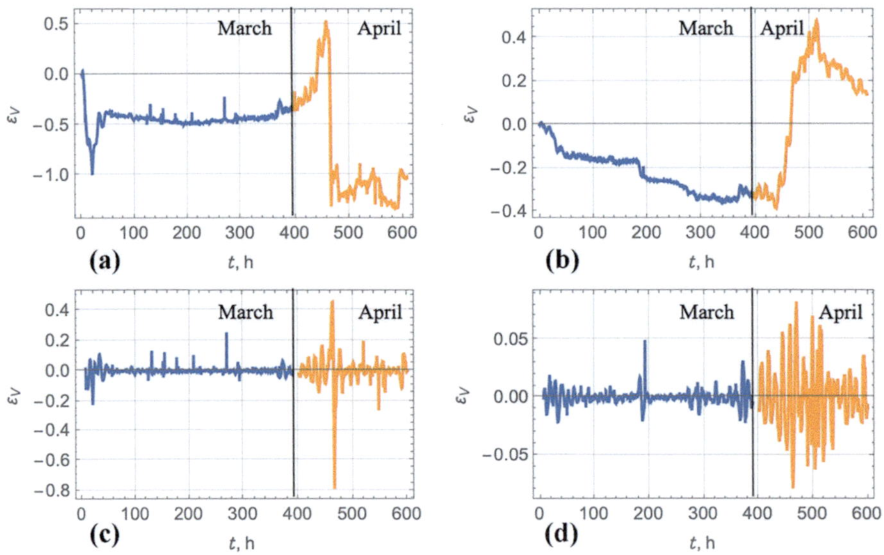

Fig. 2.29 Volumetric deformations of drift ice calculated using data of the ITb, ITc, and ITe (**a**); and ITb, ITd, and ITe (**b**) depending on time. Oscillating parts of the volumetric deformations calculated using data of ITb, ITc, and ITe (**c**); and ITb, ITd, and ITe

The Iridium messages with GPS coordinates became not regular after April 15, but it was possible to follow several ITs. The ITa communicated during the longest time and drifted around Spitsbergen to the Ice Fjord (Fig. 2.30a). Finally, it was picked up on shore near Longyearbyen (Fig. 2.31). Ice map in Fig. 2.30b from June 4, 2010, shows a field of sea with low ice concentration below 40% drifting into

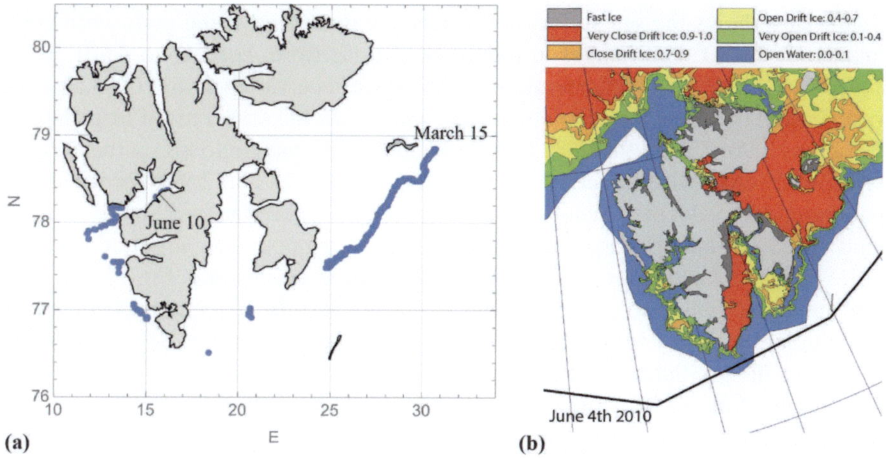

Fig. 2.30 **a** Trajectory of IT 330 from March 15 to June 10, 2010. **b** Ice conditions in the Svalbard region on June 4, 2010

Fig. 2.31 Aleksey Marchenko with ITa picked up on shore near Longyearbyen, June 10, 2010

the Ice Fjord. Previous ice maps show that this ice drifted from the region south of Spitsbergen.

2.3.4 Measurements in 2012

Four ITs were deployed on drift sea ice in April 2012 south of the Stoor Fjord and Edge Island (Yulmetov et al. 2013). Trajectories of the ITs are shown in Fig. 2.32. The ITa drifted in the SFW sub-region, and the ITb-d drifted in the SFE sub-region. The ice thickness varied from 30 to 70 cm at the points of the IT deployments. Figure 2.33 shows that ice drift from April 17 to April 30 was caused by the northeastern winds.

Figure 2.34 shows the sea surface level depending on time at the point with coordinates 77° N, 20° E. Changes in the ice conditions from April 17 to April 30 are shown in Fig. 2.35. Ice concentration at the sea surface varied from 100% near the west coast of the Stoor Fjord to open water space near the east coast of Edge Island.

Figures 2.96, 2.97, 2.98 and 2.99a show that all ITs passed loops on their trajectories in the clockwise direction. Figures 2.96, 2.97, 2.98 and 2.99b show that the maximum drift velocities measured around $t = 200$ h correlate with the maximum wind speed measured at that time (Fig. 2.33a). The maximum amplitudes of sea level were calculated at $t = 150$ h (Fig. 2.34). Fourier spectrum in Figs. 2.96, 2.97, 2.98 and 2.99c indicate the semidiurnal frequency and the second harmonic of the semidiurnal frequency in the oscillations of the IT drift velocities. Spectrum of ITa shows also the oscillations of the drift velocity at the diurnal frequency (Fig. 2.96c). Histograms of the IT drift velocities are shown in Figs. 2.96, 2.97, 2.98 and 2.99d. The mean drift velocity of ITa was 0.23 m/s, and the mean drift velocities of ITb-d were approximately 0.4 m/s. The maximum drift velocity of ITa was 1.4 m/s, and the maximum drift velocity of ITb-d was approximately 1.1 m/s. Points of the IT trajectories where the maximum drift velocities were found are shown in Fig. 2.32 by large open markers. Figures 2.97, 2.98 and 2.99e support the general trend in increasing of the IT drift velocity with decreasing sea depth, although the maximum of ITa drift velocity was reached at a depth of 150 m (Fig. 2.96e).

Figures 2.96, 2.97, 2.98 and 2.99f show that the minimum radii of curvature of the ITa trajectory was approximately 1.3 km at the mean ITs drift velocity of 0.23 m/s, and the minimum radii of curvature of the ITb-d trajectories were approximately

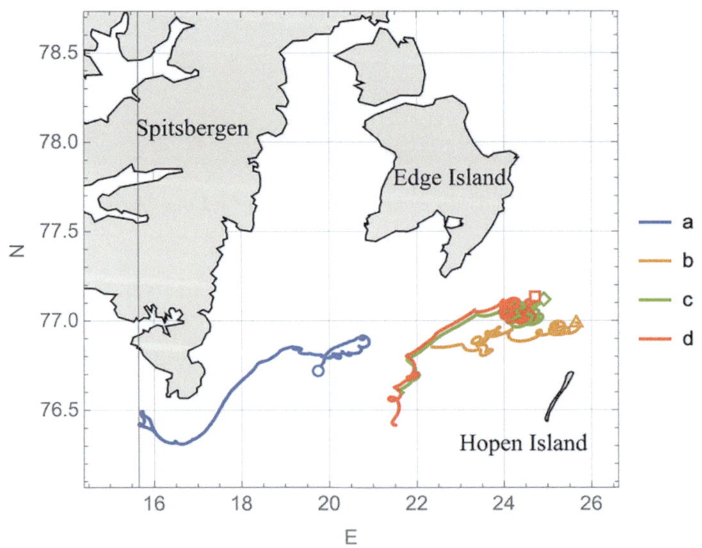

Fig. 2.32 Trajectories of ITs over the deployment period from April 17 to April 30, 2012. The initial points are shown by small open markers. Large open markers show locations with the maximum drift velocities

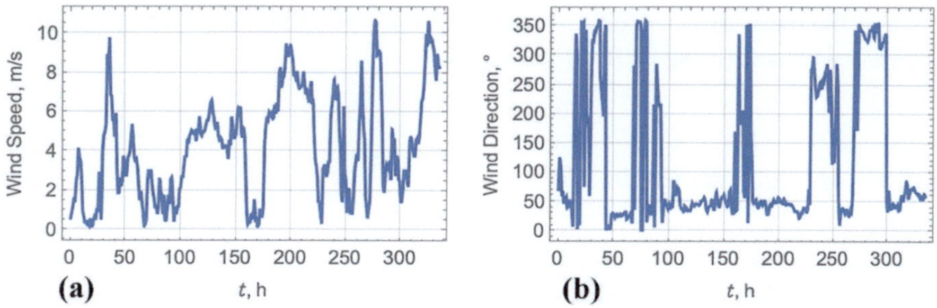

Fig. 2.33 Wind speed and direction depending on time measured on the meteorological station of Hopen Island from April 17 to April 30, 2012

Fig. 2.34 Water level depending on time at the point with coordinates 77° N, 20° E. The time is accounted from April 17, 00:00 UTC, 2013

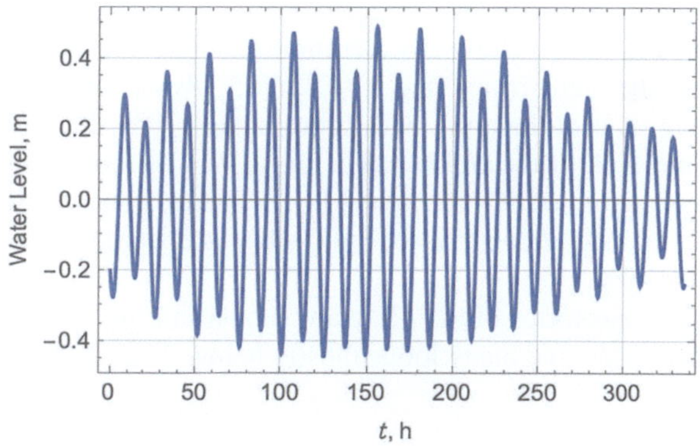

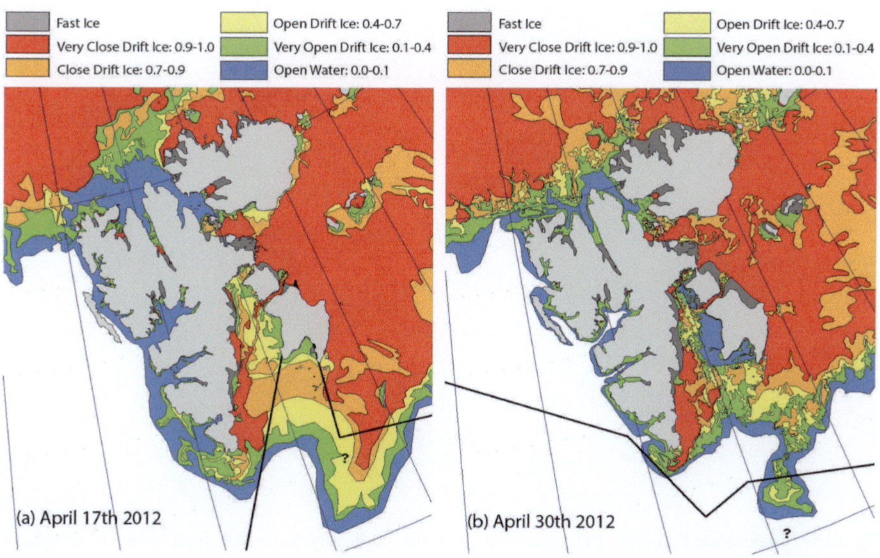

Fig. 2.35 Change of ice conditions over the deployment period from April 17 to April 30, 2012

2 km at the mean drift velocity of 0.4 m/s. Thus, the minimum radii of curvature decrease with decreasing drift velocity.

2.3.5 *Measurements in 2013*

Three ice trackers ITa-c were deployed on drift ice south of Edge Island in the beginning of May 2013. The ITa and ITc drifted along the east coast of the Stoor Fjord to the north, while the ITb initially deployed almost at the same location as the ITc drifted to the west (Fig. 2.36). All ITs remained in the SFE sub-region. The ITa-c transmitted the data until the beginning of June. Wind speed and wind direction recorded at the meteorological stations of Hopen Island and Edge Island are shown in Fig. 2.37. The wind direction changed several times from the northwestern to southeastern from May 1 to June 5. Water level depending on time is shown in Fig. 2.38 at the point with coordinates 77° N, 20° E. The changes in ice conditions during this time are shown in Fig. 2.39. The Stoor Fjord was covered by sea ice with concentration 90–100% on May 2. The ice concentration was less than 70% on June 2, and significant part of the fjord was covered by rare drift ice with concentration 10–40%.

Trajectories of the ITa-c are shown in Figs. 2.100, 2.101 and 2.102a. All trajectories include many loops mostly below 77.2° N. Figure 2.35 shows zoomed image

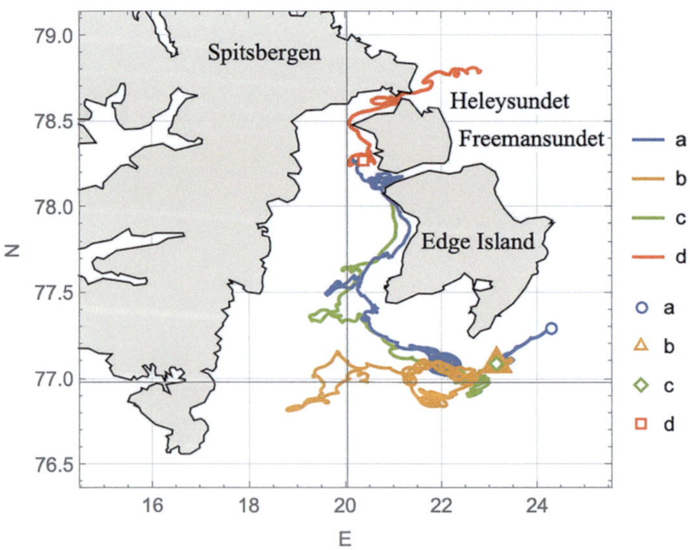

Fig. 2.36 Trajectories of ITa, ITb, and ITc over the deployment period from May 1 to June 5, 2013. Trajectories of ITd from June 20 to June 30, 2010. Straits Freemansundet and Heleysundet are indicated. Open markers show initial positions of the ITs

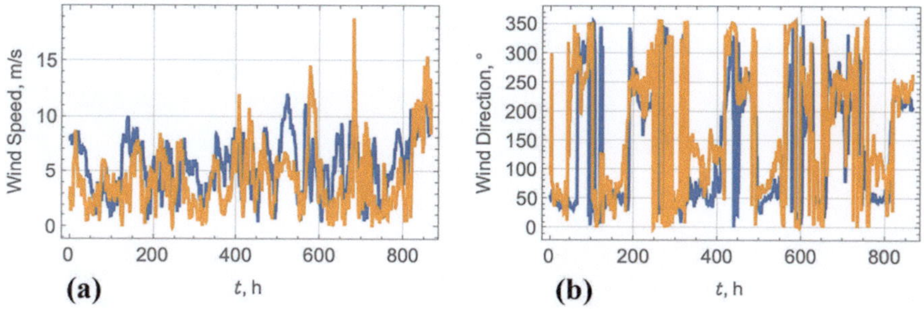

Fig. 2.37 Wind speed and direction depending on time measured at the meteorological stations of Hopen Island (blue) and Edge Island (yellow) from May 1 to June 5, 2013

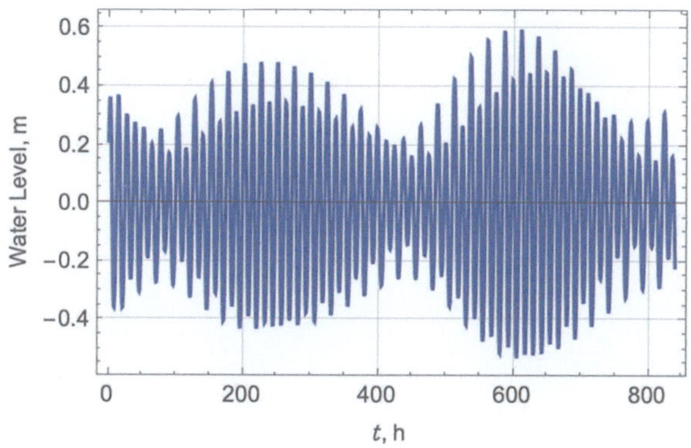

Fig. 2.38 Water level depending on time at the point with coordinates 77° N, 20° E. The time is accounted from May 1, 00:00 UTC, 2013

of the ITa trajectory around 77.07° N (a) and 78.16° N (b). Markers show the ITa positions during the semidiurnal period 12.42 h. Figure 2.40a corresponds to the maximum amplitude of semidiurnal oscillations of water level (Fig. 2.38) and wind speed lower than 10 m/s (Fig. 2.37a). Markers in Fig. 2.40a show that the ITa passed one loop during one semidiurnal period in the clockwise direction. Figure 2.40b shows the ITa trajectory near Freemansundet. One can see four small loops (1–4) on the ITa trajectory (Fig. 2.20b). The ITa passed the loops also with the semidiurnal periodicity. Loop 1 was passed in the counterclockwise direction, and loops 2–4 were passed in the clockwise direction.

Figures 2.100, 2.101 and 2.102b show that the maxima of the ITa-c drift velocities and water level amplitudes occurred at the same time around $t = 200$ h. The maximum drift velocities were calculated when the ITs passed the loops. The Fourier spectra of the ITa-c drift velocities show oscillations with the semidiurnal frequency and the second harmonics (Figs. 2.100, 2.101 and 2.102c). Fourier spectra of the ITa drift velocities show also oscillations of the third harmonic of the semidiurnal frequency (Fig. 2.100c). The maximum drift velocities of ITa,b,c were approximately 1.3 m/s

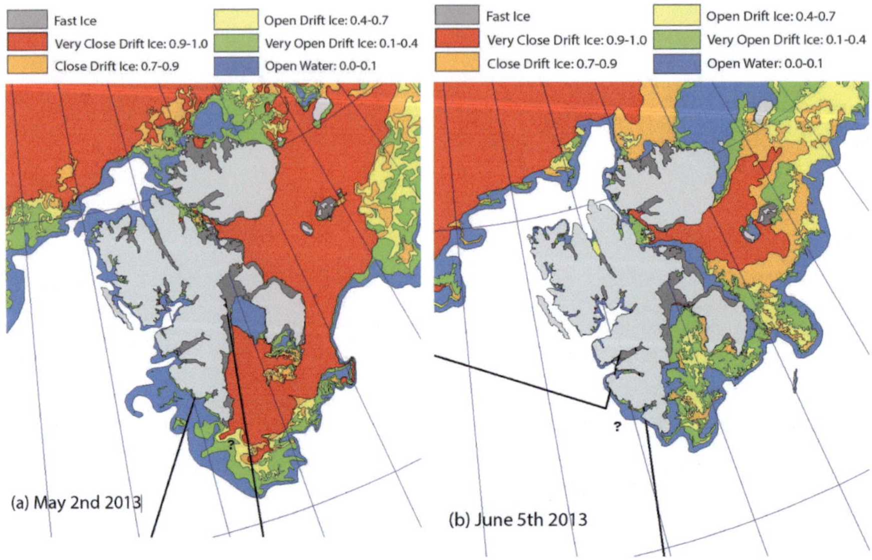

Fig. 2.39 Ice conditions on May 2 (**a**) and June 5 (**b**), 2013

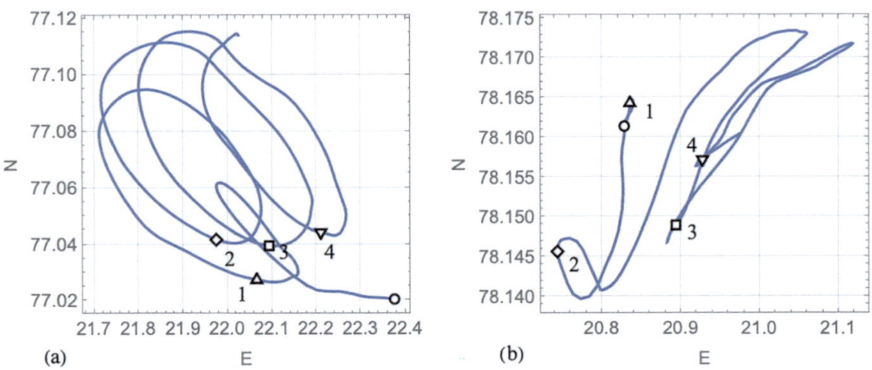

Fig. 2.40 Fragments of the ITa trajectory at $t \in (160, 216)$ h (**a**) and $t \in (944{,}965)$ h (**b**). Black circles mark the initial points of the fragments at $t_{a,0} = 160$ h and $t_{b,0} = 944$ h. Other markers mark the IT positions at $t = t_{a,0} + j\Delta t$ (**a**) and $t = t_{b,0} + j\Delta t$ (**b**), where $\Delta t = 12.42$ h and $j = 1, 2, 3, 4$

in May. The maximum drift velocity of the ITa in June was 0.94 m/s. The mean drift velocities of ITa,b,c were approximately 0.3 m/s in May. The mean drift velocity of the ITa in June was of 0.13 m/s. Histograms in Figs. 2.100, 2.101 and 2.102d show that the most frequent drift velocities were approximately 0.1 m/s for the ITa and ITc, while the most frequent drift velocity of the ITb was similar to the mean drift velocity approximately 0.3 m/s.

Figures 2.100, 2.101 and 2.102e demonstrate an increase in the IT drift velocity with decreasing sea depth. Figures 2.100, 2.101 and 2.102e show that the minimum radius of curvature of the ITa and ITc trajectories was 2 km at the mean drift velocity of 0.3 m/s, and the minimum radius of curvature of the ITb was 1.3 km in May. It is explained by a decrease on the number of loops in the trajectories of the ITa and ITc when they drifted in the region with latitudes north of 77.2° N (Fig. 2.36).

The ITd was deployed near the entrance to the Heleysundet Strait on June 20 (Marchenko and Kowalik 2017). The ITd trajectory is shown in Fig. 2.36 and Fig. 2.41 by red lines. The ice map in Fig. 2.42 shows that Heleysundet was covered by fast ice on June 20, and the strait was ice free on July 1, 2013. Very open drift ice with concentration 10–40% was located only to the east of Heleysundet in the Olga Basin on July 1. The ITd was in this region at that time. Markers 1, 2, 3, and 4 in Fig. 2.41 show the locations where the ITd drift velocity reached the maximum values (Fig. 2.103b). The highest velocity was approximately 4 m/s. The Norwegian Pilot Guide (Norwegian Hydrographic Service 2018) reports that "several vessels recorded a tidal stream velocity of 8–9 knots (4.1–4.6 m/s) in the narrowest part. Even engine vessels with high power are at risk of becoming icebound as it presses together in the sounds' funnel-shaped entrance areas". The mean ITd drift velocity was 0.3 m/s. Figure 2.103b shows that the maximum drift velocities are accompanied by very large accelerations. Figure 2.103b and the histogram in Fig. 2.103d show that drift velocities were lower than 0.2 m/s over a significant part of the ITd trajectory. Figure 2.103f shows that the minimum radius of curvature of the ITd trajectory was 1 km at a mean drift velocity of 0.3 m/s (Fig. 2.43).

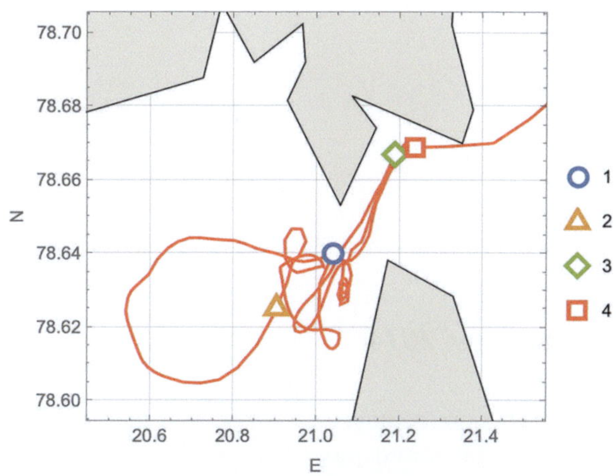

Fig. 2.41 Trajectories of ITd from June 20 to June 30, 2013

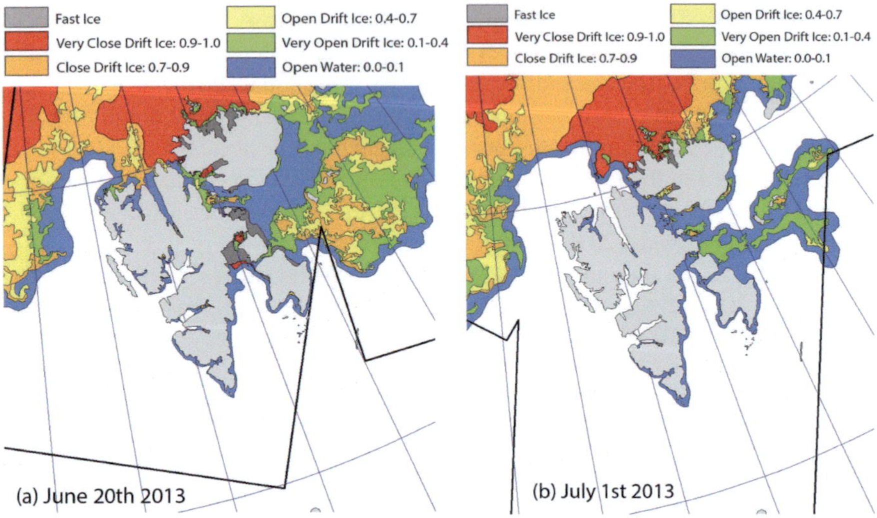

Fig. 2.42 Change in ice conditions over the deployment period of IT 4 from June 20 to July 1, 2013

Fig. 2.43 Drift speeds of IT 4 versus time. Local maxima of drift speed 1–4 were reached at the points 1–4 shown in Fig. 2.103b

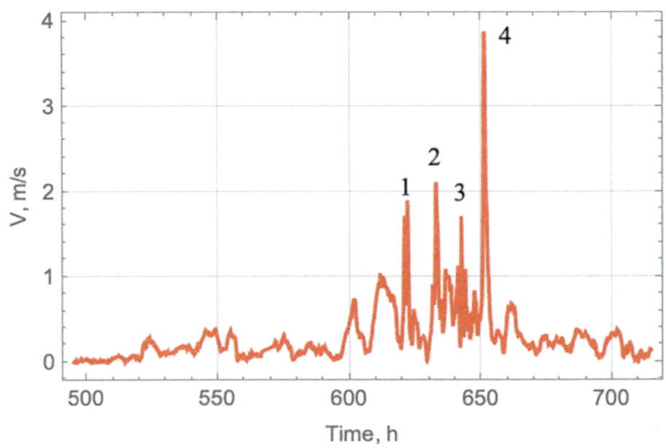

2.3.6 *Measurements in 2014*

In 2014, the ITa and ITb were deployed on a floe near the west coast of the Stoor Fjord on April 25, and the ITc was deployed on a small iceberg frozen in the floe on April 27 (Fig. 2.44). All ITs drifted in the SFW sub-region. Figure 2.45 shows the trajectories of the ITa and ITb, which are very close. The ITc was equipped with a temperature string 10 m long with 0.5 m spacing between neighboring temperature sensors. Temperature string was placed in a 2-inch well drilled in the ice near the ITc (Fig. 2.44b, c). The ITa and ITb transmitted the GPS data with a 10 min sampling interval. The ITc transmitted the GPS and temperature data with one hour sampling interval. Figure 2.46 shows the wind speed and wind direction depending on time from April 25 to May 25, 2014. The wind direction varied between the southwestern and northwestern directions. The ITa and ITb drifted to the south along the west coast of the Stoor Fjord during the first 370 h of the deployment time. Then, their drift changed to the east-northeastern directions due to the wind from the southwest. Ice conditions in the Stoor Fjord changed significantly from April 25 to May 25, 2014. The ice map in Fig. 2.47a shows ice concentration of 70–100% in the fjord on April 25. On May 23, only small area of very rare ice is seen in Fig. 2.47b in the region where the ITa and ITb drifted at that time.

Figure 2.104b show that the ITa did not move during 160 h from at $t = 120$ h to $t = 280$ h. Although the dependence of the ITa drift velocity oscillates depending on time (Fig. 2.104b), the Fourier spectrum in Fig. 2.104c does not have any spectral peaks. The histogram in Fig. 2.104d shows that the number of zero drift velocity measurements is significant. A maximum velocity of 0.5 m/s was reached when the ITa drifted to the northeast at $t = 368$ h (Fig. 2.104b). The mean drift velocity equal to 0.14 m/s is approximately close to the most frequent drift velocity in the histogram

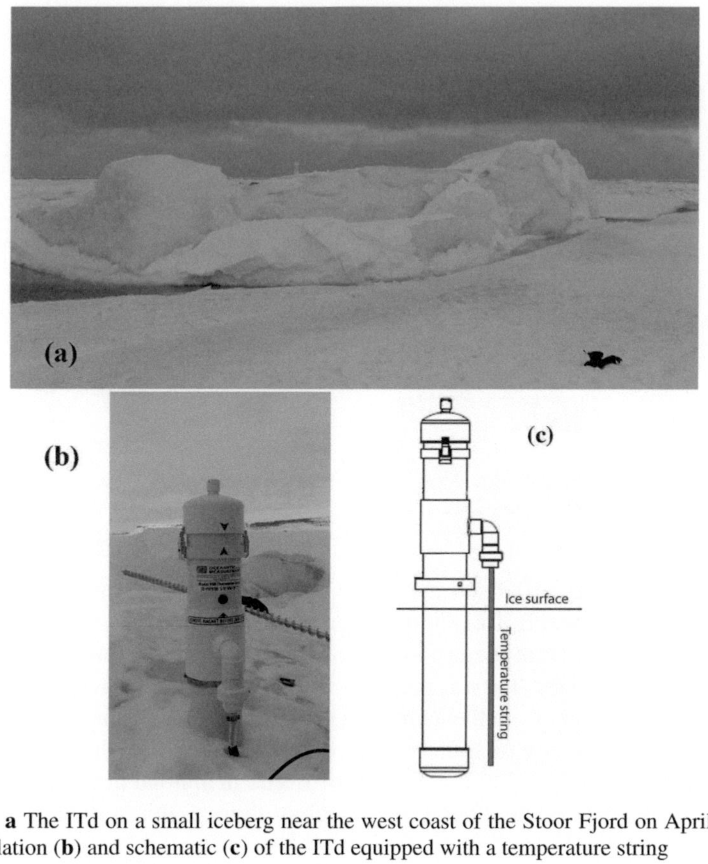

Fig. 2.44 **a** The ITd on a small iceberg near the west coast of the Stoor Fjord on April 27, 2014 (**a**). Installation (**b**) and schematic (**c**) of the ITd equipped with a temperature string

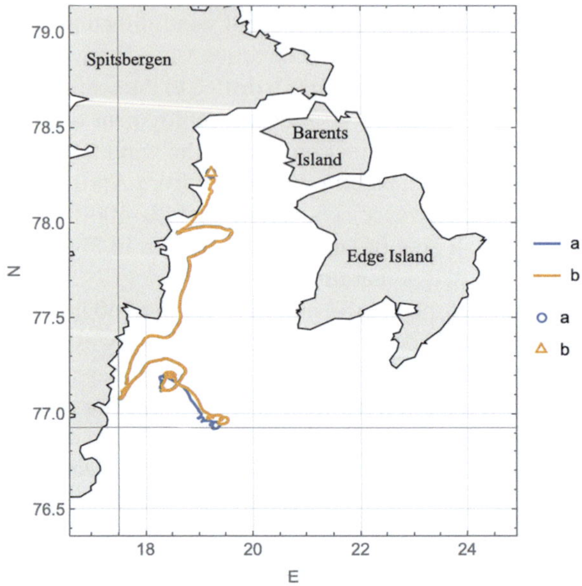

Fig. 2.45 Trajectories of the ITa and ITb from April 25 to May 24, 2014. The initial points are shown by open markers

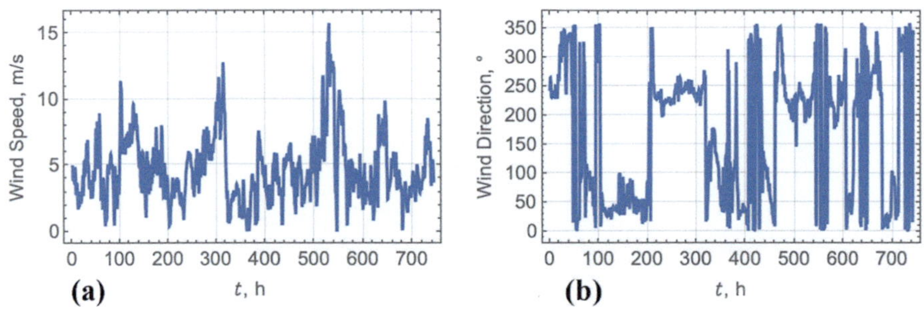

Fig. 2.46 Wind speed and direction depending on time measured at the meteorological station of Edge Island from April 25 to May 25, 2014

in Fig. 2.104d. Figure 2.104f shows that the minimum radius of curvature of the ITa trajectory is approximately 1 km at the mean drift velocity of 0.14 m/s.

Figure 2.48a shows that the distance between the ITa and ITb was $\Delta_{ab} \approx 350$ m until $t = 512$ h. Then, the floc was probably broken, and the distance increased up to 5 km on May 25. Changes in the azimuthal angle of the line connecting the ITa and ITb indicate the clockwise rotation of the floe during the first 100 h of the drift along the west coast of the Stoor Fjord (Fig. 2.24b). The azimuthal angle was changing in the clockwise and counterclockwise directions after the floe started to drift to the east at $t > 280$ h.

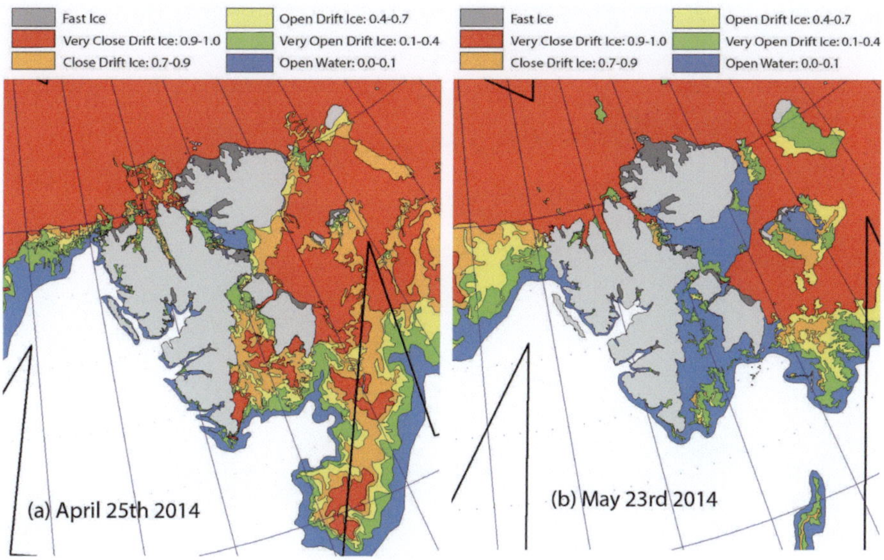

Fig. 2.47 Ice conditions on April 25 (**a**) and May 23 (**b**), 2014

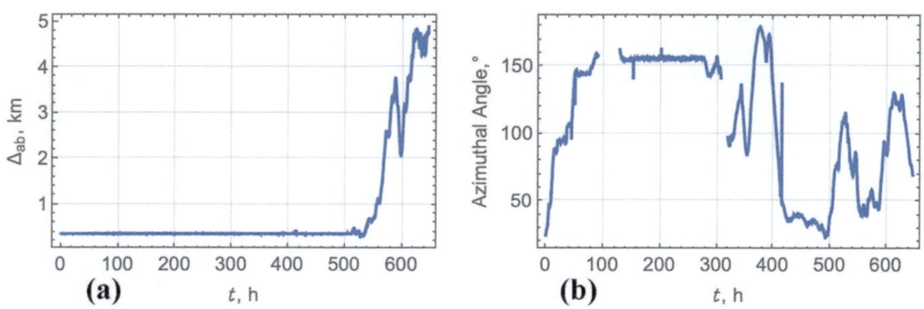

Fig. 2.48 **a** Distance between ITa and ITb versus time. **b** Azimuthal angle of the line connecting the ITa and ITb versus time

Figure 2.49 shows the drift trajectories of the ITa, ITb, and ITc since April 27, 10:00 UTC, when the ITc was deployed on the iceberg. The ITc transmitted the data only during 50 h until April 29, 09:00 UTC. Trajectories of the three ITs were similar during 45 h. Then, the iceberg trajectory became different from the trajectories of the ITa and ITb, i.e., the iceberg disconnected from the floe. It could influence the iceberg floatation and position of the ITc at the iceberg surface. Figure 2.50 shows temperature profiles measured inside the iceberg every 6 h during 48 h. Temperature profile below water level became almost linear 18 h later after the deployment. Iceberg draft was estimated at approximately 5 m. Sea water temperature below the iceberg was at the freezing point.

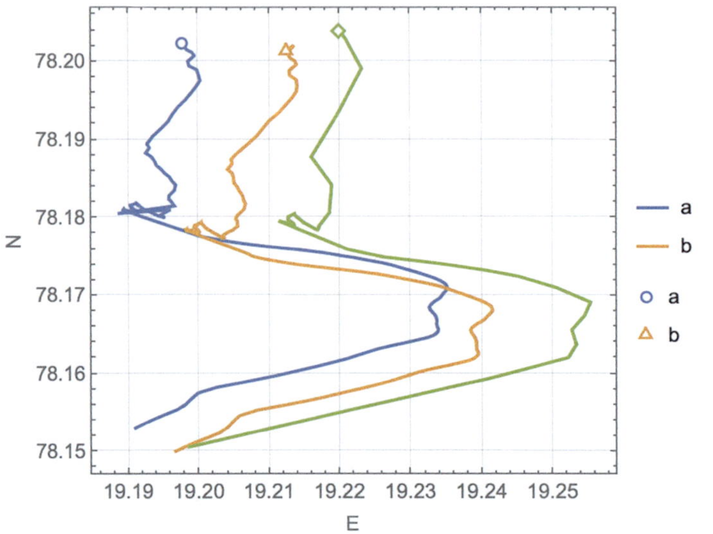

Fig. 2.49 Trajectories of ITa, ITb, and ITc at 43 h < t < 91 h. The initial points are shown by open markers

Fig. 2.50 Vertical profiles of ice temperature in the iceberg measured by IT 7 at 43 h < t < 90 h

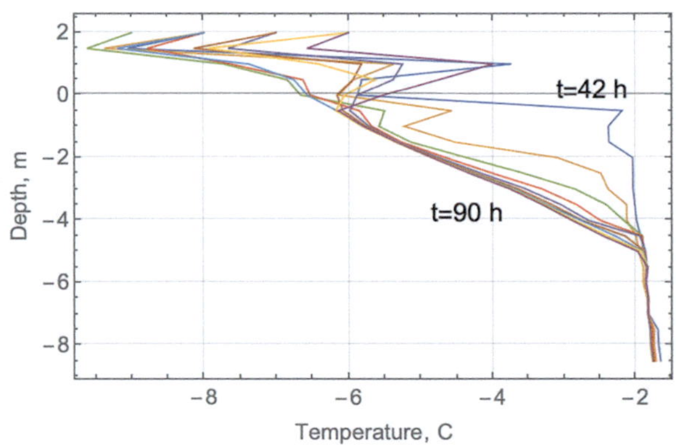

2.3.7 Measurements in 2015

The IT was deployed on multiyear floe consisting of ice rubble in marginal ice zone north of Svalbard on April 15, 16:00 UTC (Fig. 2.51). The IT was equipped with a temperature string 10 m long with 0.2 m spacing between neighboring temperature sensors. The initial ice thickness at the deployment point was approximately 8 m. The IT trajectory is shown in Fig. 2.52. Open markers show IT locations every 50 h. Figure 2.53 shows that the southwestern wind forced the IT drift to the north in the first 100 h of operation. Then, at t > 150 h, the winds of the northern direction influenced the IT drift to the south until the IT was grounded in a shallow water region near the entrance to the Wood Fjord around May 4. The ice maps in Fig. 2.54 show ice conditions in the beginning of the IT drift on April 15 (Fig. 2.54a), and on May 4 (Fig. 2.54b).

Figure 2.55a shows that the IT drift velocity reached 0.8 m/s when the wind speed was 25 m/s at $t = 200$ h. The Fourier spectrum of the drift velocity includes

Fig. 2.51 Deployment of IT on multiyear floe north of Spitsbergen

Fig. 2.52 Trajectory of IT on a multiyear floe from April 15 to May 4, 2014. Open markers show the IT locations over every 50 h

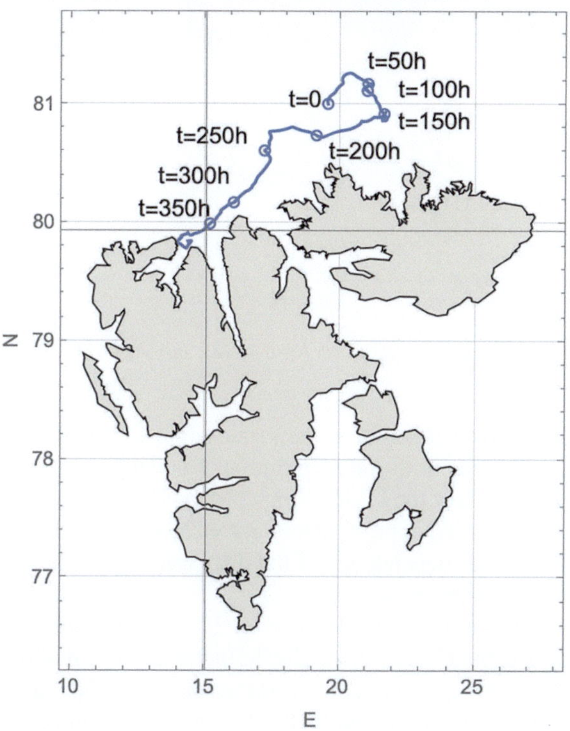

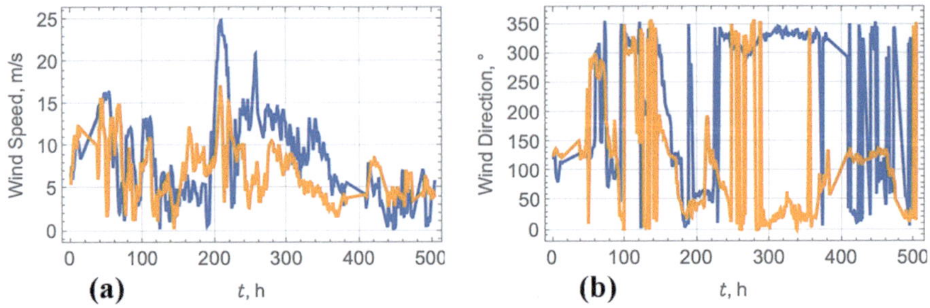

Fig. 2.53 Wind speed and direction depending on time measured on the meteorological stations of Kvitoya (blue) and Verlegenhuken (yellow) from April 15 to May 4, 2015

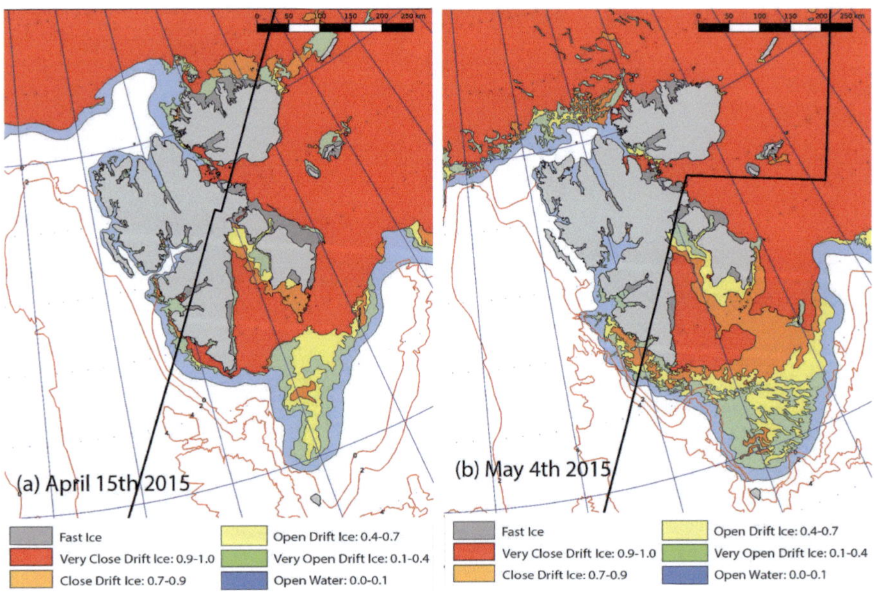

Fig. 2.54 Ice conditions on April 15 (**a**), and May 4 (**b**), 2015

small local maxima at the semidiurnal frequency and the second harmonic of the oscillations with the semidiurnal frequency. The mean drift velocity was 0.2 m/s over 470 h of the IT drift (Fig. 2.55a). The maximum drift velocity as high as 0.8 m/s was recorded approximately at $t = 200$ h when the wind speed was maximal. There is no clear dependence of the drift velocity on the sea depth since the main driving force was the wind drag.

Temperature profiles in Fig. 2.56 correspond to fast destruction of the ice keel during the IT drift at $t > 200$ h. In the first 150 h of the IT drift the water temperature below the keel was at the freezing point around -2 °C. Then, the water temperature began to increase and finally became higher than 0 °C. The initial ice draft of 8 m

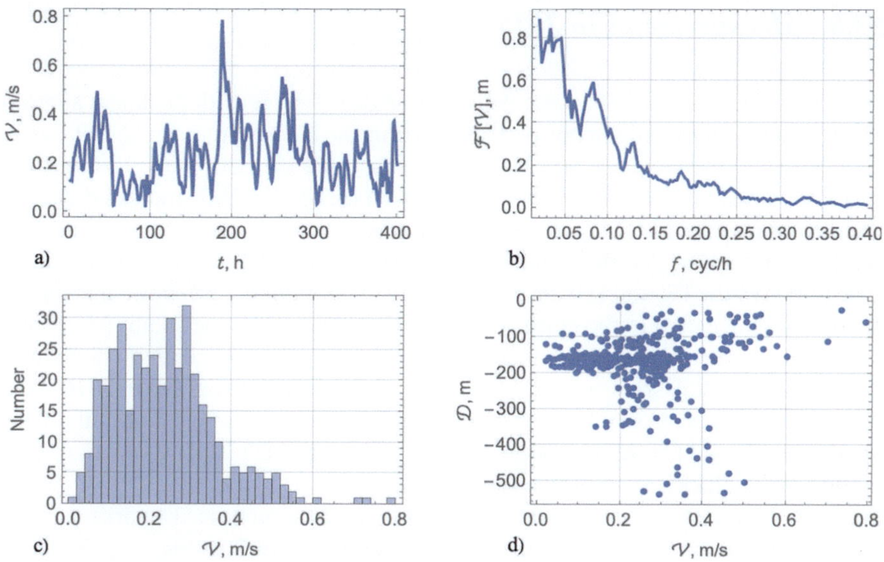

Fig. 2.55 The IT drift velocity versus time (**a**). The Fourier spectrum (**b**) and histogram (**c**) of the IT drift velocity. The IT drift velocity versus local sea depth (**d**)

decreased to 3.5 m after 350 h of the IT drift. The heat flux for melting 4.5 m of pure ice in 350 h is 1.1 kW/m^2. If the keel was not completely consolidated then the heat flux could be lower, but still, it should be high enough to melt 4.5 m of ice during 350 h. Mechanical erosion of the keel due to water flow could be an additional reason for the observed decreasing of the ice draft.

Fig. 2.56 Vertical temperature profiles inside and below the ice keel every 50 h for 350 h

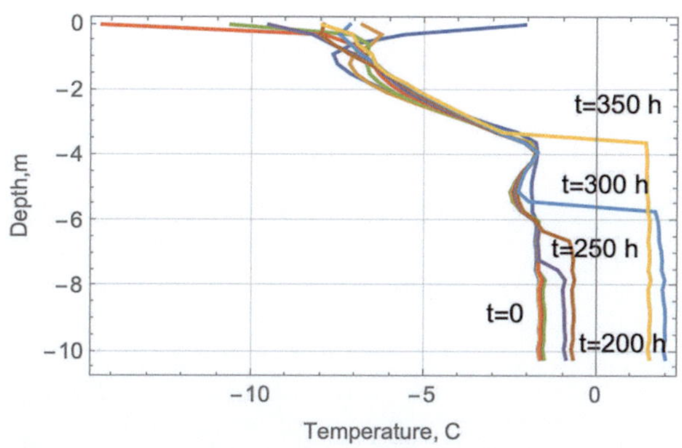

2.3.8 *Measurements in 2016*

Two ice trackers ITa and ITb were deployed on drift sea ice with thickness approximately 30 cm in the beginning of May 2016 in the BS sub-region. The ITs were equipped with anemometers and IMU and transmitted the data every 10 min from May 1 to May 23 (ITa) and from May 3 to June 6 (ITb). The IMU indicated that the ITa have fallen over on 6 May and was continuing to transmit while lying on its side. The ITa wind speed and the wind direction data were not correct since that time. Trajectories of the ITa and ITb are shown in Fig. 2.57. Figure 2.58 shows that during the first 400 h ice drift to the north was caused by winds from the south. Afterward, winds of the northern directions influenced ice drift in the inverse southern directions. Ice maps in Fig. 2.59 show very mild ice conditions in the Barents Sea in 2016. Very close drift ice existed only in the northwestern part of the Barents Sea in the Olga Basin. The ITa and ITb were deployed in this region. The ITb IMU shows dominant rotation in the clockwise direction (Fig. 2.60).

Figure 2.105b shows the ITb drift velocity depending on time. The drift velocity has three well pronounced local maxima marker 1, 2, and 3 in Fig. 2.105b. Open markers 1, 2, and 3 in Fig. 2.57 show the points where the maximum of the ITb drift velocities was estimated. The maximum drift velocities 2 and 3 approximately equal to 2 m/s were estimated at points 2 and 3 near the entrance to the Heleysundet Strait. The Fourier spectrum of the ITb has a local maximum at the semidiurnal frequency and at the second harmonic of the semidiurnal frequency (Fig. 2.105c). The histogram in Fig. 2.105d shows that the ITb drift velocities were mostly below

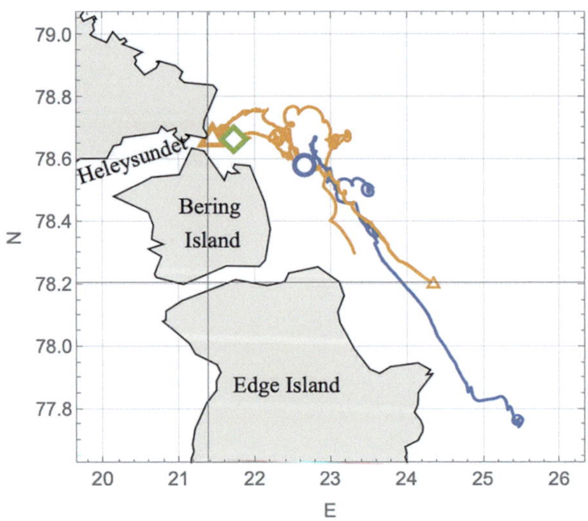

Fig. 2.57 Trajectories of the ITa from May 1 to May 23 and ITb from May 3 to June 6, 2016, are shown by blue and yellow lines, respectively. The initial points of deployments are shown by small markers. Large markers 1–3 show the points of local maxima of the ITb drift velocity

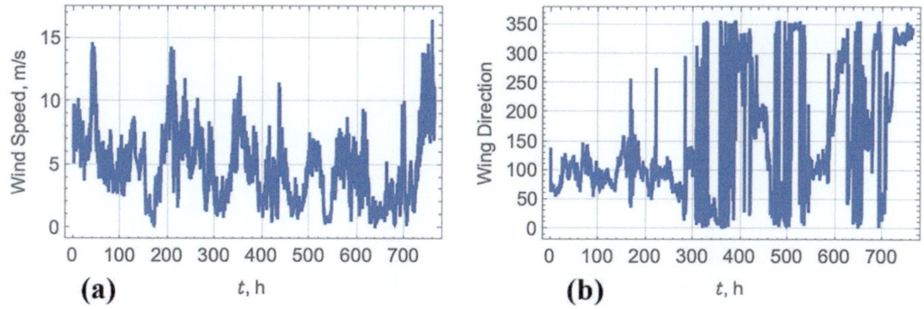

Fig. 2.58 Wind speed and wind direction measured by anemometer of ITb

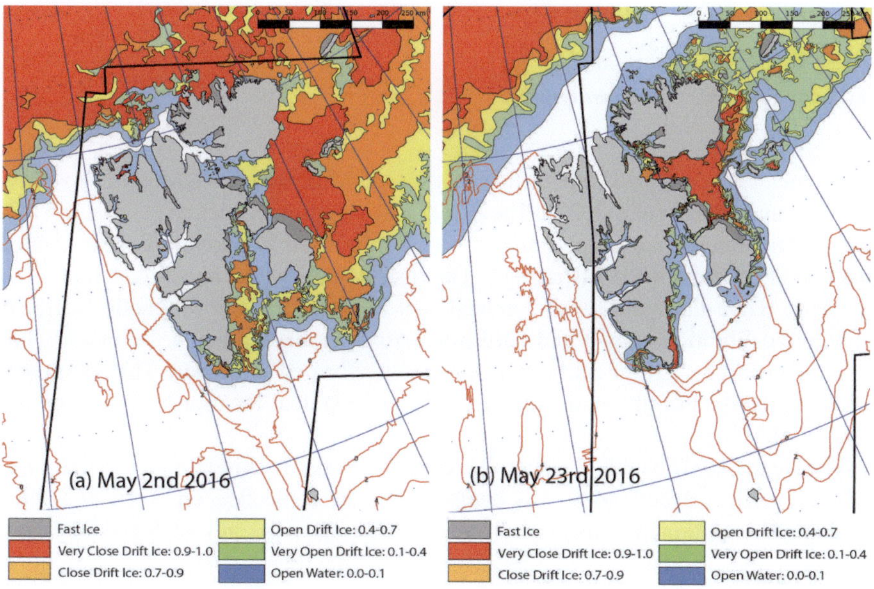

Fig. 2.59 Ice conditions on May 2 (**a**) and May 23 (**b**), 2016

0.5 m/s. The mean ITb drift velocity was equal to 0.15 m/s. Figure 2.105e shows an increase in the ITb drift velocity with decreasing sea depth. The radius of curvature of the ITb trajectory is shown depending on the drift velocity in Fig. 2.105f. The minimum radius of curvature equals to 1 km at a mean drift velocity of 0.15 m/s.

2.3.9 Measurements in 2017

The *Polarsyssel* was moored to a drift floe in the location close to 76.4° N, 22.8° E on April 24, 2017, in the SpB sub-region. The ice tracker ITa equipped with an

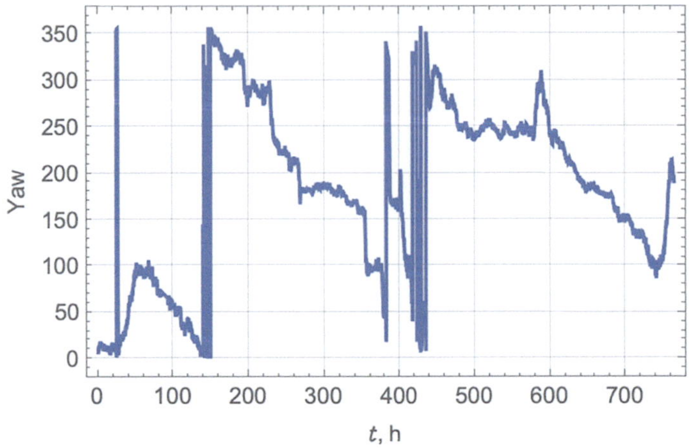

Fig. 2.60 Yaw angle of ITb versus time

anemometer was deployed on this floe (Fig. 2.61) (Marchenko and Marchenko 2019; Turnbull and Marchenko 2022). The WindDrift electronics package contained a 3-axis magnetometer and accelerometer which were used to derive tilt-compensated magnetic headings of the instrument. The instrument gives a heading of 0 degrees magnetic when the wind sensor vertical arm is pointed to the Magnetic North. A complete calibration of soft and hard iron errors has been done at the factory. For accurate headings, a brief calibration was also performed at the floe. The instrument collected the data on the wind velocity and geographical location with a sampling interval of 10 min.

The ice tracker ITb without anemometer programmed for high frequency measurement of vertical acceleration was deployed on the same floe on April 26, 2017. Vertical acceleration was measured with a sampling frequency of 5 Hz for 5 min

Fig. 2.61 The ITb on the drifting floe on April 24, 2017

one time per 2 h. The ITb also recorded the geographical position with a sampling interval of 10 min. The distance between the ITa and ITb was 25 m. The ice floe with deployed trackers ITa and ITb is shown in Fig. 2.62. The shape of the floe is shown in Fig. 2.63. The drawing was created using laser scanning and drilling data (Marchenko 2020, 2022). The maximum sail height of the floe was 2.16 m, and the maximum draft of the keel was 3.82 m. The total volume of the floe was $V_i = 3270$ m^3 and the waterplane area was $S_i = 934$ m^2.

The ITa transmitted the data from April 24 to May 05. The ITb transmitted the data from April 26 to May 2. Ice conditions and wind direction are shown in Fig. 2.64 over this time. The wind velocity changed over this period significantly. Sea ice tongue extended from Hopen Island to the south in the direction of Bear Island was clearly seen on May 26–27 and almost disappeared by May 2. An increase of wind

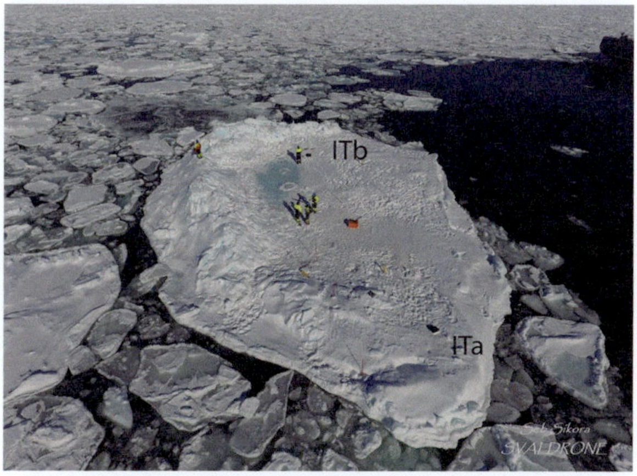

Fig. 2.62 Drifting floe where ice trackers ITa and ITb were deployed on April 24–26, 2017

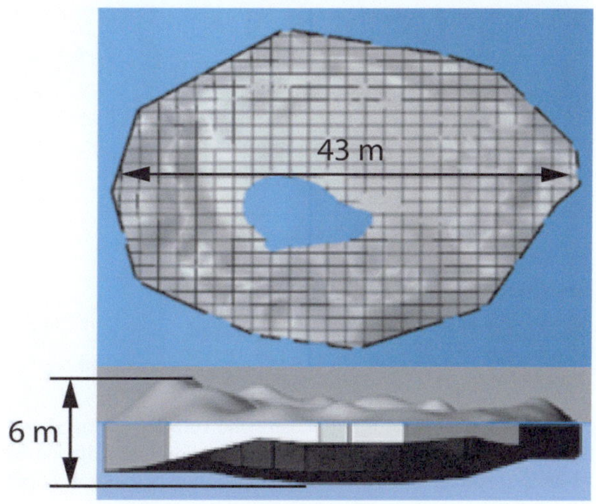

Fig. 2.63 Shape and size of the floe where ice trackers were deployed

speed on May 28 influenced ice dispersion and drift away from the Spitsbergen Bank. Figure 2.65 shows a Sentinel-1 Synthetic Aperture Radar (SAR) satellite image of the ice environment around Hopen Island taken on April 29, 2017, one day after the tracked ice floes underwent a strong dispersion to the southwest of Hopen Island. The drift track of the ITb is shown by a red line from April 26 to May 1. The yellow circled region shows a clear delineation between higher and lower concentration pack ice to the northwest and southeast, respectively. The tracked ice floes began to increase their drift speeds and disperse as they entered this region of lower concentration ice (Fig. 2.67).

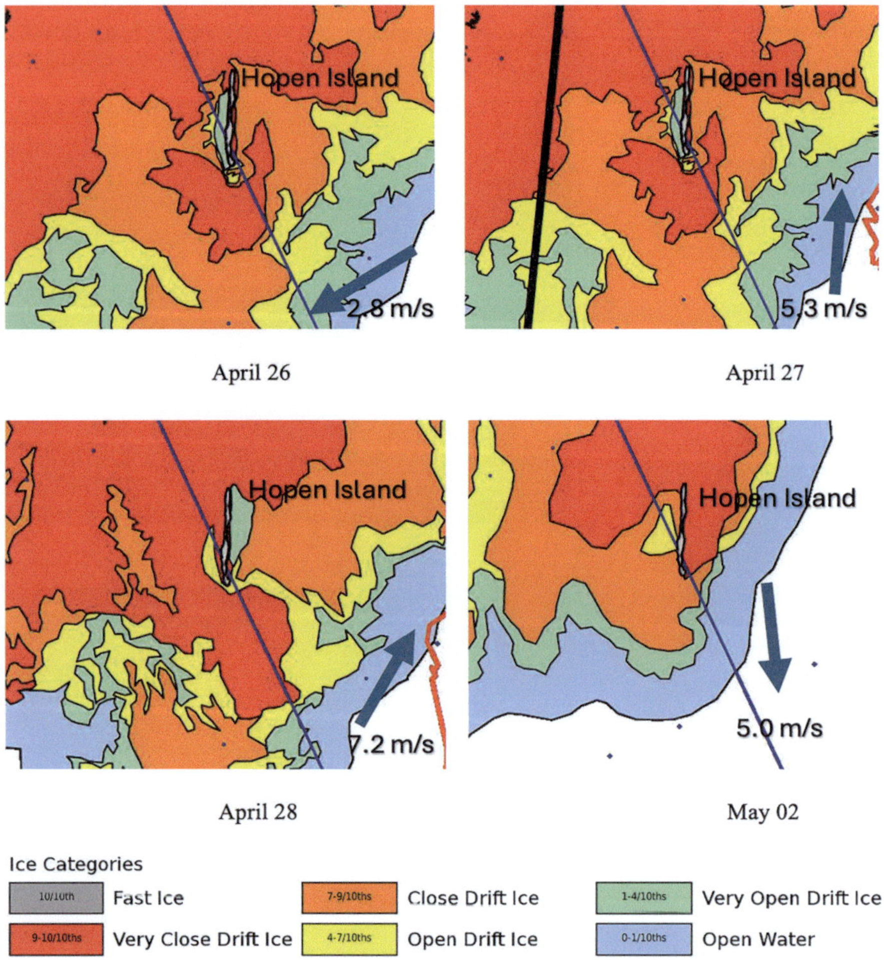

Fig. 2.64 Ice conditions and mean wind velocity averaged over 24 h in the region of the floe drift from April 26 to May 02, 2017. Colors show the drift ice with concentrations 90–100% (red), 70–90% (brown), and 40–70% (yellow). Blue color shows open water

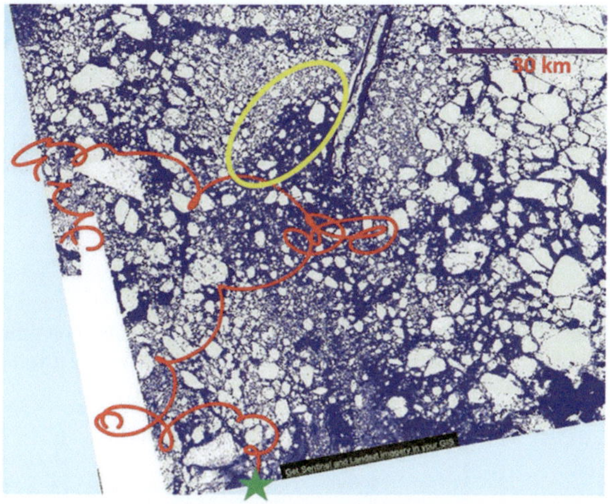

Fig. 2.65 Sentinel-1 satellite image of the ice environment around Hopen Island on April 29, 2017, with the wind and wave beacon 3840 track overlaid from April 26 to May 1; the green star marks the end of the recorded beacon trajectory

Figure 2.106a shows the ITb drift trajectory. The initial point corresponds to 12:00 UTC on April 26, and the final point corresponds to 16:00 UTC on May 02, 2017. General direction of the drift was from northwest to southeast. The resulting displacement was 44.5 km over 126 h with the mean velocity of 0.1 m/s. Figure 2.106b shows the ITb drift velocity depending on time from 00:00 UTC on April 26, 2017. The local maxima of the ITb drift velocity repeated over ~12.4 h are related to the semidiurnal tide. The maximum ITb velocity exceeded 1.6 m/s at $t = 75$ h at the point shown by the circle marker in Fig. 2.106a. The wind speed reached a local maximum of 13 m/s five hours before $t = 70$ h. The mean ITb drift velocity was 0.48 m/s.

Figure 2.106c shows that the spectrum of the ITb drift velocity has a local maximum at the semidiurnal frequency and the second harmonic of the semidiurnal frequency. The histogram in Fig. 2.106d shows that the drift velocities were mostly below 1 m/s. Figure 2.106e shows an increase in the ITb drift velocity with decreasing sea depth. From Fig. 2.106f it follows that the minimum radius of curvature of the ITb trajectory was 25 km at a minimum drift velocity of 0.48 m/s. Figure 2.68 shows the spectrum of the vertical accelerations a_z recorded by the ITb. It is obvious that the increase of the ITb drift velocities occurred at $t > 60$ h (Figs. 2.106b and 2.66a) with increasing wave energy. Wind from the north influenced opening of the lead with rare ice south of Hopen Island (Figs. 2.64 and 2.65) and excited surface waves in the lead with period of 8 s (Fig. 2.68). These waves accelerated the drift of the floe with ITb in the lead.

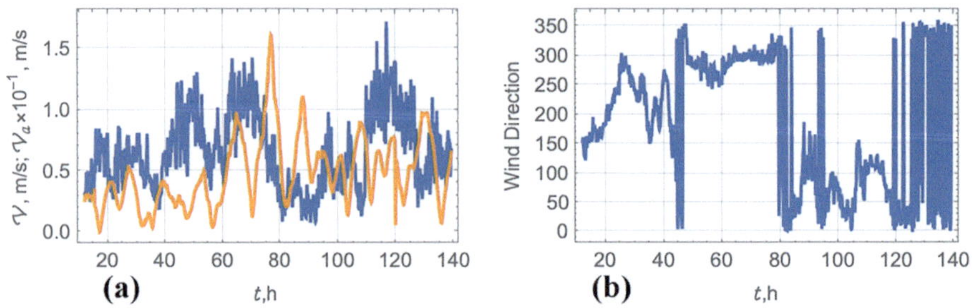

Fig. 2.66 Wind speed V_a (**a**, blue line) and wind direction, **b** depending on time measured by the ITb anemometer. The ITa drift velocity V depending on time is shown by a yellow line (**a**)

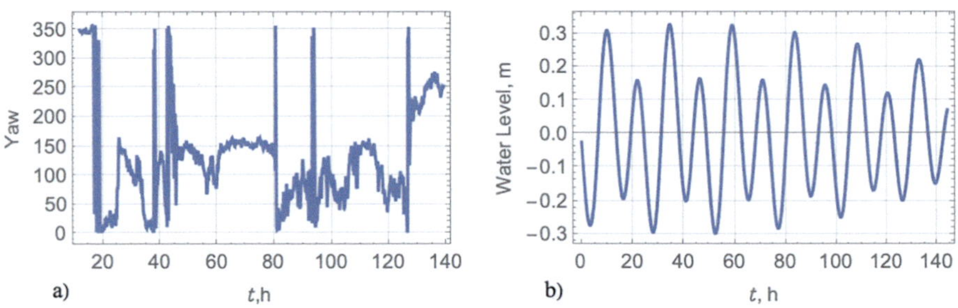

Fig. 2.67 **a** The yaw angle of the ITb depending on time. **b** Water level depending on time in the point with coordinates (23.5 E, 76.3 N)

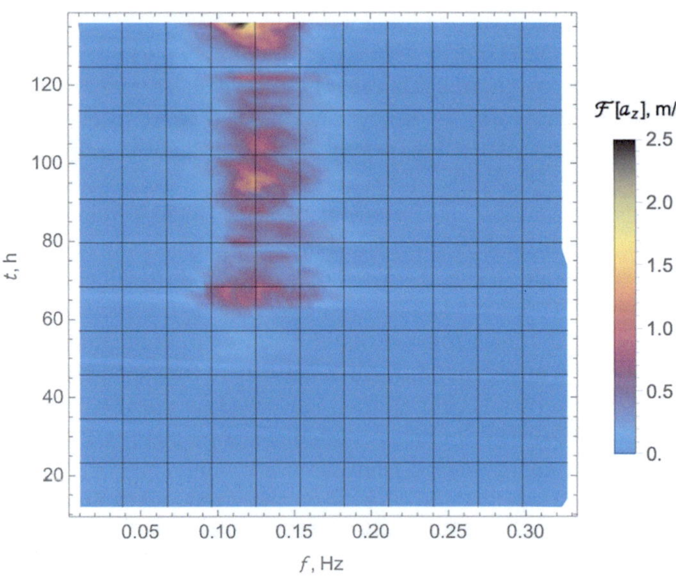

Fig. 2.68 Spectrum of the vertical accelerations of the ITb versus on time

2.3.10 Measurements in 2018

The IT deployed on April 27, 2018, transmitted the GPS data with a sampling interval 10 min and communicated well until February 27, 2019 (Marchenko et al. 2021a; Marchenko and Kowalik 2022; Kowalik and Marchenko 2023). The IT was deployed on a floe and continued to transmit the data after the floe melted. The IT drifted in the SpB, SFE, and SFW sub-regions for approximately six months (Fig. 2.69). There are 7 patterns (a-g) of the IT trajectory corresponding to the drift in the sub-regions, which are shown in different panels of Fig. 2.69. The IT passed this trajectory from April 27 to May 5. Figure 2.70 shows ice conditions on April 27 and May 3. Table 2.2 and the spectrum in Fig. 2.107c show the existence of the oscillations of the IT drift velocity at the semidiurnal frequency, a sum of the diurnal and semidiurnal frequencies, and the second and third harmonics of the semidiurnal frequency. The maximum drift velocity was 1.3 m/s, and the mean drift velocity equal to 0.7 m/s is close to the most frequent drift velocity in the histogram in Fig. 2.107d. Figure 2.107e demonstrates an increase in the IT drift velocity with decreasing sea depth. Figure 2.107f shows that the minimum radius of curvature of the IT trajectory was 1 km during very low drift velocities, and the minimum of radius of curvature of the IT trajectory was 42 m at the mean drift velocity of 0.7 m/s. We note that mostly the radii of curvature were of about 1 km.

Figure 2.108 shows the characteristics of the IT drift in the SFW from May 5 to June 30 along the blue trajectory "b" in Fig. 2.69. Table 2.2 and the spectrum in Fig. 2.108c show the existence of the oscillations of the IT drift velocity at the diurnal and semidiurnal frequencies, a sum of the diurnal and semidiurnal frequencies, and the second and the third harmonics of the semidiurnal frequency. The duration of the IT drift in SFW was much longer than the duration of the IT drift in pattern "a" (Table 2.2). Sea bottom in the SFW was much deeper than on the Spitsbergen Bank, and the dependence of the drift velocity on sea depth was not significant (Fig. 2.108e). As a result, the maximum and the mean drift velocities were 0.6 m/s and 0.2 m/s, respectively. Figure 2.108f shows that the minimum radius of curvature of the IT trajectory was 0.1 km during very low drift velocities, and the minimum of radius of curvature of the IT trajectory was 1 km at the mean drift velocity equal to 0.2 m/s.

Figure 2.109 shows the characteristics of the IT drift in the SpB from June 30 to August 16 along the green trajectory "c" in Fig. 2.69. Table 2.2 and the spectrum in Fig. 2.109c show the existence of the oscillations of the IT drift velocity at the diurnal and semidiurnal frequencies, a sum of the diurnal and semidiurnal frequencies, and the second and the third harmonics of the semidiurnal frequency. The duration of the IT drift in the SFW was much longer than along trajectory "a" and along trajectory "b" (Table 2.2). Figure 2.109e shows an increase in the drift velocity with decreasing sea depth. The maximum and the mean drift velocities were 1.7 m/s and 0.41 m/s, respectively. Figure 2.109b shows the IT drift velocity depending on time. The dependence has four local maxima 1, 2, 3, 4. Four points of the local maxima of the IT drift velocity are shown by open markers on the IT trajectory in Fig. 2.109a. Two largest maxima occurred near the southern tip of Hopen Island. Figure 2.109f shows

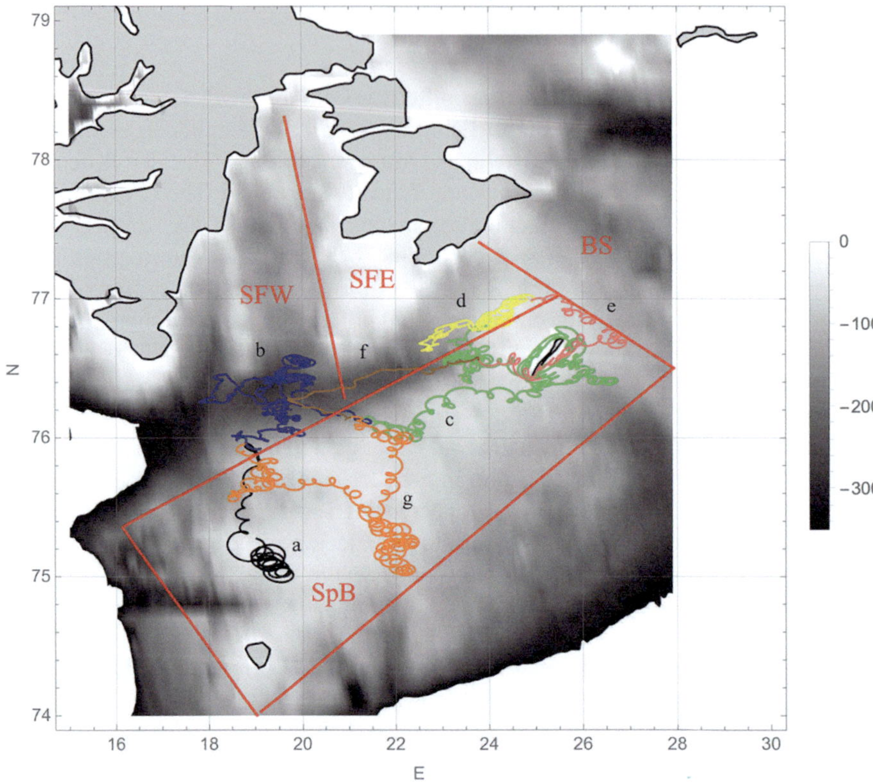

Fig. 2.69 Trajectory of the IT deployed in April 2018 in the SpB sub-region. **a** The trajectory patterns **a** (black), **c** (green), **e** (red), and **g** (orange) are in the SpB sub-region. The trajectory pattern **b** (blue) is in the SFW sub-region. The trajectory patterns "f" (brown) and "d" (yellow) are in the SFE sub-region

that the minimum radius of curvature of the IT trajectory was 0.1 km during low drift velocities, and the minimum of radius of curvature of the IT trajectory was 1 km at mean drift velocity equal to 0.41 m/s.

Figure 2.110 shows the characteristics of the IT drift in the SFE from August 16 to September 2 along the yellow trajectory "d" in Fig. 2.69. Table 2.2 and the spectrum in Fig. 2.110c show the existence of oscillations of the IT drift velocity at the diurnal and semidiurnal frequencies and the second harmonic of the semidiurnal frequency. The duration of the IT drift in the SFW was shorter than along trajectories "b" and "c" (Table 2.2). Figure 2.110e shows an increase of the drift velocity with decreasing sea depth. The maximum and the mean drift velocities were 0.9 m/s and 0.38 m/s, respectively. Figure 2.110f shows that the minimum radius of curvature of the IT trajectory was 8.2 km when the mean drift velocity was 0.38 m/s. There are three points in Fig. 2.110e where the minimum radius of curvature of the IT trajectory equals to 0.1, 1, and 3 km when the mean drift velocity was 0.38 m/s, but most of

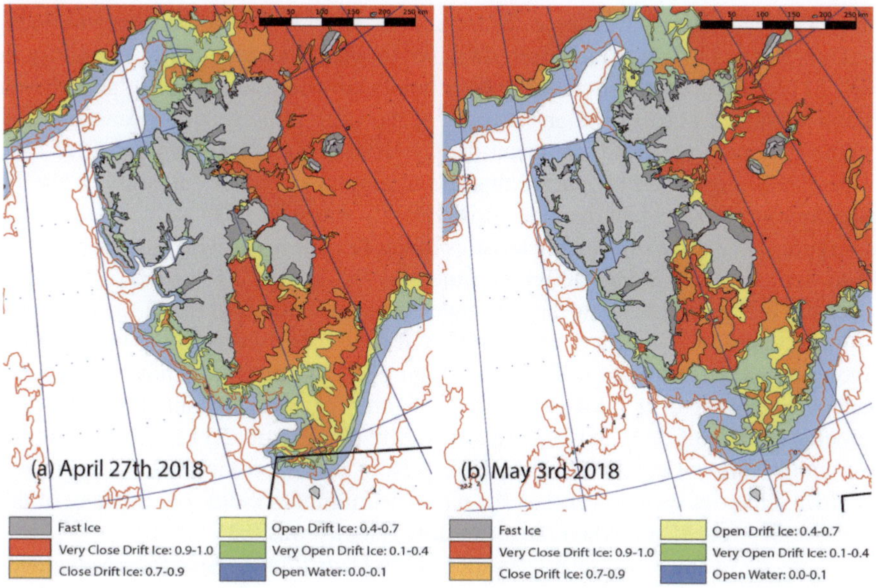

Fig. 2.70 Ice conditions on April 27 and May 3, 2018

Table 2.2 Duration of IT drift (Δt), minimum sea depth $D_{\min}$, existence of diurnal (f_D), semidiurnal (f_{SD}) and multiple ($f_D + f_{SD}$, $2f_{SD}$) frequencies in the spectrum of IT drift velocity, the mean ($\langle V \rangle$) and the maximum ($V_{\max}$) drift velocities along the IT trajectory patterns (a–g). Symbol HI means that maximum velocity was measured near Hopen Island

	Location	Δt, h	$D_{\min}$, m	f_D	f_{SD}	$f_D + f_{SD}$	$2f_{SD}$	$\langle V \rangle$, m/s	$V_{\max}$, m/s
a	SpB	226	20	–	v	v	v	0.7	1.3
b	SFW	1350	100	v	v	v	v	0.2	0.6
c	SpB	1150	20	v	v	v	v	0.41	1.7; HI
d	SFE	400	40	v	v	–	v	0.38	0.9
e	SpB	300	20	–	v	–	v	0.4	1.4; HI
f	SFE	200	70	–	v	–	v	0.25	0.6
g	SpB	950	50	v	v	v	v	0.43	0.85

the points in Fig. 2.110e have vertical coordinates corresponding to the radius of curvature larger than 8.2 km.

Figure 2.111 show the characteristics of the IT drift in the SpB from September 2 to September 15 along the pink trajectory "e" in Fig. 2.69. Table 2.2 and the spectrum in Fig. 2.111c show the existence of the oscillations of the IT drift velocity at the semidiurnal frequency and the second harmonic of the semidiurnal frequency. The duration of the IT drift in the SpB was shorter than the IT drift along trajectories "b" and "c" (Table 2.2). Figure 2.111b shows the IT drift velocity depending on time. The dependence has three local maxima 1, 2, 3. Three points of the local maxima of the IT drift velocity are shown by open markers on the IT trajectory in Fig. 2.111a. Figure 2.69 shows that these points are near the southern tip of Hopen Island. Figure 2.111e shows an increase of the drift velocity with decreasing sea depth. The maximum and the mean drift velocities were 1.41 m/s and 0.42 m/s, respectively. Figure 2.111f shows that the minimum radius of curvature of the IT trajectory was 1 km when the mean drift velocity was 0.42 m/s.

Figure 2.112 shows the characteristics of the IT drift in the SFE from September 15 to September 23 along the brown trajectory “f” in Fig. 2.69. Table 2.2 and the spectrum in Fig. 2.112c show the existence of the oscillations of the IT drift velocity at the semidiurnal frequency. The duration of the IT drift in the SFW was shortest among all IT trajectories in Table 2.2. Figure 2.112e does not show any visible dependence of the drift velocity on the sea depth. The maximum and the mean drift velocities were 0.6 m/s and 0.25 m/s, respectively. Figure 2.112f shows that the minimum radius of curvature of the IT trajectory was approximately 0.1 km when the drift velocities were lower than 0.1 m/s. The minimum radius of curvature of the IT trajectory was 85 km when the mean drift velocity was 0.25 m/s.

Figure 2.113 shows the characteristics of the IT drift in the SpB from September 23 to November 1 along the orange trajectory “g” in Fig. 2.69. Table 2.2 and the spectrum in Fig. 2.113c show the existence of the oscillations of the IT drift velocity at the diurnal and semidiurnal frequencies, a sum of the diurnal and semidiurnal frequencies and the second harmonic of the semidiurnal frequency. The maximum drift velocity was 0.85 m/s, and the mean drift velocity 0.43 m/s is close to the most frequent drift velocity in the histogram in Fig. 2.113d. Figure 2.113e demonstrates an increase in the IT drift velocity with decreasing sea depth. Figure 2.113f shows that the minimum radius of curvature of the IT trajectory was 0.1 km during very low drift velocities, and the minimum of radius of curvature of the IT trajectory was 10 km when the mean drift velocity was 0.43 m/s. We note that most of the radii of curvature were of about 1 km.

The diameters of loops D passed by the IT on the trajectory patterns “a” and “g” can be estimated using the formula $\langle V \rangle = \pi f_{SD} D$, where f_{SD} is the semidiurnal frequency. Assuming $f_{SD} = 1/12.4\ \mathrm{h}^{-1}$, we find $D \approx 10$ km on the trajectory “a”, and $D \approx 6$ km on trajectories “d” and “g”. One can see from Fig. 2.69 that the diameters of the loops on the trajectory “a”, where the sea depth was 20–30 m, are larger than the diameters of the loops on the trajectory “g”, where sea depth varied in the range from 50 to 100 m. Fragments of the IT trajectories “a” and “g” shown in Fig. 2.71 confirms that these estimates are reasonable values of the sizes of the loops.

2.3.11 *Measurements in 2019*

Field works on drift ice south of Spitsbergen and Hopen Island were performed in April–May 2019 during the cruise of the *Polarsyssel* organized within the study

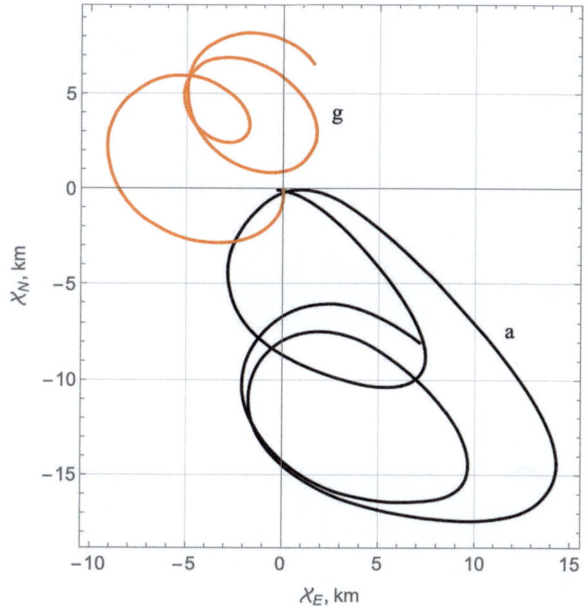

Fig. 2.71 Fragments of the IT trajectories "a" and "g" in metric coordinates

program of the Arctic Technology Department at the UNIS. The IT was deployed at a bergy bit drifting to the southeast of Hopen Island (Fig. 2.72) (Marchenko et al. 2020, 2021b). Laser scanning was performed from several locations onboard the *Polarsyssel* with laser scanner Riegl-Vz 1000 (Fig. 2.73). The horizontal sizes of the bergy bit varied within 15–25 m. The volume of the above water part of the bergy bit was estimated as $V_{fb} = 600\text{–}650\ \mathrm{m}^3$, and the mean free board was $h_{fb} = 1.75$ m. The bergy bit keel was investigated visually with a ROV. It had a conical shape with a maximal draft of 16 m. The volume of bergy bit was estimated with the formula $V_i = \rho_w V_{fb}/(\rho_w - \rho_i)$, where $\rho_w \approx 1020\ \mathrm{kg/m^3}$ is sea water density, and $\rho_i \approx 916\ \mathrm{kg/m^3}$ is density of ice. From (2.1) it follows that $V_i = 5884\text{–}6375\ \mathrm{m}^3$, and the bergy bit mass is $M_i = 5.39\text{–}5.84 \cdot 10^3$ t. Maximal length of the sail is estimated from Fig. 2.73 about $L = 23$ m, and the ratio M_i/L^3 is 0.44–0.48.

The buoy Oceanetic Model 703-1901 uses a VectorNav VN-100 Inertial Measurement Unit (IMU) to measure acceleration of ice floes due to water waves. Acceleration is measured in the three axes True North, East, and Down at a selectable rate and duration. A Fast Fourier Transform (FFT) of each axis is programmed in C, and the real and imaginary FFT results are broadcasted over the Iridium satellite network. Along with the FFT results, the buoy measures its yaw or heading angle relative to the True North at the beginning of each sampling session, so that rotation of the buoy and ice floe may be monitored. The buoy also measures GPS position every 30 min. All raw acceleration data and FFT results are recorded on an SD card.

The buoy may be remotely programmed with the following parameters:

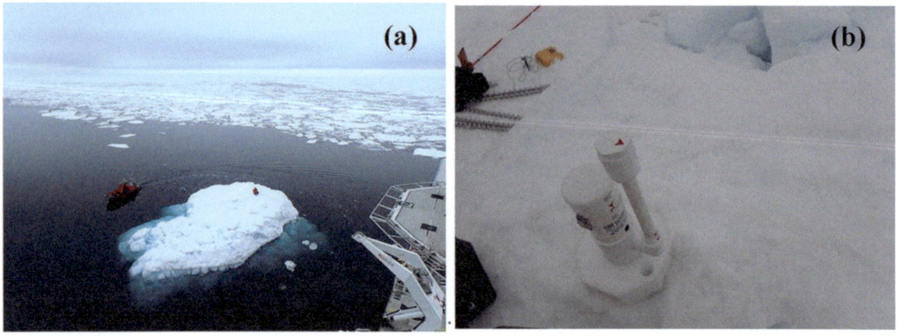

Fig. 2.72 **a** Deployment of ice tracker on the bergy bit on April 27, 2019. **b** Ice tracker deployed on the bergy bit

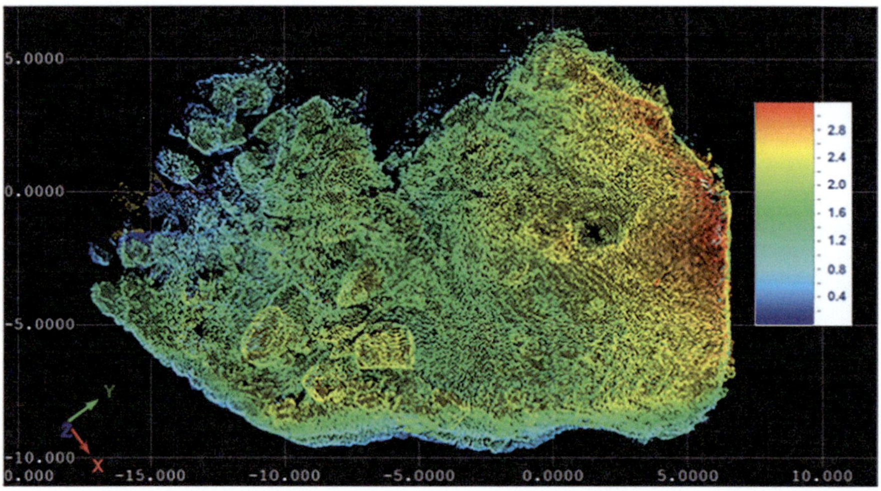

Fig. 2.73 Bergy bit surface after laser scanning from the *Polarsyssel*

- Transmit up to 160 FFT bins, selectable in groups of 5 from bins 0–239 (default bins 30–99)
- FFT length 512 or 1024 (default 1024)
- IMU sample rate of 200, 250, 400, or 500 ms (default 500 ms)
- Number of IMU samples of 512, 1024, 1536, or 2048 (default 2048)
- Period of sample collection from 1 to 24 h (default 4 h)
- IMU data are generated at 800 Hz sampling frequency and can be averaged over 4, 100, 200, or 400 samples (default 200).

The VN-100 uses its internal magnetometer, gyroscope, and accelerometers to determine its orientation relative to the True North and uses GPS location in conjunction with the Magnetic World Model. As such, the VN-100 is sensitive to magnetic

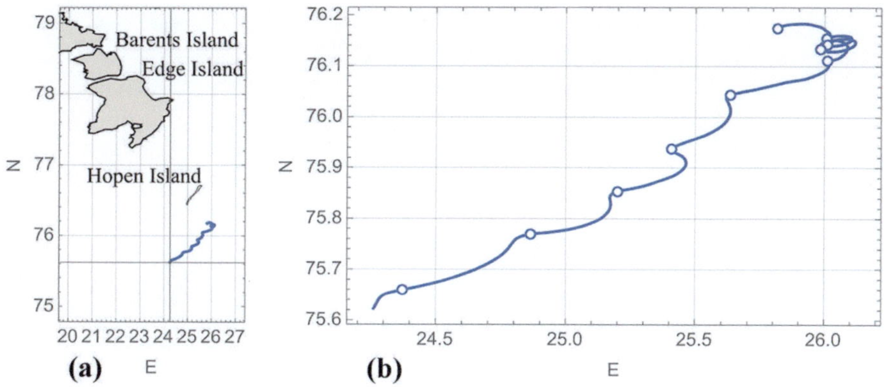

Fig. 2.74 **a** The IT trajectory near Hopen Island from April 27, 18:00 UTC, to May 2, 14:30, UTC, 2019. **b** The IT trajectory with open markers showing the IT locations over every semidiurnal period of 12.42 h starting from the initial position

fields, and is therefore mounted on a boom away from the body of the buoy. The VN-100 is calibrated at the factory for soft and hard iron effects and does not require a calibration step for the magnetometer. Figure 2.72b shows the IT on the bergy bit. The default settings were used with the period of sample collection equal to 1 h. It means that raw data were recorded during the interval of 512 s two times per hour. Then, 70 complex FFT amplitudes of the North, East and Down accelerations were transmitted via Iridium for each interval of 512 s.

The IT trajectory deployed on the bergy bit are shown in Fig. 2.74. General direction of the IT drift was to the southwest. The trajectory had several lopes and "corner" points with high curvature because of combined action of the semidiurnal tide and wind drag. Open circles show in Fig. 2.74b the IT locations over the semidiurnal period M_2 equal to 12.42 h.

Figure 2.75 shows ice conditions in the region of the IT deployment from April 26 to May 2, 2019. Figure 2.76 shows the wind velocity depending on time over the same period. Winds from the west influenced the ice drift to the east during the first 20 h of the IT drift (Fig. 2.74). During the next 20 h, the wind speed was low, and the bergy bit was passing tidal ellipses with a drift velocity of 0.1–0.2 m/s (Fig. 2.77). Then, the wind speed increased, and wind direction changed to the north (Fig. 2.76). It forced the drift of the bergy bit to the southwest (Fig. 2.74). Figure 2.77 shows the locations where the IT drift velocity had local maxima and minima. Open circles show that maximum and minimum drift velocities occurred in the locations with the maximum and minimum radii of curvature of the IT trajectory, respectively.

The Fourier spectrum of the IT drift velocity in Fig. 2.114c has two local maxima at the semidiurnal frequency and the second harmonic of the semidiurnal frequency. The histogram in Fig. 2.114d shows that the most frequent drift velocity was approximately 0.2 m/s, and the maximum of the velocity was 0.6 m/s. The mean drift velocity was equal to 0.27 m/s. Figure 2.114e shows an increase in the IT drift velocity with decreasing depth. Figure 2.114f shows that the minimum radius of curvature of the

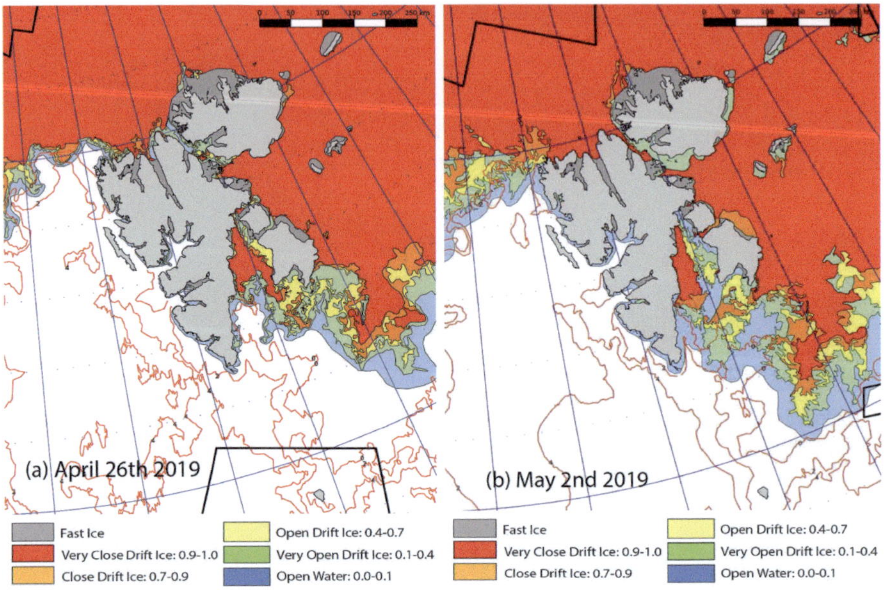

Fig. 2.75 Ice conditions from April 26, 18:00 UTC, to May 2, 14:30 UTC, 2019

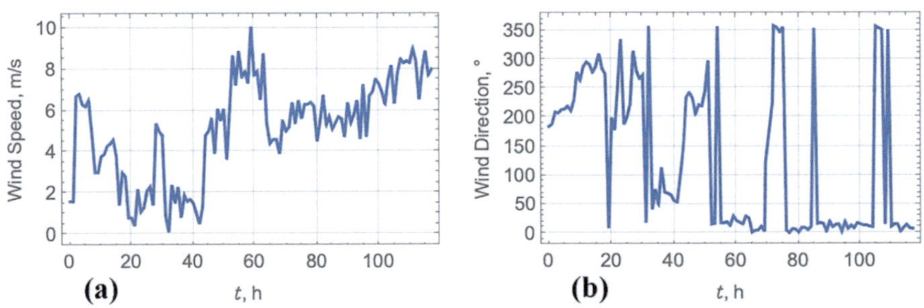

Fig. 2.76 Wind speed (**a**) and direction (**b**) depending on time measured on the meteorological station of Hopen Island from April 27, 18:00 UTC, to May 2, 14:00 UTC, 2019

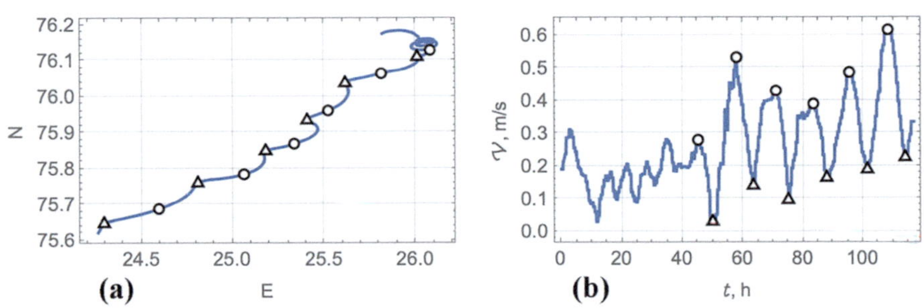

Fig. 2.77 The bergy bit trajectory with open markers showing locations of maximum (circles) and minimum (triangles) drift velocities. **b** Drift velocity of IT depending on time

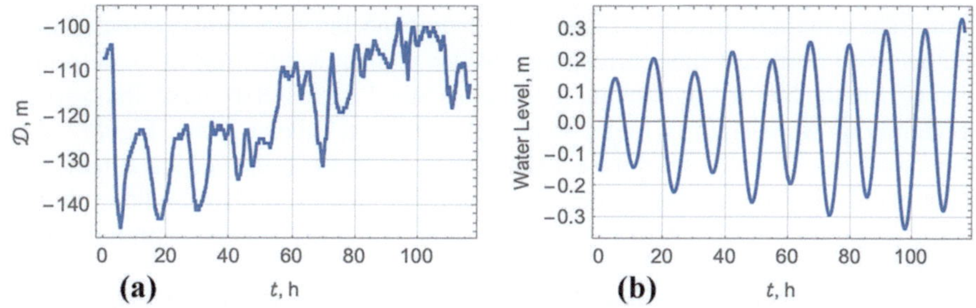

Fig. 2.78 **a** Sea water depth along the IT trajectory. **b** Tide elevations of sea level in the region of the IT drift

IT trajectory was 11.8 km at the mean drift velocity of 0.27 m/s. Figure 2.77b shows an increase in the mean IT drift velocity and in the amplitudes of the oscillations of the IT drift velocity with time. Figure 2.78 shows that it can be explained by a decrease in sea depth and the tidal amplitude with time.

The absolute IT acceleration is shown in Fig. 2.79a depending on time. From Fig. 2.77b it follows that the representative amplitude of the IT drift velocity was about 0.2 m/s per 5 h. Hence, it gives an acceleration of 0.01 mm/s^2. Figure 2.79a shows accelerations of the same order. Figure 2.79b shows the heading angle of the IT depending on time. During the first 60 h of the drift motion, the IT heading angle changed four times by 200°–300°. It can be explained by the interaction of the bergy bit with the surrounding floes. Then, changes of the heading angle became smoother. Smooth temporal oscillations of the heading angle can be explained by hydrodynamic interaction of the bergy bit with water (Marchenko et al. 2019).

Figure 2.80 shows spectra of the horizontal and vertical accelerations of the IT transmitted via Iridium. The frequency of most energetic oscillations varied within 0.08–0.01 Hz. It corresponds well to the swell periods 10.0–12.5 s in the Barents Sea. The inverse FFT was applied to the transmitted spectra of accelerations. Figure 2.81a shows the spatial distribution of the IT accelerations on the horizontal plane. There is a dominant direction of the oscillating surge motion in each time pointed out in the

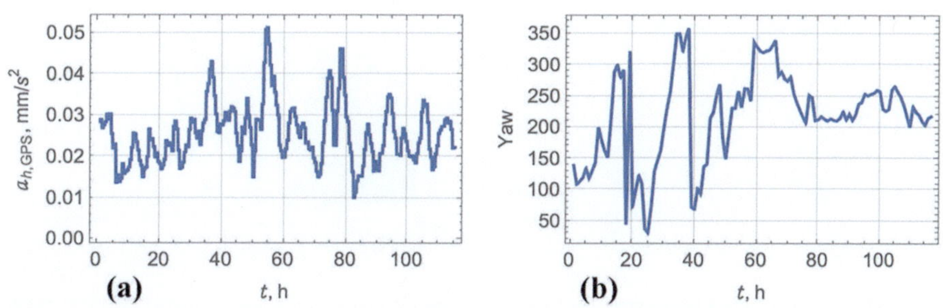

Fig. 2.79 **a** Absolute acceleration of IT depending on time computed using the data on GPS locations recorded with a sampling interval of 30 min. **b** Yaw of IT depending on time

figure. The dominant directions changed with the time, but were mostly oriented from southwest to northeast that corresponds to the dominant direction of swell propagating from the North Atlantic into the Barents Sea. Figure 2.81b shows that vertical accelerations were slightly larger than the horizontal accelerations. Figure 2.81b also shows that an increase of the acceleration amplitudes in the vertical direction is accompanied by an increase of their horizontal amplitudes. It also can be explained by the increasing of the horizontal orbital water velocities in swell with increasing of the swell amplitude.

Figure 2.82 shows the maximum absolute horizontal and vertical accelerations depending on time. Blue lines and yellow lines correspond to two consequent intervals of 512 s within the same hour. Figure 2.83 shows the mean absolute horizontal and vertical accelerations depending on time calculated for every interval of 512 s. The maximum and the mean horizontal accelerations are much higher than the accelerations shown in Fig. 2.79a. The ratios of the mean horizontal and vertical accelerations of the bergy bit caused by waves are shown in Table 2.3.

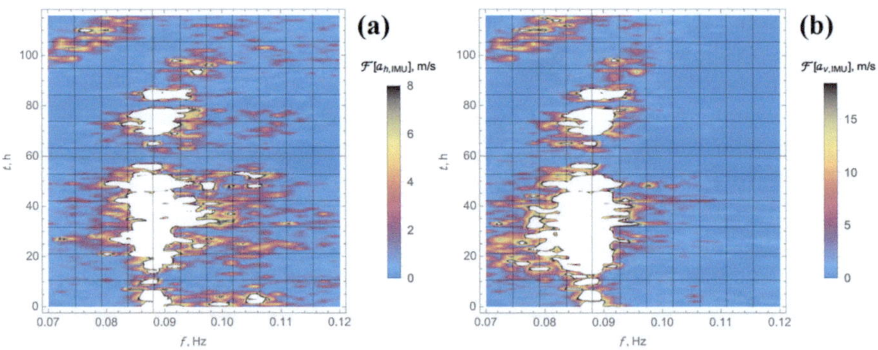

Fig. 2.80 Spectra of the horizontal (**a**) and vertical, **b** accelerations recorded by the IMU depending on time

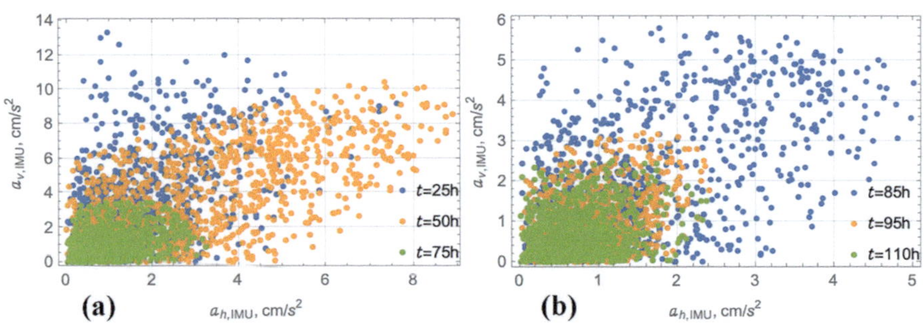

Fig. 2.81 Reconstructed horizontal (HA) and vertical (VA) accelerations recorded by IMU buoy. The time corresponding to the color of points is indicated

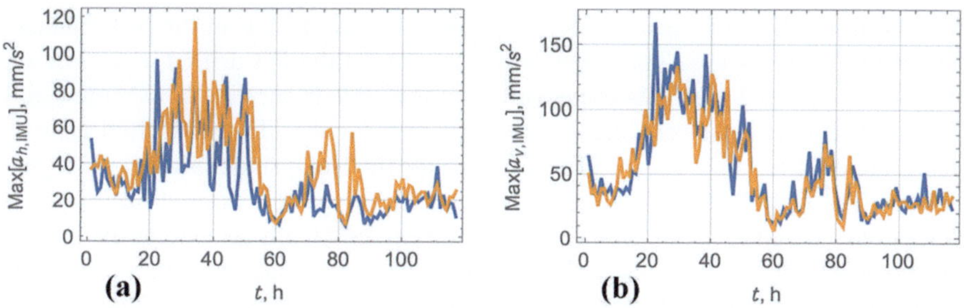

Fig. 2.82 Maximum horizontal (**a**) and vertical, **b** accelerations of the IT measured with 2 Hz frequency versus time

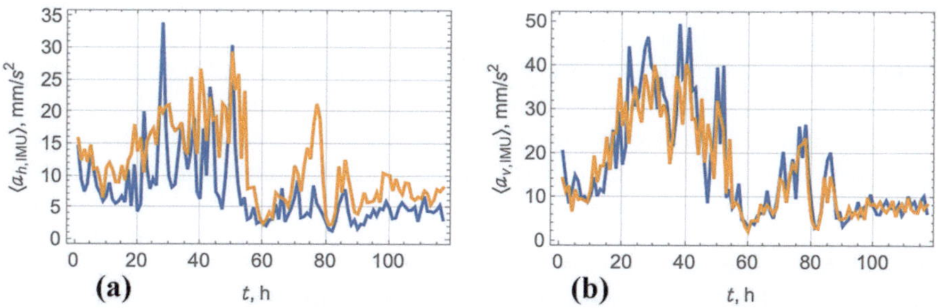

Fig. 2.83 Mean horizontal (**a**) and vertical, **b** accelerations of IT measured with 2 Hz frequency versus time

Table 2.3 Ratios of the mean horizontal and vertical accelerations of a bergy bit measured with IMU buoy

Time, h	25	50	75	85	95	110
$\langle HA\rangle/\langle VA\rangle$	0.5	0.8	0.9	0.8	1.0	1.0

2.4 Concluding Remarks on Ice Drift

The main results of the IT data processing are summarized in Tables 2.4, 2.5, 2.6 and 2.7, Figs. 2.84, 2.85, 2.86 and 2.87. Figure 2.84 was constructed based on the data in Tables 2.4, 2.5, 2.6 and 2.7. Figure 2.84a shows that the maximum number of harmonics 4 in the spectral composition of the IT drift velocity was calculated for the ITs with the largest time of deployment greater than 900 h. The diurnal frequency was presented in the spectra of the drift velocities of these ITs. Triangle markers in Fig. 2.84a show that the number of harmonics in the drift velocities of the ITs in SpB increased proportionally to the ITs deployment time. The third harmonics of the semidiurnal frequency was calculated only for the ITs drifted in the BS. Figure 2.84b does not demonstrate any dependence of the meant drift velocities on the deployment time of the ITs. The mean drift velocities of the ITs drifted in the SpB were larger than 0.4 m/s. The mean drift velocities of the ITs that drifted in the SFE changed between 0.2 and 0.4 m/s. The mean drift velocities of the ITs drifted in the BA and SFW were about 0.2 m/s. This value corresponds well to the earlier data (Vinje 1985).

The maximum drift velocities were calculated for the ITs that drifted near the Heleysundet Strait. In 2013, the ITd drifted inside the Heleysindet Strait with a speed of 4 m/s. The ITd (2013) and ITa (2016) drifted with a speed of around 2 m/s near the Heleysundet Strait from the side of the Stoor Fjord and from the side of the Barents Sea, respectively. The Norwegian Pilot Guide (2018) reported: "Several vessels have recorded a tidal stream speed of 8–9 knots in the narrowest part". A stream speed of 8–9 knots means 4.1–4.6 m/s. The high drift speeds were found near the southern tip of Hopen Island. The ITs drifted there with the speeds of 1.6 m/s (2017), 1.7 m/s (IT trajectory c 2018), and 1.4 m/s (IT trajectory e 2018). It adds valuable information to the Norwegian Pilot (2018) reporting on the Hopen Island region:" The tidal stream, which changes at high and low water, rapidly rotates around the island with unusual strength and in some places, especially around the southern and northern points, it sets up a violent sea which can be dangerous to smaller vessels".

The maximum IT drift velocities of 1.3–1.4 m/s were calculated from the IT data in the SpB, SFW and SFE sub-regions. The maximum IT drift velocities measured

Table 2.4 **Barents Sea** sub-region. Duration of IT drift (Δt), minimum sea depth $D_{\min}$, existence of diurnal (f_D), semidiurnal (f_{SD}), and multiple ($f_D + f_{SD}$, $2f_{SD}$) frequencies in the spectrum of IT drift velocity, the mean ($\langle V \rangle$) and the maximum ($V_{\max}$) drift velocities along the IT trajectory patterns "a–g"

	Δt, h	$D_{\min}$, m	f_D	f_{SD}	$f_D + f_{SD}$	$2f_{SD}$	$3f_{SD}$	$\langle V \rangle$, m/s	$V_{\max}$, m/s
2008a	550	60	v	v	–	v	–	0.23	0.55
2010a	482	100	–	v	–	v	–	0.17	0.47
2010b	758	50	–	v	–	v	v	0.19	0.6
2010c	809	<50	–	v	–	v	–	0.19	0.6
2010d	619	50	–	v	–	v	v	0.17	0.58
2010e	763	<50	–	v	–	v	v	0.19	0.6
2016a	764	<50	–	v	–	v	–	0.15	2.23

Table 2.5 **Spitsbergen Bank** sub-region. The same characteristics as in Table 2.4. Symbol HI means that the maximum velocity was measured near Hopen Island

	Δt, h	$D_{\min}$, m	f_D	f_{SD}	$f_D + f_{SD}$	$2f_{SD}$	$3f_{SD}$	$\langle V \rangle$, m/s	$V_{\max}$, m/s
2008b	589	<50	v	v	–	v	–	0.41	1.36
2017	226	50	–	v	–	v	–	0.4	1.6
2018a	764	20	–	v	v	v	–	0.7	1.3
2018c	1150	<50	v	v	v	v	–	0.4	1.7; HI
2018e	300	<50	–	v	–	v	–	0.4	1.4; HI
2018g	950	50	v	v	v	v	–	0.4	0.85

Table 2.6 **Stoor Fjord West** sub-region. The same characteristics as in Table 2.4

	Δt, h	D_{min}, m	f_D	f_{SD}	$f_D + f_{SD}$	$2f_{SD}$	$3f_{SD}$	$\langle V \rangle$, m/s	V_{max}, m/s
2012a	276	50	v	v	–	v	–	0.23	1.4
2014a	713	<50	–	–	–	–	–	0.19	0.6
2018b	1350	100	v	v	v	v	–	0.2	0.6

Table 2.7 **Stoor Fjord East** sub-region. The same characteristics as in Table 2.4

	Δt, h	D_{min}, m	f_D	f_{SD}	$f_D + f_{SD}$	$2f_{SD}$	$3f_{SD}$	$\langle V \rangle$, m/s	V_{max}, m/s
2012b	204	<50	–	v	–	v	–	0.37	1.1
2012c	234	<50	–	v	–	v	–	0.43	1.1
2012d	265	<50	–	v	–	v	–	0.42	1.1
2013a	1243	<50	v	v	v	v	–	0.22	1.33
2013b	1285	<50	v	v	v	v	–	0.33	1.27
2013c	802	<50	–	v	–	v	–	0.28	1.3
2018d	400	<50	v	v	–	v	–	0.38	0.9
2018f	200	70	–	v	–	v	–	0.25	0.6

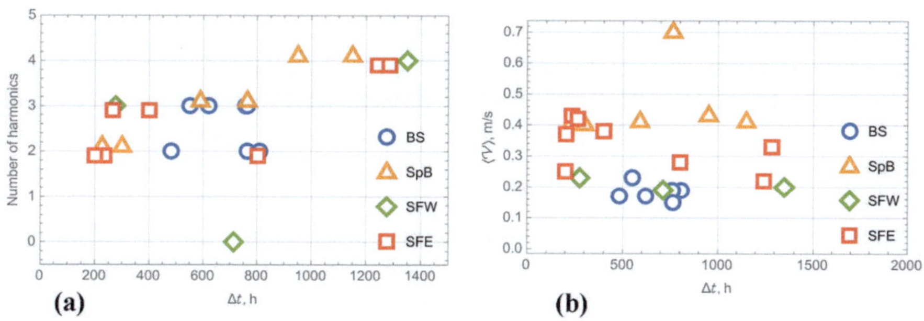

Fig. 2.84 Number of harmonics is the spectra of the IT drift velocities (**a**) and the mean IT drift velocities, **b** depending on the deployment time Δt

in the BS sub-region were 0.6 m/s. The maximum drift IT velocity equal to 1.4 m/s in the SFW correlated with increasing wind velocity, while the maximum drift velocities measured in the BS, SpB, and SFE sub-regions were associated mostly with the tidal currents.

Figure 2.85 demonstrates the general trend to increasing the maximum IT drift speed with decreasing sea depth in the BS, SpB, and SFE sub-regions. Several green dots in Fig. 2.85a with drift velocities exceeding 1 m/s correspond to the IT drift near the Heleysundet Strait in 2016. Several yellow and green dots with drift velocities exceeding 1.4 m/s correspond to the IT drift near Hopen Island in 2017 and 2018. Several blue dots with drift velocities exceeding 0.6 m/s in Fig. 2.85c correspond to the IT acceleration by strong wind in 2012. An increase in the drift velocity with

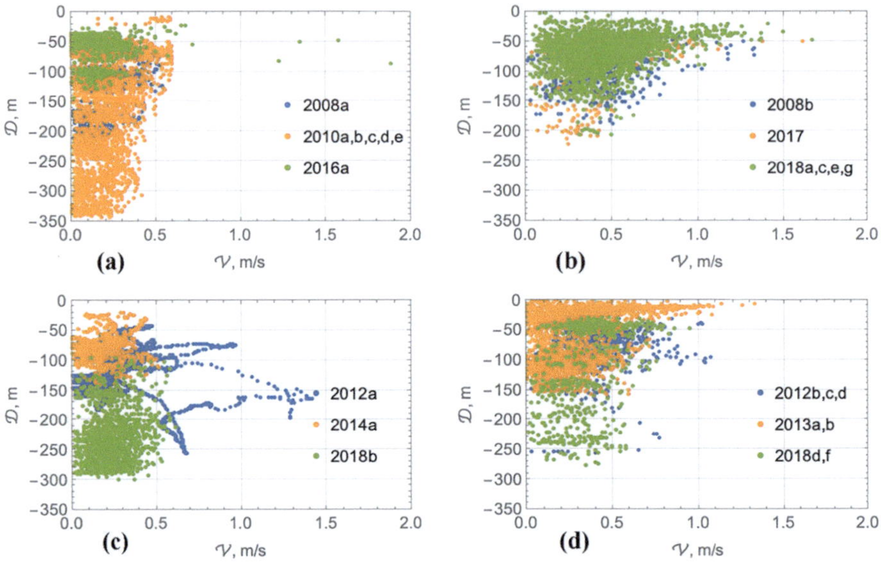

Fig. 2.85 The IT drift velocity depending on the local sea depth in the BS (**a**), SpB (**b**), SFW (**c**), and SFE (**d**) subregions, respectively

decreasing sea depth is best pronounced in the SpB (Fig. 2.85b). Figure 2.85d shows a very similar tendency with slightly smaller maximal drift velocities. An increase of drift velocities with decreasing sea depth is less pronounced in the BS (Fig. 2.85a). The dependence of the IT drift velocities on the sea depth is absent in the SFW (Fig. 2.85c).

Lines in Fig. 2.86 show the minimum radii of curvature of the IT trajectories. Open markers show the minimum radii of curvature at the mean IT drift velocities. Figure 2.86a, c show that all values of R at the mean IT drift velocities are of 1 km in the BS and SFW sub-regions, and change from 1 to 3 km in the SFE. The values of R at the mean IT drift velocities change from 50 to 10 km in the SpB (Fig. 2.86b). This means that drift ice in SpB is more mobile compared to the other regions.

Comparison of the dependences of the minimum radii of curvature on the drift velocity for two IT deployed on icebergs in 2009 and 2019 with the ITa deployed on drift ice in BS in 2008 is shown in Fig. 2.87. The radii of curvature of the ITs deployed on icebergs are similar to each other and to the larger radii of curvature of the IT deployed on drift ice at $V > 0.2$ m/s. This can be explained by the greater influence of the iceberg's inertia on its drift compared to level ice, the drift of which depends to a greater extent on wind and water drag forces.

An assessment of data from NOAA (2025) on the floating buoys and those fixed on ice installed by the UNIS in the western part of Barents Sea shows that they both demonstrated the possibilities of ice and surface water drift into the Ice Fjord from the Barents Sea. This motion is supported by the sea current system around

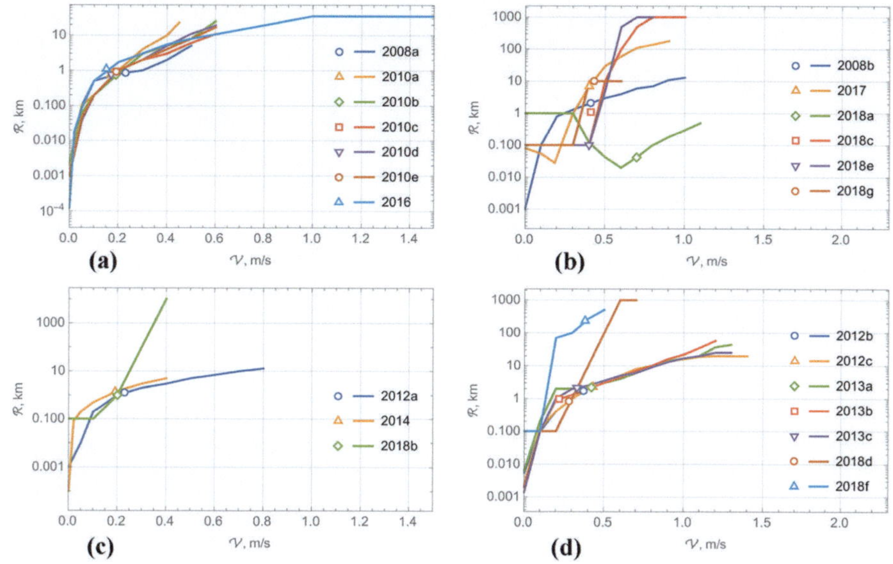

Fig. 2.86 Bottom envelopes of the IT radii of curvature in the BS (**a**), SpB (**b**), SFW (**c**), and SFE (**d**) sub-regions

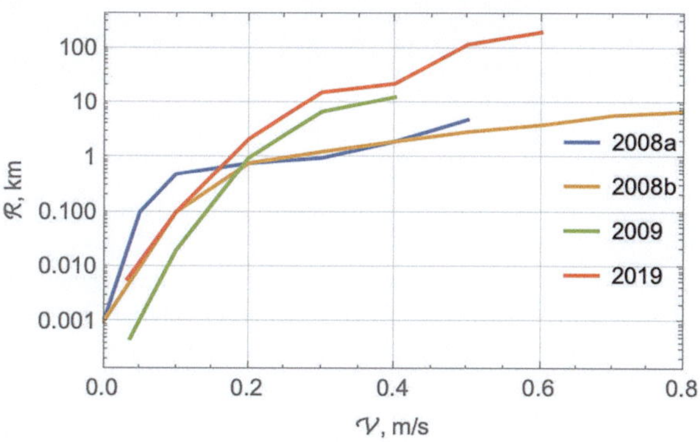

Fig. 2.87 Bottom envelopes of the radii of curvature of the iceberg trajectories in 2009 and 2019 in comparison with the radius of curvature of the IT deployed on drift ice in 2008

Spitsbergen and local counterclockwise circulation of sea water inside the Ice Fjord (Marchenko and Marchenko 2015).

The mean drift speed of ITs calculated using the GPS data with a six-hour sampling interval is higher than the mean speed of floating buoys (FB) calculated using the GPS data with the same sampling interval. In the examples considered here, the mean speed of the ITs was estimated within 0.2–0.4 m/s, while the mean speed of floating buoys was within 0.1–0.2 m/s. The maximum drift speed of the ITs is higher than the maximum drift speed of the FBs by 30%. The maximum speed of the ITs calculated using GPS data with a 6-h sampling interval is lower than the maximum speed of

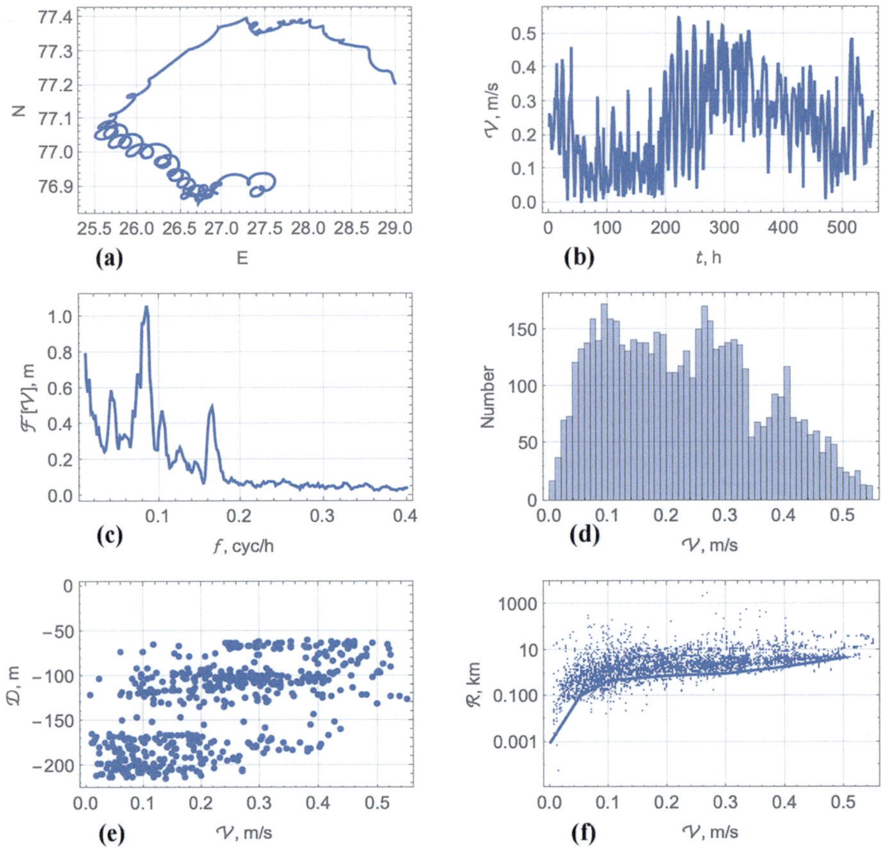

Fig. 2.88 2008a. **a** The IT trajectory in the BS from May 9 to May 31. **b** The IT drift velocity depending on time. **c** Fourier spectrum of the IT drift velocity. **d** Histogram of the IT drift velocity. **e** The IT drift velocity depending on local sea depth. **f** Radius of curvature of the IT drift trajectory depending on the drift velocity

the ITs calculated using a 10- or 20-min sampling interval. The difference reached 20–30% in the considered examples (Marchenko and Marchenko 2015).

Appendix

See Figs. 2.88, 2.89, 2.90, 2.91, 2.92, 2.93, 2.94, 2.95, 2.96, 2.97, 2.98, 2.99, 2.100, 2.101, 2.102, 2.103, 2.104, 2.105, 2.106, 2.107, 2.108, 2.109, 2.110, 2.111, 2.112, 2.113 and 2.114.

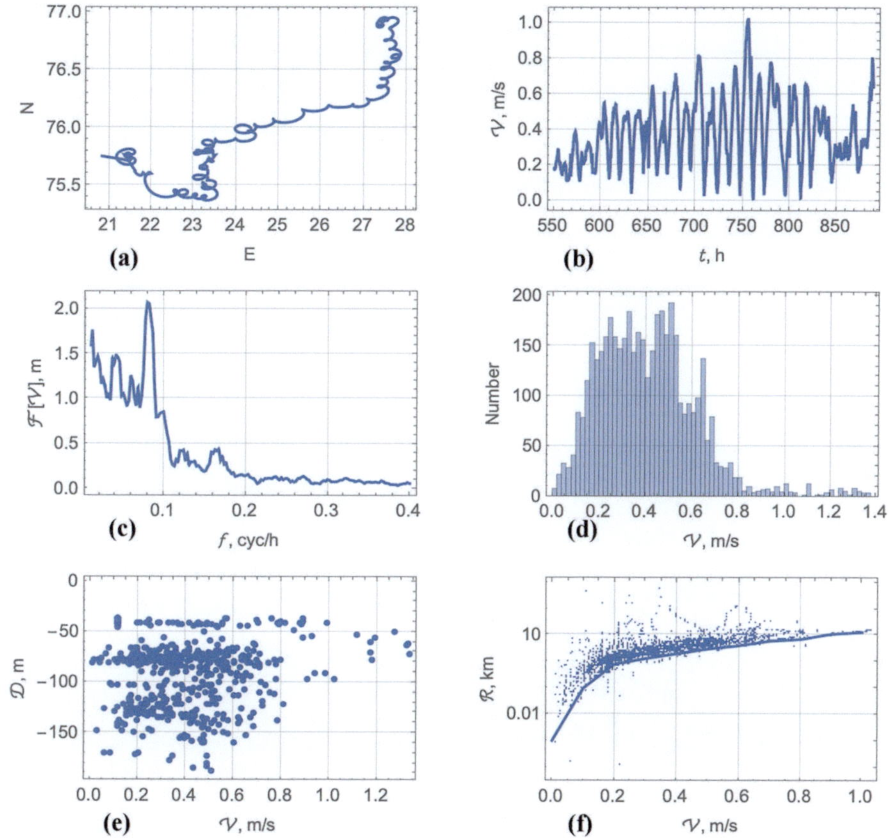

Fig. 2.89 2008b. **a** The IT trajectory in the SpB from June 1 to June 20. **b** The IT drift velocity depending on time. **c** Fourier spectrum of the IT drift velocity. **d** Histogram of the IT drift velocity. **e** The IT drift velocity depending on local sea depth. **f** Radius of curvature of the IT drift trajectory depending on the drift velocity

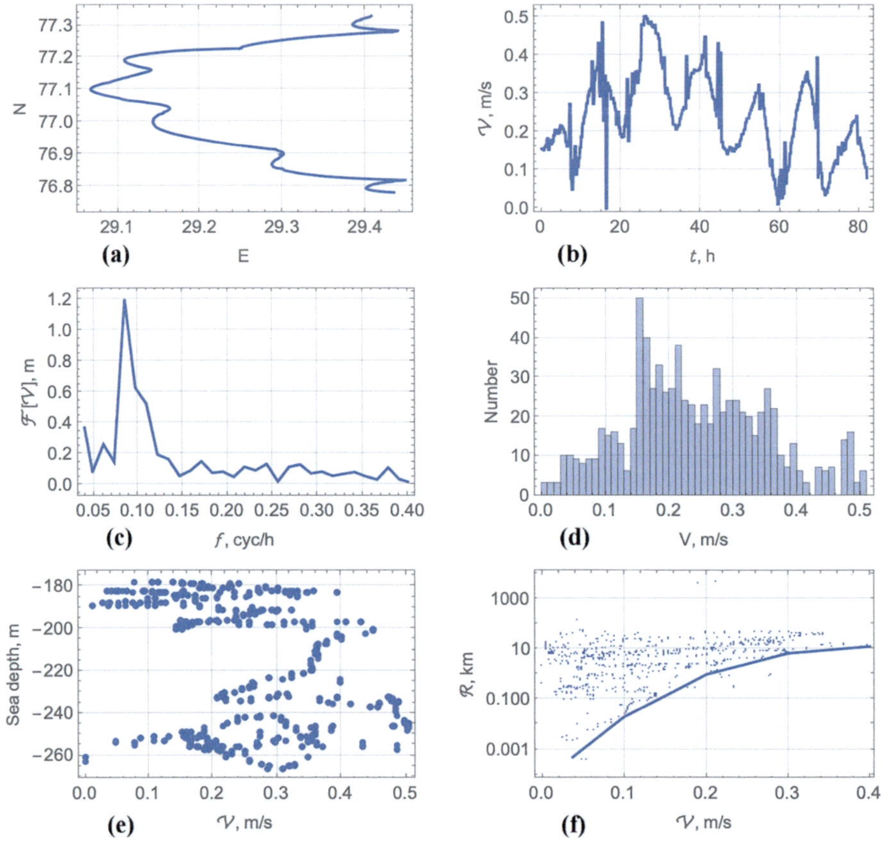

Fig. 2.90 2009. **a** The IT1 trajectory in the BS from May 2 to May 5. **b** The IT1 drift velocity depending on time. **c** Fourier spectrum of the IT1 drift velocity. **d** Histogram of the IT1 drift velocity. **e** The IT1 drift velocity depending on local sea depth. **f** Radius of curvature of the IT1 drift trajectory depending on the drift velocity

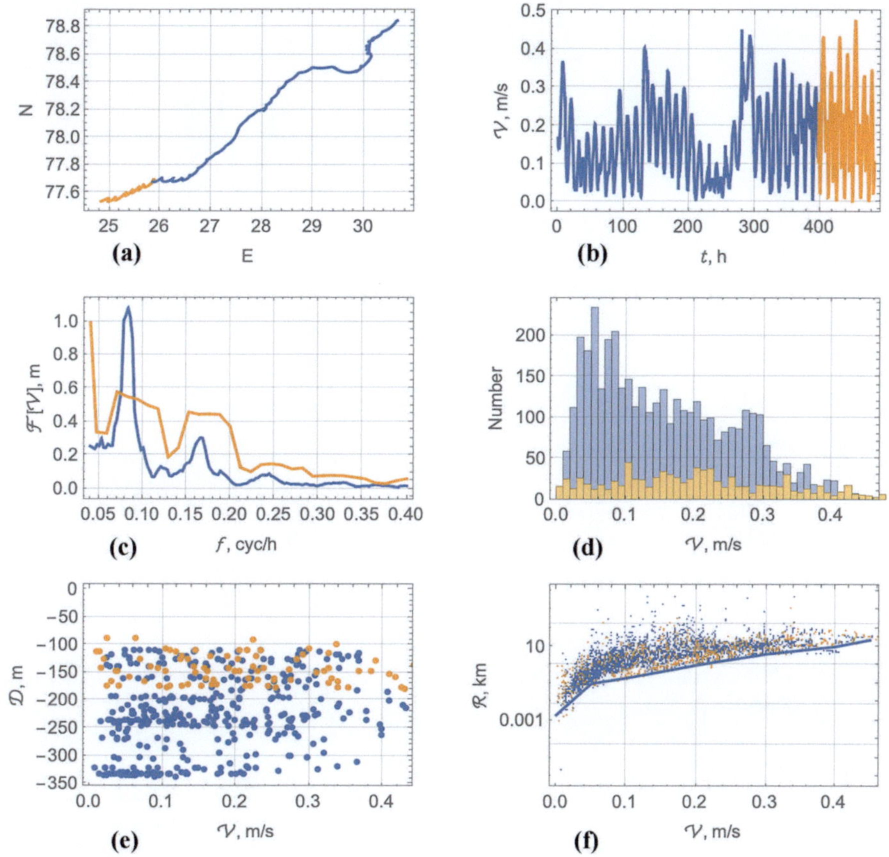

Fig. 2.91 2010a. **a** The IT trajectory in the BS from March 15 to April 15. **b** The IT drift velocity depending on time. **c** Fourier spectrum of the IT drift velocity. **d** Histogram of the IT drift velocity. **e** The IT drift velocity depending on local sea depth. **f** Radius of curvature of the IT drift trajectory depending on the drift velocity. Characteristics of the IT drift are shown by blue color on March 15–31, and by yellow color on April 1–15

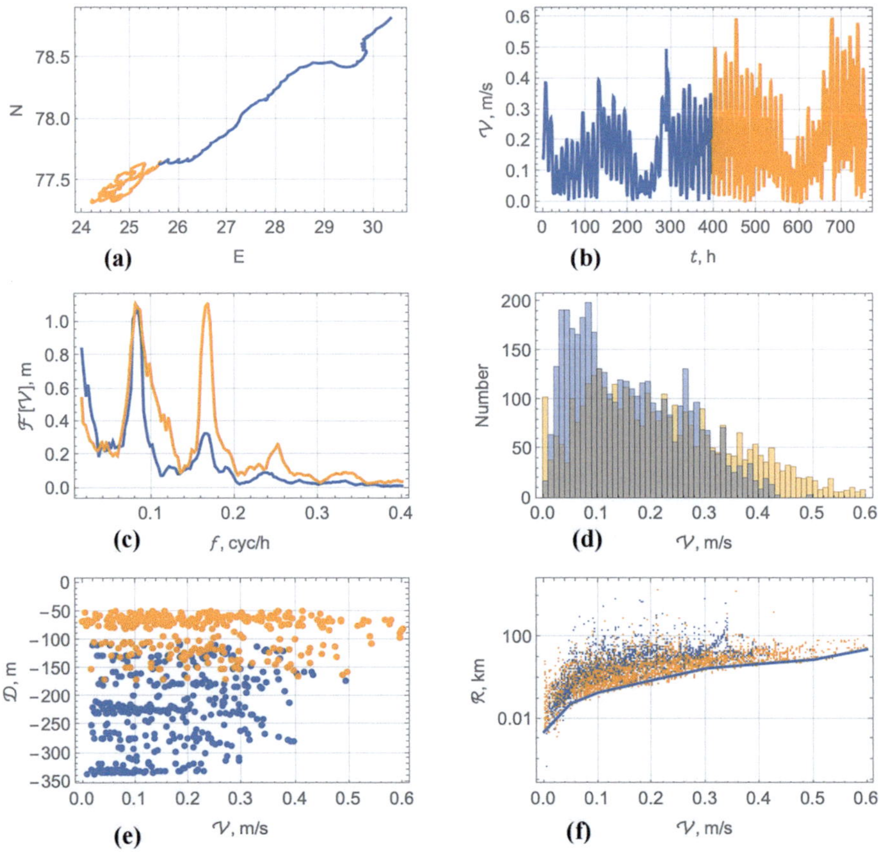

Fig. 2.92 2010b. **a** The IT trajectory in the BS from March 15 to April 15. **b** The IT drift velocity depending on time. **c** Fourier spectrum of the IT drift velocity. **d** Histogram of the IT drift velocity. **e** The IT drift velocity depending on local sea depth. **f** Radius of curvature of the IT drift trajectory depending on the drift velocity. Characteristics of the IT drift are shown by blue color on March 15–31, and by yellow color on April 1–15

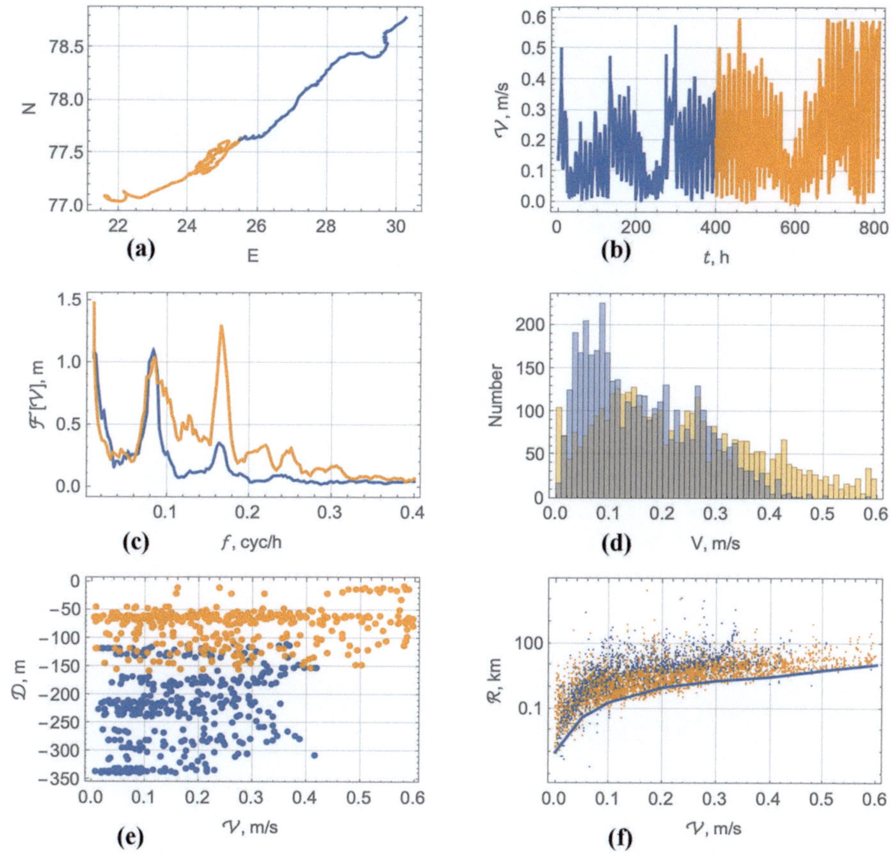

Fig. 2.93 2010c. **a** The IT trajectory in the BS from March 15 to April 15. **b** The IT drift velocity depending on time. **c** Fourier spectrum of the IT drift velocity. **d** Histogram of the IT drift velocity. **e** The IT drift velocity depending on local sea depth. **f** Radius of curvature of the IT drift trajectory depending on the drift velocity. Characteristics of the IT drift are shown by blue color on March 15–31, and by yellow color on April 1–15

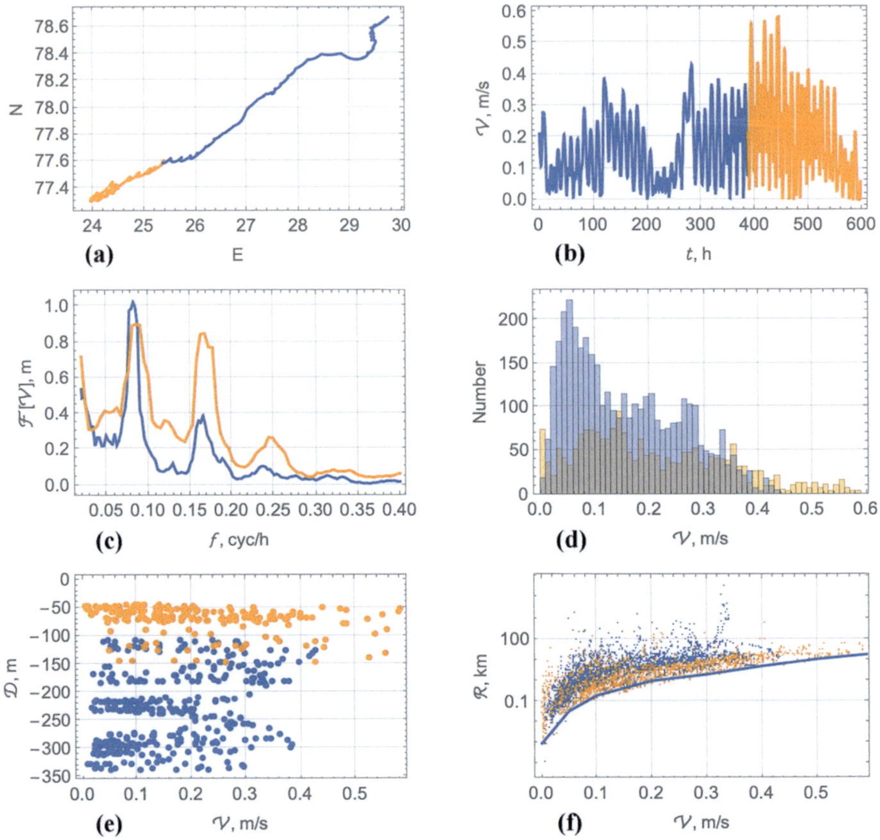

Fig. 2.94 2010d. **a** The IT trajectory in the BS from March 15 to April 15. **b** The IT drift velocity depending on time. **c** Fourier spectrum of the IT drift velocity. **d** Histogram of the IT drift velocity. **e** The IT drift velocity depending on local sea depth. **f** Radius of curvature of the IT drift trajectory depending on the drift velocity. Characteristics of the IT drift are shown by blue color on March 15–31, and by yellow color on April 1–15

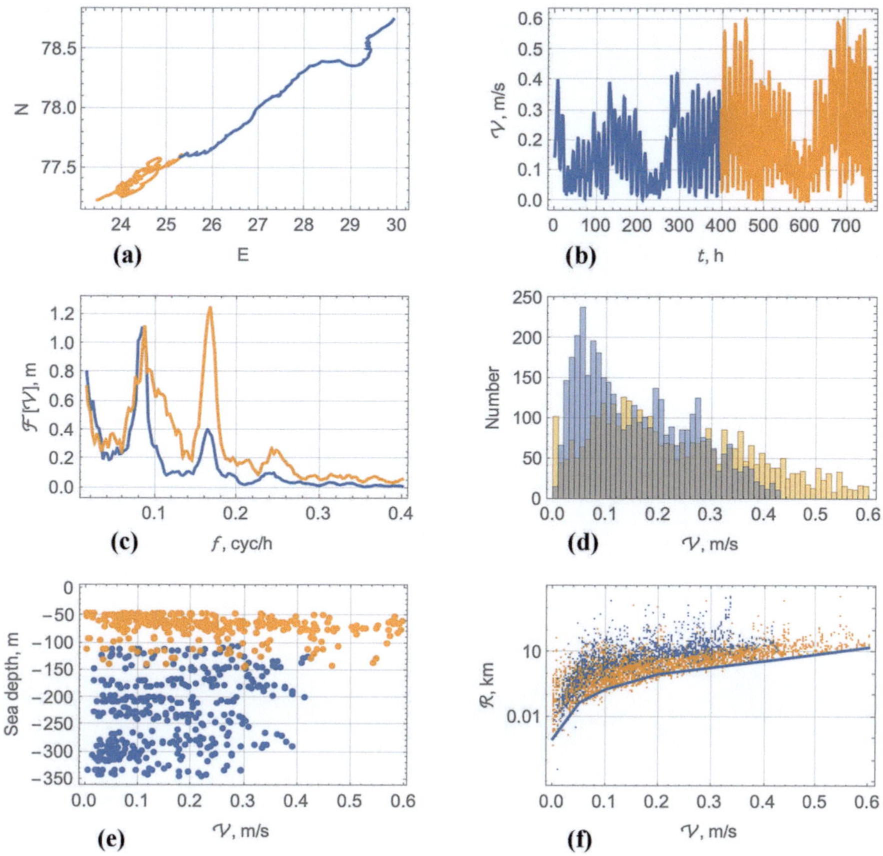

Fig. 2.95 2010e. **a** The IT trajectory in the BS from March 15 to April 15. **b** The IT drift velocity depending on time. **c** Fourier spectrum of the IT drift velocity. **d** Histogram of the IT drift velocity. **e** The IT drift velocity depending on local sea depth. **f** Radius of curvature of the IT drift trajectory depending on the drift velocity. Characteristics of the IT drift are shown by blue color on March 15–31, and by yellow color on April 1–15

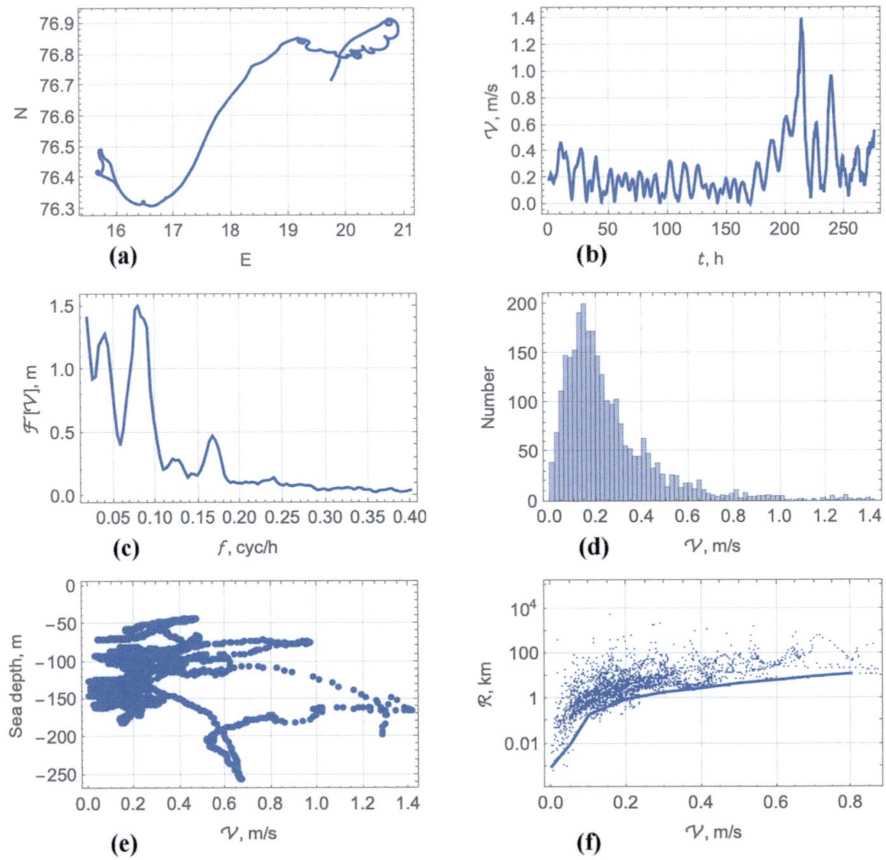

Fig. 2.96 2012a. **a** The IT trajectory in the SFW from April 17 to April 29. **b** The IT drift velocity depending on time. **c** Fourier spectrum of the IT drift velocity. **d** Histogram of the IT drift velocity. **e** The IT drift velocity depending on local sea depth. **f** Radius of curvature of the IT drift trajectory depending on the drift velocity

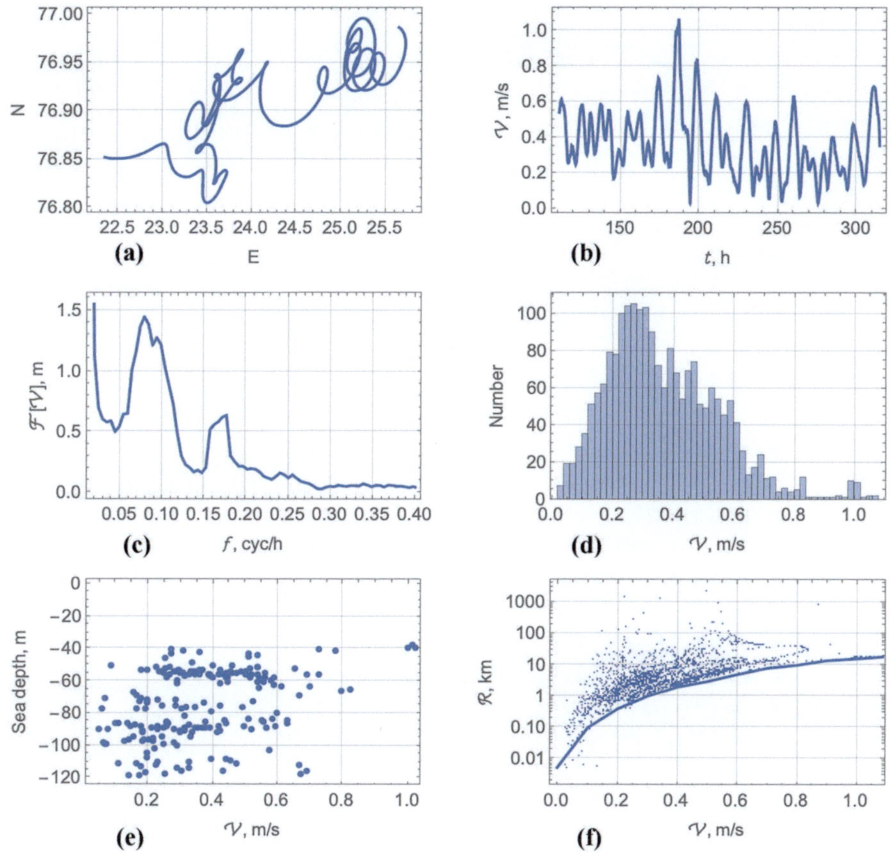

Fig. 2.97 2012b. **a** The IT trajectory in the SFE from April 22 to April 30. **b** The IT drift velocity depending on time. **c** Fourier spectrum of the IT drift velocity. **d** Histogram of the IT drift velocity. **e** The IT drift velocity depending on local sea depth. **f** Radius of curvature of the IT drift trajectory depending on the drift velocity

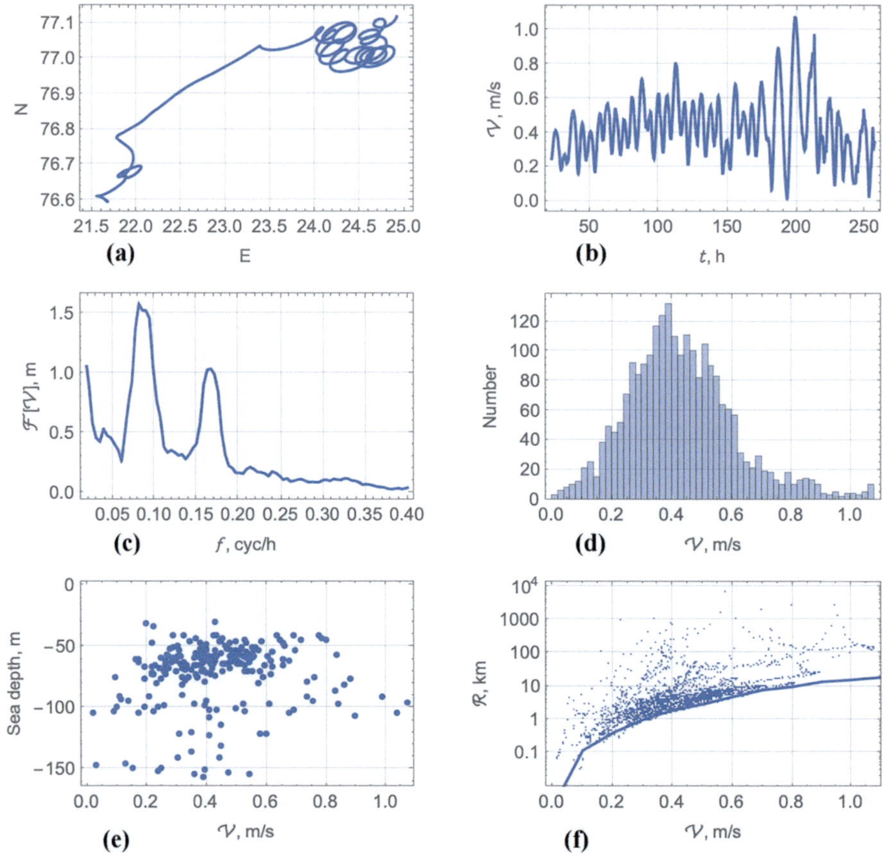

Fig. 2.98 2012c. **a** The IT trajectory in the SFE from April 18 to April 28. **b** The IT drift velocity depending on time. **c** Fourier spectrum of the IT drift velocity. **d** Histogram of the IT drift velocity. **e** The IT drift velocity depending on local sea depth. **f** Radius of curvature of the IT drift trajectory depending on the drift velocity

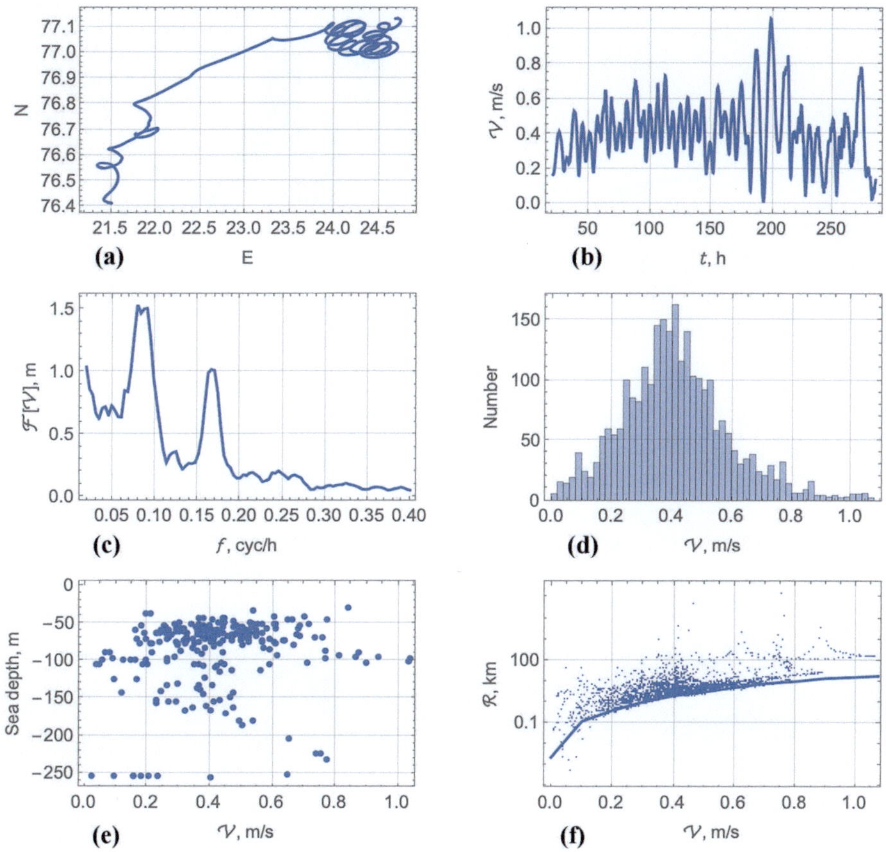

Fig. 2.99 2012d. **a** The IT trajectory in the SFE from April 18 to April 29. **b** The IT drift velocity depending on time. **c** Fourier spectrum of the IT drift velocity. **d** Histogram of the IT drift velocity. **e** The IT drift velocity depending on local sea depth. **f** Radius of curvature of the IT drift trajectory depending on the drift velocity

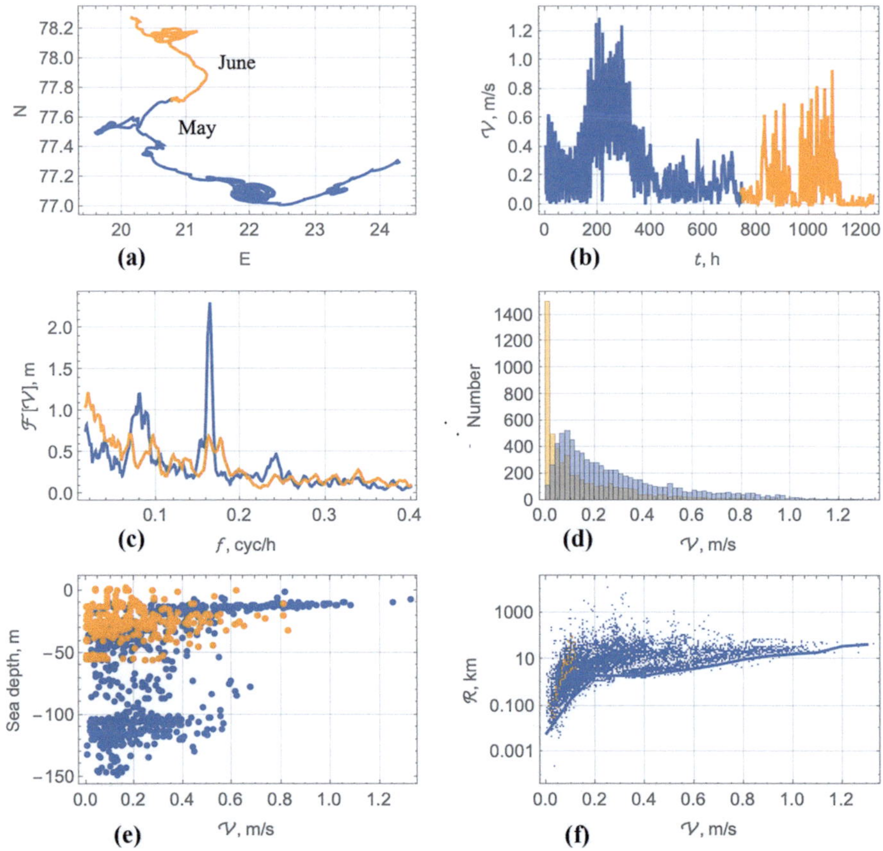

Fig. 2.100 2013a. **a** The IT trajectory in the SFE from May 1 to June 21. **b** The IT drift velocity depending on time. **c** Fourier spectrum of the IT drift velocity. **d** Histogram of the IT drift velocity. **e** The IT drift velocity depending on local sea depth. **f** Radius of curvature of the IT drift trajectory depending on the drift velocity. Blue and yellow colors correspond to the measurements in May and June

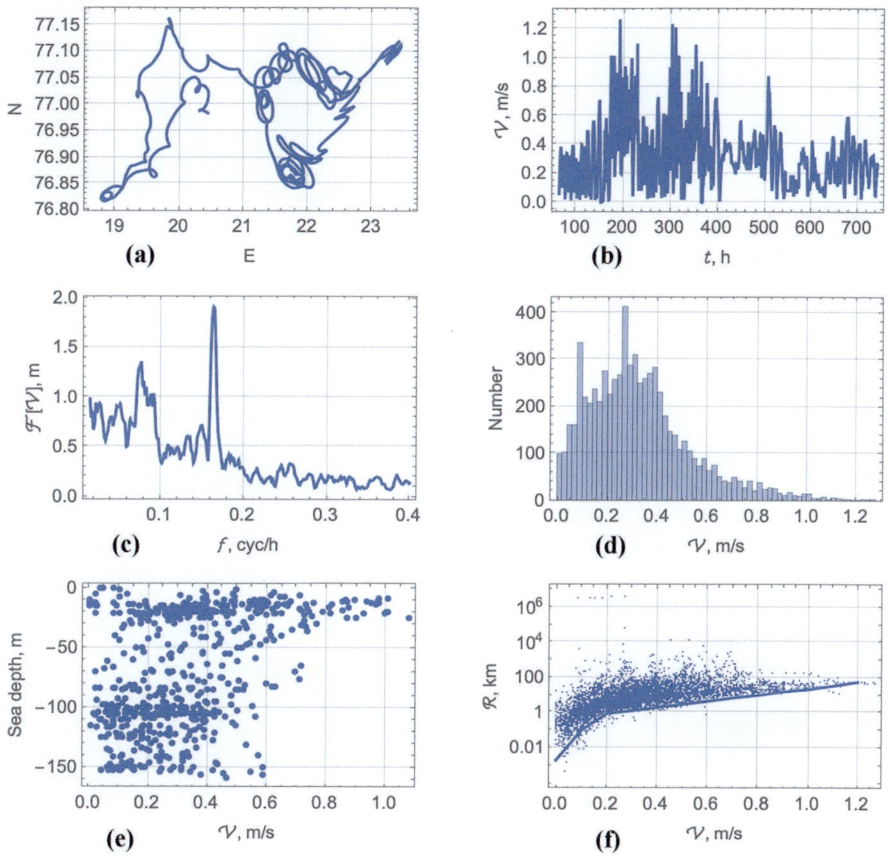

Fig. 2.101 2013b. **a** The IT trajectory in the SFE from May 3 to May 31. **b** The IT drift velocity depending on time. **c** Fourier spectrum of the IT drift velocity. **d** Histogram of the IT drift velocity. **e** The IT drift velocity depending on local sea depth. **f** Radius of curvature of the IT drift trajectory depending on the drift velocity

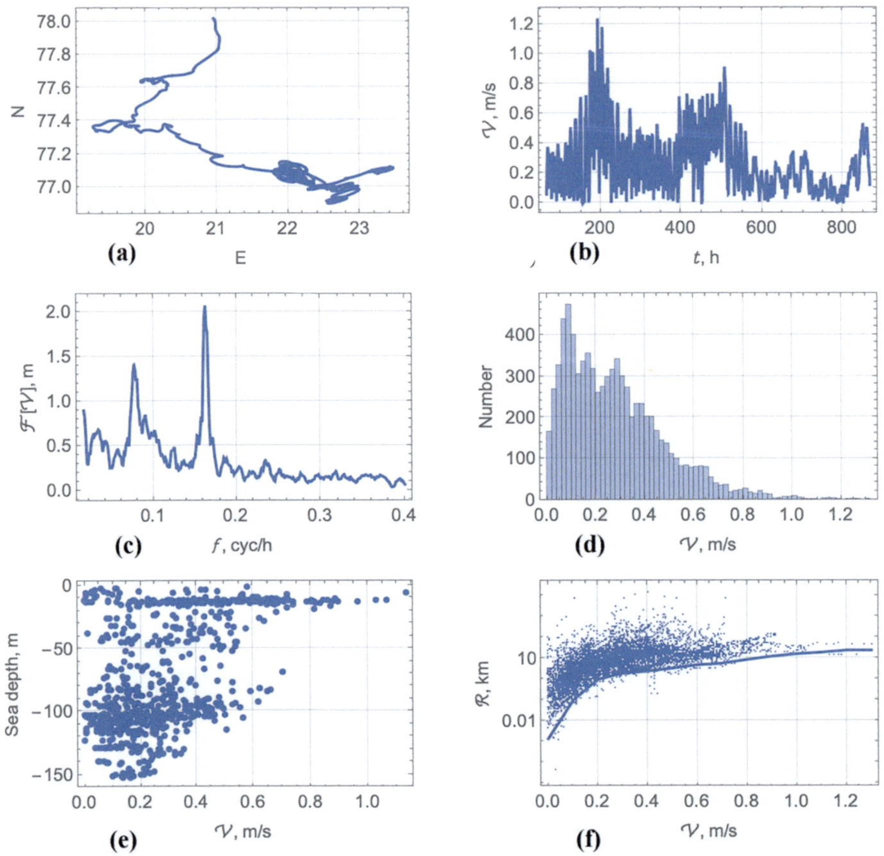

Fig. 2.102 2013c. **a** The IT trajectory in the SFE from May 3 to June 6. **b** The IT drift velocity depending on time. **c** Fourier spectrum of the IT drift velocity. **d** Histogram of the IT drift velocity. **e** The IT drift velocity depending on local sea depth. **f** Radius of curvature of the IT drift trajectory depending on the drift velocity

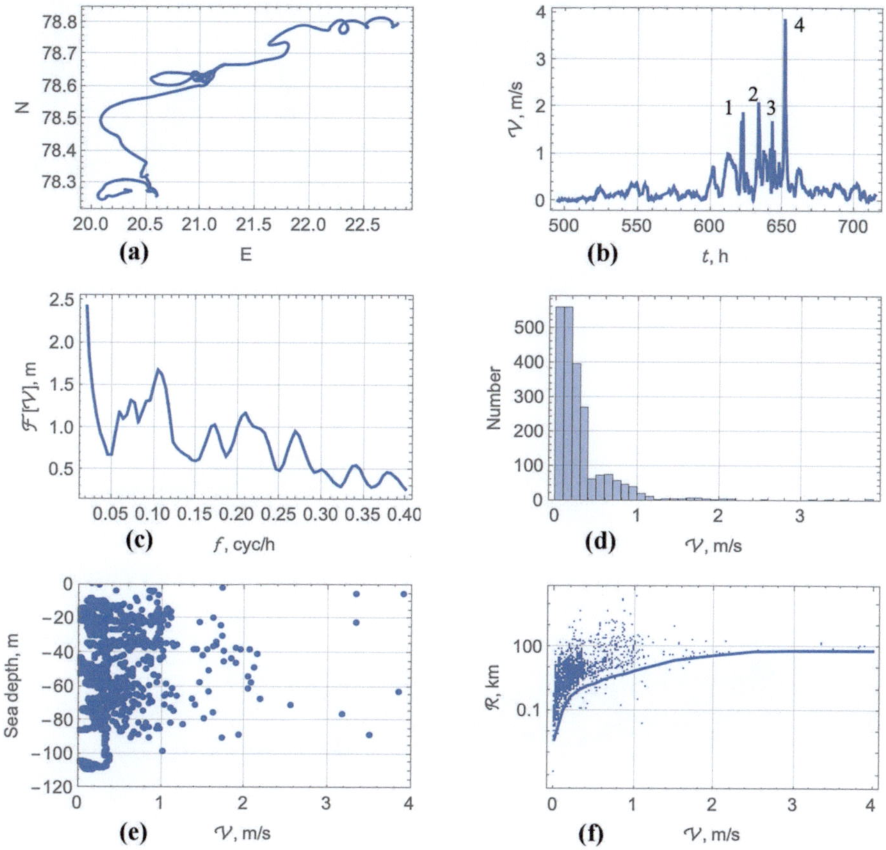

Fig. 2.103 2013d. **a** The IT trajectory in the SFE from June 20 to June 29. **b** The IT drift velocity depending on time. **c** Fourier spectrum of the IT drift velocity. **d** Histogram of the IT drift velocity. **e** The IT drift velocity depending on local sea depth. **f** Radius of curvature of the IT drift trajectory depending on the drift velocity

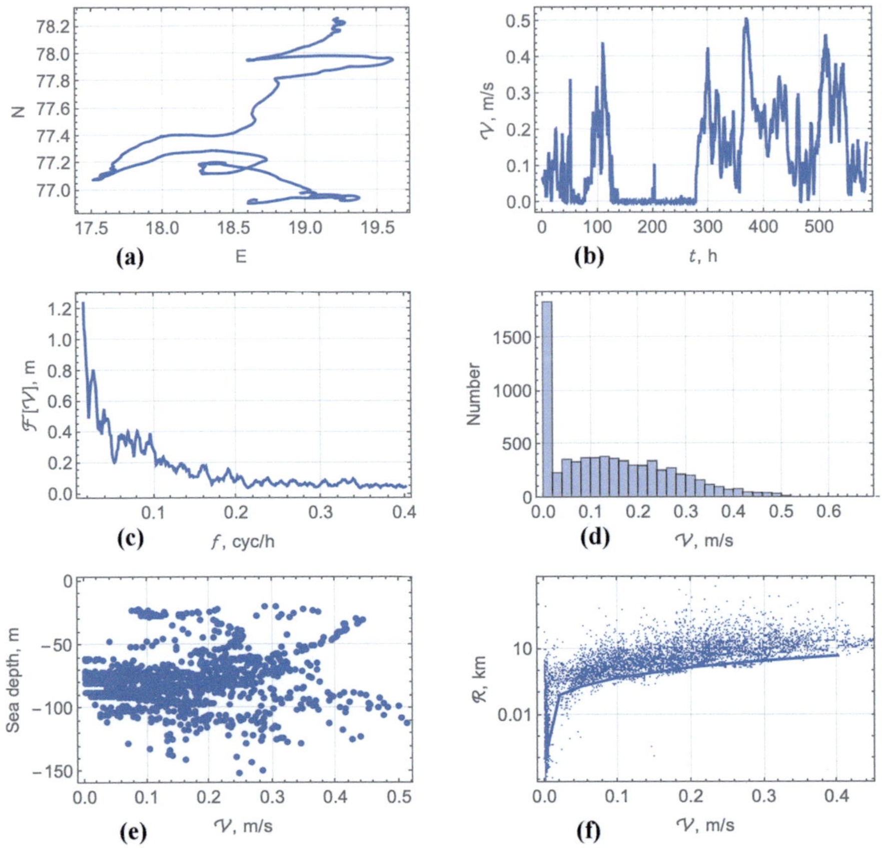

Fig. 2.104 2014a. **a** The ITa trajectory in the SFW from April 30 to May 25. **b** The ITa drift velocity depending on time. **c** Fourier spectrum of the ITa drift velocity. **d** Histogram of the ITa drift velocity. **e** The ITa drift velocity depending on local sea depth. **f** Radius of curvature of the ITa drift trajectory depending on the drift velocity

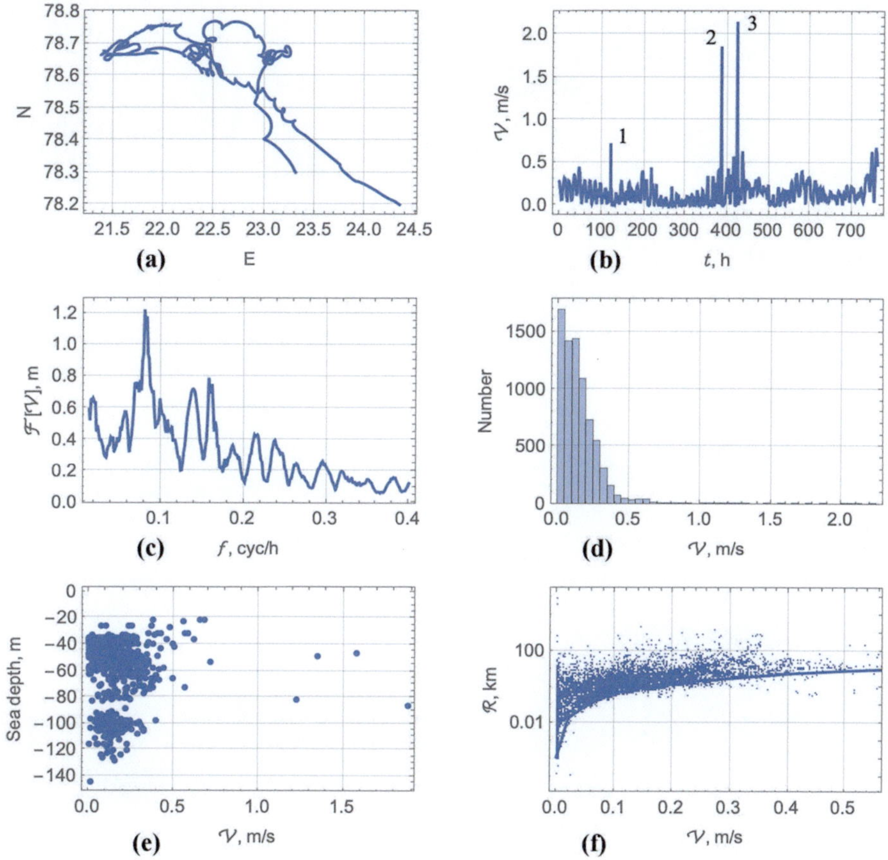

Fig. 2.105 2016b. **a** The ITb trajectory in the BS from May 3 to June 4. **b** The ITb drift velocity depending on time. **c** Fourier spectrum of the ITb drift velocity. **d** Histogram of the ITb drift velocity. **e** The ITb drift velocity depending on local sea depth. **f** Radius of curvature of the ITb drift trajectory depending on the drift velocity

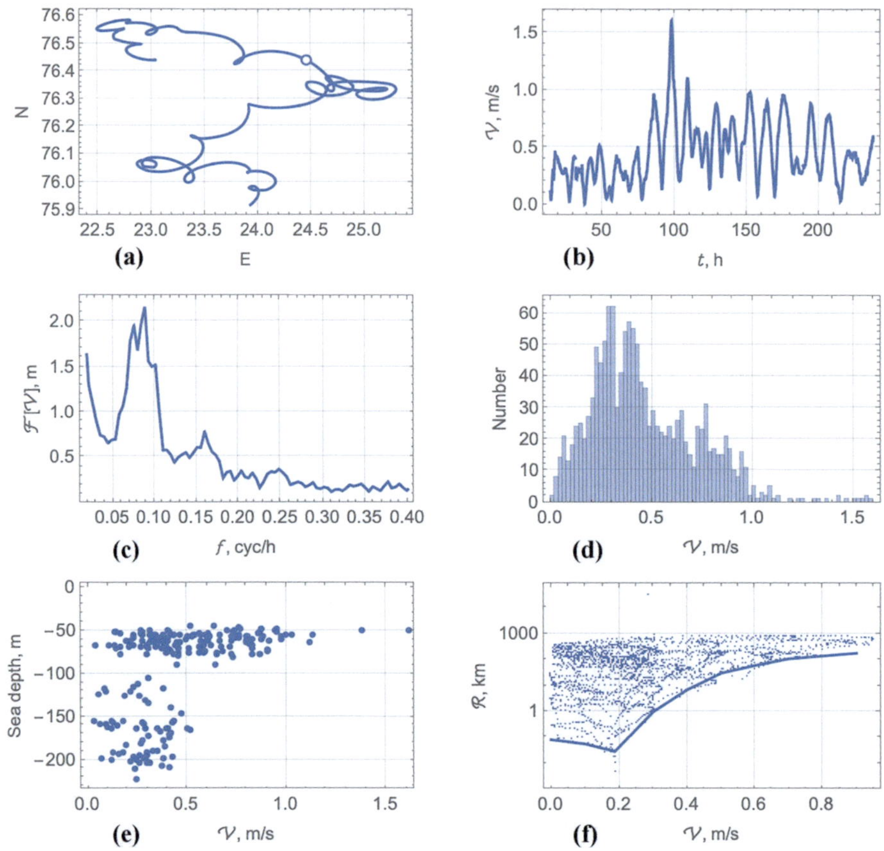

Fig. 2.106 2017b. **a** The ITb trajectory in the SpB from April 25 to May 5. **b** The ITb drift velocity depending on time. **c** The Fourier spectrum of the ITb drift velocity. **d** Histogram of the ITb drift velocity. **e** The ITb drift velocity depending on local sea depth. **f** Radius of curvature of the ITb drift trajectory depending on the drift velocity

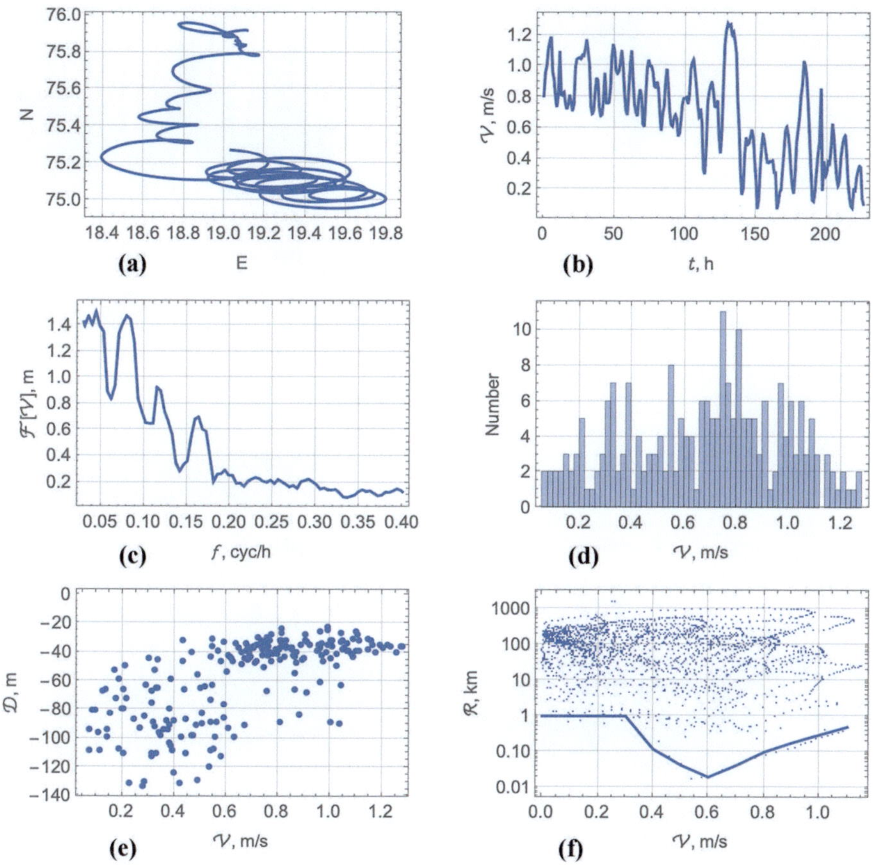

Fig. 2.107 2018a. **a** The IT trajectory in the SpB from April 27 to May 5. **b** The IT drift velocity depending on time. **c** Fourier spectrum of the IT drift velocity. **d** Histogram of the IT drift velocity. **e** The IT drift velocity depending on local sea depth. **f** Radius of curvature of the IT drift trajectory depending on the drift velocity

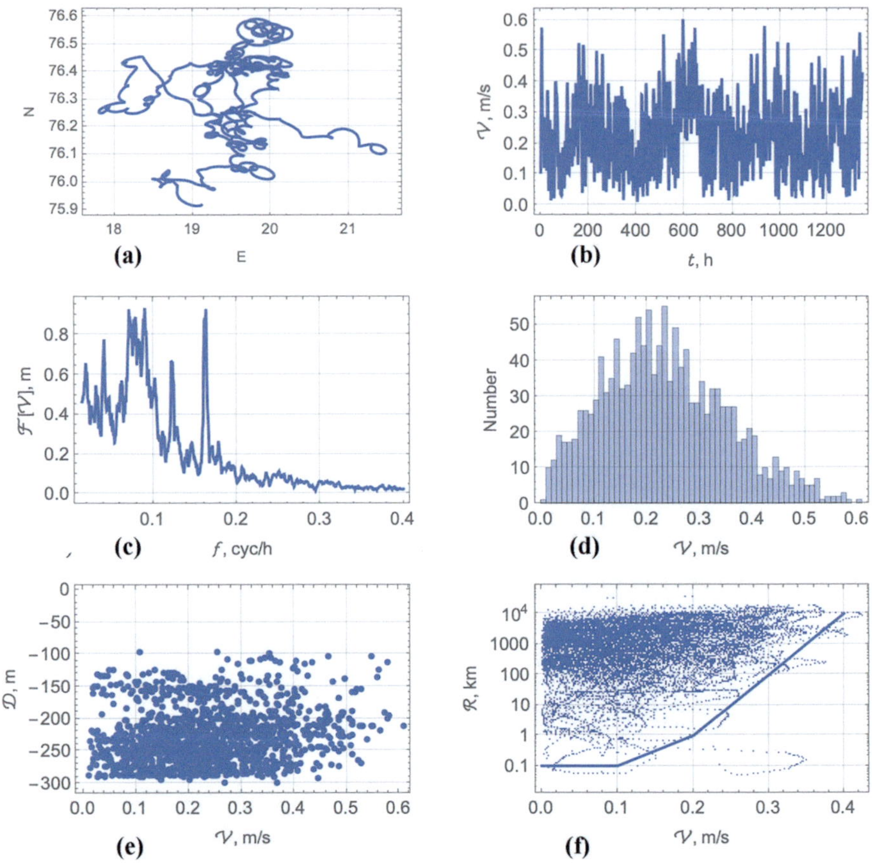

Fig. 2.108 2018b. **a** The IT trajectory in the SFW from May 5 to June 30. **b** The IT drift velocity depending on time. **c** Fourier spectrum of the IT drift velocity. **d** Histogram of the IT drift velocity. **e** The IT drift velocity depending on local sea depth. **f** Radius of curvature of the IT drift trajectory depending on the drift velocity

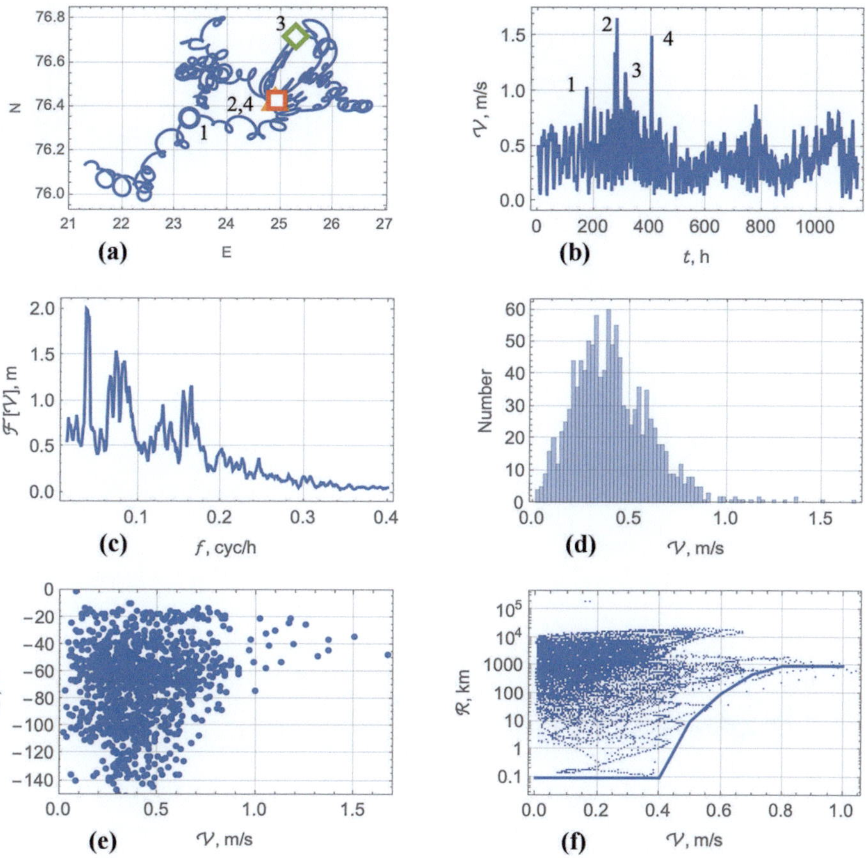

Fig. 2.109 2018c. **a** The IT trajectory in the SpB from June 30 to August 16. **b** The IT drift velocity depending on time. **c** Fourier spectrum of the IT drift velocity. **d** Histogram of the IT drift velocity. **e** The IT drift velocity depending on local sea depth. **f** Radius of curvature of the IT drift trajectory depending on the drift velocity

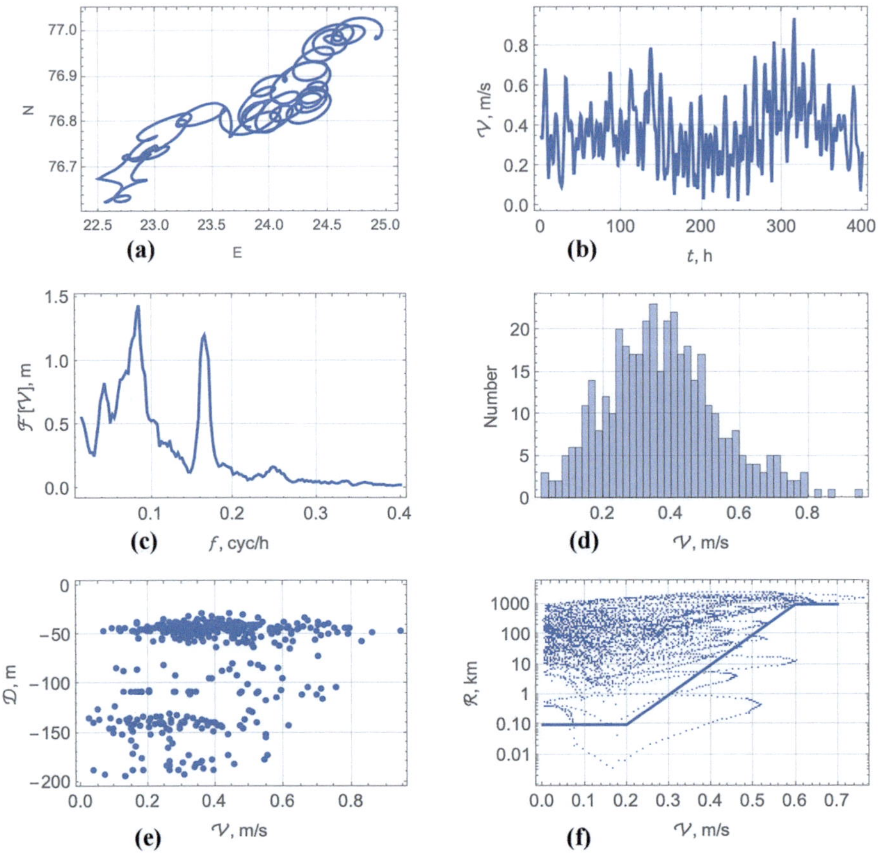

Fig. 2.110 2018d. **a** The IT trajectory in the SFE from August 16 to September 2. **b** The IT drift velocity depending on time. **c** Fourier spectrum of the IT drift velocity. **d** Histogram of the IT drift velocity. **e** The IT drift velocity depending on local sea depth. **f** Radius of curvature of the IT drift trajectory depending on the drift velocity

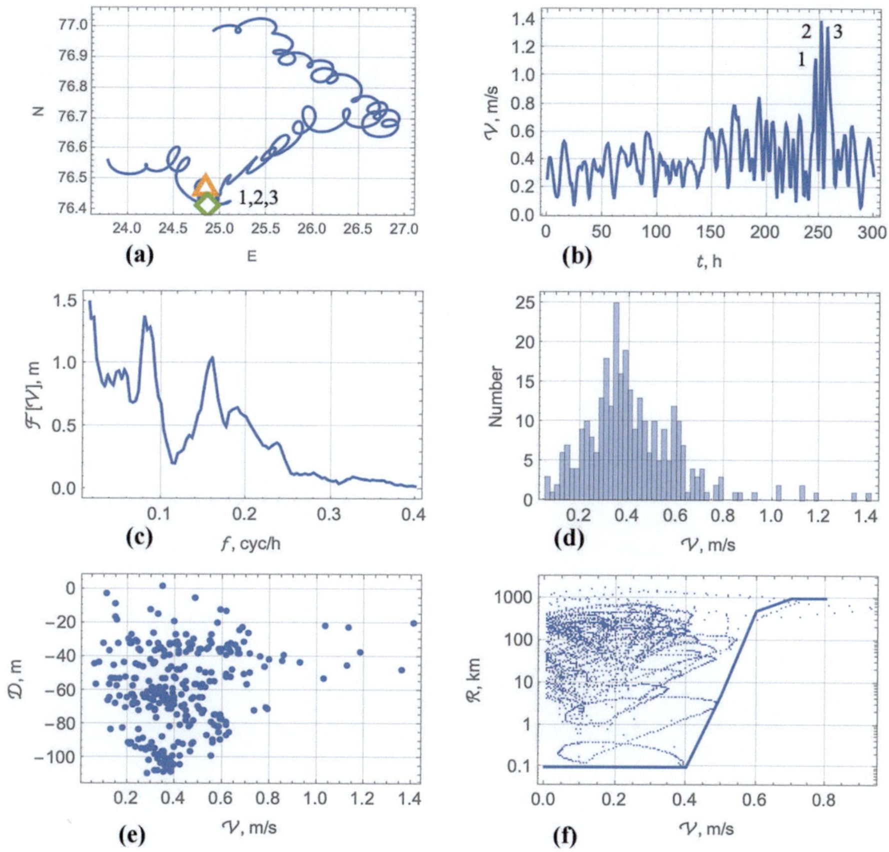

Fig. 2.111 2018e. **a** The IT trajectory in the SpB from September 2 to September 15. **b** The IT drift velocity depending on time. **c** Fourier spectrum of the IT drift velocity. **d** Histogram of the IT drift velocity. **e** The IT drift velocity depending on local sea depth. **f** Radius of curvature of the IT drift trajectory depending on the drift velocity

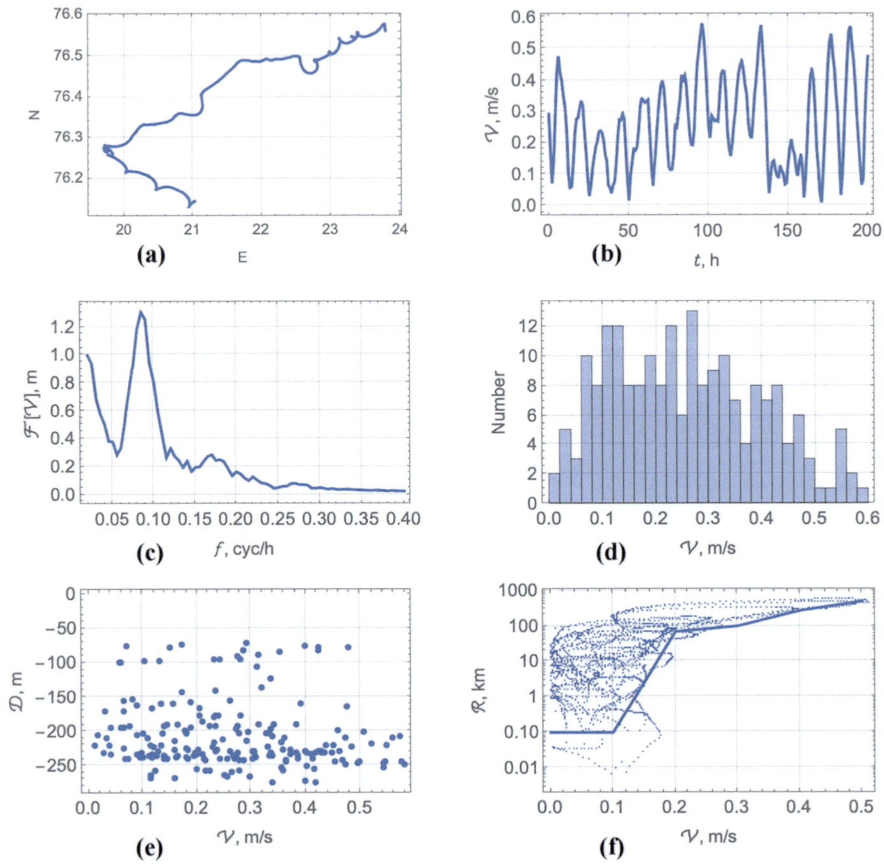

Fig. 2.112 2018f. **a** The IT trajectory in the SFE from September 15 to September 23. **b** The IT drift velocity depending on time. **c** Fourier spectrum of the IT drift velocity. **d** Histogram of the IT drift velocity. **e** The IT drift velocity depending on local sea depth. **f** Radius of curvature of the IT drift trajectory depending on the drift velocity

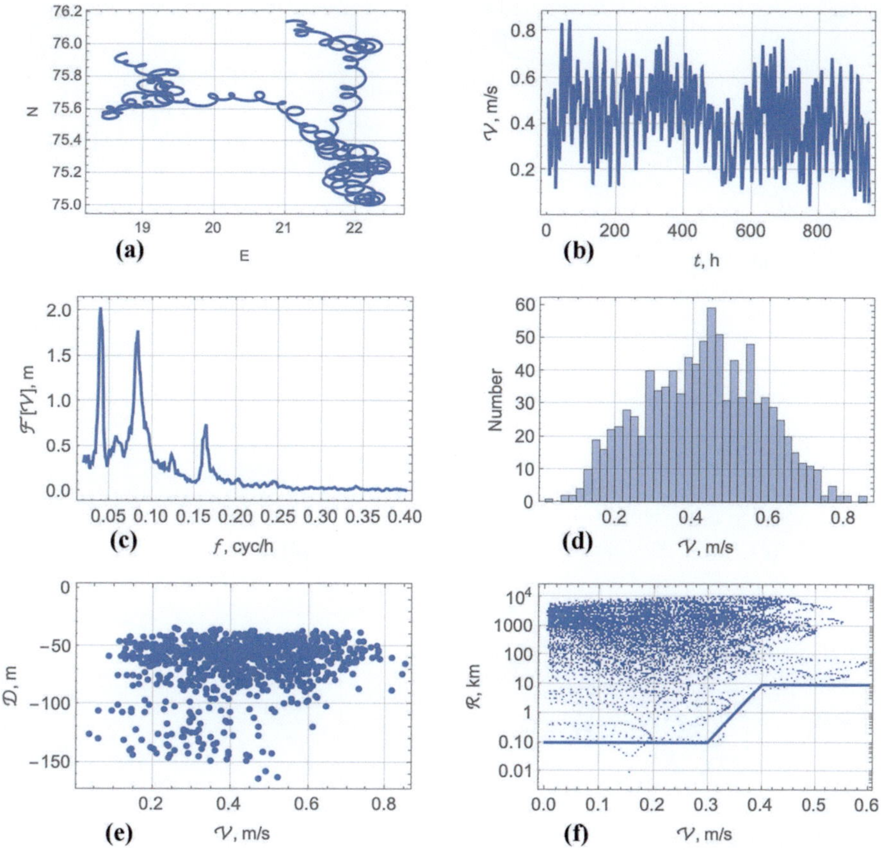

Fig. 2.113 2018g. **a** The IT trajectory in the SFE from September 23 to November 1. **b** The IT drift velocity depending on time. **c** Fourier spectrum of the IT drift velocity. **d** Histogram of the IT drift velocity. **e** The IT drift velocity depending on local sea depth. **f** Radius of curvature of the IT drift trajectory depending on the drift velocity

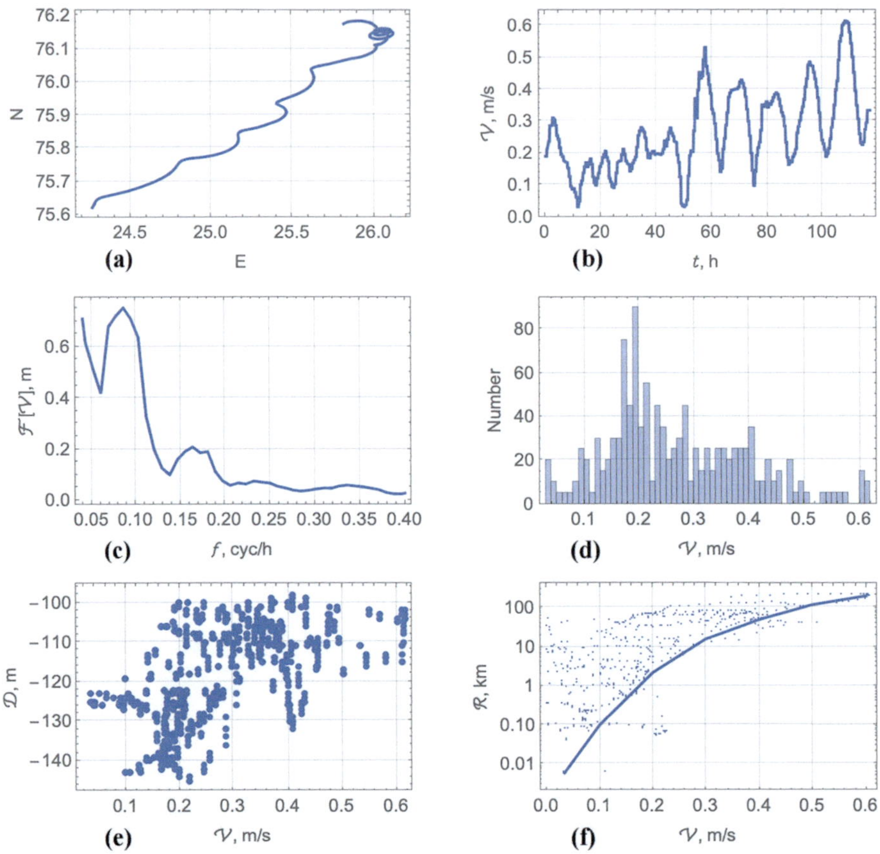

Fig. 2.114 2019. **a** The IT trajectory in the SpB from April 27 to May 2. **b** The IT drift velocity depending on time. **c** Fourier spectrum of The IT drift velocity. **d** Histogram of the IT drift velocity. **e** The IT drift velocity depending on local sea depth. **f** Radius of curvature of the IT drift trajectory depending on the drift velocity

References

Albanov VI (1917) Southward, to the Franz Josef Land, Petrograd, Tipografia Morskogo Ministerstva

Barr W (1975) South eo Zemlya-Frantsa-Iosifa—Cruise of Sv Anna and Albanovs Sledge Journey, 1912–14. Canadian Slavonic Papers-Revue Canadienne des Slavistes 17:567–595

Buzin I, Nesterov A, Gudoshikov YP et al (2020) Use of the GPS beacons for the aims of drift study of the ice cover and icebergs in 2012–2017. In: Conference: XVII glaciological symposium. AARI, St. Petersburg, Russia

Divine DV, Steer A, Granskog MA et al (2025) Data on physical properties of sea ice in the northern Barents Sea and adjacent Arctic Basin from the Nansen Legacy Project. Geosci Data J 12:e70001

Dmitriev NY, Nesterov AV (2007) Iceberg drift in the Barents Sea according to the observation data and simulation results. The Seventeenth International Offshore and Polar Engineering Conference. International Society of Offshore and Polar Engineers, Lisbon, Portugal

Frolov IE, Gudkovich ZM, Karklin VP et al. (2009) Climate change in Eurasian Arctic shelf seas. Centennial ice cover observations. Springer Praxis Books. Springer-Verlag Berlin and Heidelberg GmbH & Co., 166 p.

Keghouche I, Bertino L, Lisaeter KA (2009) Parameterization of an iceberg drift model in the Barents Sea. J Atmosph Oceanic Technol 26:2216–2227

Kowalik Z, Marchenko A (2023) Tidal motion enhancement on Spitsbergen Bank, Barents Sea. J Geophys Res Oceans 128: e2022JC018539

Marchenko AV, Kulyakhtin SA, Eik K (2010) Icebergs drift in the Barents Sea: data analysis of ice tracking buoy and numerical simulations. In: 20th IAHR symposium on ice. Lahti, Finland

Marchenko AV, Kvamstad B, Fjørtoft K et al (2011) Characteristics of ice drift in the Western Barents Sea reconstructed by the data of the trackers deployed on drifting ice in 2008 and 2010. In: Port and ocean engineering under arctic conditions, Montreal, Canada

Marchenko A, Kvamstad B, Fjørtoft KE, Holmen J (2013) Characteristics of ice drift in the western Barents Sea. Fram Forum 34–38

Marchenko A, Kowalik Z (2017) Investigation of ocean currents in navigational straits of Spitsbergen. In: Marine navigation. Marine navigation and safety of sea transportation. CRC Press

Marchenko A, Marchenko N (2019) Characteristics of ice drift and waves on Spitsbergen-banken. In: The 25th International conference on port and ocean engineering under arctic conditions. Delft, the Netherlands

Marchenko A, Diansky N, Fomin V (2019) Modeling of iceberg drift in the marginal ice zone of the Barents Sea. Appl Ocean Res 88:210–222

Marchenko A, Zenkin A, Marchenko N et al (2020) Monitoring of 3D motion of drifting iceberg with an ice tracker equipped with accelerometers. In: The 25th IAHR international symposium on ice. Trondheim, Norway

Marchenko A, Turnbull I, Kowalik Z et al (2021a) Properties of drift ice and sea currents on Spitsbergen Bank. In: The 26th International conference on port and ocean engineering under arctic conditions. Moscow

Marchenko AV, Morozov EG, Kozlov IE, Frey DI (2021b) High-amplitude internal waves southeast of Spitsbergen. Continental Shelf Res 227:104523

Marchenko A (2022) Thermo-hydrodynamics of sea ice rubble. In: Tuhkuri J, Polojärvi A (eds) IUTAM Symposium on physics and mechanics of sea ice. Springer International Publishing, pp 203–223

Marchenko A, Kowalik Z (2022) Tidal wave—Elliptic island interaction above the critical latitude. J Phys Oceanogr 53:683–698. https://doi.org/10.1175/JPO-D-22-0018.1

Marchenko NA, Marchenko AV (2015) Sea currents and ice drift in western part of Barents Sea. A comparison of data from floating and fixed on ice buoys. In: The 23rd international conference on Port and ocean engineering under arctic conditions (POAC 2015). Trondheim

Marchenko N (2020) Marginal ice zone (MIZ) field investigation in the Western Barents Sea in April 2017–2019. In: The 25th IAHR international symposium on ice. Trondheim

Mironov EU (2004) Ice conditions in the Greenland and Barents seas and their long-term forecast. St. Petersburg, AARII [in Russian]

Nansen F, Sverdrup ON (1897) Fram over polhavet: den norske polarfaerd 1893–1896, H. Aschehoug

NOAA (2025) Global drifter program. https://www.aoml.noaa.gov/global-drifter-program/. Accessed Sept 2025

Norwegian Hydrographic Service (2018) The Norwegian pilot guide. Saling direction. Svalbard and Jan Mayen. Third Edition Stavanger, NHS and NPI

Pease CH, Turet P, Pritchard RS (1995) Barents Sea tidal and inertial motions from Argos ice buoys during the Coordinated Eastern Arctic experiment. J Geophys Res Oceans 100:24705–24718

Petermann A (1865) Spitzbergen und die arktische central-region: Eine reihe von aufsätzen und karten als beitrag zur geographie und erforschung der Polar-regionen, Gotha, Justus Perthes

Rabault J, Sutherland G, Gundersen O, Jensen A, Marchenko A, Breivik Ø (2020) An open source, versatile, affordable waves in ice instrument for scientific measurements in the Polar Regions. Cold Reg Sci Technol 170:102955

Rabault J, Müller M, Voermans J et al (2023) A dataset of direct observations of sea ice drift and waves in ice. Sci Data 10:251

Spring W (1994) Ice data acquisition summary report. Mobil Research and Development Corporation, Dallas, Texas

Turnbull ID, Marchenko AV (2022) Deformation of an ice pack influenced by waves and topographic enhancement of tidal motion near Hopen Island in the Barents Sea. Cold Reg Sci Technol 194:103463

Vinje T (1978) Sea ice conditions and drift of Nimbus-6 buoys in 1977. Norsk Polarinstitutt Årbok 1977:10

Vinje TE (1979) Sea ice conditions and drift of Nimbus-6 buoys in 1978. Årbok 1978. Norsk Polarinstitutt, Oslo

Vinje T (1985) Physical environment, Western Barents Sea. Drift, composition, morphology and distribution of the sea ice fields in the Barents Sea Oslo, Norsk polarinstitutt

Vinje T, Kvambekk AS (1991) Barents Sea drift ice characteristics. Polar Res 10:59–68

Yulmetov R, Marchenko A, Løset S (2013) Ice drift and sea current analysis in the Northwestern Barents Sea. In: Proceedings of the international conference on port and ocean engineering under arctic conditions, Espoo, Finland

Zubakin GK (2006) Ice formations of the seas of Western Arctic, St. Petersburg, AARI [in Russian]

Zubakin GK, Gudoshnikov YP, Naumov AK et al (2007) Results of investigations of icebergs, glaciers and their frontal zones in the northeastern part of the Barents Sea. Dalian Univ Technol Press, Dalian

Chapter 3
Internal Waves

3.1 General Notes and Introduction

Arctic regions, especially in the Arctic seas of Russia are characterized by a wide shelf and low surface temperatures. Ice cover is an important factor for generating internal waves because the Garrett-Mank spectrum of internal waves is partly maintained by generating inertial oscillation by wind and further transfer of energy to higher frequences of the spectrum (Sugiyama et al. 2009). Ice cover prevents wind forcing at the surface. Low salinity due to ice melting and river discharge is a characteristic feature of surface waters. In some regions warm and saline Atlantic waters spread in the deep layers of the Arctic Ocean. This spreading combined with the freshened surface waters facilitate strong stratification, which is an important property for internal wave existence and generation.

Internal tides transform the velocity field of the barotropic tide so that relatively low tidal velocities form a velocity field with higher horizontal velocities and amplitudes of the vertical displacements, which are confined to the regions of sharp changes in the bottom topography.

It is known that tidal internal waves are generated in the ocean when currents of the barotropic tide flow over irregularities of bottom topography (Prinsenberg et al. 1974; Baines 1982; Garrett and Kunze 2007). The vertical components of currents, which appear when the tidal currents flow over submarine slopes, cause periodical displacements of isopycnal surfaces. The main energetic frequency of internal tide is the frequency corresponding to the barotropic tide M_2 with a period of 12.4 h.

Many properties of internal tides depend on latitude. Let us show this on the example of a differential equation for vertical velocities w caused by internal waves (LeBlond and Mysak 1978):

$$\frac{d^2w}{dz^2} + \frac{N^2(z)}{g}\frac{dw}{dz} + \frac{N^2(z) - \omega^2}{\omega^2 - f^2}k^2 w = 0 \tag{3.1}$$

A. Marchenko et al., *Geophysical Hydrodynamics of Currents, Waves, and Ice in the Atlantic Arctic*, https://doi.org/10.1007/978-3-032-14122-4_3

This Eq. (3.1) is deduced for a flat bottom. This is an equation with the boundary conditions for the vertical velocity, which we assume zero at the surface and bottom. Here, k is the wavenumber. The denominator in the third term of this equation is important for the problems we discuss here. As the latitude increases the difference between the M_2 tidal frequency and the Coriolis parameter decreases (ω is the frequency of the M_2 tide). Approximately at latitude $\phi = 74°30'$, this difference becomes equal to zero, which excludes oscillatory solutions of Eq. (3.1) in the linear formulation.

Thus, latitudes close to $74°30'$ are critical for the internal tides with a period of 12.4 h (M_2 frequency). However, the process of their generation at high latitudes does not differ from a similar process at low latitudes. Namely, the currents of the barotropic tide flow over underwater slopes and periodically displace isopycnal surfaces up and down due to the appearance of vertical components of these currents near submarine slopes (Baines 1982; Konyaev and Sabinin 1992). The differences between the behavior of internal tides at lower and higher latitudes appear because in the region of super-critical latitudes, internal wave with frequency M_2 becomes a forced wave over a submarine slope and is not be able to spread from the slope as a free propagating wave as is the case at lower latitudes. Rotation prevents its spreading from the slope.

A description of one of the first observations of internal tides at high latitudes is given in a well-known monograph by Krauss (1966). The author describes the measurements during several days from an anchored ship at a station in the Barents Sea and shows an example of the existence of internal tides north of 75° N. Despite the theoretical result that free internal waves with a tidal period cannot exist north of the critical latitude (Krauss 1966), such waves were observed (Konyayev et al. 2000, Levine et al. 1985; Parsons et al. 1996; Padman et al. 1992; Plueddemann et al. 1998).

A decrease in the efficient Coriolis frequency is one of the possible mechanisms that make possible the existence of free semidiurnal internal tides of the M_2 period. This decrease can appear due to the background negative (anticyclonic) vorticity. If the efficient Coriolis frequency becomes smaller than the M_2 tidal frequency, the existence of internal tides north of the theoretical critical latitude becomes possible. It was shown by Kozubskaya et al. (1999) that negative vorticity up to $5.5 \cdot 10^{-6}$ s^{-1} observed in the ocean can shift the efficient critical latitude up to the latitudes of Spitsbergen (78° N).

It was shown in Vlasenko et al. (2003) that at high latitudes nonlinearity should be high to allow generation of internal waves otherwise, nonlinear effects are suppressed by rotation. The authors demonstrate that such generation is possible if the problem has a nonlinear formulation. They suggested a numerical nonlinear model, which allows generation of internal tide at high latitudes and its breaking into packets of short-period internal wave during a specific phase of internal tide, when conditions for its breaking are formed.

Semidiurnal internal tides are frequently observed over underwater slopes. It was shown by Pisarev (1988, 1991, 1992), Kurkina and Talipova (2011) that the semidiurnal oscillations are observed in the Arctic regions over underwater slopes, while only packets of short-period waves are observed far from the slopes over relatively flat bottom. This is confirmed by a model presented by Kurkina and Talipova (2011) and by observations reported by Morozov and Paka (2010). Influence of the critical latitude on tidal internal waves of frequency M_2 (period 12.4 h) is shown on the example of moored measurements. The critical latitude for internal waves M_2 is latitude 74.5°N. North of this latitude the temperature spectra reveal only the S_2 peak (period 12.0 h), while the current spectra show two peaks M_2 and S_2.

The problem of existence of internal tides near the critical latitude has been analyzed by Morozov and Pisarev (2023) based on a numerical model developed by Vlasenko (1992). As a result of calculations and analysis, a map of the amplitudes of semidiurnal internal tidal waves in the Arctic seas has been constructed. The maximum amplitudes of internal tidal waves occur over underwater slopes.

Very large amplitudes of internal tides were found in the Kara Gates. Internal waves in the Kara Gates Strait were studied on the basis of moored measurements and towed CTD-profilers in the scanning mode. Internal waves turned out to be extremely large and their dynamics is close to the similar waves in the Strait of Gibraltar. Waves with heights of up to 40 m are directed towards the Barents Sea. The opposing current east of the strait decreases the wavelength, and, consequently, the amplitude of internal waves increases.

3.2 Influence of Critical Latitude (74°30′ N)

Let us consider how critical latitudes influence internal tides north of the critical latitude (74°30′ N). We will analyze the results of modeling the internal waves in the region of the Arctic shelf and discuss the numerical experiments with varying parameters reported by Morozov and Pisarev (2002). We used a numerical model developed by Vlasenko (1992), Vlasenko and Hutter (2002), and Vlasenko et al. (2005).

We shall choose basic parameters and change them comparing the new solution with the basic solution. Let us specify a strong stratification corresponding to a warm and freshened upper layer with density $\sigma_t = 23.5$ and denser water at the bottom ($\sigma_t = 27.5$) with a sharp density interface at a depth of 10 m corresponding to the Brunt–Väisälä frequency $N = 0.06$ cycle/s. The mean Brunt–Väisälä frequency over the whole depth (60 m) is 0.04 cycle/s. Such density stratification can appear when the temperature and salinity at the surface are equal to 7 °C and 30 psu, and the corresponding values at the bottom are equal to 1 °C and 34 psu, respectively. The tidal flow was specified so that the altitude of the surface elevation at the coast was equal to 1 m. The maximum tidal velocities over the slope were approximately 20 cm/s. The density contour lines obtained in the model calculation after eight periods of semidiurnal oscillations are shown in Fig. 3.1.

Analysis of the vertical structure of the generated waves shows that the vertical motions of water particles occur in-phase, while the largest deviations of the isopycnals occur in the vicinity of the sharp pycnocline and near the sharp deepening of

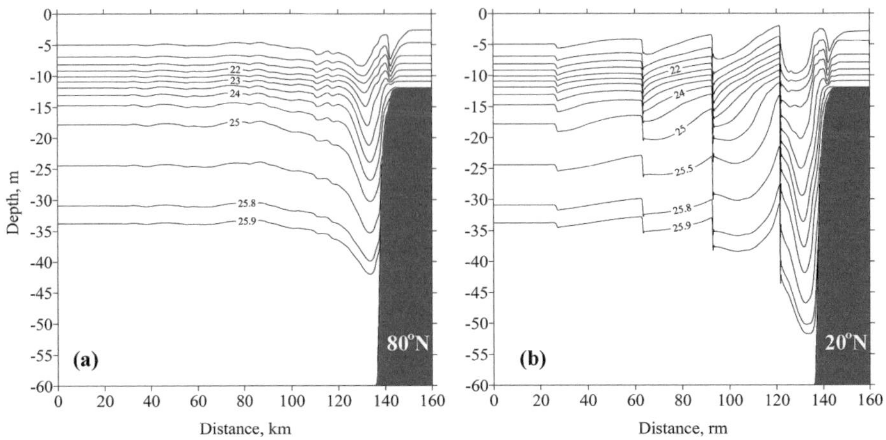

Fig. 3.1 Field of the isopycnal surfaces at different latitudes (all other parameters are the same). The slope of the bottom is 6 10^{-3}, and the Brunt–Väisälä frequency at the depth of the density interface is equal to 6 10^{-2} cps: **a** latitude 80°, **b** latitude 20°

the bottom, where they can reach 10 m. In this numerical experiment, the latitude of the region was equal to 80° N (Fig. 3.1a).

We performed calculations of the perturbations of the density field by internal waves at the same parameters but at latitude 20º. The corresponding graphs are shown in Fig. 3.1b. The region at 20° is very far from the critical latitude; therefore, intense internal tides are developed here, which propagate from the slope with a lesser decay.

High amplitude forced internal waves of tidal period (12.4 h) develop over the slope but they do not propagate from the slope due to the restriction of such propagation north of 74.5° N. However, a shorter wave packet is formed every 12.4 h that runs from the slope because there is no restriction for the existence of free propagating waves of shorter periods. Evolution of this wave packet is shown in Fig. 3.2. Three patterns of the wave perturbations of the density field are shown with a time step of 1/20 of the tidal period 12.4 h, which is 0.62 h. The distance traveled by the short-period internal wave during this period is ~2.1 km. Hence, the wave speed is $c = 0.94$ m/s, which is a reasonable value for the internal wave speeds. The wave length of the short-period internal wave estimated from the figure is 333 m. Hence the wave period is $T = L/c = 354$ s ≈ 5 min, which is several times greater than the period corresponding to the mean Brunt–Väisälä frequency.

One can see from Fig. 3.2 that the fluctuations in the short-period wave are in phase over the depth. We assume that the wave is of the first mode. We can estimate the velocity of internal wave from a simple dispersion relation.

$$c = \frac{NH}{\pi} = \frac{0.04 \times 60}{3.14} = 0.76\text{m/s}. \tag{3.2}$$

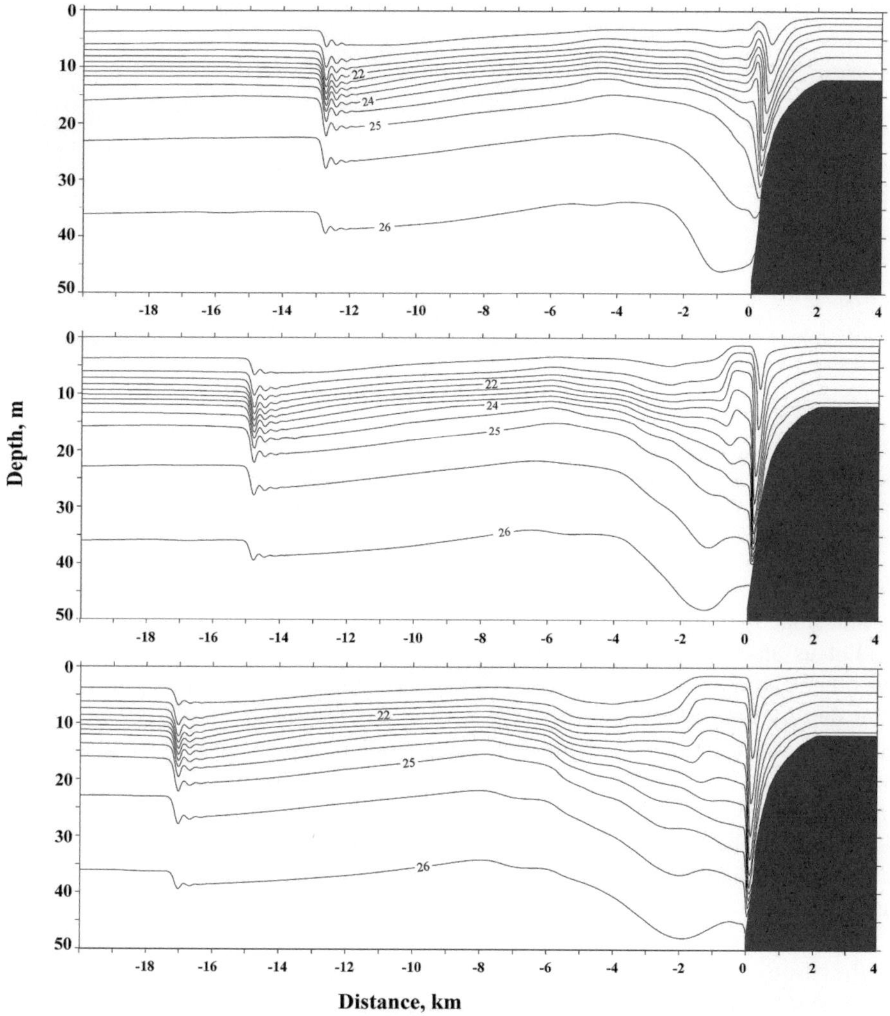

Fig. 3.2 Evolution of the density field shown in Fig. 3.1a based on model simulations. The variations in the density field are shown from top to bottom with a time step of 0.62 h

Then, the velocity of the short-period internal wave appears close to the one measured from the model simulation.

Let us analyze how parameters of the simulation influence the resulting density field. First, we change the slope and make the slope of the bottom 1 10^{-3} (Fig. 3.3b) instead of 6 10^{-3} (Fig. 3.3a). Comparison of two simulations are shown in Fig. 3.3. All other parameters of simulation are the same. One can see that density perturbations over the slope are smaller in the case of a gentler slope of the bottom. This is quite reasonable because in the case of a steep slope greater amplitudes of vertical displacement and greater vertical velocities are generated when tidal currents encounter a

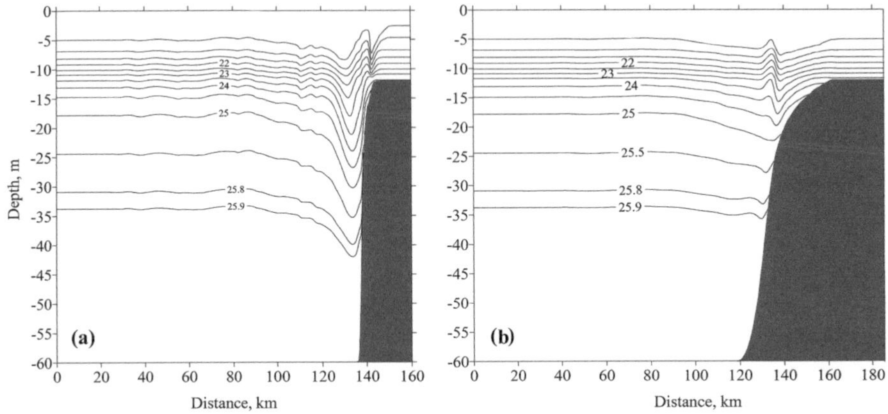

Fig. 3.3 Comparison of variations in the density field caused by different slope. The slope of the bottom is 6 10^{-3} **a** and 1 10^{-3} **b**; the Brunt–Väisälä frequency is the same at the depth of the density interface (6 10^{-2} cps), tidal velocities and latitude are the same 80° N

submarine slope. Tidal currents induce longer excursions of water particles over the gentle slope than those over a steeper one.

Let us also analyze how the density field changes if we change stratification all other parameters remaining the same. The simulations are shown in Fig. 3.4. The density near the bottom at 60 m was assumed the same 1.027 g/cm^3. In the previous cases the density at the surface was 1.023 g/cm^3. In this simulation we assumed that it is 1.007 g/cm^3, which corresponds to a lower salinity in the upper layer caused, for example, by ice melting or river discharge. This surface density is realistic in the northern seas of Russia. Simulation with stronger stratification revealed the displacements of isopycnals became smaller due to the resistance of stratification to the tidal forcing over a sloping bottom.

We also note an obvious fact. The greater the mass flux transported by the currents of the barotropic tide through a section parallel to the slope the greater the amplitudes of the internal tide.

3.3 Spectral Peaks of Semidiurnal Tides in the Arctic Ocean

Measurements on moorings north of the critical latitude allow us to analyze the influence of the critical latitude. We analyzed data from several moorings in the Arctic Ocean. A chart of these moorings is shown in Fig. 3.5.

Measurements at several moorings in the Arctic Ocean show that internal tides are low. A mooring was deployed in 2002 in the Canada Basin at 73°30' N, 137° W (Timmermans et al. 2007). Vertical motions with periods close to f and M_2 with amplitudes 10–20 m were recorded at a depth of 2500 m. Moored measurements were carried out almost exactly at the North Pole from April 2001 to April 2008. The

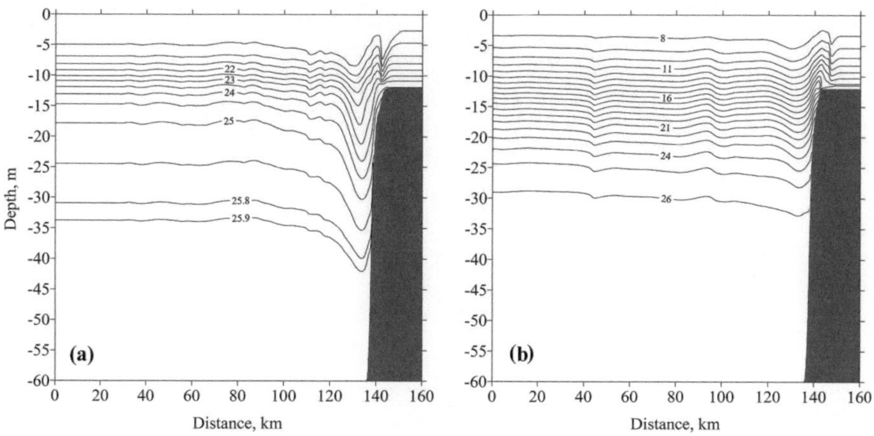

Fig. 3.4 Comparison of variations in the density field caused by different stratification. The density field in **a** is the same as in the previous simulations with the slope equal to 6 10^{-3} **a**; the Brunt–Väisälä frequency in **a** is the same at the depth of the density interface (6 10^{-2} cps), tidal velocities and latitude are the same 80° N in both panels. Density stratification in panel **b** is stronger. Water density at the surface was specified 1.007 g/cm^3, which at the bottom it remained 1.027 g/cm^3

moorings were replaced every year. As expected, peaks at the M_2 frequency were almost absent. Possible generation of internal tides occurred close to the ice surface owing to uneven ice surface at the boundary with water. Judging from the vertical temperature gradient the vertical amplitudes of the internal tide on all moorings did not exceed 10 m.

Twelve moorings were deployed at the eastern coast of Greenland in 1987–1989. Some of the moorings were located over the continental slope, while the others were in the open sea. At the slope the amplitudes of internal tides were estimated at 35–40 m. In the open sea, the amplitudes at 200–215 m over the oceanic depth of 3600–3700 m were in the range 15–20 m. The peaks near the slope were at the M_2 and S_2 frequencies, while only the S_2 peaks were found over the flat bottom.

Moored measurements near the Lomonosov Ridge were carried out from August 1995 to August 1996. Three moorings were deployed in the region (Fig. 3.5). The depths of the ocean over the slope were ~1700 m.

Internal tides of the M_2 frequency (12.42 h) cannot exist at these latitudes; however, this restriction is not related to the internal tides of the S_2 frequency (12.0 h). The measurements on the northern mooring (81° N) confirm this rule. The spectra of velocity and temperature fluctuations at a depth of 712 m are shown in Fig. 3.6. Two peaks at frequencies of M_2 and S_2 related to the currents of the barotropic tide are found on the velocity spectra. Only the peak at the S_2 frequency was found on the temperature spectra. Insignificant peaks were found on the temperature spectra at several deep levels of the measurements because the moorings were deployed on the slopes but not on a flat bottom.

Let us consider temperature and velocity measurements on moorings almost exactly at the North Pole. The moorings were deployed from 2011 to 2008. There

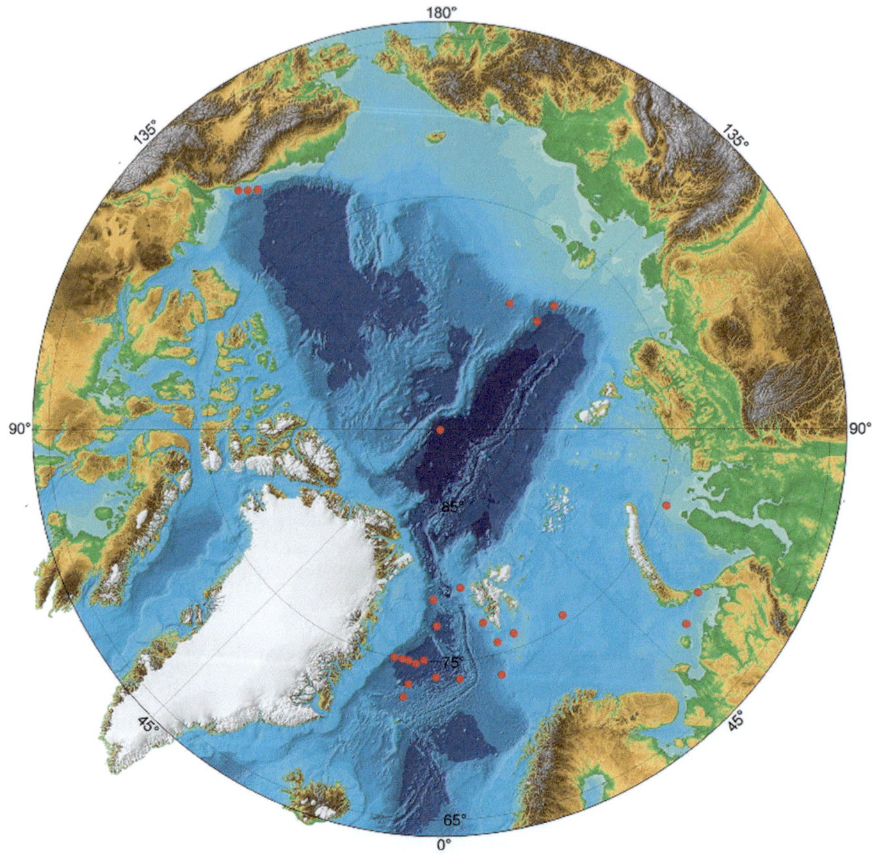

Fig. 3.5 Chart of moorings in the Arctic Ocean. Data from these moorings were used in the study

were 6 sequential deployments longer than one-year each with change of the equipment. We will consider continuous measurements on one mooring from April 2006 to November 2008. The depths of the measurements were from 88 to 4306 m over the relatively flat ocean depth close to 4315 m. The moorings were deployed in the Amundsen Basin close to the Lomonosov Ridge but not on the slopes of the ridge. Fifteen instruments measuring velocity, temperature, pressure, and salinity were attached to the mooring wire. Vertical profile of the mean temperature over 718 days are shown in Fig. 3.7. A temperature increase is seen at depths of approximately 300 m because a warm and saline jet of Atlantic waters flows closely to the North Pole over the slopes of the Lomonosov Ridge.

Spectra of temperature and velocity fluctuations from a mooring at the North Pole at a depth of 1029 m are shown in Fig. 3.8. Spectral peaks on the velocity spectra at the M_2 and S_2 frequencies are clearly seen. However, no peaks are seen at tidal frequencies on the temperature spectra, which reflect temperature fluctuations related to internal tides. Vertical temperature gradient at 1029 m as seen from Fig. 3.7 is

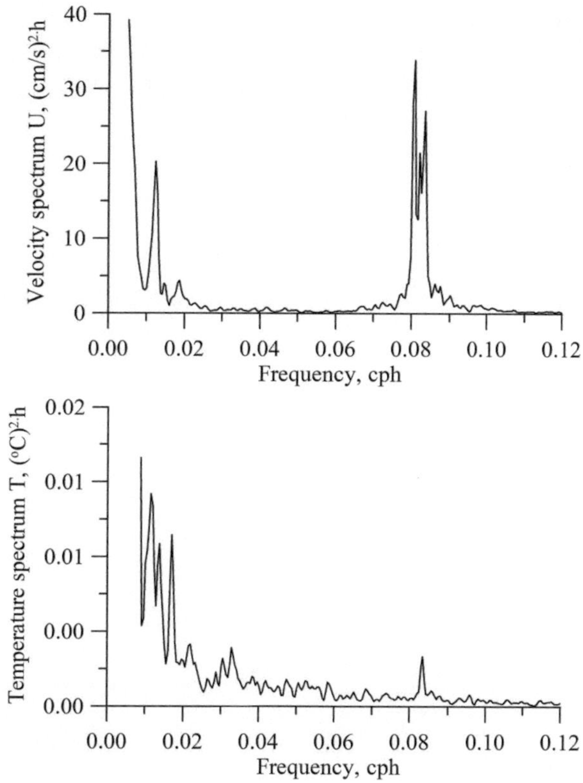

Fig. 3.6 Velocity (top panel) and temperature (bottom panel) spectra from a mooring deployed near the Lomonosov Ridge (81°04’ N, 138°54’ E). Measurements at a depth of 712 m over the ocean depth 1712 m

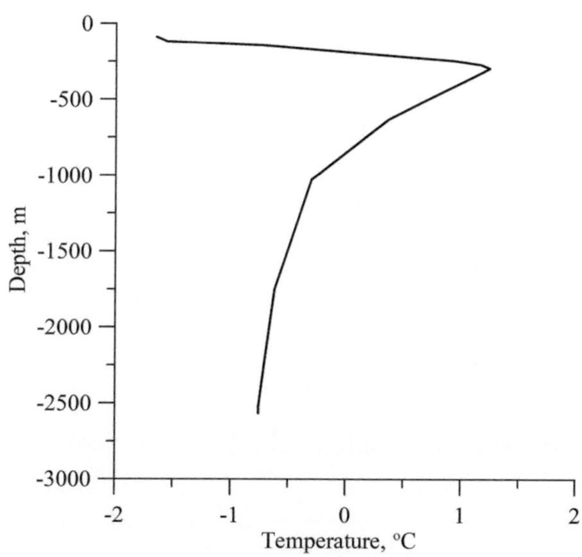

Fig. 3.7 Temperature profile almost at the North Pole (89°20′ N, 77°07′ E)

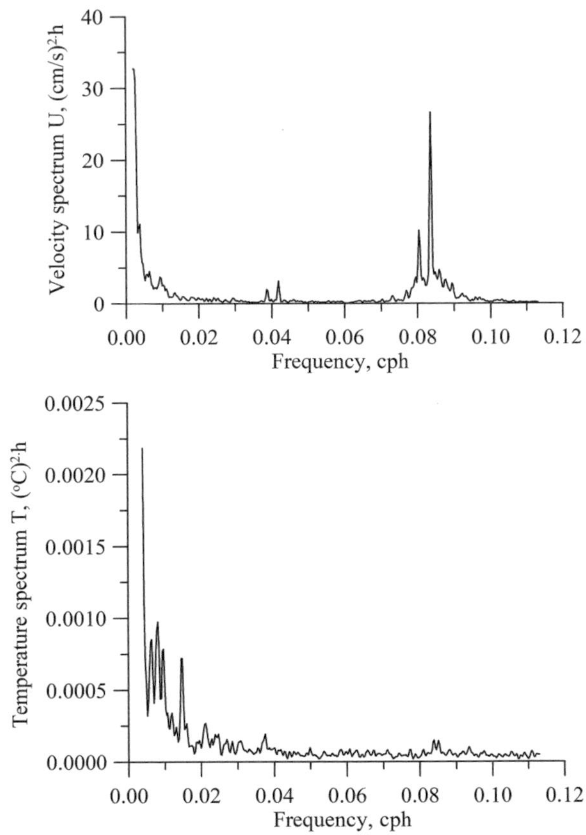

Fig. 3.8 Velocity (top panel) and temperature (bottom panel) spectra from a mooring deployed near the North Pole (89°20′ N, 77°07′ E). Measurements at a depth of 1029 m over the ocean depth 4315 m

far from zero. Slight peaks of low confidence are poorly visible at the S_2 frequency (0.0833 cph, 12.0 h period) on the temperature spectra (Fig. 3.8).

Fluctuations of velocity and temperature change in the upper layer at a depth of 118 m. Horizontal fluctuations with a period of 12.0 h became stronger. The peak at a frequency of the S_2 tide is more reliable, while the peak at M_2 frequency became lower. Only insignificant peaks at the semidiurnal frequency S_2 on the temperature spectra could be found because the latitude of the North Pole is almost critical for internal tides with the S_2 period (Fig. 3.9).

3.4 Internal Tide in the Barents Sea

The geographical location of the arctic shelf in the Barents Sea at latitudes north of 70° N is close to the critical latitudes. Publications in (Morozov et al. 2003a, 2003b; Morozov and Pisarev 2002, 2004; Pisarev 1988, 1996) report the results of studies of internal waves in this region. Internal tide cannot exist close and north of the critical

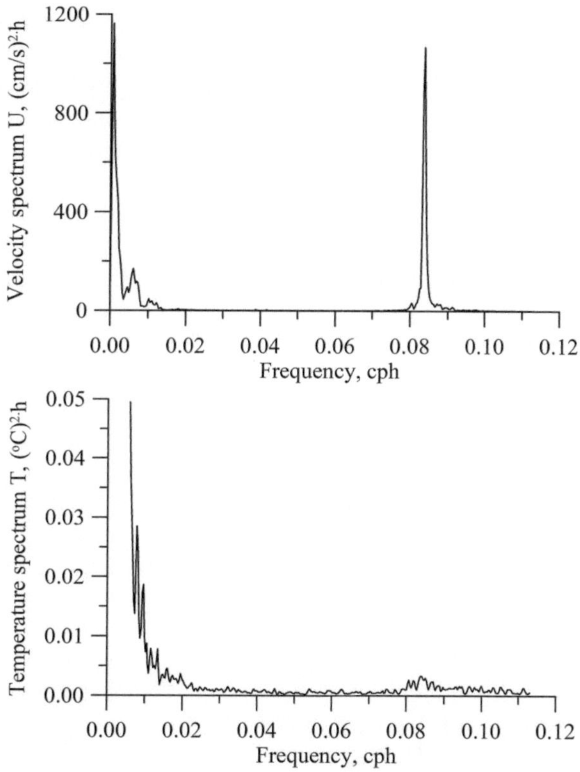

Fig. 3.9 Velocity (top panel) and temperature (bottom panel) spectra from a mooring deployed near the North Pole (89°20′ N, 77°07′ E). Measurements at a depth of 118 m over the ocean depth 4315 m

latitude as a progressive internal wave over a flat bottom. Measurements in many regions in the northern Russian seas reveal generally low amplitudes of internal tides excluding some regions in the Barents Sea.

We applied the numerical model developed by Vlasenko (1992), Vlasenko and Hutter (2002) to study the generation and propagation of internal tidal waves in the Barents Sea and review the approach used in (Morozov et al. 2002a) and review our publication (Morozov et al. 2003c). Characteristic regional oceanographic parameters were used as the input data for the model. Although the model is two dimensional, the equation for the V-component of velocity normal to the x, z-plane is introduced in the model to account for rotation. However, the V-component is considered constant. Internal wave is considered flat and propagating normal to the continental slope.

We specify the density field undisturbed by internal waves, which is characterized by the distribution of the Brunt–Väisälä frequency $N(z)$. A long barotropic tidal wave propagates from the open sea to the continental slope. The tidal currents flowing over the continental slope obtain a vertical component. The periodically changing vertical components of these currents stimulate vertical oscillations of isopycnal surfaces, and internal waves of the tidal period are generated due to this mechanism. Thus, the problem of calculating the parameters of the baroclinic wave perturbations is solved based on the parameters of the barotropic tide, stratification, and bottom topography. The model simulations allow us to estimate the amplitudes of internal tides.

To evaluate the 3D distribution of the amplitude, we used the approach based on the well-known strip theory. This theory is used in aerodynamics to calculate the airflow around the wing of an airplane. If the variations of properties along the transversal coordinate are significantly smaller than along the longitudinal coordinate, the study region is divided into strips with approximately equal properties across the strips and sharply changing along the strips. In our case, the strips for the study were located across the isobaths so that the entire basin of the sea was divided into strips 20–100 km wide and 100–300 km long. Within each strip we specified the model topography of the bottom, characteristic stratification, and currents of the barotropic tide.

Water transport by the barotropic tidal currents in each zone (strip) was specified by recalculating them into the stream function. The currents of the barotropic tide were calculated from the satellite data from the NASA database at Oregon State University (USA). We used the TPXO.9 global inverse tidal model with the Ocean Tidal Prediction Software (OTPS) algorithm (http://volkov.oce.orst.edu/tides/otps.html) based on the satellite data assimilation (Egbert and Erofeeva 2002). The ellipses of tidal currents in characteristic locations of the Arctic seas are shown in Fig. 3.10.

In each of the zones for calculation, we specified the unperturbed density field by means of the Brunt–Väisälä frequency distribution. The density stratification was specified on the basis of the Russian–US Oceanographic Atlas of the Arctic Ocean (Atlas 1998). We used the climatic data averaged approximately over 50 years of observations. The sea bottom was specified on the basis of the ETOPO database (http://web.ngdc.noaa.gov/mgg/global/seltopo.html).

In the model calculations, we used a work field 300 km long with a horizontal step of 50 m and 20 vertical levels. The time step was equal to 2 s. The coefficients of the horizontal eddy viscosity and density diffusivity were specified as 3–4 $m^2 s^{-1}$, and the corresponding vertical coefficients were equal to 0.0001 $m^2 s^{-1}$. Calculations of internal wave propagation in the Arctic basin near the critical latitude require that the rotation dispersion does not suppress the nonlinearity of the waves. Therefore, we had to choose such a small horizontal step.

After dividing the sea basin into a large number of calculation zones (strips) and using profiles of vertical density stratification and tidal currents in each of the zones, we obtained a pattern of the distribution of internal tide amplitudes in the region in the summer period. The calculations were corrected by individual measurements at moored stations in the open sea. The simulated distribution in the Barents Sea is shown in Fig. 3.11a. Different gray tones correspond to the amplitudes of vertical displacement associated with internal tides.

A few regions with large amplitudes of internal tides were distinguished. The regions with the largest amplitudes were found in the Kara Gates Strait. Here, the depth of the sea in the fractures of the ridge crossing the strait is 120–160 m; while the wave amplitude reached 40 m and the corresponding wave height (double amplitude or vertical displacement) was very close to the thickness of the water column from the surface to the bottom. This is an extraordinary effect observed only in some straits,

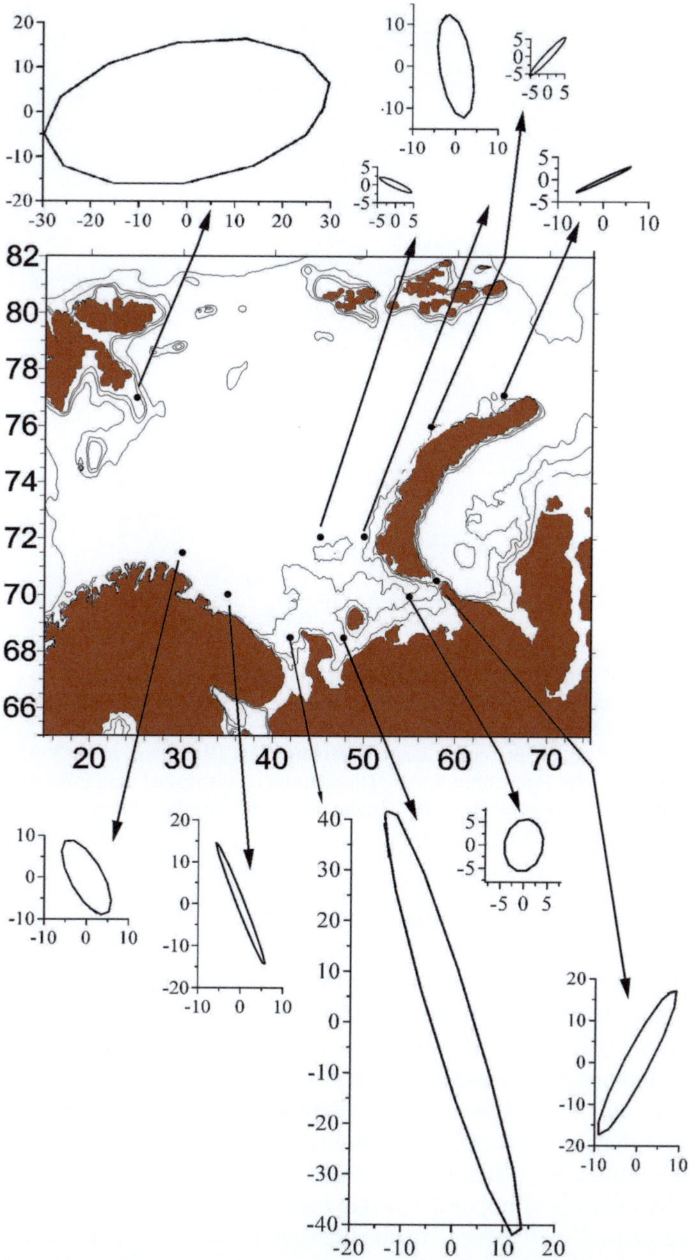

Fig. 3.10 Ellipses of the barotropic tide currents (cm/s) at some characteristic points of the Barents Sea based on the TPXO.9 global inverse tidal model with satellite data assimilation (Egbert and Erofeeva 2002). The isobaths of 25, 50, 100, and 500 m are shown

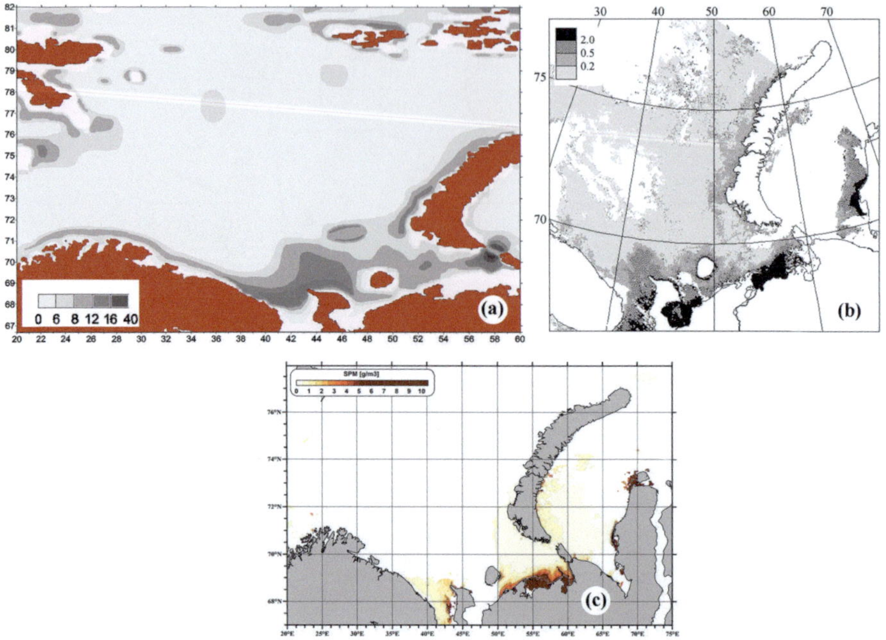

Fig. 3.11 **a** Chart of amplitudes (meters) of internal tides in the Barents Sea based on the model simulation; **b** Distribution of the suspended matter based on the satellite SeaWiFS color scanner data in July–August 1998 **b**. Gradations of the suspended matter concentration (>2.0; 0.5, and 0.2 mg/l) are shown in the inset. White color denotes missing data. (Burenkov et al. 2001). Used with permission of the Pleiades Publishing, Ltd.; **c** Distribution of suspended matter based on SeaWiFS color scanner data on September 20, 2006. Gradations of the suspended matter concentration (0–10 g/m^3) are shown in the inset

for example, in the Strait of Gibraltar. This stimulates intense generation of internal waves.

The amplitudes of the internal tides (about 10 m) along the northern slope of the Kola Peninsula (35–40° E) are related to the steep continental slopes along the coast and strong barotropic tide. The large amplitudes of internal tides between the Kola and Kanin peninsulas and around the Kanin Peninsula (44° E) are caused by significant currents of the barotropic tide in this region even though the continental slope is flatter than in the northern part of the Kola Peninsula.

Large amplitudes of the internal tide found near western coast of Novaya Zemlya are caused by the joint effect of the strong barotropic currents and steep submarine slopes of the coastal shelf close to the inclination of internal tide characteristics. The large amplitudes in the northern parts of the Barents Sea over steep submarine slopes are caused by the physical mechanism of internal tide generation at supercritical latitudes, where a very steep slope is not necessary for the generation of strong internal tides as at southern latitudes. Large amplitudes (40 m) of internal tide in the northwestern part of the Barents Sea were confirmed by field measurements in

2019. The results are reported in the Chapter, section: High-amplitude internal waves southeast of Spitsbergen.

Interesting results were found by comparing the chart of internal tide amplitudes and the chart of water turbidity in the surface layer based on the satellite SeaWiFS (Sea-viewing Wide Field-of-view Sensor) ocean color scanner data. A chart of water turbidity (Fig. 3.11b) was constructed using a special algorithm for determining the concentration of suspension matter in the Barents Sea waters on the basis of the SeaWiFS data (Burenkov et al. 2001). A comparison of these charts demonstrates similar features in the distribution of internal tide amplitudes and suspension matter concentration in shallow waters. The correlation between the internal tide amplitudes with water turbidity at shallow depths appears because strong bottom currents induced by the internal waves facilitate stirring up bottom sediments and their mixing in the entire water column (Gjevik et al. 1994). The general features, which are well seen on the chart, are increased turbidity in the region of the Kanin Peninsula, where the currents of the barotropic tide and the amplitudes of the internal tide are large. In the shallow water regions close to the strait connecting the Barents and White seas (Gorlo Strait) increased turbidity is caused only by the strong currents of the barotropic tide over the shallow depths of the strait. The eigen function for the horizontal velocities in the internal tides has maxima at the bottom and at the surface; therefore, internal waves should cause relatively high bottom currents over the shelf, thus increasing turbidity raising suspended particles from the bottom, which is seen in the satellite images of the shallow sea.

Near the Kara Gates Strait the turbidity is high. However, the strait was covered with ice and we could not get satellite data on turbidity. High turbidity is also observed along the northwestern coast of Novaya Zemlya where we estimated large amplitudes of the internal tide. The observed correlation between the charts is well justified in the shallow regions. In the center of the sea, both the suspended matter concentration and the amplitudes of internal tides are small. Here the concentration of suspension matter can be determined by the other factors.

Thus, the main conclusions of the numerical experiments are the following. In the region of the Arctic coast, the barotropic tide generates an internal tide, whose influence most strongly manifests itself near the steep bottom topography. The internal tide, even in the regions located south of the critical latitude, almost does not develop into a free propagating internal wave. The fluctuations of the isopycnal surfaces with significant amplitudes are observed only near the bottom slopes. It is likely that here the internal waves lose their energy, which is transferred to the energy of short-period waves.

The model simulations indicate that the bottom slopes characteristic of the region can lead to the generation of forced internal tide. Since the stratification and bottom slopes are irregular, the internal wave field should also have a patchy structure. The maximum amplitudes would be observed at the places where the inclination of the characteristic lines for the internal tides randomly coincides with the bottom slope. In order to form large amplitudes of the internal tide, the bottom slopes should not necessarily be as steep as they should be in the regions of low and mid-latitudes. Because the inclination of the curves of the characteristic equation for the internal waves in the Arctic is significantly smaller, large amplitudes of the internal tide can be generated even at small bottom inclinations. The proximity of the critical latitude for the internal tide, which is determined not only by the latitude of the place (approximately 74°30′ N) but also by the vorticity of the currents, can play a significant role here.

3.5 Internal Tide in the Kara Gates

The Strait of Kara Gates connects the Barents and Kara seas. The Barents Sea is a relatively warm sea in Northern Europe because the warm water from the Atlantic Ocean is transported into the sea and warms it. The southern part of the sea never freezes. The Kara Sea is a very cold sea. The sea is ice-free only in August–October. A warm flow exists from the Barents Sea to the Kara Sea due to the density difference. A shallow sill crosses the strait. The minimal depth of the sill is 30 m, but there are fractures in the sill, which are 100 m deep.

Moored measurements of currents and temperature in the Strait of Kara Gates were performed in September–October 1997. After the experiment we published the results in (Morozov et al. 2003a, 2003b). Three moorings with currents and temperature meters were deployed in the strait for a few days at 70°22′ N, 57°38′ E; 70°18′ N, 57°58′ E; 70°17′ N, 58°11′ E (Fig. 3.12). The fourth mooring was deployed at a distance of 200 km southwest of the strait at 69°43′ N, 51°02′ E.

Later, internal tides in the Kara Gates were analyzed by Morozov et al. (2008, 2017) on the basis of moored measurements, numerical modeling, towed CTD, and satellite images. A current from the Barents Sea to the Kara Sea was recorded based on the mooring measurements with velocities ranging from 6 to 25 cm/s. The maximum velocity reached 50 cm/s. A bottom reverse flow from the Kara Sea with a mean velocity of 11 cm/s was recorded by a deep instrument in a fracture of the sill across the strait.

Moored temperature measurements in 1997 show that in the spring phase of the tide the vertical peak to peak displacement of isothermal surfaces reached 60 m, while the total depth at the measurement site is 230 m. The amplitude of internal tides on a mooring located 200 km southwest of the Strait was 12 m.

Based on the data of the previous studies (Morozov et al. 2003a, 2003b, 2008, 2017), the structure of the internal tides in the Kara Gates is similar to that in the Strait of Gibraltar and Bab el Mandeb Strait. Internal tides are generated due to a strong barotropic tide on the slopes of the sill that crosses the strait (Morozov et al. 2002b). A warm current from the Barents Sea to the Kara Sea generated by sea level difference intensifies the internal tide propagating to the west (Morozov et al. 2017). Superposition of the current and barotropic tide flow forms a complex dynamic situation in the Kara Gates. Interaction with the currents intensifies internal tides.

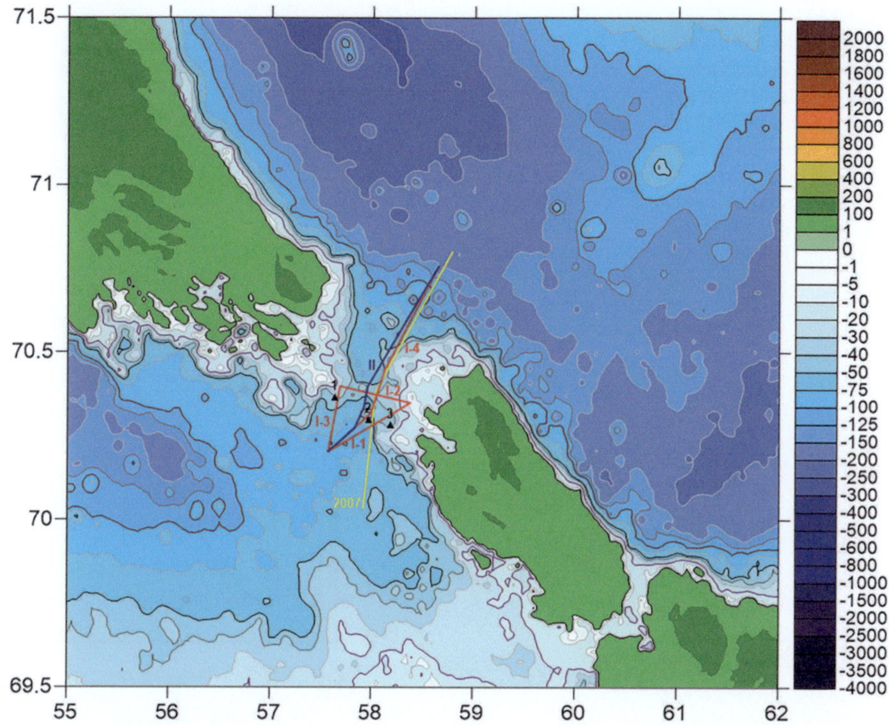

Fig. 3.12 Bottom topography in the Kara Gates based on ETOPO1. Ship tacks with towed CTD are shown: yellow line denotes the tack in 2007, the other lines show tacks in 2015. The moorings in 1997 are shown with black triangles

The field studies were continued in 2007, 2015, and 2016. A scheme of the field works in the strait is shown in Fig. 3.12. Moon phases are important in the investigation of internal tides. The measurements in 2007 and 2015 coincided with the spring tide. The ellipses of the baroclinic tides M_2 and S_2 on the dates of towed measurements across the strait were calculated on the basis of the data assimilation of satellite altimetry using the OTIS algorithm of the Oregon State University (http://volkov.oce.orst.edu/tides/otps.html) (Egbert and Erofeeva 2002). The ellipses are shown in Fig. 3.13.

On August 28–29, 2015, towed measurements in the scanning regime were carried out in the strait with a towed CTD-profiler IDRONAUT 320-plus by the ship moving at a speed of 6 knots. The instrument was periodically lowered and raised so that the records were made almost from the surface to the bottom. The cycle from the appearance at the surface to the next one lasted 5 min. The horizontal distance between the points where the instrument reached the surface was several hundred meters (Morozov et al. 2017). A scheme of the ship tacks is shown in Fig. 3.12.

A temperature section was plotted on the basis of towed measurements (Fig. 3.14). The figure shows a snapshot of the displacements of isotherms along the cruise tack.

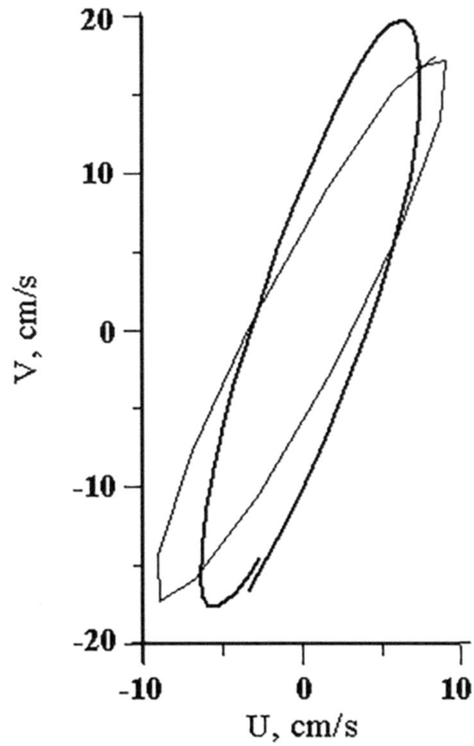

Fig. 3.13 Ellipse of tidal currents at the M_2 frequency on mooring in the Kara Gates calculated from the measurements in 1997 is outlined with a thick line. Ellipse of tidal currents calculated from the satellite data (M_2 and S_2) is shown with a thin line (Morozov 2006)

Fluctuations of the isotherms caused by internal waves are seen in the plot. The Barents Sea is on the left side. A hydraulic jump (deepening of deep isotherms) is seen on the Kara Sea side. In the western part of the section one can see fluctuations of the isotherm typical for the internal tide. Internal tide is best pronounced in the fluctuations of isotherms 1.0–2.5 °C. Judging from the distances between the maximum depths of isotherms the wave nearest to the sill is 12 km long, the next is 15 km, and the third is 18 km. The internal tide wavelength increases because the ocean becomes deeper.

The ship was moving from the Barents Sea opposite to the propagation of internal tide in the Barents Sea. On the Kara Sea side the wavelengths are longer: 25–27 km. Their amplitudes are smaller than on the Barents Sea side.

Numerical modeling using a fully non-hydrostatic model developed by Vlasenko (1992), Vlasenko and Hutter (2002) was used to study the influence of the currents on internal tides. In this calculation, we used the data of topography from the tacks of our towed measurements in the region. Stratification was taken from the CTD data. The amplitude of barotropic tidal currents was assumed equal to 12 cm/s on the basis of satellite altimetry (Egbert and Erofeeva 2002). Simulations were carried out using a domain 200 km long with a horizontal step of 75 m and 20 vertical levels. The time step was approximately equal to 1.5 s. We specified the coefficients of

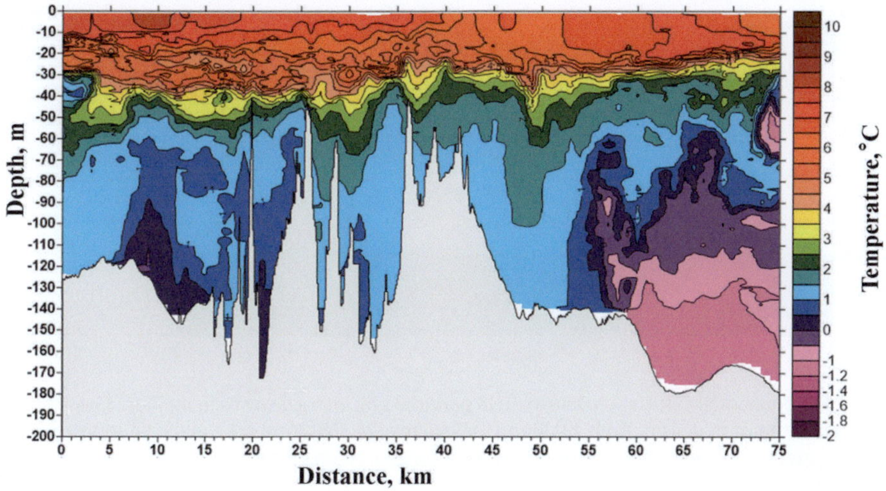

Fig. 3.14 Distribution of temperature over the towed CTD section during the ship motion from the Barents to the Kara Sea. Isotherms are shown with an interval of 0.5 °C. Gray color shows the bottom profile (below) (Morozov et al. 2017). Used with permission of the Pleiades Publishing, Ltd

horizontal eddy viscosity and density diffusivity as 8 m^2/s over the ridge and 2 m^2/s beyond the ridge over the flat bottom in the model. The coefficients of vertical turbulent viscosity and density diffusion were set at 0.001 m^2/s over the ridge and 0.0001 m^2/s beyond the ridge. A small horizontal step in these calculations allows us to increase the non-linearity, which suppresses the dispersion due to strong rotation at high latitudes.

In the model, we introduced a mean flow from the Barents Sea with a velocity of 12 cm/s, which is directed to the northeast and occupies the entire depth. In the model, the direction of the flow is from left to right. A periodical barotropic tidal flow was superimposed on this current. Periodical changes in the tidal horizontal flow with an amplitude of 9 cm/s induce an internal wave propagating in both directions from the sill. The estimates of barotropic tide current velocities were calculated from the satellite data on the day of the towed CTD section. Close values of model tidal velocities were reported by Kagan and Timofeev (2015).

The perturbations of the density field induced by the propagating internal tide after three tidal periods (36 h) of calculation are shown in Fig. 3.15. The fluctuations of the density field are not symmetrical relative to the sill that crosses the strait.

It is clear from the physical considerations that a stronger internal tide should be observed in the Barents Sea because the southwestern direction of the wave propagating from the sill in the strait is opposite to the current. This is confirmed by the numerical simulation. When internal tide propagates opposite to the current its amplitude increases due to the decrease in the wavelength and conservation of energy over one period. The leading edge of the wave is flat and the trailing edge is steep. Owing to strong non-linear effects and the mean current opposite to the

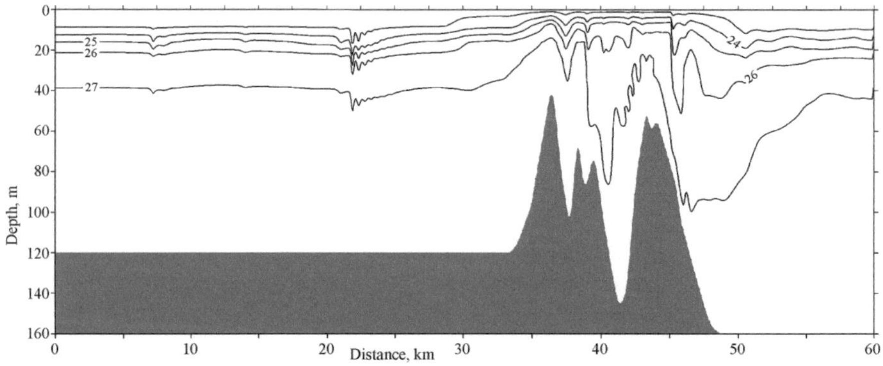

Fig. 3.15 Numerical calculation density field perturbed by internal waves in the Kara Gates Strait. The contour lines of density from 1.023 to 1.027 g/cm^3 (24, 25, 26, and 27 in the figure) perturbed by the current in the flow and internal waves are shown based on the numerical simulations. Bottom is shown with gray color

wave propagation internal bore is formed east and west of the sill. The isopycnals sharply deepen to 10–15 m forming a bore. A packet of short-period waves follows the bore. These smaller scale waves and the bore induce vertical motions, which manifest themselves at the surface. These high-frequency waves cause spatial inhomogeneities in the vertical motions, which manifest themselves at the surface as bands of ripples in the convergence zones and bands without ripples in the divergence zones. The wavelength is approximately 17–18 km. Calculation of the wavelength from the dispersion relation yields that the wavelength is 23 km. The difference appears due to the Doppler effect.

A sharp deepening of isopycnals is observed on the Kara Sea side. It is a hydraulic jump that appears on the lee side of a flow beyond an obstacle. According to the numerical simulations, the 1.027 g/cm^3 isopycnal deepens from 40 to 100 m. A similar but slightly weaker pattern is observed in the data of towed measurements (Fig. 3.14).

Wave packets of short-period internal waves directed to the Barents Sea are seen on a satellite image. The length of the wave packets is 1–2 m. The distance between packets is 4–6 km. High contrasts on the images are caused by the formation of intense convergence of vertical flows, in which backscattering is strongly enhanced over wind ripples and wave breaking (Morozov et al. 2017).

3.6 Short-Period Internal Waves in an Arctic Fjord in Spitsbergen

Short-period internal waves are observed everywhere in the ocean in addition to the tidal internal waves of high energy (Sabinin and Serebryany, 2005). There are no evident sources of short-period internal waves in the ocean, and there are no clear narrow peaks on the spectra of oceanographic characteristics in the frequency range of short-period internal waves. On the other hand, any perturbation in a stratified fluid generates internal waves in a wide frequency range. In this section, we will review our publication (Morozov and Marchenko 2012).

The existence of internal waves with the M_2 period north of the critical latitude over a flat bottom is theoretically impossible; however, many authors demonstrate that such waves exist over submarine slopes. The authors of papers (Morozov et al. 2003c; Pisarev 1991, 1992; Kurkina and Talipova 2011) show that semidiurnal oscillations are observed in the Arctic regions over submarine slopes and only packets of short-period internal waves are recorded over relatively flat bottom far from the slopes. This is also confirmed by model calculations in (Vlasenko et al. 2003).

Hazel (1972) constructed the curves of neutral stability for plane parallel flows. The curve of neutral stability has one maximum for the flows, in which the density and velocity vary similarly with depth and have a transition layer (i.e., interface) (Fig. 3.16). This curve shows that the greatest rate of perturbation growth corresponds to the dimensionless wave number $k = 0.5$.

When a long-period wave breaks (Hazel 1972), waves with wavenumber $k = \alpha/2$ are generated with the maximum probability, where α is the wave number equal to the inverse value of a specific size. Long waves and short wave are not likely to appear. According to the Hazel's curve, not all scales in the perturbation scale have the same rate of amplitude growth under the condition that the flow reaches the instability state (for example, at Richardson numbers $\mathrm{Ri} < 0.25$). The perturbations with the dimensionless wavenumber $k = 0.5$ obtain energy faster than the other perturbations.

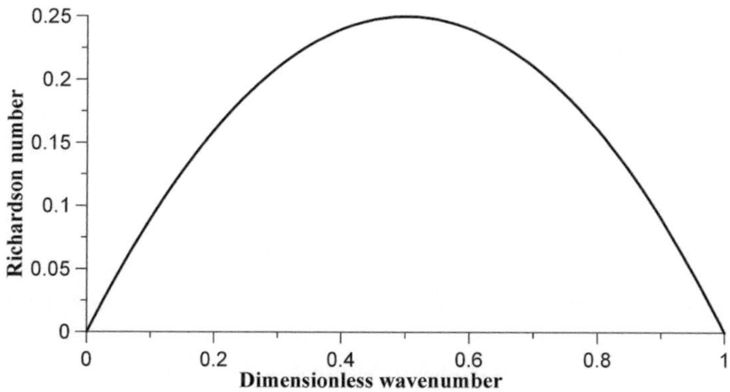

Fig. 3.16 Stability boundary of the flow according to Hazel (1972)

If the wavenumber is close to zero or unity, very low values of the Richardson number are required for the instability appearance. More recent laboratory experiments on the development of this instability of long internal waves that cause wave breaking in a near node region are described in Kalinichenko (2005).

In August and September 2008 and 2009, in March 2008 and 2010, and in May 2010, experiments were carried out in Spitsbergen to measure internal waves in the Van Mijen Fjord. The amplitude of the water level at the coastline reached 2 m. In winter, measurements were carried out from the ice cover (Marchenko et al. 2010, 2011, 2013). We used temperature and salinity profilers SBE 39, SBE 19 (in the mooring mode), and current meters ADV and Aquadopp. In the summer, the instruments were fixed on a mooring with a surface buoyancy. Assuming that the ocean depth H and the Brunt–Väisälä frequency N are constant, a dispersion relation is given in (Tareev 1966) which is a good approximation for long internal waves to estimate their phase velocity:

$$c = \frac{\omega}{k} = \frac{NH}{\pi} \tag{3.3}$$

The depth at the mooring location (H) was approximately 15 m. The characteristic value of the Brunt–Väisälä frequency (N) in the study region was $N = 0.07\ \mathrm{s}^{-1}$ (the period is 90 s). In summer, the study region is characterized by a sharp salinity interface in the depth range 2–6 m (6 PSU) and by a moderate temperature gradient. The measurements of the vertical profiles were carried out using the SBE 19plus profiler. The time variability of the temperature fluctuations was measured by the SBE 39 instruments fixed on a mooring. In the summer of 2009, the measurements were made at 4 and 6 m depths over a submarine slope where the depths sharply increase from 3–4 m to 15 m.

Let us estimate the possible periods of short-period waves in the packets based on the results obtained by Hazel in the laboratory experiments (Hazel 1972). The perturbations with dimensionless wavenumber $k = 0.5$ obtain energy faster than the other perturbations. Let us take the depth $H = 12$ m as the characteristic scale when the flow in the fjord breaks due to instability. Then, the critical dimensional wavenumber ($k = 0.5$) would be equal to 1/24 m^{-1}. It is related to the internal waves that are formed during the perturbation of the flow caused by internal tide and obtain energy faster than the other perturbations (dimensionless wave number $k = 0.5$). The frequency of long internal waves is estimated from relation (3.4):

$$\omega = \frac{NHk}{\pi} = \frac{0.0712}{24\pi} = 0.0111\mathrm{s}^{-1} \tag{3.4}$$

Thus, the characteristic wave period in the packets of waves in our study region is 565 s, or approximately 9 min. Since our calculations are only estimates, we should consider that the characteristic period is within a few minutes.

The amplitude of the vertical oscillations recorded by each of the instruments is approximately half a meter. A significant spectral peak corresponds to a frequency of 0.18–0.19 cycle/min (approximate period is 5 min).

It was shown in (Sugiyama et al. 2009) on the basis of numerical experiments that the Garrett–Munk spectrum is maintained only under the condition that the energy is transferred to the ocean under forcing at frequencies close to the inertial frequency due to wind and at a frequency of the semidiurnal tide. When both energy sources are present, nonlinear interactions between internal waves lead to the fact that the resulting spectrum becomes close to the Garrett–Munk spectrum. In this case, the ratio of the incoming energy at the inertial and semidiurnal frequencies is not principally important. If we exclude one of the sources, the level of the resulting spectrum would be lower than the GM spectrum.

The moored Garrett–Munk spectrum, which can be obtained experimentally from the measurements at moored buoy stations, is written as (Konyaev and Sabinin 1992):

$$MS(\omega, z) = C_M f \frac{\sqrt{\omega^2 - f^2}}{N(z)\omega^3}, \tag{3.5}$$

where, C_M is a coefficient (204 m^2/h), f is the inertial frequency, ω is the wave frequency, and $N(z)$ is the Brunt–Väisälä frequency. The expression becomes simpler for high frequency waves

$$MS(\omega, z) = C_M \frac{f}{N(z)} \omega^{-2}, \tag{3.6}$$

and the spectrum is approximated by the inverse relation of the squared frequency. It is worth noting that, in this case, this is the spectrum of the vertical displacements; therefore, the measured values should be recalculated into vertical displacements.

The spectra of vertical displacements are shown in Fig. 3.17 (the spectra are in the logarithmic scale). The background Garrett–Munk spectra are also shown in this figure for summer and winter. The stratification in winter is weaker (the Brunt–Vaisala frequency is in the denominator); therefore, the upper line corresponds to winter. The theory of the Garrett–Munk spectrum was developed for conditions which are not observed in the Arctic seas under ice (shallow depths, high latitudes, etc.); however, we give these formal dependencies as reference functions.

The levels of the three winter spectra are notably lower than the summer spectra, which are determined by joined effect of tides and wind. An increase in the spectral densities in summer is also related to the sharp increase in the summer stratification, which enhances the nonlinearity of wave generation. It is worth noting that the three winter spectra are very similar in their level of the spectral densities, although each was measured using different methods: the winter measurements in 2008 were carried out by the SBE 39 instrument (temperature), while instruments SBE 39 (temperature) and SBE 19 (with sensors of temperature, conductivity, and pressure) were used in the winter measurements in 2010.

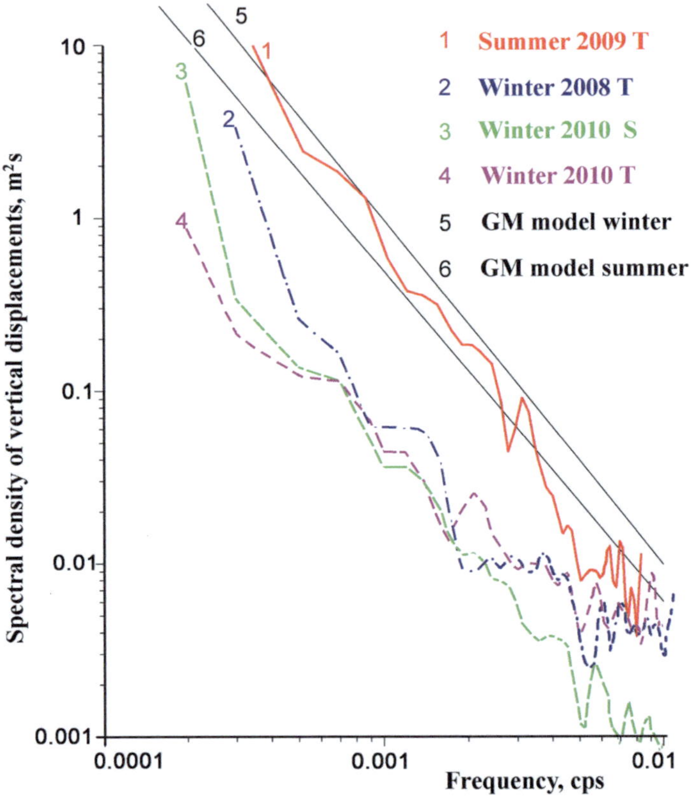

Fig. 3.17 High-frequency spectra of the amplitudes of vertical displacements based on the moored measurements in the fjord. (1) August 2009 (two moored temperature sensors), (2) March 2008 (one temperature sensor under ice), (3) March 2010 (salinity sensor under ice, (4) March 2010 (temperature sensor under ice), (5) the Garrett–Munk spectrum (1972) for winter conditions, and (6) the Garrett–Munk spectrum for summer conditions (Morozov and Marchenko 2012). Used with permission of the Pleiades Publishing, Ltd

The spectral level of the fluctuations of vertical displacements caused by internal waves in winter is significantly less than in summer. When the surface of the fjord is covered with ice, the wind stress is not transferred to the short-period internal oscillations, and the spectral level of internal waves decreases. Strong summer stratification enhances the nonlinearity of wave generation processes.

3.7 High-Amplitude Internal Waves Southeast of Spitsbergen

Many publications related to the generation of internal tides discuss the phenomenon that internal tides transform the velocity field of the barotropic tide (Garrett and Kunze 2007). In the regions of sharp changes in the bottom topography low velocities of the barotropic tide form a velocity field with higher horizontal velocities and amplitudes of the vertical displacements. Internal tide in the ocean is generated when tidal currents overflow submarine slopes. This overflow causes the appearance of vertical velocity components, which periodically displace isopycnals. The most energetic component of the internal tide is the M_2 wave with a period of 12.4 h.

It is well known that semidiurnal internal tidal of the M_2 frequency cannot exist north of the critical latitude (74°30′ N) (Vlasenko et al. 2003; Morozov and Paka 2010). Despite the fact that it is impossible to explain the existence of the semidiurnal internal tide of the M_2 period (12.4 h) at polar latitudes north of 74.5° N within the linear theory of internal waves in the ocean of constant depth (Krauss 1966), internal waves of the M_2 period exist in the Arctic Ocean and can be observed in field studies (Kozubskaya et al. 1999). These waves cannot exist over the flat ocean bottom and zero background vorticity (Krauss 1966; Konyaev et al. 2000).

Nevertheless, the currents of the barotropic tide flow over submarine slopes and periodically displace the isopycnals due to the existence of vertical components of tidal currents over the slopes. But the generation process of internal tides at the northern latitudes differs from that at low and mid-latitudes (Baines 1982; Konyaev and Sabinin 1992). The difference is related to the fact that in the region of supercritical latitudes the internal tide is a forced wave and cannot freely propagate from the slope as it does at lower latitudes. This statement is related only to the flat bottom.

However, internal tides were repeatedly recorded over the underwater slopes in the Arctic region (Parsons et al. 1996; Plueddemann 1992; Levine et al. 1985; Pisarev 1996; Kozubskaya et al. 1999; Rippeth et al. 2017). The measurements from an autonomous drifting buoy over the Nansen Basin, Yermak Plateau, and Greenland Sea in 1987–1988 gave much data on the semidiurnal tides close to the critical latitudes (Padman and Dillon 1991; Padman et al. 1992; Plueddemann et al. 1998).

In the previous work of one of the authors it was shown that forced internal tides are strong over the slopes of the southeastern Spitsbergen (Morozov 2006). The amplitudes (peak-to-peak particle displacements) of forced internal tide based on the model simulations were as high as 36 m. Such high amplitudes are caused by strong tidal forcing. The amplitudes of the barotropic tide velocities in this region are 20 cm/s (Morozov 2006). This region is also known for high rates of barotropic to baroclinic tidal energy conversion in the M_2 band (Rippeth et al. 2017). Modeling of high amplitude internal waves was performed in (Wood and Grue 2003).

The objective of this study is to estimate the properties of forced internal tides southeast of Spitsbergen north of the critical latitude (74.5° N). Internal tides are generated over the slopes of submarine ridges, continental slopes, and seamounts (Morozov 1995). This effect intensifies when barotropic tidal currents are normal to a submarine ridge. It is obvious that the stronger the barotropic tidal currents, the greater the amplitude of internal tide. The analysis is based on field observations, model simulations, and satellite images. We review our publication (Marchenko et al. 2021).

On April 24–27, 2019, we carried out field work in the region of the Barents Sea southeast of Spitsbergen approximately at 76°10′ N, 25°50′ E. The region was over the bottom slope of Hopen Island at a distance of 30 km from the island over a depth of 98–110 m (Fig. 3.18). The region of measurements was located in the marginal ice zone. A CTD cast was made each hour from the "Polarsyssel" vessel that kept its position near a drifting ice floe. The drift of the ice floe was recorded using a satellite transponder set on the floe. A scheme of the floe drift is shown in Fig. 3.19. Some irregularities in the drift are caused by the ship motion that could slightly displace the floe in attempt to make ice-free space for CTD casts. The ship was keeping its position near the ice flow. The motion of the ship was recorded using GPS navigation. From time to time the ship had to move to keep closer to the ice flow. The drift of the ice floe in the figure is combined with the graphs of the tidal ellipses based on the model simulations using the satellite altimetry measurements processed by the TPXO9.1 global tidal model (Oregon State University, USA) (Egbert and Erofeeva 2002).

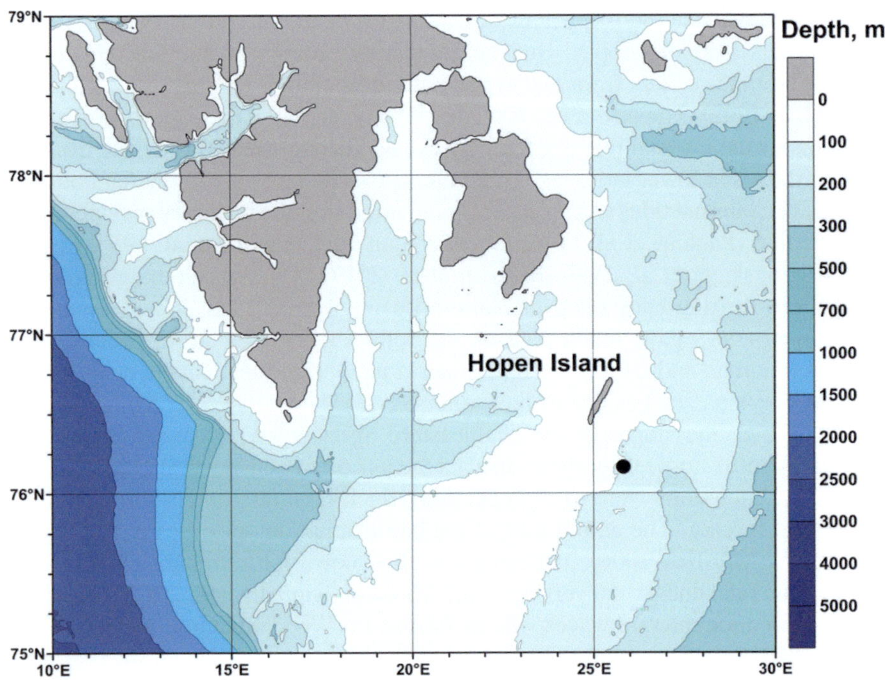

Fig. 3.18 Map of the study site southeast from Svalbard. Location of measurements is shown with a black

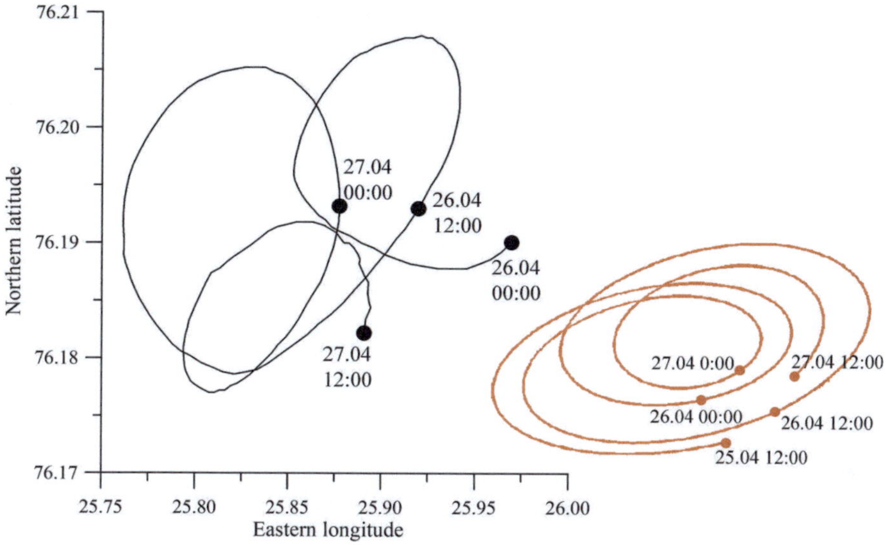

Fig. 3.19 Drift of the ice floe (black) and tidal ellipses based on the TPXO9.1 model (brown) including the M_2, S_2, N_2, K_2, K_1, O_1, P_1, Q_1, MM, MF, M_4, MN_4, MS_4, $2N_2$, and S_1 tidal components. The reference time moments are indicated (Marchenko et al. 2021)

During the time of field work from April 25 to April 27, 2019, 47 CTD casts were made using the SBE 19 plus profiler. The instrument was lowered down to the bottom; the lowest measurements were made at a distance of 20 cm from the seabed. The bottom topography used in this research was taken from the GEBCO digital database. It is important to note that the measurements were carried out during the neap-tide. The full moon was on April 19, and the third quarter was on April 27. This means that the field measurements were performed in the period of low tidal currents.

Let us consider CTD profiles in the study region. Examples of temperature and salinity profiles are shown in Fig. 3.20. We note strong time variations in the vertical profiles. Two examples of temperature and salinity profiles taken on April 26 at 10–00 and 14–00 are shown in Fig. 3.20. Such strong variations are caused by internal motions generated over the bottom slope. Salinity variations in time are more representative to analyze internal wave perturbations since salinity is generally gradually increasing with depth, while temperature profiles have maxima in the middle of the water column. The time evolution of isohalines is shown in Fig. 3.21. A two-day time record allows us to distinguish a semidiurnal internal wave. Depressions of isohalines are clearly seen at 17:00, 07:30, 18:00, 06:30. The amplitude of oscillations reached 50 m.

Let us consider model simulations. The numerical model is based on the primitive hydrodynamic equations. The equations are nonlinear. We take into account turbulent exchange and density diffusion (Vlasenko 1992; Morozov et al. 2002b). The system of equations describes a two-dimensional (x, z) flow in a continuously stratified

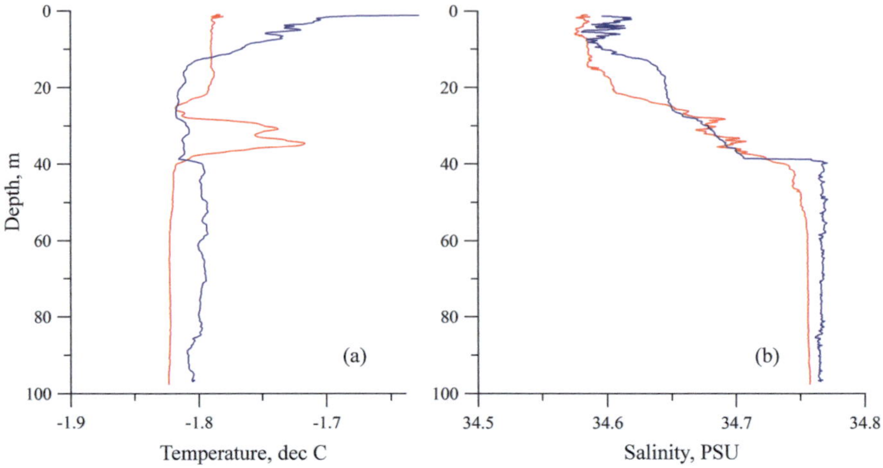

Fig. 3.20 Temperature **a** and salinity **b** profiles on April 26 at 10–00 (red) and 14–00 (blue) (Marchenko et al. 2021)

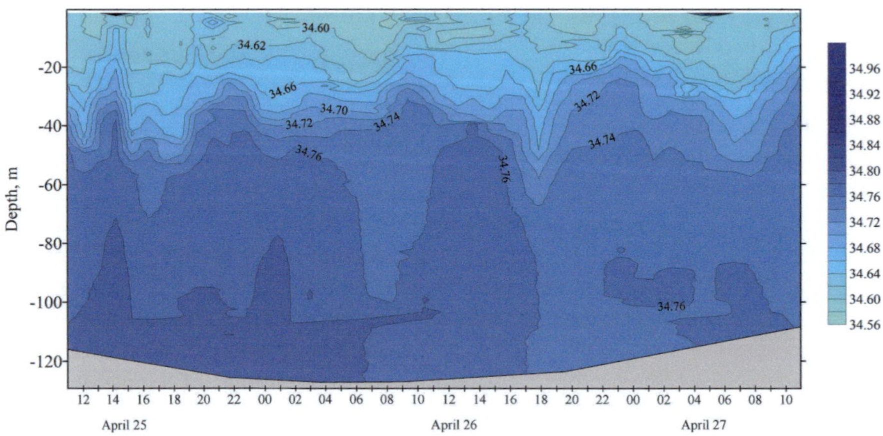

Fig. 3.21 Time evolution of salinity (Marchenko et al. 2021)

rotating ocean of variable depth. The barotropic tide flowing over the slope displaces isopycnic surfaces up and down and generates internal waves propagating in both directions from the slope, which is the source of perturbations.

The density field not perturbed by internal waves corresponds to the specified distribution of the Brunt-Väisälä N (z). A long tidal barotropic wave propagates from the open part of the sea to the slope of Hopen Island. The amplitudes of the barotropic tide velocity were specified as 20 cm/s based on the simulations using the TPXO.9 model for this region and the dates of the experiment. Thus, the numerical model calculations allow us to estimate the amplitude of internal waves in the immediate vicinity of the generation region.

We selected a working domain for the model calculations 120 km long with a horizontal step of 40 m and 20 vertical levels. The time step was 15 s. Such values completely satisfy the Courant-Friedrichs condition for the stable operation of the numerical model. The coefficients of horizontal viscosity and density diffusion were set equal to 4.0 m^2/s, and the corresponding vertical coefficients were equal to 0.0001 m^2/s. Smaller values led to numerical instability due to the occurrence of strong nonlinear waves at large amplitudes, and at large values of the coefficients, wave motion occurred in an excessively viscous medium, which reduced the nonlinear effects of internal wave.

The model simulations showed that from five to seven periods of the semidiurnal internal tide of model calculations are needed to get the estimates of the density field corresponding to the pattern of developed wave motion.

A model calculation of the wave motions of isopycnals is shown in Fig. 3.22. The figure shows perturbations of density. We show the initial density field with black lines and the resulting density field perturbed by internal waves is shown with the red lines. The model shows that immediately above the slope at a depth of 100 m, the oscillations of isopycnals exceed 50 m; they rapidly decrease with the distance from the slope because only forced internal tides of the M_2 period can exist over irregularities of the bottom topography.

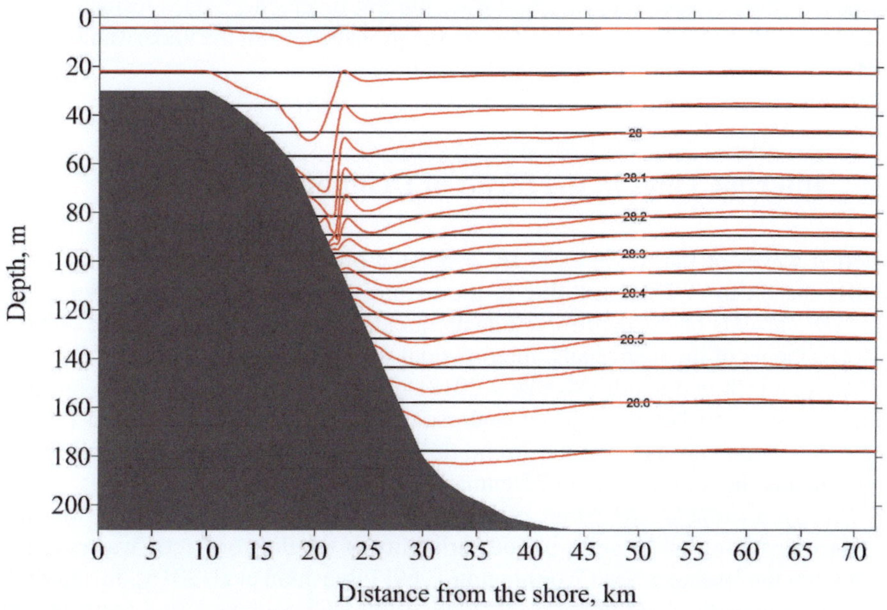

Fig. 3.22 Density perturbations over the slope based on the model simulations. The black lines show the initial state of rest, and red lines show perturbations (Marchenko et al. 2021)

Thus, the region near Spitsbergen in the Norwegian Sea is known for the high amplitude internal tides generated due to the interaction between the currents of the barotropic tide and bottom topography north of the critical latitudes (74°30′ N). In the end of April 2019, we performed field measurements in the region near Hopen Island in the marginal ice zone. We analyze a series of CTD-casts to estimate the vertical displacements of water particles caused by internal waves generated over the continental slope. The vertical motions of water particles are clearly seen from the salinity variations in time. Peak to peak displacements caused by forced internal tides are as high as 50 m. Generation of strong internal tides is also shown based on numerical modeling. Generation of internal tides over the continental slope is confirmed by satellite images of their surface manifestation. The M_2 internal tidal waves are forced waves over the slopes of the bottom topography. According to the theoretical studies, such wave cannot propagate in the northern latitude as free propagating waves. Since the energy of the barotropic tide is permanently transformed to internal motions, but internal waves cannot carry this energy from the slopes, the internal waves of higher frequency are generated that can propagate north of the critical latitudes. The development of forced internal tides in the vicinity of the M_2 critical latitude is accompanied by generation of short-wave trains of internal solitary waves, whose surface manifestations are seen on the satellite images. As seen from the satellite images of the region they are typically observed in the form of rank-ordered narrow concentric strips having either an enhanced radar backscatter when the sea ice was accumulated in surface current convergence zones or as dark strips of a lower backscatter in the surface current divergence zones formed due to the propagation of internal waves. The waves propagate from the underwater slopes generally in the northern direction to the shallow regions.

3.8 Internal Tides in the Strait of Denmark

Internal waves of the tidal period are generated on the sill of the transverse ridge across the Strait of Denmark due to the barotropic tide currents overflowing the sill. The currents of the barotropic tide are incident on the slope and cause vertical displacements of the isopycnals. Internal tidal waves propagate to both sides of the sill to the north and south (Morozov 2018). A similar effect is observed in other straits (Kagan and Timofeev 2005; Androsov et al. 1993). In the Strait of Denmark, there is a mean current from the Arctic to the Atlantic in this region, which flows over the threshold of the Strait of Denmark.

The pycnocline in the Strait of Denmark is subjected to significant vertical displacements caused by short-period variability of the flow of Arctic waters to the south into the Irminger Sea (Worthington 1969; Filyushkin et al. 2013). In addition, tidal currents induce additional oscillations of the thermocline with a semi-diurnal period. The currents of the barotropic tide acquire a vertical component above uneven sloping bottom, which leads to the vertical displacements of the pycnocline. Periodic (12-h) displacements of the pycnocline up and down relative to the average position lead to the formation of progressive internal waves that propagate to both sides of the slope, while the front of the internal waves is located approximately parallel to the direction of the sill.

Such a mechanism of the formation of internal tidal waves has been many times discussed in the literature (Morozov 1995; Garrett and Kunze 2007). Numerous measurements of internal waves over the slopes of the bottom topography show that internal waves are more energetic over the underwater slopes than in the regions with a relatively flat bottom (Morozov 2018). In this paper, we will try to estimate the amplitudes of the vertical displacements of water particles caused by internal tides based on direct measurements at the six-day mooring and using a numerical model.

In July 2018, we carried out investigations in the Strait of Denmark to study the flow of Arctic waters into the Atlantic and the water exchange between the two oceans. We performed CTD/LADCP casts using an SBE 911 probe and two velocity profilers RDI QM 150 kHz (down-looking) and WHS 300 kHz (up-looking). In addition, a mooring station was deployed with instruments of different types. A map of the work area is shown in Fig. 3.23. The goal of the work was to estimate the amplitudes of internal waves in the Strait of Denmark. We review our publication (Morozov et al. 2019a).

A submerged mooring was deployed in the Strait of Denmark at coordinates 66°00′ N, 27°30′ W at a depth of 610 m. The station performed measurements from July 22 to July 28, 2018. In addition to the devices for studying suspended matter, temperature and pressure sensors were installed at the station on two RDRduet-6000 instruments at depths of 355 and 505 m and an SBE 37 SMP-ODO temperature and

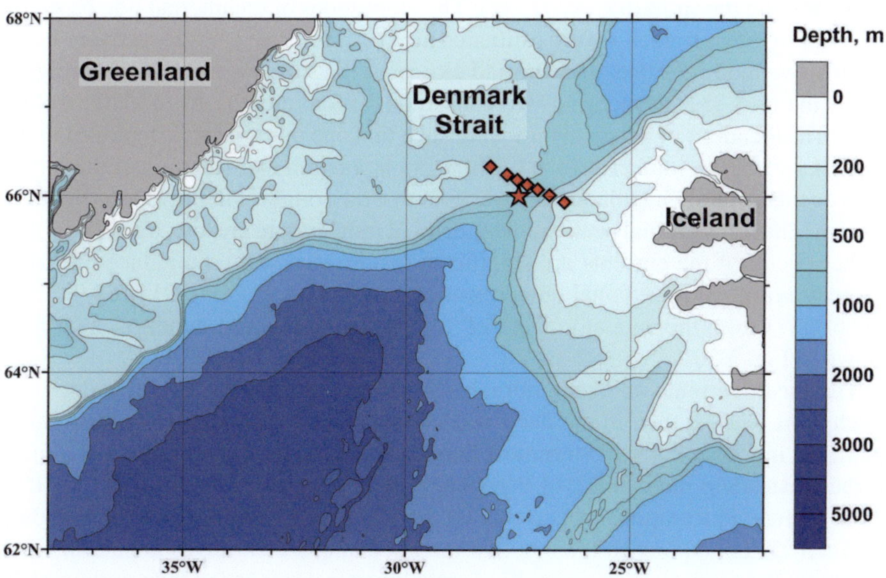

Fig. 3.23 Bathymetric map of the study site. Red star shows the location of the mooring on July 22–28, 2018. Diamonds show the CTD-stations of the section across the strait

pressure meter at a depth of 555 m. The station was supported by buoyancy consisting of 21 aluminum and 5 plastic fishery floats. Before deploying the mooring, the floats were tested at a depth of 150 m. The fishery floats were supposed to operate at a depth of 55 m. However, the currents in the strait were so strong that they periodically submerged the buoyancy; the upper temperature sensor moved between the depths of 330 and 531 m (instead of the planned position at 355 m). Thus, the upper sensor was submerged by more than 200 m relative to its position on the vertical wire. The buoyancy was displaced even more (from 55 m to approximately 350 m), which led to the destruction of the plastic floats and to a decrease in its buoyancy ability. In this case, the entire system of three temperature sensors moved in the range of ± 100 m, and the lower sensor periodically changed the depth of position from 537 to 592 m, instead of the planned 505 m.

Nevertheless, if the temperature measurements from three sensors are available and the depth of each of them at the time of measurement is also known, it was possible to construct the depth of the vertical oscillations of the isotherms. Oscillations of isotherms 0.5, 0.8, 1.0, 2.0, 3.0, 4.0, and 5.0 °C are shown in Fig. 3.24. It is seen from the figure that the range of oscillations of isotherms with a period of 12 h corresponding to the propagation of internal tidal waves, was in the range of 30–50 m, and the maximum displacements of isotherms, 2 and 3 °C, were close to 100 m.

Measurements at the moorings were also performed in this region before. In 1986–1999, 22 long-term moorings were deployed in the Strait of Denmark. The measurements were carried out by the scientists from the Ministry of Agriculture, Fisheries, and Food of the UK (MAFF). The moorings were deployed south of the ridge crossing the strait (65° N, 31° W). The amplitudes of the internal tide based on the mooring measurements were estimated at 30 m. Here, we analyze measurements on only five moorings that were located closest to the sill. A map of the moorings is shown in Fig. 3.25. The moorings were installed for four months, from March to July 1990. The locations of the moorings were such that one mooring was deployed in the deepest part of the southern slope of the strait at a depth of 2200 m, and the others were on the slope of Greenland. The westernmost mooring was deployed at a depth of 1080 m. The semi-diurnal peaks in the spectra of temperature are low, while the peaks in the spectra of the currents are significant (exceed the 95% confidence interval). A spectrum of the meridional component of currents on mooring 914 (64.913° N, 30.668° W) from the data at a depth of 1155 m over an ocean depth of 2005 m is shown in Fig. 3.26.

The mean currents and semidiurnal fluctuations of currents over the slope are much more intense than in the deep-water part. The average current at the bottom over four months on the westernmost buoy is 46 cm/s (the flow is directed almost to the west along the isobath of the slope), and at a greater depth (2180 m) it is about 5 cm/s. Fluctuations of currents on the slope and in the deep part are shown in Fig. 3.27. In addition to the fact that the fluctuations themselves are stronger on the slope, the semi-diurnal variability is much better pronounced than at the depth, where variations with a period of several days prevail.

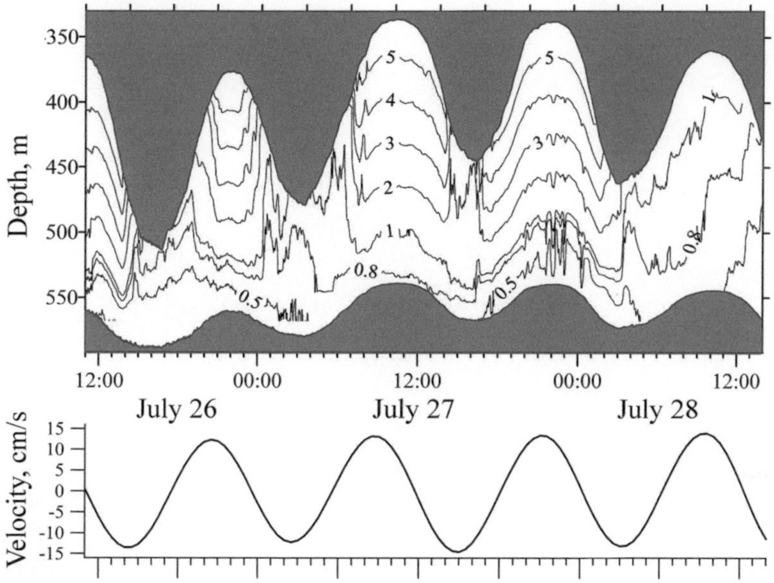

Fig. 3.24 Fluctuations of isotherms: 0.5, 0.8, 1.0, 2.0, 3.0, 4.0, and 5.0 °C based on the mooring data (top panel). The intervals of depths, in which there were no measurements due to the periodic deepening of the mooring leading to the depth fluctuations of the entire system, are shaded in gray. The bottom panel of the figure shows the time variations in the velocity of the barotropic tide (meridional component) based on the TPXO7.2 model. Adopted from (Morozov et al. 2019a). Used with permission of the Pleiades Publishing, Ltd

The time on the x axis is laid off in hours from 12:00 March 9 to 12:00 March 29, 1990.

We used the model developed by V. Vlasenko (Vlasenko and Hutter 2002; Vlasenko 1992) to calculate the characteristics of the internal tide. The model requires the following input data: barotropic tide flow velocities, stratification, and bottom topography. The velocity of the barotropic tide currents was calculated using satellite altimetry measurements processed by the TPXO7.2 global tidal model (Oregon State University, USA) (Egbert and Erofeeva 2002). The graphs of the tidal ellipses based on the model simulations are shown in Fig. 3.28.

Knowing the amplitudes of the barotropic tide, let us return to the analysis of the mooring data; the mooring was subject to strong currents, and therefore the instruments changed the depth. If we compare the velocity graphs of the barotropic tide and the deepening of the entire line, which is supported by buoyancy whose holding force is insufficient for the existing load, it turns out that the maximum depths (almost the "lying" position of the mooring) correspond to the maximum tidal currents directed to the south. The vertical position of the mooring corresponds to the maximum current velocities directed to the north.

This seeming contradiction (it seems that the period of deepening of the whole system should be 12, not 24 h) is explained by the presence of the mean current

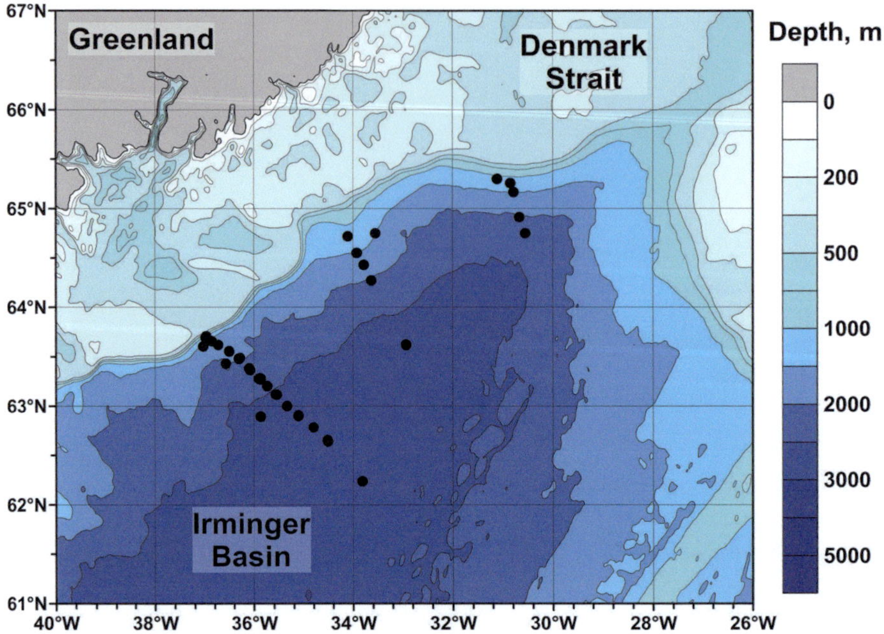

Fig. 3.25 Chart of moorings south of the Denmark Strait

Fig. 3.26 Spectral density of oscillations of the meridional component of velocity measured at a depth of 1155 m over the ocean depth 2005 m. The mooring was located at 64.913° N, 30.668° W. Adopted from (Morozov et al. 2019a). Used with permission of the Pleiades Publishing, Ltd

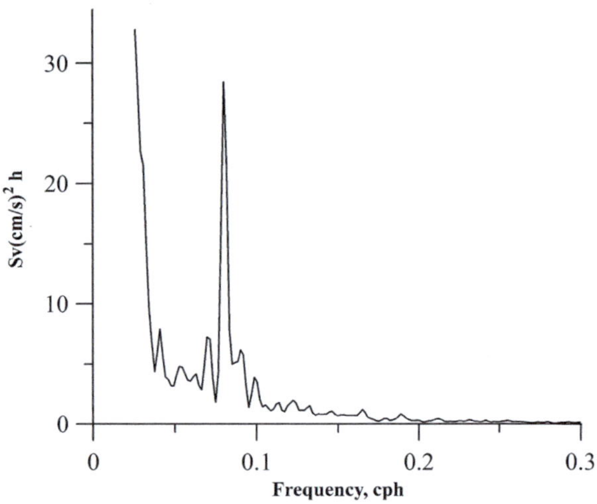

directed to the south, whose speed is close to the maximum speed of the tidal currents. During the phase of the barotropic flow in the southern direction, the speeds double due to the addition of the barotropic tide and the mean flow. When the tidal current is directed to the north, it compensates for the average flow to the south and the overall

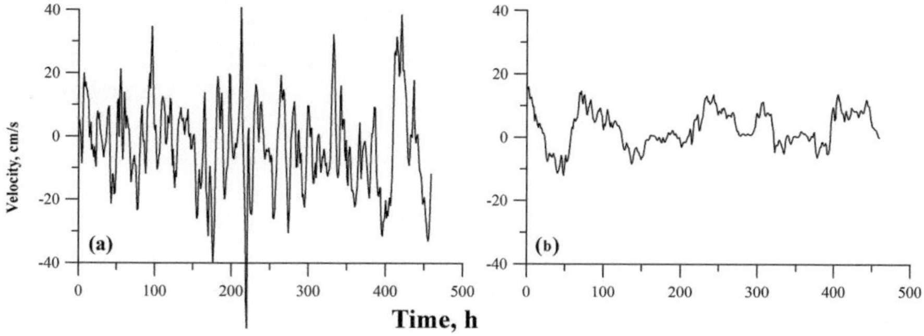

Fig. 3.27 Fluctuations of meridional components of velocity on the moorings: **a** mooring 924 at (65.298° N, 31.115° W), the current meter at a depth of 1062 m over the ocean bottom at a depth of 1080 m; **b** mooring 913 at (64.750° N, 30.550° W) the current meter at a depth of 2180 m over the ocean bottom at a depth of 2200 m

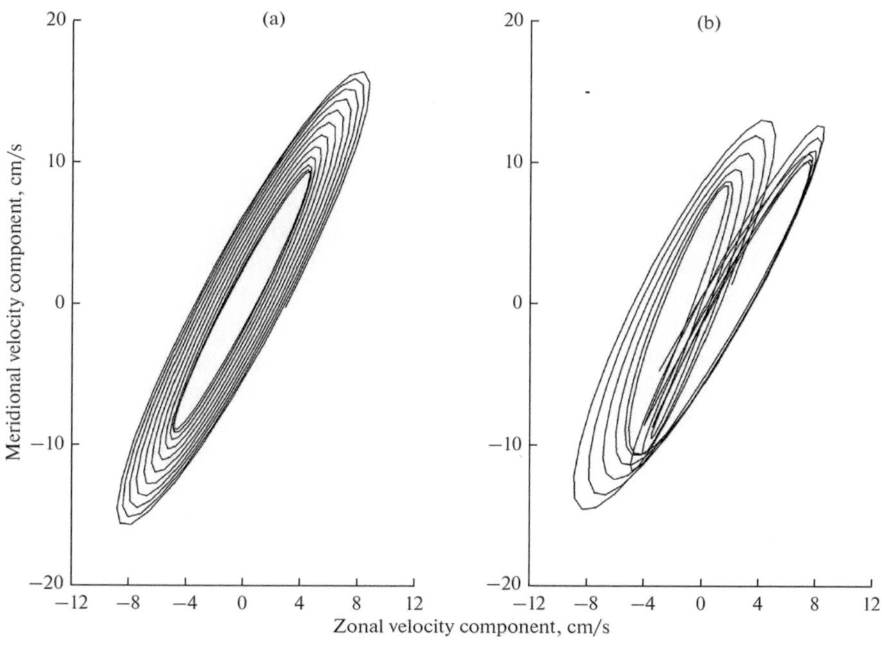

Fig. 3.28 Graphs of the ellipses of tidal currents on July 22–28 based on TPXO7.2 model. Left panel: two tidal harmonics M2 and S2. Right panel: harmonics N2, K2, K1, O1, P1, Q1, MF, MM, M4, MS4, MN4 are added

velocity becomes small, which makes it possible for the mooring station to get a vertical position.

A CTD cast at station 5795 was performed before deploying the mooring. The profiling was made at the low tide phase (based on the model). The average speeds based on the LADCP profile were: U = –37 cm/s (direction to the west), V = –14 cm/s (to the south). The tide velocities according to the model at the moment when the instrument reached the bottom were: U = –0.5 cm/s, V = –1 cm/s. After recovering the mooring a CTD cast at station 5884 was performed. The average speeds based on the LADCP profile were: U = –25 cm/s, V = –42 cm/s. The profiling was made at the high-water phase of the tide. The tide velocities were: U = –9 cm/s, V = –15 cm/s. If the mean currents and tidal currents are directed approximately in the same direction, they strongly deepen the buoyancy of the mooring. If they are directed in the opposite directions, the tidal currents weaken the total current and the mooring assumes an almost vertical position.

We used a numerical model developed by V. Vlasenko (Vlasenko 1992) to model internal motions in the study region. The barotropic tide deforms the isopycnic surfaces over the irregularities of the bottom topography and generates internal waves propagating in both directions from the source of perturbations. If the lateral boundaries of the region are located far from the sources so that the leading front of the internal wave reaches the lateral boundaries not earlier than after several periods of the tidal cycle, then during this time a quasi-periodic motion mode would be established in the vicinity of the non-uniform bottom topography, which can be interpreted as steady wave oscillation regime.

The bottom topography in the strait and the depth profiles in the direction normal to the coast were taken from the Smith & Sandwell digital topography databases. The actual depth data was taken from the Internet site: (http://web.ngdc.noaa.gov/mgg/global/seltopo.html). The ocean stratification (unperturbed state) was taken from the averaged CTD measurement data in cruise 71 of the R/V *Akademik Mstislav Keldysh*. We believe that a barotropic tidal wave is incident on the sill of the transverse ridge of the Strait of Denmark. Due to the interaction with large inhomogeneities of the ocean floor, the barotropic tide currents acquire a vertical component, and this leads to the generation of baroclinic perturbations. The formulation of the problem is reduced to determining the field of generated internal waves based on the known parameters of stratification, bottom topography, and the maximum velocity of the barotropic tide.

We put forward a task to construct a mathematical model of baroclinic tide generation under the condition of stable vertical stratification of fluid in the Strait of Denmark based on the full hydrodynamic equations, taking into account nonlinearity of the wave process and the turbulent exchange. The density field that is not perturbed by internal waves, which corresponds to the distribution of the Brunt-Väisälä frequency $N(z)$, is specified. A long tidal barotropic wave propagates from the open part of the sea to the sill of the ridge. Tidal currents caused by this wave encounter an obstacle in the form of a bottom slope and acquire vertical components. Periodically varying vertical components of the currents cause vertical oscillations of isopycnic surfaces, and due to this mechanism, internal tidal waves are generated. Thus, according to the specified parameters of the exciting wave, stratification and bottom topography, the problem of calculating the parameters of baroclinic wave disturbances is solved. Numerical model calculations allow us to estimate the amplitude of internal waves and their energy fluxes in the immediate vicinity of the generation region.

The motions of water particles caused by internal waves occur along the characteristic lines. The group velocity vector is also directed along the same lines. Characteristic lines extend from the surface to the bottom reflecting from the surface and from the bottom. One full horizontal distance between two tangencies of the bottom or surface corresponds to the scale of the internal wavelength. When an ensemble of internal wave perturbations is generated, beams of characteristic lines depart from irregularities in the bottom topography. If the number of sources of perturbations over an uneven slope is large, a whole ensemble of such ray disturbances is generated. The vertical components of the perturbation wave vectors propagating up and down compensate each other. Thus, a standing wave or a vertical mode is formed. Obviously, the greater the mass transported by the barotropic tide through a section normal to the coast, the greater the amplitude of the internal tide.

We selected a working domain for the model calculations 720 km long with a horizontal step of 200 m and 20 vertical levels. The maximum depth of the sill in the Strait of Denmark is 650 m. The time step was 12 s. Such values completely satisfy the Courant-Friedrichs condition for the stable operation of the numerical model. The coefficients of horizontal viscosity and density diffusion were set equal to 2.0 m^2/s, and the corresponding vertical coefficients were equal to 0.0001 m^2/s. The numerical problem at these coefficients of viscosity and diffusion was solved at the edge of stability. Smaller values led to numerical instability due to the occurrence of strong nonlinear waves at large amplitudes, and at large values of the coefficients, wave motion occurred in an excessively viscous medium, which reduced the nonlinear effects of internal wave propagation that we are interested in.

We specify a constant current to the south throughout the entire water column at a speed of 15 cm/s and superimpose a periodical barotropic tidal current on the mean current. In the northern part of the strait, the vertical components of the flow through the ridge cause the ascent of isopycnals, and in the southern part they descend as a hydraulic jump. On the northern slopes of the ridge, the average current is directed towards the internal tidal wave, which makes the wavelength shorter, which leads to an increase in its amplitude. In the southern part, where the mean current and the tidal internal waves are directed to the south, the internal tide wavelength becomes larger, the amplitude of oscillations of isopycnals becomes smaller. This effect was previously observed in the Strait of Kara Gates (Morozov et al. 2003b; 2008).

The model simulations showed that if we start modeling from the unperturbed state of the ocean, then in order to get the estimates of the density field corresponding to the pattern of developed wave motion, it is sufficient to perform calculations during the physical time from five to six periods of the semidiurnal internal tide. After that, the calculated wave field would evolve according to the motions caused by the passage of an internal wave over a given topography of the ocean floor. In this case, changes in stratification will be caused by internal waves. Thus, it is possible to trace the generation of the internal wave over the underwater slope.

A model calculation of the wave motions of isopycnals is shown in Fig. 3.29. The model shows that immediately above the slope the oscillations of isotherms exceed 50 m; they rapidly decrease with the distance from the slope. The model estimates and the measured fluctuations of isotherms have similar values in the range of 30–50 m.

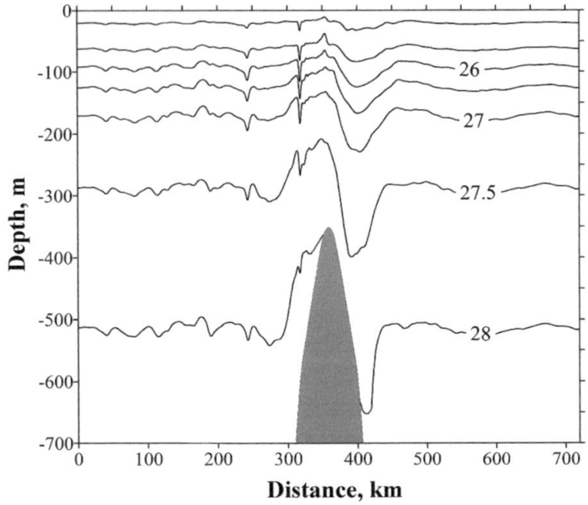

Fig. 3.29 Model simulation of the oscillations of isopycnals caused by internal waves. Density at the ocean surface is specified as 1.0253 kg/dm^3. Contour lines of density 1.027, 1.028 kg/dm^3 and others are denoted as 27 and 28, respectively. The density contour lines are drawn with an interval of 0.005 kg/dm^3. The gray color shows the sill of the ridge in the strait. The right part of the figure corresponds to the southern part of the strait

It was shown on the basis of the estimates of isotherm oscillations based on the measurements on a mooring in the Strait of Denmark and numerical simulations that the height of the semidiurnal internal tides near the sill of the transverse ridge across the Strait of Denmark exceeds 50 m. The amplitude decreases to 30 m at distance of about 100 km from the sill. In the northern part of the strait, internal waves are stronger, since the tidal wave propagates opposite to the mean current. In the southern part of the strait, internal waves and currents are directed to the south, which increases the wavelength and reduces their amplitude.

3.9 Internal Waves in the Region of the Akselsundet Strait of Spitsbergen

(This section was written in cooperation with S.V. Pisarev)

Tidal internal waves are generated in the ocean when barotropic tidal currents flow over topographic irregularities and cause periodic vertical displacements of isopycnal surfaces. Observations in the ocean indicate that tidal internal waves with a period of M_2 are most energetic. The stronger the currents of the barotropic tide, the greater the amplitudes of the generated internal waves. Latitude 74.5° is critical for linear internal waves with a semidiurnal tide period of 12.4 h, since at this latitude the tidal frequency becomes equal to the Coriolis parameter. In the flat bottom regions in the absence of eddy currents, the propagation of internal tides is not allowed by theoretical estimates. One of the conditions for the existence of semidiurnal tidal waves north of the critical latitude is a sloping bottom. Barotropic tide currents over a sloping bottom obtain a vertical component and generate an internal tide that exists as a forced wave only above the slope. The wave amplitude rapidly decreases as the wave propagates beyond the slope over a flat bottom.

Let us consider measurements on moorings in terms of identifying local properties of internal tidal waves north of the critical latitude. Three moorings were deployed at the mouth of the Van Mijen Fjord in the southern part of West Spitsbergen Island (Fig. 3.30). The fjord cuts deeply inland for about 80 km. From the ocean side, the fjord is almost completely blocked by Akseløya Island, which is about 8 km long. There are two straits to the north and south of the island. The northern strait is named after Akseløya Island (Akselsundet). The width of the strait is about 1200 m. The maximum depth of the strait based on the navigation chart is about 30 m [http://toposvalbard.npolar.no]. Two straits are located south of Akseløya Island (Mariasundet), which surround Maria Island (Mariaholmen), one 200 m wide and 2 m deep, the second is 500 m and 12 m, respectively. Due to the tide, very strong currents up to 3 m/s exist during spring tides. The Akselsundet Strait is navigable. Cargo ships with coal mined in the mine in the village of Svea far in the fjord can pass through this strait. Due to strong currents, the strait almost never freezes in winter, but ice is transported out of the fjord through the strait (Marchenko et al. 2015).

In August 2010, during expedition work on the Norwegian vessel *Haakon Mosby*, three moorings were deployed in the Van Mijen Fjord. We review our publication (Morozov and Pisarev 2023) The deployment of moorings was carried out within the framework of the project of the Norwegian Scientific Council for Norwegian-Russian cooperation in the field of researching the nature of Svalbard "Internal hydraulic processes in the Arctic fjord" (Scientific supervisors of the project were I. Fer from the Geophysical Institute of the University of Bergen and S.V. Pisarev from the Shirshov Institute of Oceanology of the Russian Academy of Sciences). A scheme

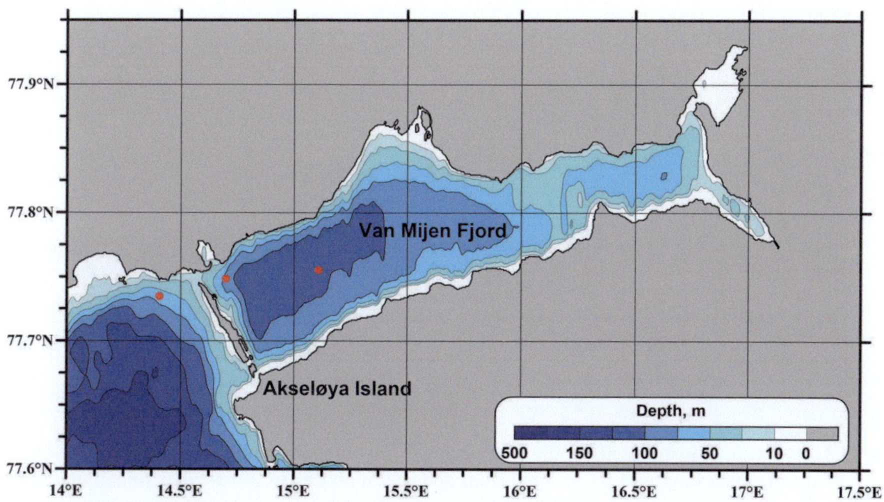

Fig. 3.30 Bottom topography in the region of the Van Mijen Fjord and scheme of moorings (red dots). Moorings are numbered from west to east (1, 2, 3)

of the location of moorings is shown in Fig. 3.30. Two stations were installed to the east and west of the Akselsundet Strait, and the third one was in a deep depression (about 115 m) in the middle of the fjord. The operation of all instruments of the three moorings began at 17:00 UTC on August 9, 2010, and ended at 09:00 UTC on September 5, 2011. Measurements on the instruments were carried out with an interval of 1 h.

Several temperature, salinity, and current meters were fixed on the wire of each of the three moorings: SBE37, SBE39, Aanderaa RCM7, Seaguard WLR. The holding buoyancy of each mooring was eight Vitrovex glass spheres. The bottom depths at the location of moorings were 71 m at mooring 1, 70 m at mooring 2, and 115 m at mooring 3.

The currents in the strait caused strong deviation of the mooring wire from the vertical position. Under normal tidal currents, fluctuations in depth were within 1–2 m. During the strongest spring tidal currents, the instruments lowered deeper. The maximum depth was noted on mooring 2: instead of the planned 36 m depth of the instrument during strong spring tidal currents, the current almost completely deepened the entire mooring wire to the bottom, and the instrument ended up at a depth of 68 m (Fig. 3.31). The moorings deepened at intervals of 14 days due to the spring-neap variations of the tide.

The maximum tides do not have to correspond exactly to the new moon and full moon. They can occur 1–3 days before or after the new moon and full moon [https://www.seatemperatu.re/europe/norway/svalbard/tides.html]. The maximum deepening of the mooring caused by strong tidal currents may, in turn lag behind the time of the maximum tidal currents, although they should occur at close times.

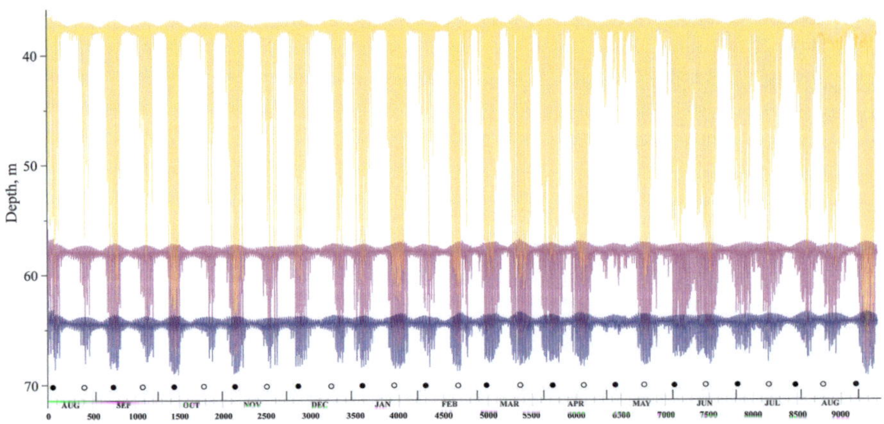

Fig. 3.31 Maximum instrument depths on mooring 2 (instruments at nominal depth 36 m: yellow line; 57 m: magenta line; and 63 m: blue line) and moon phases in 2010–2011 (black and open circles). The numbers along the abscissa indicate hours from the start of work. The nominal depth indicates the depth of the instrument that is not submerged by strong currents (Morozov and Pisarev 2023). Used with permission of the Pleiades Publishing, Ltd

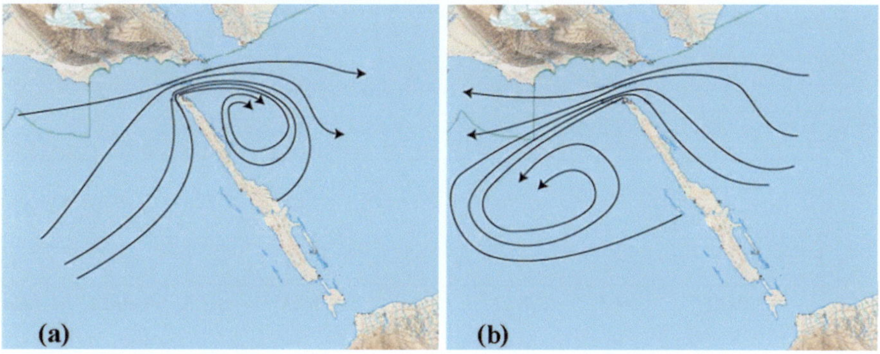

Fig. 3.32 Scheme of tidal currents (streamlines) in the Akselsundet Strait in the phases of **a** high tide and **b** low tide (Marchenko et al. 2015)

Mooring 2 was located closer to the strait (2.5 km east) than mooring 1 (5 km west) from the strait, mooring 3 was installed 12 km east of the strait. Note that the jet of a strong current in the strait continues at the outflow of the current from the strait, while at the inflow of the current into the strait, water collects from a small surrounding water region and does not form a jet before flowing into the strait. After six hours, the direction of the flow changes, and the jet flows at high speed in the other direction from the strait, forming an eddy. A scheme of currents in the strait is shown in Fig. 3.32. Current meters AADI RCM7 were deepened almost to the bottom along with the entire wire of the mooring during strong currents and the device recorded data from varying depth.

Due to the described property of the jet, strong deepening of the mooring (by tens of meters) occurred with an interval of about 12.4 h, while the direction of tidal currents changed with a period of about 6.2 h. Figure 3.33 shows examples of changing the position of the instrument on moorings 1 and 2 over time. When the mooring was deepened by the current from the east side of the strait, the western mooring was not deepened. It is notable that the eastern mooring was deepened stronger because it was located closer to the strait and was affected by a jet with a higher flow velocity.

The current velocities recorded by the current meters at a nominal depth of 36 m behaved similarly; during the period of strong currents, they were deepened almost to the bottom. The graphs show that strong currents on the eastern mooring were directed to the east (positive values of the east component), similar currents with a lower speed on mooring 1 were westward (negative). The moments of strong currents on different moorings were shifted by 6 h (Fig. 3.34).

Temperature variations based on the measurements on moorings show that in winter the entire water layer cools almost to the temperatures close to freezing (Fig. 3.35). The temperature of the entire layer down to the bottom was –1.87 °C. At salinities of 34.3–34.4 PSU, the freezing point is about –1.89 °C. In the summer months, positive temperature stratification was observed.

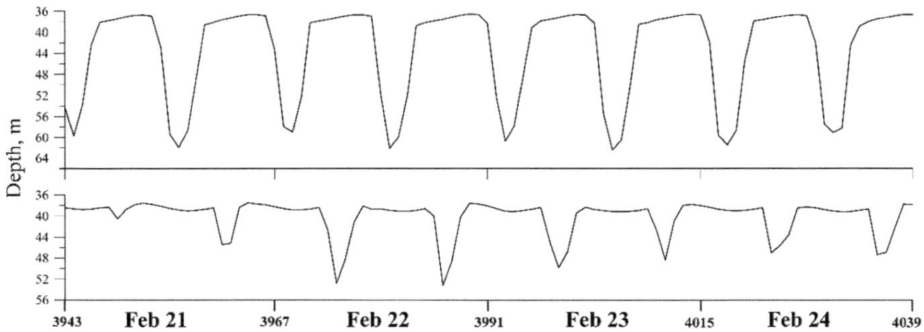

Fig. 3.33 Deepening of moorings 1 (bottom panel) and 2 (top panel) based on the sensors at a nominal depth of 36 m. Numbers 3943, 3967… show hours from the start of measurements (Morozov and Pisarev 2023). Used with permission of the Pleiades Publishing, Ltd

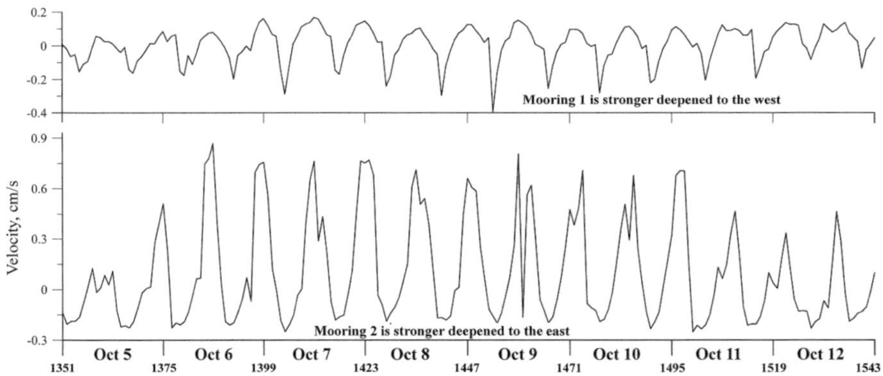

Fig. 3.34 Zonal velocities on mooring 1 (top panel) and mooring 2 (bottom panel) based on the current meters at a nominal depth of 36 m. Numbers 1351, 1375… show hours from the start of measurements (Morozov and Pisarev 2023). Used with permission of the Pleiades Publishing, Ltd

The study site was located in the zone of relatively warm waters transported from the south by the West Spitsbergen Current, which is a continuation of the North Atlantic and Norwegian currents. The open ocean does not freeze in winter. The entire fjord freezes by January, except for the part that is close to the strait. In some warm winters, a significant part of the fjord remains with open water. To the west, the current brings water with changing thermohaline properties to Akseløya Island. This is seen from the changes in temperature and salinity based on the data of moorings.

In the summer periods, the temperature of the upper layer was higher than that of the deeper layers. In autumn and spring, such stratification may change to warmer waters in the deeper layers. This is due both to the advection of waters with other properties from the ocean, and to the cooling of water from the surface and convection. This effect was especially notable during the autumn months when the surface water cools.

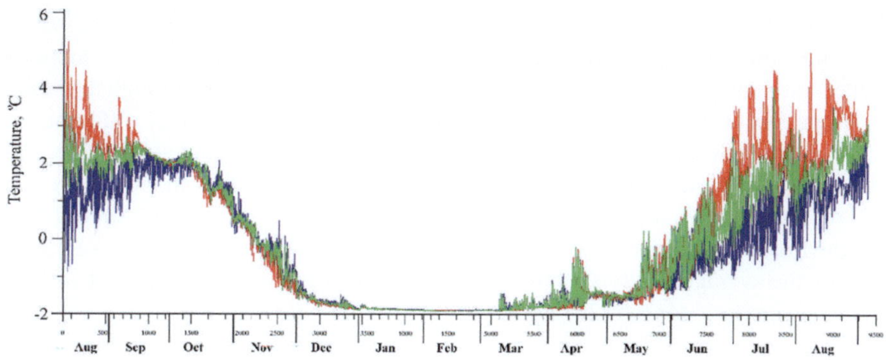

Fig. 3.35 Temperature variations on mooring 2 during the year. Sensors at nominal depths of 16 m (red line), 36 m (green line), and 63 m (blue line). Small tics and numbers from 0 to 9500 indicate hours from the beginning of the mooring operation (Morozov and Pisarev 2023). Used with permission of the Pleiades Publishing, Ltd

Let us consider the effect of stratification variations in autumn, when the temperature in the upper layer becomes lower than in the lower layer. This is illustrated in Fig. 3.36.

The higher temperature in the lower layer is compensated in terms of density by salinity and does not lead to unstable stratification. We show this in Tables 3.1 and 3.2 on the example of measurements on two dates, August 20, and November 1, when

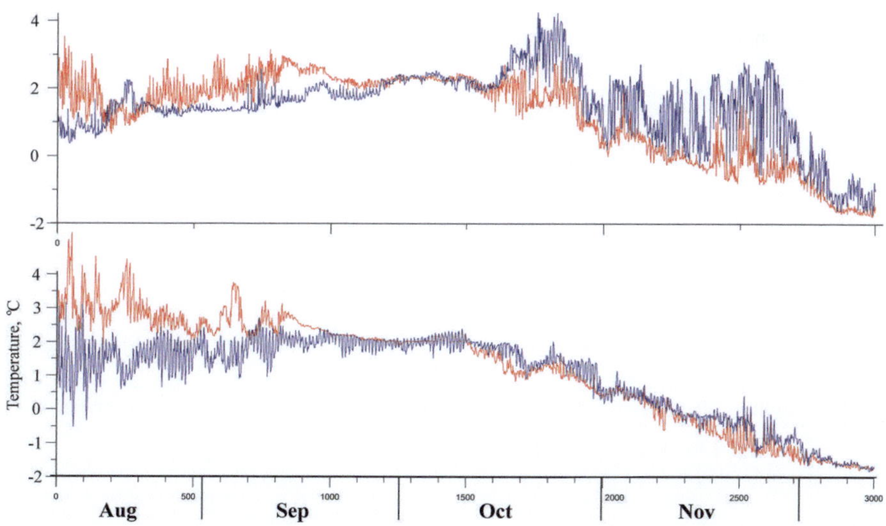

Fig. 3.36 Temperature variations in the upper layer (red curve) and lower layer (blue) on mooring 1 (top panel) and mooring 2 (bottom panel). Top panel: nominal depths 37 m and 57 m; bottom panel: nominal depths 16 m and 57 m (Morozov and Pisarev 2023). Used with permission of the Pleiades Publishing, Ltd

the moorings were not deepened by a strong current. Calculations show that when the temperature increases with depth, higher salinity in the deeper layers supports stable stratification. At the same time, despite the strong exchange through the strait, the properties of the waters to the west and east of the strait were very different.

Let us estimate the balance of water entering the Van Mijen Fjord at high tide. The area of the fjord is 480 km^2. The rise of the water level during high and low tides is $\pm$ 0.75 m. Thus, during six hours of high tide, $480 \times 10^6 \times 1.5\,\text{m}^3$ of water should flow into the fjord through the Akselsundet Strait. The width of the Akselsundet Strait is 1000 m, the average depth is 18 m. From here we can estimate the transport with an average velocity V for 6 h (21,600 s):

$$480 \times 1.5 \times 10^6 m^3 = 1000 \times 18V \times 21600\, m^3$$

We find that the mean flow velocity in the strait over 6 h is $V = 1.85$ m/s, which is consistent with the maximum flow velocity of the current jet in the spring period, which is about 3 m/s. The maximum transport by the tidal current in the strait is estimated as 0.05 Sv.

Temperature fluctuations are not very indicative for the study of internal waves, since strong currents periodically deepened the entire line of measuring instruments. Therefore, the measurements were not at the same depth. Due to the change in stratification and vertical temperature profile, it is difficult and even impossible to judge

Table 3.1 Mooring 1. Characteristics of temperature, salinity, and density with changing stratification

Nominal depth, m	Parameter	Time 250 h or August 20	Time 1680 h or November 1
37	T	1.25	1.93
	S	34.75	34.02
	σ	27.98	27.33
57	T	1.54	2.36
	S	34.65	34.16
	σ	28.02	27.55

Table 3.2 Mooring 2. Characteristics of temperature, salinity, and density with changing stratification

Nominal depth	Parameter	Time 250 h or August 20	Time 1680 h or November 1
16	T	4.10	0.87
	S	31.87	33.18
	σ	25.39	26.68
57	T	8.17	0.93
	S	33.46	33.52
	σ	26.32	27.15

the vertical displacements of water particles in internal waves. When the stratification and vertical temperature profile change, there is a period when the temperature is almost the same throughout the entire layer. A similar effect is observed during winter cooling. Nevertheless, the spectrum of temperature fluctuations reflects the main property of fluctuations determined by semidiurnal tidal internal waves (Fig. 3.37). Mooring 3 was deployed far from the strait, and the tidal internal wave should decay over such a distance in the region of supercritical latitudes. This is exactly what the temperature spectrum on mooring 3 shows: the peak at frequency M_2 is weak. Salinity stratification persists even in winter; therefore, the salinity spectra calculated over the entire series of measurements show confidence peaks at the frequency of the semidiurnal tide M_2 (Fig. 3.37).

Oscillations of isotherms are more informative; these fluctuations are equivalent to the vertical oscillations of water particles. Let us calculate the changes in the depths of isotherms based on the available measurements on moorings at several depths, bearing in mind that we know from the pressure sensor at what depth the temperature measurements were taken at each moment of time, while the mooring was deepening under the influence of currents.

One can see from Fig. 3.38 that changes in stratification with time lead to the variation in the depths of isotherms within the considered depth range and their locations are replaced by isotherms of other temperatures. This figure shows the fluctuation of isotherms during a period when typical stratification of the summer period was maintained and warm water was located higher than cold water. Let us

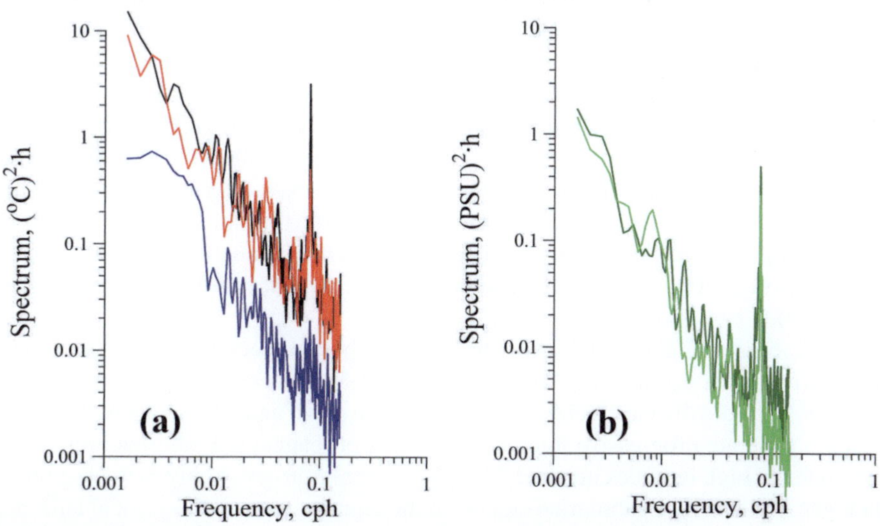

Fig. 3.37 Temperature spectra **a** from temperature measurements at a nominal depth of 64 m, mooring 1 (black line), 37 m, mooring 2 (red line), and 65 m, mooring 3 (blue line). Salinity spectrum **b** at mooring 1 at a nominal depth of 57 m (dark line), 37 m (light line) (Morozov and Pisarev 2023). Used with permission of the Pleiades Publishing, Ltd

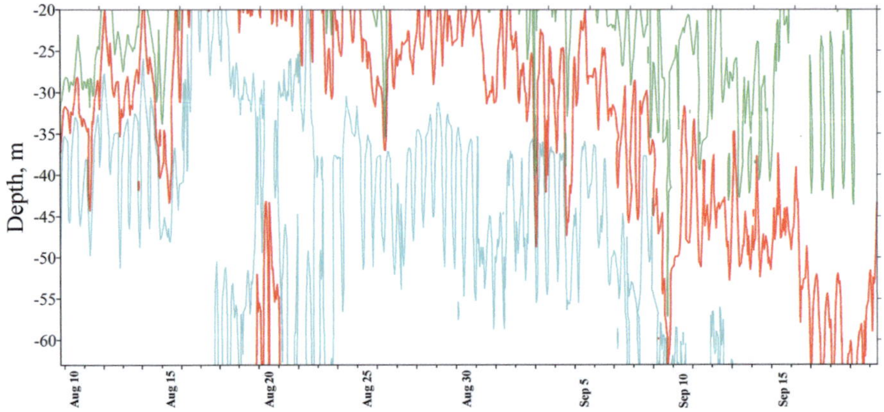

Fig. 3.38 Fluctuations of isotherms on mooring 1 in the period from August 9 to September 19, 2010. Isotherm 2.5 °C: green line, 2.0 °C: red line, 1.5 °C: blue line (Morozov and Pisarev 2023). Used with permission of the Pleiades Publishing, Ltd

isolate the 2 °C isotherm, digitize the depths of this isotherm in time and calculate the spectrum of vertical displacements of water particles from it, assuming that the molecular and turbulent heat exchange are small (that is, heating and cooling of water particles is insignificant during the considered time period of about 20 days) from August 10 to 30, 2010.

Figure 3.39 shows the oscillations of the 2 °C isotherm in time in the depth range 18–65 m. Visually, the amplitude of the oscillations of the isotherm with a period of 12 h can be estimated as 8–9 m with a displacement of water particles of about 16–18 m. The depth fluctuations of the 2 °C isotherm were subjected to bandpass filtering so that semidiurnal fluctuations remained. The result is shown by the red line in Fig. 3.39. Isotherms fluctuate with an amplitude of about 7–9 m. The initial and final sections of the series were cut off due to bandpass filtering. The remaining oscillations after filtering were apparently caused by the formation of internal waves on a sloping bottom. The spectrum calculated from the fluctuations of the isotherm over a period of 500 h shows a reliable peak at the frequency of the semidiurnal tide (Fig. 3.40).

Generation of tidal internal waves on a sloping bottom occurs because barotropic tidal currents run up the slope and induce vertical displacements of the isopycnal surfaces. Periodic displacements of the pycnocline down and up relative to the mean position generate progressive internal waves that propagate in both directions from the slope. At high latitudes north of 74.5° N, internal tides exist only in the region of their generation above submarine slopes. In the regions north of this critical latitude, the linear theory of internal waves over a flat bottom does not allow the existence of solutions for waves with tidal frequency M_2 over a flat bottom, since this frequency is smaller than the local Coriolis frequency (LeBlond and Mysak 1978).

Therefore, waves decay over a flat bottom. Semidiurnal internal tides are often observed over submarine slopes (Marchenko et al. 2021; Pisarev 1988, 1991, 1992,

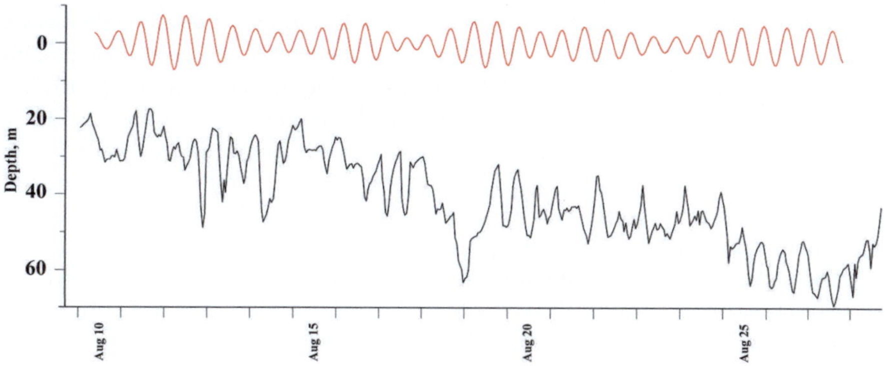

Fig. 3.39 Fluctuations of the 2 °C isotherm depth in the period of August 10–25, 2010, when this isotherm was in the depth range 18–65 m (black line). Isotherm oscillations after band-pass filtering, which left only oscillations with a period of 12.4 h (red line). Data from mooring 1 (Morozov and Pisarev 2023). Used with permission of the Pleiades Publishing, Ltd

Fig. 3.40 Spectrum of fluctuations of the 2 °C isotherm depth in the period August 10–25, 2010 based on mooring 1 data (Morozov and Pisarev 2023). Used with permission of the Pleiades Publishing, Ltd

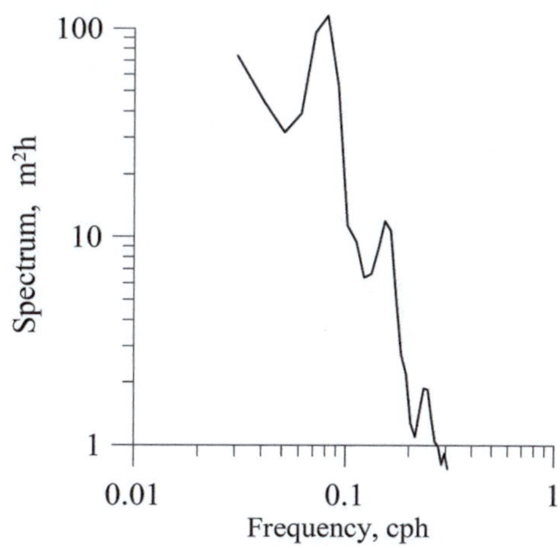

1996). It was shown in (Pisarev 1992, 1996; Kurkina and Talipova 2011) that in the Arctic regions near submarine slopes, internal oscillations with a period of the local barotropic tide are observed, whereas far from the slopes, only packets of short-period waves are observed over a relatively flat bottom. This is confirmed by the model presented in (Kurkina and Talipova 2011; Morozov and Paka 2010).

Velocity of barotropic tide currents in the strait was estimated from measurements on a mooring. Only the time interval was considered when the mooring was located vertically, and the currents were not very strong. The ocean stratification (undisturbed state) was taken from the data of CTD measurements and the temperature and salinity profile recorded by the instruments on the mooring. Unperturbed density field by internal waves has been specified with the corresponding Brunt-Väisälä frequency distribution $N(z)$. A long tidal barotropic wave propagates from the open part of the sea, onto the underwater slope of the strait. The tidal currents caused by this wave, encounter an obstacle in the form of a slope and obtain a vertical component. This leads to the generation of baroclinic disturbances. The formulation of the problem is reduced to determining the field of generated internal waves from the known parameters of stratification, bottom topography, and barotropic tide velocity.

Measurements of temperature, salinity, and currents in the region of the strongest tidal currents of the Akselsundet Strait near the mouth of the Van Mijen Fiord in the southern part of West Spitsbergen Island showed that currents, when flowing over the underwater transverse sill in the strait, generate intense internal waves of the semidiurnal tidal period. Waves are forced over the underwater slope and decay propagating from the strait and slope. Displacements of water particles in internal waves reach 20 m at depths above the slope of about 70 m. Tidal currents in the strait reach a speed of 3.5 m/s and form an intense jet as they flow out of the strait. When flowing into the strait, water is collected from the water basin near the strait and does not form a concentrated jet when flowing. It is formed when flowing out of the strait in its eastern or western parts. An intense jet of currents during spring tides causes that the entire line of the mooring with instruments and buoyancy almost lies on the bottom. The instruments were equipped with pressure sensors, which made it possible to calculate the fluctuations of isotherms in time, despite such a vertical motion of the entire mooring. At a distance of 12 km from the strait, the semidiurnal internal wave was no longer recorded at the mooring.

3.9.1 Sea Ice Evolution and Internal Wave Generation Due to a Tidal Jet in a Frozen Sea

This Section is Written Together with K. Filchuk, Z. Kowalik, and I. Ryzhov

We investigated the influence of tidal currents on sea ice in Spitsbergen fjords, which may cause rapid decrease of the ice thickness due to erosion and melting of the ice. The effect was studied in-situ near the narrow channel connecting the Van Mijen Fjord and Lake Vallunden. The strong jet-like tidal currents in the strait driven by semidiurnal tide continue into the lake preventing ice freezing along a narrow strip during high tide and relatively warm weather. Understanding the formation of open water regions or regions with thin ice is important for the safe transportation on ice. We estimate conditions and representative time over which strong tidal current influences ice thickness along a narrow strip in solid ice. Changes of tidal phase and decrease in air temperature influence freezing of the strip in one-two days. While the tidal flow leaves the strait it overflows a shallow bar and generates internal lee waves propagating downslope and mixing the water. Tidal forcing of internal waves was measured using pressure gauges and by scanning the ice surface during flood and ebb phases. Internal waves were measured using three types of CTD instruments and an ADCP current meter. The generation of wave packets occurs every tidal cycle when the current flows into the lake, but no generation occurs during the ebb phase of the tide because the currents over the bar slope are low. Parameters of internal waves were estimated. Model simulations confirm generation of internal wave train by the tidal current descending downslope. In this section we review publication (Morozov et al. 2019b).

Short-period internal waves generated by a discharge water flow from rivers were analyzed in (Nash and Moum 2005; Stashchuk and Vlasenko 2009; Osadchiev 2018). Nash and Moum (2005) observed internal waves generated by the outflow of the Columbia River to the ocean, while Stashchuk and Vlasenko (2009) modeled this process. Osadchiev (2018) made observations of wave generation by small mountainous rivers flowing into the Black Sea. A model of such generation by the tidal flow was suggested by Maxworthy (1979). Theoretical and laboratory modeling of such processes was considered in (Wood and Grue 2003; Carr et al. 2008; Fructus et al. 2009). Propagation of wave trains from bottom topography has been demonstrated by satellite imagery in the Andaman Sea (Osborne and Burch 1980) and by direct measurements in Massachusetts Bay (Haury et al. 1979) and by (Farmer and Smith 1980) over the sill in Knight Inlet in Canada. Generation of packets of short-period internal waves by tidal currents in the Strait of Gibraltar and surface manifestations of these packets have been reported in many publications. Watson and Robinson (1990) present a review of such observations in 20 papers.

Formations of open water regions or regions with thin ice are typical phenomena for ice covered waters. Prediction and pointing out of such regions is of practical importance because of the safe navigation in the regions of floating ice. There are several locations in Spitsbergen fjords where such phenomena are recorded. Typical examples are related to the open water formation in navigational straits Akselsundet and Heleysundet shown in Fig. 3.41. Strong currents in the straits prevent ice formation. Maximal currents recorded in the Akselsundet Strait ranged from 2.5 to 5 m/s (The Norwegian Pilot 2016; Marchenko and Kowalik 2017). Regions of circular shapes known as Murdoch-hole are systematically formed in a specific place of the Temple Fjord (Fig. 3.41). The Murdoch-hole effect is associated with strong local currents (Johansen 2000; Svalbardposten 2011).

It is known that regions with thin ice may appear in the Tempel Fjord, at the entrance to the Dickson Fjord and in Van Mijen Fjord, and near Akseloya. In February 2007, the track with sledges of 10 tons weight fell through the ice near Svea. The ice thickness was only 37 cm in the place where the accident has happened, while the mean thickness of ice in the fjord was greater than 50 cm (Svalbardposten 2007). In 2017, a tourist group fell through the ice crossing the Tempel Fjord in the same place where they travelled safely one day before (Svalbardposten 2017). It shows that local ice conditions can change relatively fast depending on the weather conditions, sea currents, and internal waves driven by tides.

Charts and aerial photo of the region are shown in Fig. 3.42. It was found that the maximum speed of the tidal current directed along the fjord reaches 50 cm/s in the ice season, but it does not have a significant influence on the ice thickness. The maximum speed of the sea current measured at the entrance into Lake Vallunden (Fig. 3.42a) in September 2010 exceeded 1.5 m/s. Numerical simulations and field

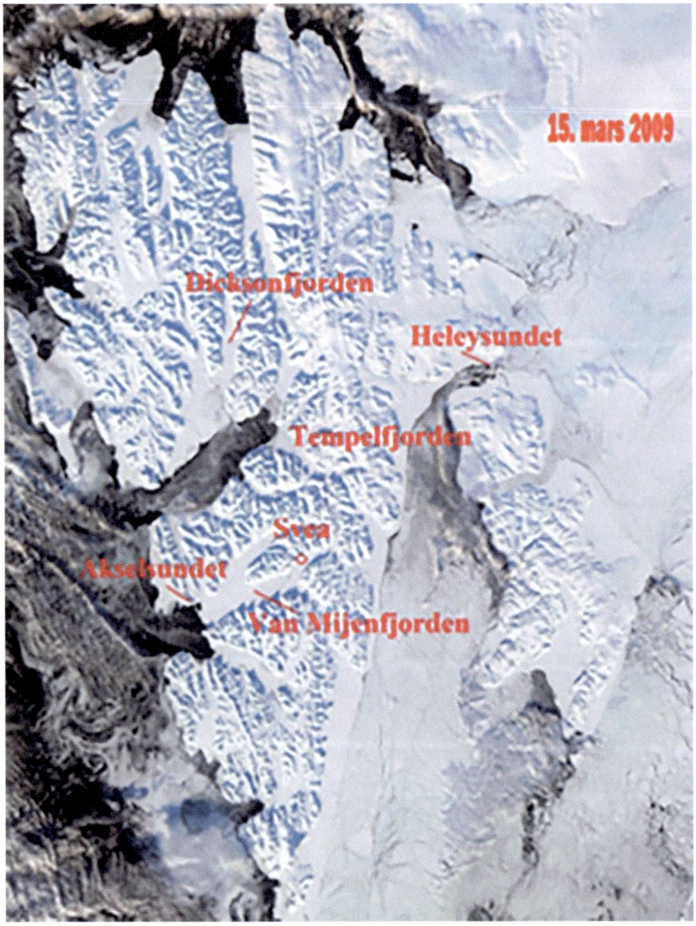

Fig. 3.41 Satellite image of Spitsbergen (UNIS archive) on March 15, 2009. Numerous field works were performed in the shallow area located in the end of the Van Mijen Fjord, which is called Braganzavagen (Marchenko and Morozov 2013; Wrangborg et al. 2013; Shestov et al. 2015)

measurements showed that the maximum speed of sea current at the entrance from the strait into the lake can reach 2.5 m/s in the ice-free season (Marchenko and Morozov 2013). Strong tidal currents develop in the channel because the amplitude of the tidal elevation in the Van Mijen Fjord varies between 0.6 and 2 m, while in the lake it does not exceed 0.5 m (Marchenko and Morozov 2013; Kowalik et al. 2015). Field observations show that the strait from the fjord into the lake is sometimes ice free in the winter season, and sometimes covered by thinner ice in comparison with the ice in the fjord. In winter 2013, A. Marchenko fell through the ice in the water during ice thickness measurements near the entrance into the strait at the fjord side even though the mean ice thickness in the fjord was 70 cm.

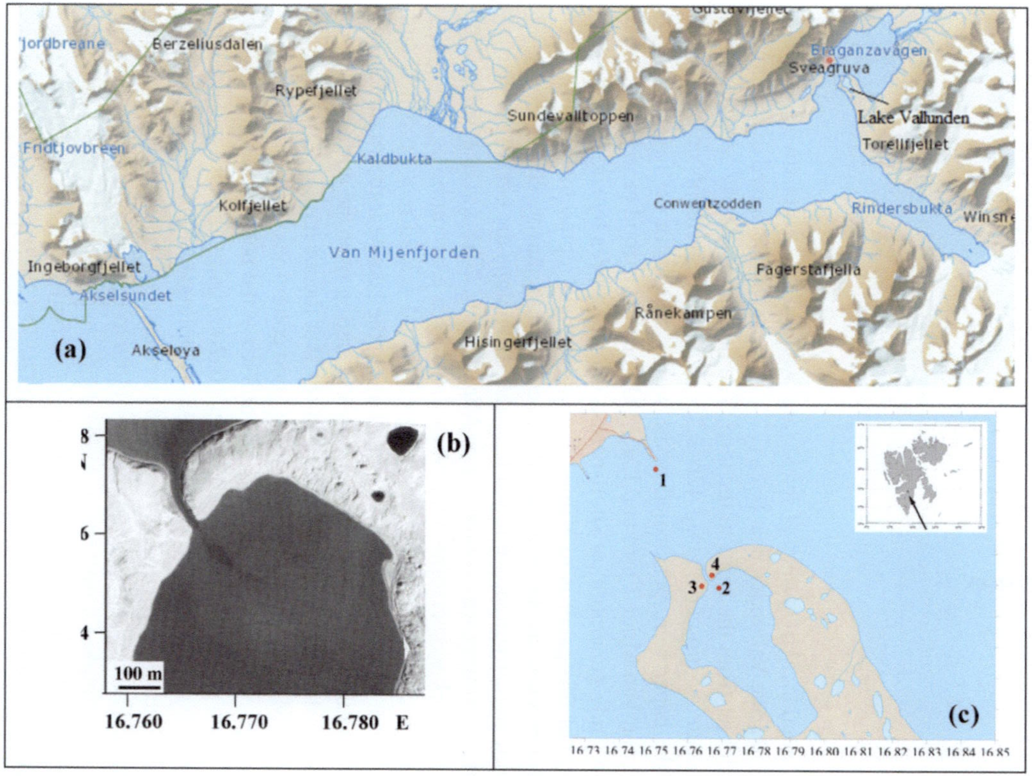

Fig. 3.42 **a** Chart of the Van Mijen Fjord; **b** Aerial photograph of the lake and channel during the phase of tidal current flowing into the lake (aerial image s2011_25163_00668_13 on July 26, 2011, 7:30 AM CET, courtesy of the Norwegian Polar Institute); **c** Chart of the location. Northern and eastern coordinates are indicated at the axes. Spitsbergen map is shown in the inset. Red dots indicate the following locations: (1) pressure gauge in the fjord; (2) pressure gauge in the lake; (3) location of time laps camera Reconix and laser scanner Riegl Vz-1000; (4) summer photo in Fig. 3.53 (Morozov et al. 2019b)

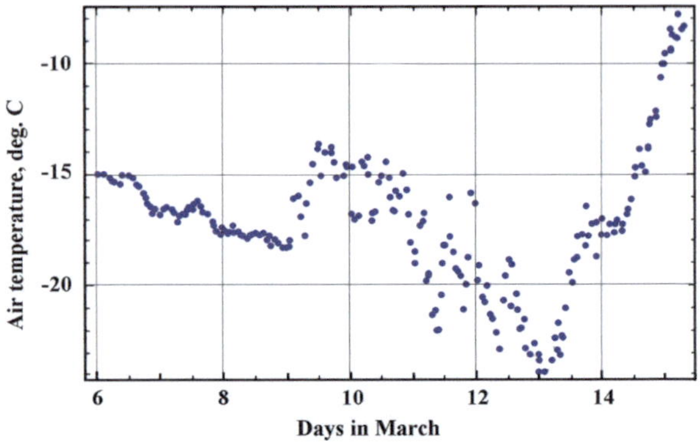

Fig. 3.43 Record of the air temperature in March 2018 from the Svea meteorological station over the period of the field works. (Morozov et al. 2019b)

We investigated local dynamics of sea ice and water in the strait between the Van Mijen Fjord and Lake Vallunden near Svea (77°53′ N, 16°46′ E) caused by the inflow of a strong tidal jet from the fjord to the lake. The tidal jet is clearly seen on aerial photographs made in summer (Figs. 3.42b, 3.53). Field measurements near the strait described in the paper were performed in the winter season for the first time. The strait between the Van Mijen Fjord and Lake Vallunden is a good example of "field laboratory" to study generation of short-period internal waves by surface currents. The motivation of this paper is to estimate conditions and representative time over which strong tidal currents influence ice thickness along a narrow strip in the solid ice and generate high-frequency internal lee wave downslope the bar.

Our recent experiment was conducted on March 7–16, 2018 in Lake Vallunden. During the field work we performed investigations of sea ice morphology and standard oceanographic measurements including CTD and sea current profiling, and point records of the water temperature and salinity in the lake near the strait. Record of the air temperature on the Svea meteorological station over the entire period of the field works is shown in Fig. 3.43. One can see that the air temperature dropped below –20 °C on March 12–13 providing very good conditions for the ice growth.

We have analyzed sea level data measured at Svea at the head of the fjord over a period close to two years starting on September 10, 2007 using the data collected by Caline (2010). The strongest tidal constituent M_2 (amplitude 51 cm) was very stable during the entire measuring period. The second strongest constituent S_2 varied in the periodic fashion. The cycle of the variability is close to the semiannual period with the maximum amplitude (close to 30 cm) for the S_2 occurring in winter (January) and in summer (July), while the minima (close to 10 cm) are in April and October. The timing points towards the variability are related to the equinox and solstice. Superposition of the two components with the largest amplitudes, i.e., M_2 and S_2 resulted in the new oscillations with the semidiurnal period and monthly period. The monthly temporal variability is manifested either as the upper or the lower envelope of the diurnal signal. The resulting amplitude from these two envelopes is changing with the fortnightly period from the spring tide (sum of the amplitudes), to the neap tide (difference between the amplitudes).

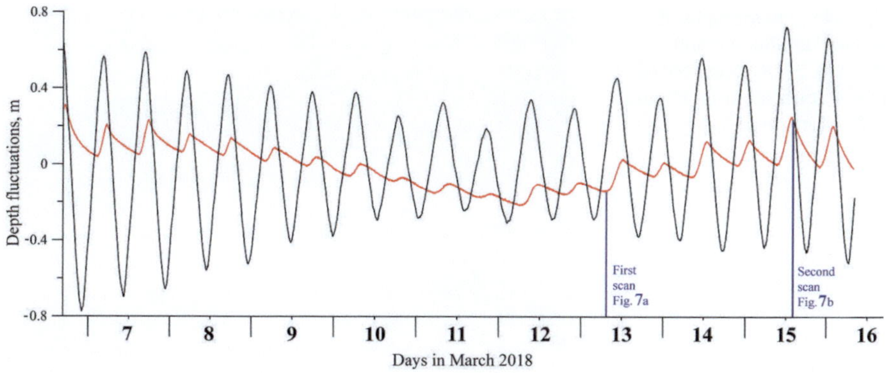

Fig. 3.44 Tidal variations in the water elevation in the fjord (black line) and lake (red line) (Morozov et al. 2019b)

During our latest measurements in 2018 the sea level in the fjord and in the lake were recorded by two pressure gauges (SBE 37 and SBE 39) deployed at locations 1 and 2 shown in Fig. 3.42c. Measurements were performed with a sampling interval of 2 min. Figure 3.44 shows the recorded oscillations of the sea level in the fjord and in the lake over a time span close to 10 days. The main period corresponds to the semidiurnal tides. In the figure the linear interaction of semidiurnal constituents M_2 and S_2 results in the semidiurnal period and fortnightly (spring-neap) period of amplitude variation, which is about half-month. Full Moon was on March 3, last quarter was on March 9, and new Moon was on March 17. The record given in Fig. 3.44 shows that the amplitude of the semidiurnal tide in the lake (red line) is smaller than the amplitude of the water level oscillations in the fjord at point 1 (Fig. 3.42c). Ice cover decreases water exchange between the fjord and lake. This effect of exchange decrease appears only in the winter time; it is explained by a constriction of the channel near the fjord and at the sides due to the ice formation (Marchenko and Morozov 2013).

Let us consider Sea ice morphology. It was observed that the strait from the fjord into the lake was ice free on March 7–10, then the surface of the strait was covered by slush and finally the strait was frozen on March 14. In the beginning of the field work on March 7 it was found that the ice-free channel in the strait continued into the lake as a narrow strip (Fig. 3.45, top panel). The strip became covered by slush and then frozen on March 8. The strip region remained frozen until the end of the field works, but the ice cover there was thinner (Fig. 3.45, bottom panel). Dynamics of slush in the strait and strip was investigated using time laps images of Reconix camera mounted at location 3 in Fig. 3.42c. Removing of slush and thin ice from the water surface inside the strip by the semidiurnal water flow is clearly seen on the images taken with a time interval of 1 min on March 7–9. Thus, ice erosion by water flow is a physical process causing the cleaning of the channel and strip from the ice. The erosion increases as the air temperature and the water flow speed increase.

Fig. 3.45 Photographs of the lake on March 7 and March 8, 2018. Recovery of ADCP at Location 1. (Photos by E. Morozov)

Drilling studies showed uneven distribution of the ice thickness near the strip. The chart of ice thickness distribution is shown in Fig. 3.46a. The ice thickness around the strip sharply increases from 25 cm at the ice edge to 40 cm at the distance of about 1 m from the ice edge on March 8. The ice thickness was about 60 cm at the distance of about 8 m from the ice edge. In the continuation of the narrow strip the ice thickness was almost 120 cm, while the mean ice thickness in the lake was about 70 cm. We explain the growth of ice thickness by the transport of small ice pieces and freezing water under the ice cover by the strong current and their accumulation under ice where the current speed decreases. The ice thickness in the strip reached 20–25 cm on March 8, and after March 9 it was save to walk over the frozen strip.

Laser scanning of the ice surface in the lake and fjord was performed to find the actual changes of the ice surface elevation over the tidal cycle. We used terrestrial scanner Riegl ZV-1000 providing 3D point cloud (www.riegl.no). The accuracy (conformity with actual value) and precision (degree to which further measurements show the same results) of the scanning are 8 and 5 mm, respectively, at a maximum

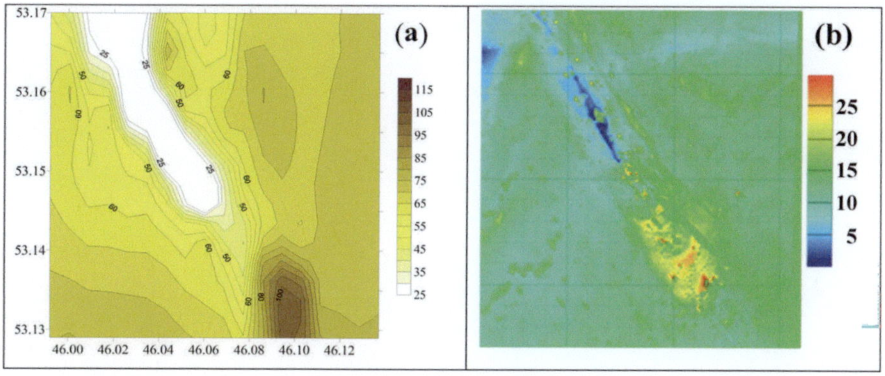

Fig. 3.46 Chart of ice thickness (cm) around the strip on March 8 **a** and scan of the ice surface near the frozen strip on March 15 **b**. The height scale is in cm. The numbers along the vertical axis in **a** indicate minutes of latitude 77° N, the numbers along the horizontal axis indicate minutes of longitude 16° E (Morozov et al. 2019b)

distance of 1400 m. Several scans were performed from the hill on the western side of the strait connecting the lake and fjord (scan location 3 is shown in Fig. 3.42c). Snow surface of the lake and fjord was captured by the scanner as a point cloud consisting of several millions of points with coordinates in project/scanner coordinate system. Point clouds were processed using the RiScan software and colored with respect to the elevation using relative color bar (in meters). The mean level of the measured surface of the fjord and lake was chosen as the zero reference level.

Two scans with the most clearly seen elevation difference are shown in Fig. 3.47. These scans were made almost in the opposite phases of the tide during low and high water (Fig. 3.44). The first scan was performed at 09:21 UTC on March 13 2018 (low water tide). The second scan was performed at 15:08 UTC on March 15 2018 (high water tide). The fjord is located on the upper (northern) side of the plots; the lake is on the lower (southern) side. The mean level of the measured surface of the fjord and lake was chosen as the zero reference level. The scan in Fig. 3.47a shows that the ice surface in the lake near the strait was about 20 cm below the reference level when the tide elevation in the lake was at the local minimum (Fig. 3.44). The scan in Fig. 3.47b shows that the ice surface in the lake near the strait was about 20 cm above the reference level, and the ice surface in the fjord near the lake was about 10 cm below the reference level when the tide elevation in the lake was at the local maximum (Fig. 3.44).

The scan in Fig. 3.47b shows that ice surface in the middle of the strait is uplifted by the tidal flow. It looks like a crest on photographs from March 15. The ice has significant bending deformations across the strait. The scan in Fig. 3.47a does not show this effect because the strait surface was covered by slush on March 12, when the scan was performed.

Figure 3.46b shows ice surface around the frozen strip from the scan made on March 15. It is clear that the ice surface in the continuation of the strip was 20–25 cm

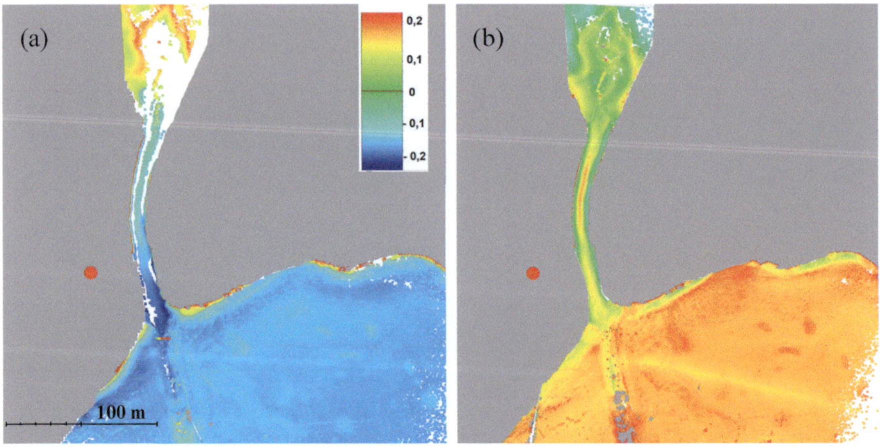

Fig. 3.47 Deviation of the ice surface from the mean level at minimal **a** and maximal **b** water levels in the lake. Color scale indicates relative deviation in meters. Land is shown with gray color. Points 1 and 2 show locations of ADCP deployments. Locations of laser scanner are shown with red dots. Distance scale (100 m) is shown in the left panel. The color scale of heights is in cm (Morozov et al. 2019b)

above the ice surface inside the strip. It can be explained by the freezing of locally overflowing water over the ice surface at the very end of the strip during the time when the strip was ice free. Buoyancy force applied to the thick ice also influences the elevation of the ice surface, but it cannot be localized over the area with a scale of 2–3 m.

Let us consider currents in the tidal jet. The ADCP profiler (RDI Sentinel, 1238 Hz) was deployed in the tidal jet on March 8, 2018, before the ice-free strip of water got frozen. Average ensemble interval of 2 min, bottom tracking mode, bin size of 5 cm, and blanking distance of 49 cm were programmed in the deployment. The ADCP was strongly fixed at the surface on a wooden "bridge" across the jet. Actual measurements of sea current profile were performed from a depth of 88 cm to the bottom. A sketch of the deployment is shown in Fig. 3.48. A section of the velocity along the strait and velocity records in Bin 001 (88 cm depth), Bin 003 (1.03 m depth), Bin 005 (1.13 m depth), Bin 0030 (2.4 m depth), and Bin 050 (3.4 m depth) are shown in Fig. 3.49 together with synchronous records of the water surface elevation in the fjord and in the lake (Locations 1 and 2 in Fig. 3.42c).

On March 8 the instrument was placed at Location 1 (Fig. 3.47a) almost at the end of the strip (Fig. 3.44). The depth at this location was approximately 4 m. During the intense tidal flow, the recorded velocities exceeded 30 cm/s almost in the entire water column, but water speed reduction in the boundary layer below the ice was stronger than near the seabed. On March 9, the instrument was replaced at Location 2 (Fig. 3.47a) almost in the beginning of the strip. The depth at this location was approximately 1.1 m. The maximum water velocities exceeded 30 cm/s. Such strong velocities were recorded almost in the entire water column. Water speed reduction

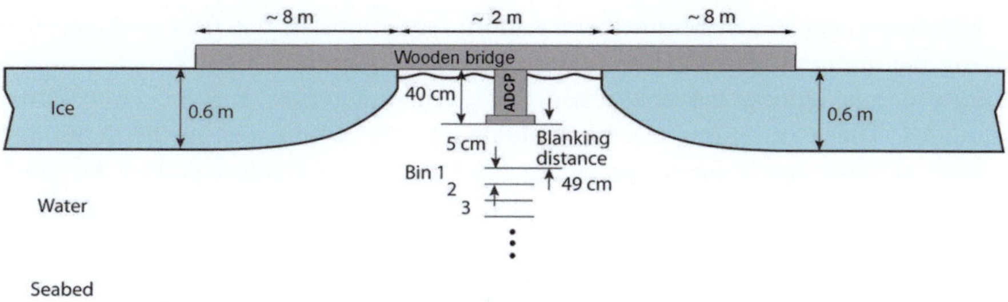

Fig. 3.48 Sketch of the ADCP deployment (Morozov et al. 2019b)

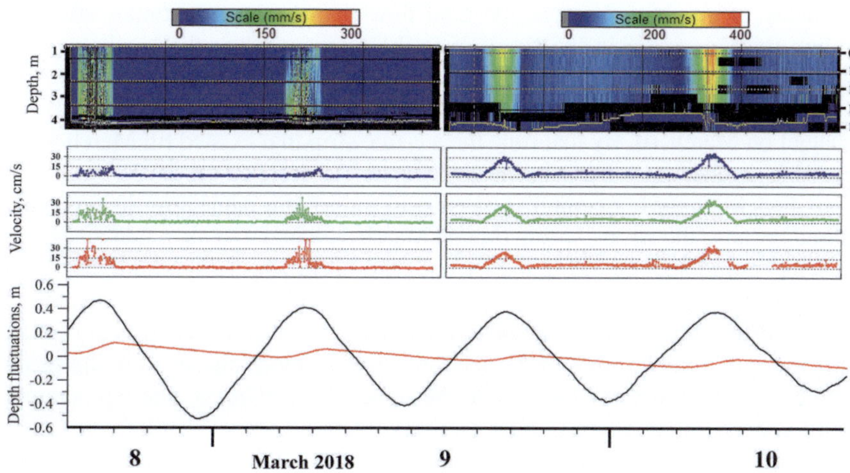

Fig. 3.49 Time scan of the velocity (cm/s) along the jet (meridional component) measured by ADCP (top panel) and several records from different ADCP bins. Tidal level measured by pressure gauges in the fjord (black line) and lake (red line) demonstrates the timing of the tidal forcing (bottom panel) (Morozov et al. 2019b)

in the boundary layer near the seabed is clearly seen; the maximum velocities were recorded in the first bins.

These high velocities were observed only at the flood tide during the time interval when the tidal flow was directed into the lake. At the ebb tide the velocities were low because water was flowing from the lake. However, our summer observations show that the velocities in the strait are almost the same at the two phases of the tide. Because our measurements were made in the lake approximately at 50 m from the strait, the jet that continued the flow in the channel was found in the lake only during the flood phase of the tide. We believe that during the ebb phase, the water in the lake converges into the strait from different directions and the jet could be observed on the other end of the strait in the fjord. Tidal graphs in Fig. 3.49 demonstrate that the maximum velocities are observed at the time of the maximum tide in the lake when the sea level differences between the fjord and the lake are the greatest.

Measurements of temperature and salinity variations in the fjord indicate that during the flooding phase of the tide the water temperature slightly increases and salinity drops. During the ebb phase the salinity increases and the temperature decreases. These variations can be explained by the differences in the depth and volume of water in the open part of the fjord and in the shallow end of the fjord. During the flood phase, water is transported in the fjord from its wide and deep part in the west. During the ebb phase the water is transported back from the narrow and shallow (up to 1 m depth) eastern part of the fjord (Braganzavågen) where strong freezing occurs; hence the water temperature decreases and its salinity increases owing to the salinification of water during freezing due to brine rejection from the ice. The maximum and minimum tidal elevations in the lake occur approximately with a delay of one hour because the tidal flow into the lake continues after the maximum tidal elevation in the fjord up to the moment when the levels become equal at both ends.

A CTD (conductivity, temperature, depth) section was occupied along the strip of the tidal jet on March 10. By that time the ice-free strip that existed till March 9 has frozen. This could happen because the air temperature dropped to –20 °C unlike the previous days when the temperature was close to zero. At the same time the tidal flow decreased because the strong spring tide on March 3 changed to neap tide on March 10. A temperature section along the frozen strip is shown in Fig. 3.50. The tidal stream from the strait was flowing under ice. We see the warmer water transported by the tidal flow in the right part of the figure (closer to the strait) whereas the colder water is located in the left part of the section, which is more distant from the strait. The strongest variability is observed in the upper layer, whereas the temperature distribution deeper than 1.5 m is more uniform.

We compared the measured water temperatures and salinities with the θ, S-plot of the sea water freezing at the atmospheric pressure from (Fofonoff and Millard 1983) (https://www.mt-oceanography.info/Utilities/freeze.html). The measured values and the freezing graph are shown in Fig. 3.51. All the measured values on the θ, S-plot are close to the freezing temperatures but still warmer. The moored measurements data (SBE-19) that extend in a line parallel to the freezing graph show water masses with slightly different temperature and salinity. These variations result from the appearance of slightly different waters transported by the tidal jet. The CTD-profile measurements occupy a region close to the freezing line. They were measured almost at the same tidal phase but at different depths.

The strong tidal current inflowing into the lake, bottom topography, and density stratification in the narrow strip of the water are preconditions for the generation of the internal lee waves. To define the properties of these oscillations two instruments (SBE-19 and SBE-37) were deployed in the tidal jet at the depths of 1.35 and 1.55 m, respectively. The instruments operated simultaneously with a time sampling of 20 s. The maximum fluctuations of temperature were observed during strong currents from the strait. This occurred because the jet overflows a shallow (1.5 m) stone bar located immediately at the end of the strait. The high local velocities at the constriction caused instability of the flow over the slope and generation of internal lee waves. Knowing the vertical temperature gradient between the two temperature sensors

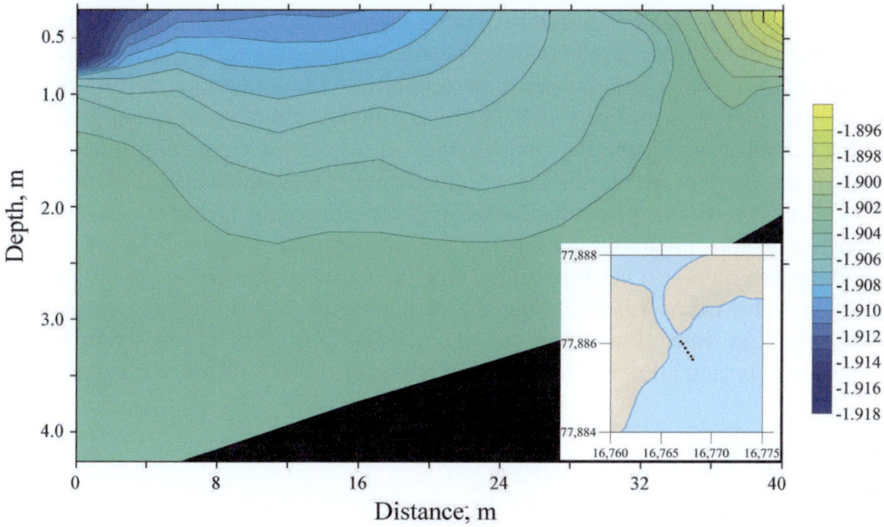

Fig. 3.50 Temperature section under ice along the strip line and its continuation. The black color shows the bottom profile. Temperature color scale is on the right. A chart of the locations of CTD casts is in the inset (Morozov et al. 2019b)

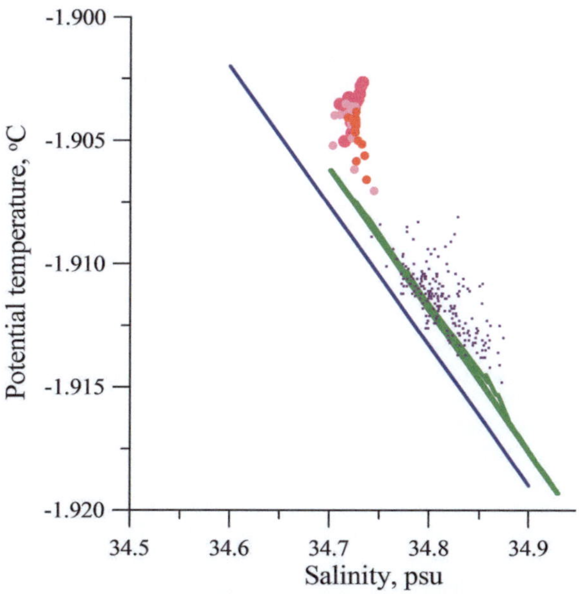

Fig. 3.51 Freezing graph based on the equation from (Fofonoff and Millard 1983) (blue line). Green line is based on the moored measurements using the SBE 19 instrument. Violet dots are from the moored SBE 37 instrument. Dark red, light red and pink dots are from CTD profiles (Morozov et al. 2019b)

it was possible to recalculate temperature fluctuations into vertical displacements. The spectra of temperature fluctuations at two instruments reveal reliable peaks at a frequency of 0.0718 cyc/min, which corresponds to 14 min period (Fig. 3.52).

The fluctuations with a period of 14 min are likely caused by lee waves generated by the current overflowing the shallow bar. Lee waves are frequently found in the ocean when currents overflow a topographic feature (Mackinnon 2013; Wright et al. 2014). Lee waves are generally found in the deep ocean; however, the physical processes responsible for lee wave generation are the same in the shallow water. They are generated owing to the instability of the shear current overflowing underwater obstacles. The Brunt-Väisälä frequency in the region (N) where relatively warm and less saline water flows in to the lake is quite high: 0.01 cps. Laboratory and numerical experiments of (Hazel 1972; Kalinichenko 2005) and also our experiments

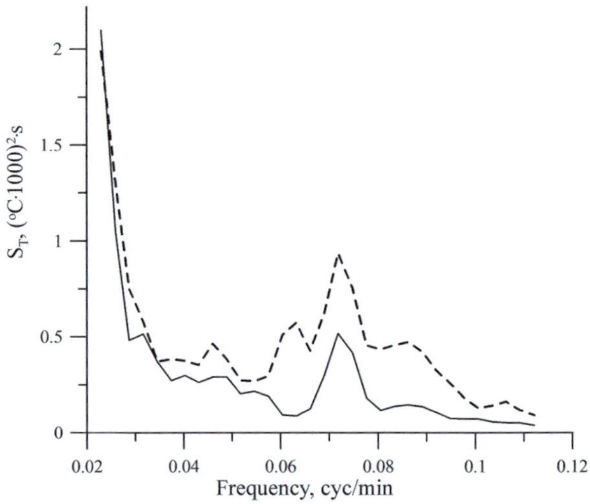

Fig. 3.52 Spectra of temperature fluctuations measured by two instruments SBE 19 at 1.35 m (solid line) and SBE 37 at 1.55 m (dashed line) (Morozov et al. 2019b)

Fig. 3.53 Manifestation of internal waves at the surface of the lake (photo by E. Morozov, September 29, 2009, 7:40 PM CET). The photo was taken from the point near the channel approximately in the southern direction, see Fig. 3.42c. Photo by E. Morozov

in the Van Mijen Fjord (Morozov and Marchenko 2012; Marchenko and Morozov 2016; Marchenko et al. 2010) indicate that the dominating periods of short-period internal waves are equal to a few minutes. They correspond to an internal wave with a dimensionless wavenumber 0.5. The perturbations with the dimensionless wavenumber equal to 0.5 obtain energy faster than the other perturbations (Hazel 1972; Kalinichenko 2005). If we assume that the Brunt-Väisälä frequency is $N = 0.01$ cps and the characteristic size is the depth of the lake ($H = 5$ m) the dimensional wavenumber becomes $k = 0.5/H = 0.1\ \text{m}^{-1}$.

The period of waves with $k = 0.1\ \text{m}^{-1}$ is equal to approximately 10 min. It is seen from the observations that the spectral peak corresponds to the temperature fluctuations with a period of 14 min, which is reasonable because the Brunt-Väisälä frequency is variable and the bottom is not even. The most likely process of internal wave generation is due to the instability of the current, which occurs over the sloping bottom where the tidal jet overflows the stony bar. The internal waves radiate from this point forming a fan-shaped structure as seen in the summer photo (Fig. 3.53).

Wave packets appear with a periodicity of 12 h as the strong tidal current from the fjord appears. We performed band filtering of the time series to separate the fluctuations with a period of 14 ± 3 min. The temperature fluctuations from the measurements by two instruments reveal perturbations coinciding in time with the inflowing tidal current. In order to recalculate the temperature fluctuations into vertical displacements we calculated the vertical temperature gradient for each time moment of sampling by subtracting the temperatures measured at two instruments. Then we divided the difference by the distance between the sensors (20 cm).

Each deviation from the zero mean value of the band-filtered time series of temperature fluctuations was then divided by the vertical temperature gradient to find the vertical displacements. After excluding unreasonable peaks, we obtained a time series of vertical displacements shown in Fig. 3.54. The mean amplitude of vertical displacements in each of the three time series exceeded 10 cm, but the maximum ones were as high as 30 cm. Wave packets coincide in time with the maximum velocities of the tidal jet associated with the maximum sea level difference between the fjord and lake shown in the bottom panel. The structure of the 14-min internal wave train on a larger scale is shown in Fig. 3.55.

The energy densities of internal waves (E_{IW}) were estimated according to (Torgrimson and Hickey 1979; Lozovatsky et al. 2003). The density of the kinetic energy of the horizontal components of internal waves was determined as the sum of squared amplitudes of velocity. The total energy density of internal waves was calculated using the following formula:

$$\mathrm{E}_{IW} = 0.25\rho(u^2 + v^2 + N^2\varsigma^2) \tag{3.7}$$

Here, u and v are the horizontal velocity components, and ζ is the vertical displacements caused by internal waves.

The velocities in the channel connecting the lake and the fjord according to the measurements in (Marchenko and Morozov 2013) are of the order of $U = 1$ m/s. The kinetic energy density (per cubic meter) of the flow in the channel can be estimated

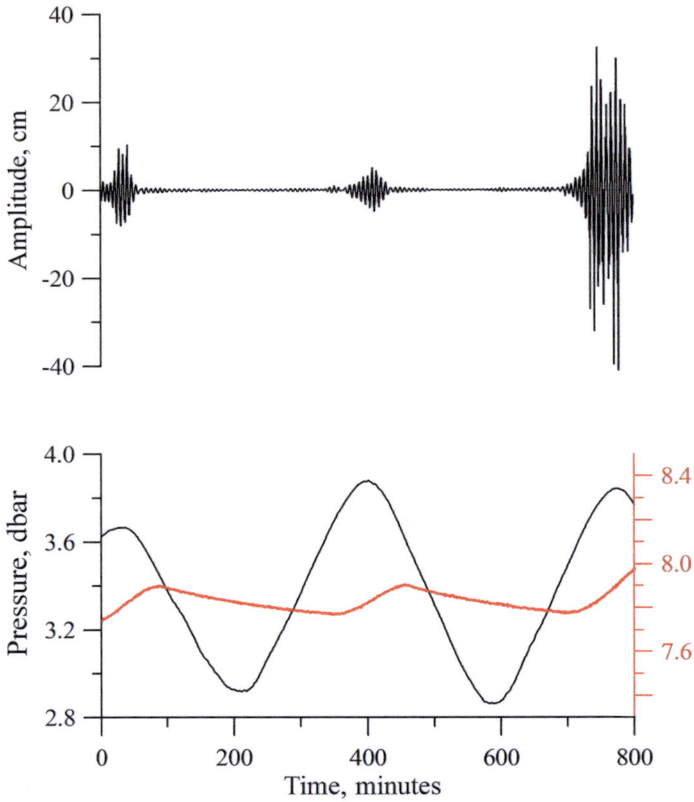

Fig. 3.54 Wave packets of internal waves of 13.9 min period generated by the tidal jet over the submarine slope (top panel). Tidal fluctuations of depth caused by tide based on the data of pressure time gauges in the fjord (black line) and lake (red line) are shown to illustrate the forcing (bottom panel) (Morozov et al. 2019b)

as ($\rho = 1027$ kg/m^3):

$$\mathrm{E}_{KIN} = 0.25\rho(U^2) = 257\mathrm{J/m}^3 \tag{3.8}$$

The energy density of internal waves at the point of measurements at a distance of approximately 50 m from the channel is estimated according to formula (3.7) as (we neglect transversal velocities *u*):

$$\mathrm{E}_{IW} = 0.25\rho(u^2 + v^2 + N^2\zeta^2) = 0.25\rho(v^2 + N^2\zeta^2) = 3.7\mathrm{J/m}^3 \tag{3.9}$$

Here, $v = 0.12$ m/s, N $= 0.01$ cps, $\zeta = 0.2$ m. The transversal component of velocity ($u \sim 0$ m/s) is small compared to the longitudinal component v. The energy of vertical displacements is small compared with the horizontal fluctuations, which also decrease with the distance from the channel. The energy of internal waves

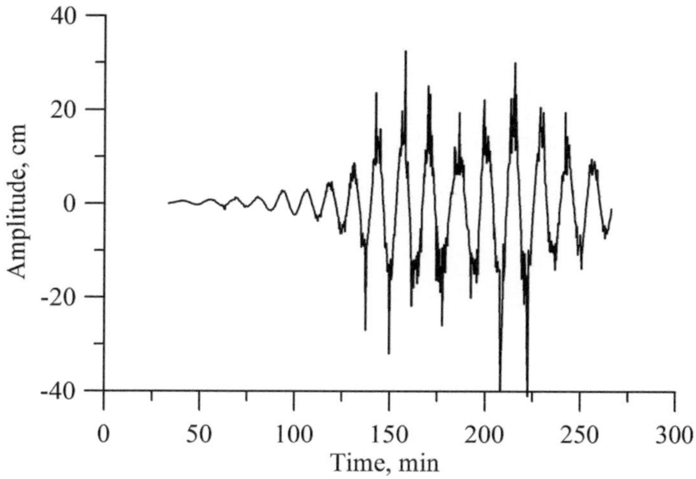

Fig. 3.55 Wave train of internal wave with a period of 13.9 min (Morozov et al. 2019b)

decreases almost by two orders of magnitude compared to the energy of the tidal flow in the channel, which is not surprising because of the radial spreading of the flow from the channel and energy transfer to ice motion and other processes rather than only to internal waves.

Let us compare field observations of internal waves with the results of numerical modeling. A two-dimensional numerical model developed by Vlasenko (2002), Vlasenko (1992) was applied for analysis. The model is based on the full nonlinear equations of hydrodynamics, which take into account wave processes, turbulent exchange, and density diffusion. We consider a two-dimensional (x, z) flow in continuously stratified rotating ocean of variable depth. A third coordinate is introduced to take rotation into account, but the variations along this coordinate are not considered.

Since internal waves occur under ice and the vertical displacements of the surface are negligibly small, the "rigid lid" condition at the surface is assumed. The calculations start from the state of rest when the fluid at the bottom is motionless, and the isopycnals are horizontal. The density field unperturbed by internal waves is specified based on our CTD-casts, and the corresponding distribution of the Brunt-Väisälä frequency is introduced. The density at the surface was specified as 1.02755 g/cm^3. The domain for model simulations was 800 m long with a horizontal step of 0.5 m and 20 vertical levels. The time step was 0.62 s.

It is important that during the phase when the tidal current flows to the lake it has a form of a jet due to the high velocities of the current outflowing from the strait. The jet is continuing in the lake. There is no jet in the lake in the opposite phase of the tide because the water converges to the stream in the channel from the vicinities of the entrance to the channel. Hence, in the model the maximum velocity of the tidal flow during the inflowing current is taken as 12 cm/s, while that of the outflowing current was only 2 cm/s.

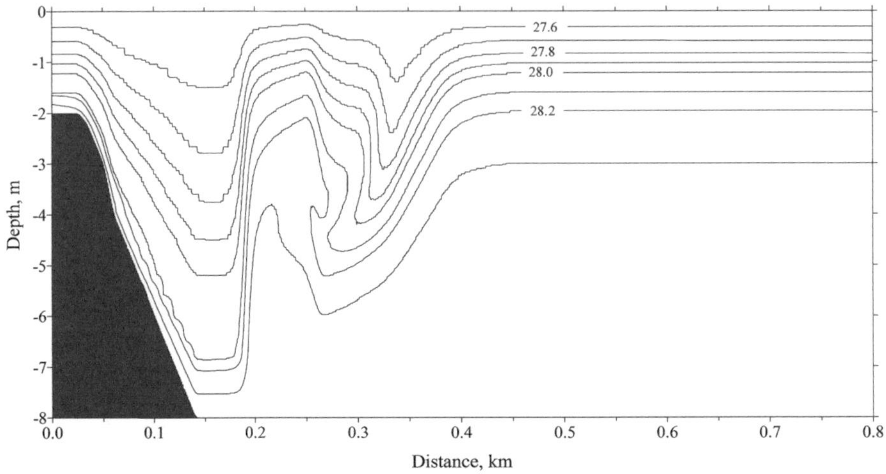

Fig. 3.56 Density field perturbed by internal waves. The internal wave pulse starts to overturn. The time corresponding to this snapshot is 0.25 of the tidal period after the inflow currents began to increase from zero. Contour lines of density from 1.0276 to 1.0283 g/cm^3 are shown based on the numerical simulations. The bottom is shown with gray color (Morozov et al. 2019b)

When the tide propagates into the lake, the fast tidal flow descends down the slope as the gravity current and a pulse of internal waves is generated down the slope. This pulse has a steep trailing edge and a flatter leading edge. In addition to the phase speed of the wave the pulse is transported by the tidal current. The currents in the upper layers of the pulse exceed the translation velocity of the pulse and it begins to overturn. This is seen in Fig. 3.56; this figure shows the density field perturbed by internal waves.

As the wave overturns, a depression of isopycnals is formed ahead of the wave pulse. The depression is followed by another front, which leads to the formation of internal oscillations of shorter period. When the maximum velocity of the tidal flow is reached, a packet of shorter internal waves is formed that follows this new front. The distribution of isopycnals is shown in Fig. 3.57. The length of short internal waves is close to 50 m, and their period is approximately 15 min, which we observed in the measurements.

Formation of the depression before the internal wave pulse is also seen on the time series of density measured by the SBE 19 instrument at a depth of 1.5 m. Figure 3.58 shows graphs of the time series of density during the formation of a wave packet. One can see in Fig. 3.58 that density sharply decreases before the maximum tidal velocities are reached (which coincides with the maximum difference in tidal elevation) and then a packet of short period oscillation appears.

Thus, we can conclude that we studied a tidal jet flowing from the Van Mijen Fjord to Lake Vallunden during 10 days in the ice season. Sea ice evolution around the jet was determined by the water–ice erosion and heat exchange between the water and air. In the beginning of the field works tidal currents were stronger and the air

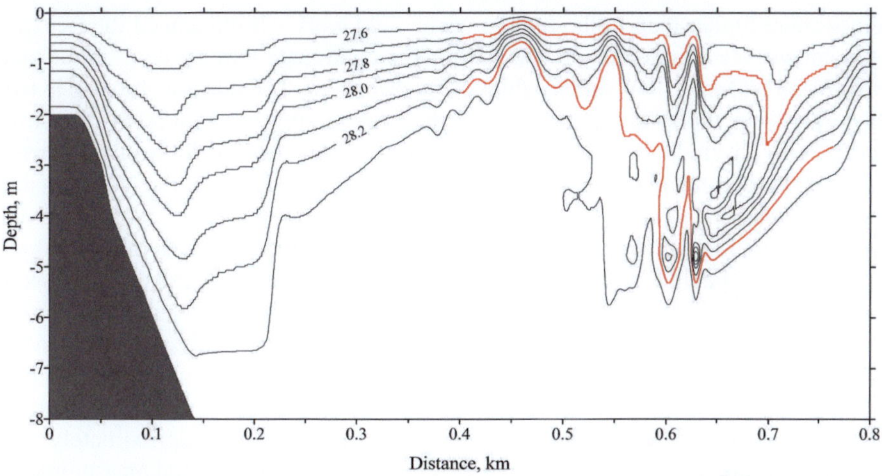

Fig. 3.57 Density field perturbed by internal waves at the moment of maximum currents of the tide. Wave packet of short internal waves is formed in the upper layer. The tidal jet and the wave propagate from left to right. Contour lines of density from 1.0276 to 1.0283 g/cm^3 are shown based on the numerical simulations. Fluctuations of density 1.0277 and 1.0282 are shown with red color. The depression and high frequency wave perturbations are highlighted with the red color. The time corresponding to this snapshot is 0.50 of the tidal period after the inflow currents began to increase from zero. The bottom is shown with gray color (Morozov et al. 2019b)

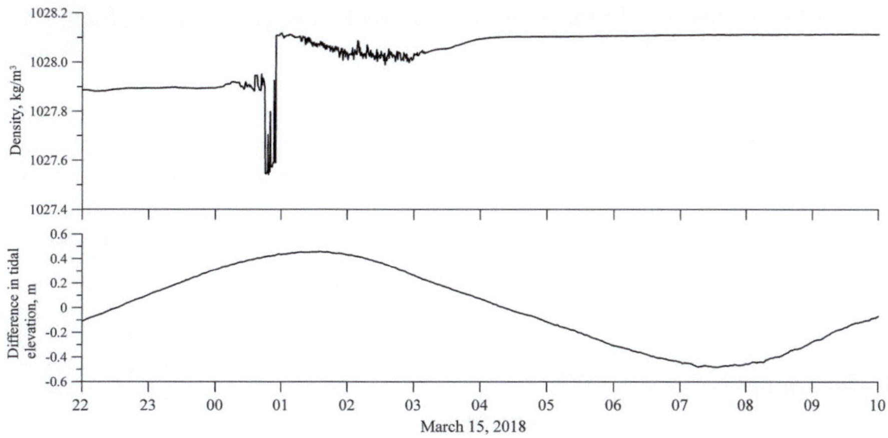

Fig. 3.58 Time series of density (top panel) and difference in the sea surface elevation due to tide between fjord and lake (bottom panel) showing depression of isopycnals before the onset of internal wave train. Time scale is shown in the bottom panel (Morozov et al. 2019b)

temperature higher than in the middle of the field works. It influenced the freezing of water surface around the jet. The strip in the continuation of the strait in the lake was frozen on March 8–9, and ice thickness reached 20–25 cm inside the strip in two days. The strip width was about 2 m, and the strip length was about 50 m. The strait surface was covered by solid ice on March 14. It is explained by the stronger current in the strait in comparison with the continuation of the tidal jet in the lake. Outside the strip, sea ice thickness was about 70 cm. It is found that after the flow passes the strait, the tidal jet overflows a shallow bar and generates packets of internal lee waves downslope. Forcing of internal wave generation was measured using pressure gauges in the fjord and in the lake. Forcing was also studied by means of scanning the ice surface elevation. Pulses of waves were generated every tidal period when the tidal current flows into the lake. The measured period of internal waves was 14 min and their amplitude was up to 20–30 cm over a total depth of 5–6 m. Numerical modeling confirmed the existence of short-period internal waves approximately with the same parameters as the measured waves. Numerical modeling and measurements of temperature and salinity show that as the speed of tidal currents increases the wave starts to overturn and a packet of short period internal waves appears.

Tidally driven currents below ice with water speed of about 1 m/s may lead to strong erosion of the sea ice and reduction of the ice thickness. The region around the strong current may not be frozen in relatively warm weather and then covered by thin ice when the air temperature drops below –20 °C. These changes may occur in few days or even in 1–2 days depending on the characteristics of tidal current and weather conditions. In the regions with smaller current speed (~0.5 m/s) the erosion is weaker and the influence of atmosphere cooling is stronger. Nevertheless, proximity of regions with strong current may influence ice thickness in the regions with smaller currents. Periodical generation of short-period (14 min) internal wave trains accompanies propagation of tidal jet under ice downslope when a tidal current overflows a stone bar and influences heat exchange over the water column. Wave packets with a period of approximately 14 min are generated with a time interval of 12.4 h but not generated during the oppositely directed currents because the flow is not strong. Numerical model supports the field experiment.

References

Androsov AA, Voltsinger NE, Kagan BA, Salusti ES (1993) Residual tidal circulation in the Strait of Messina. Izv Atmos Ocean Phys 29(4):543–552

Atlas (1998) Joint U.S.-Russian Atlas of the Arctic Ocean: oceanography atlas for the winter and summer period (CD-ROM). Environmental Working Group. National Snow and Ice Data Center (NSIDC), Boulder, Colorado

Baines PG (1982) On internal tide generation models. Deep-Sea Res 29(3):307–338

Burenkov VI, Ershova SV, Kopelevich OV, Sheberstov SV, Shevchenko VP (2001) An estimate of the distribution of suspended matter in the Barents Sea waters on the basis of the SeaWiFS satellite ocean color scanner. Oceanology 41(5):622–628

Caline F (2010) Coastal-sea-ice action on a breakwater in amicro-tidal inlet in Svalbard. PhD thesis, Norwegian University of Science and Technology. p 221

Carr M, Fructus D, Grue J, Jensen A, Davies PA (2008) Convectively induced shear instability in large amplitude internal solitary waves. Phys Fluids 20:126601

Egbert GD, Erofeeva S (2002) Efficient inverse modeling of barotropic ocean tides. J Atmos Ocean Tech 19(2):183–204. https://doi.org/10.1175/1520-0426(2002)019%3c0183:EIMOBO%3e2.0.CO;2

Farmer DM, Smith JD (1980) Generation of lee waves over the sill in knight inlet. In: Freeland HJ, Farmer DM, Levings CD (eds) Fjord oceanography. NATO conference series (IV Marine Sciences), vol. 4. Springer, Boston, MA. https://doi.org/10.1007/978-1-4613-3105-6_18

Filyushkin BN, Moshonkin SN, Myslenkov SA, Zalesny VB, Kozhelupova NG (2013) Modeling of long-term and seasonal variability of bottom current water flow in the Denmark Strait. Oceanology 53(6):725–736

Fofonoff P, Millard RC (1983) Algorithms for computation of fundamental properties of seawater. UNESCO Techn Pap Mar Sci 44:53

Fructus D, Carr M, Grue J, Jensen A, Davies PA (2009) Shear-induced breaking of large internal solitary waves. J Fluid Mech 620:1–29. https://doi.org/10.1017/S0022112008004898

Garrett C, Kunze E (2007) Internal tide generation in the deep ocean. Annual Rev Fluid Mech 39:57–87

Garrett C, Munk W (1972) Space-time scales of internal waves. Geophys Fluid Dyn 3(3):225–264

Gjevik B, Nøst E, Straume T (1994) Model simulations of the tides in the Barents Sea. J Geophys Res 99(C2):3337–3350

Haury LR, Briscoe MG, Orr MH (1979) Tidally generated internal wave packets in Massachusetts Bay, U.S.A.; preliminary physical and biological results. Nature 278:312–317

Hazel P (1972) Numerical studies of the stability of inviscid stratified shear flows. J Fluid Mech 51(5):39–61

Johansen GE (2000) Kapp Murdoch Polynyen. En våge i en arktisk fjord. Bachelor project in Geofysik. Københavns Universitet, Bohr Instituttet, Geofysisk Afdeling

Kagan BA, Timofeev AA (2005) Dynamics and energetics of surface and internal semidiurnal tides in the White Sea. Izv Atmos Oceanic Phys 41(4):844–850

Kagan BA, Timofeev AA (2015) Modeling of the stationary circulation and semidiurnal surface and internal tides in the Strait of Kara Gates. Fund Appl Hydrophys 8(3):72–79

Kalinichenko VA (2005) Development of a shear instability in nodal zones of a standing internal wave. Izv Fluid Dyn 40(6):956–964

Konyaev KV, Plueddemann A, Sabinin KD (2000) Internal tide on the Yermak Plateau in the Arctic Ocean. Izv Atmos Oceanic Phys 36(4):500–509

Konyaev KV, Sabinin KD (1992) Waves in the Ocean. Gidrometeoizdat, Leningrad, p 272. [in Russian]

Kowalik Z, Marchenko A, Brazhnikov D, Marchenko N (2015) Tidal currents in the western Svalbard Fjords. Oceanologia 57:318–327

Kozubskaya GI, Konyaev KV, Plueddemann A, Sabinin KD (1999) Internal waves at the slope of Bear Island from the data of the Barents Sea Polar Front Experiment (BSPF-92). Oceanology 39(2):147–154

Krauss W (1966) Interne Wellen. Gebrüder Borntraeger, Berlin–Nikolasee

Kurkina OE, Talipova TG (2011) Huge internal waves in the vicinity of the Spitsbergen Island (Barents Sea). Natural Hazards Earth System Studies 11:981–986

LeBlond PH, Mysak LA (1978) Waves in the ocean. Elsevier oceanographic series. Elsevier, Amsterdam, p 602

Levine MD, Paulson CA, Morison JH (1985) Internal waves in the Arctic Ocean: comparison with lower latitude observations. J Phys Oceanogr 15:800–809

Lozovatsky ID, Morozov EG, Fernando HJS (2003) Spatial decay of energy density of tidal internal waves. J Geophys Res 108(C6):3201–3216

Mackinnon J (2013) Mountain waves in the deep ocean. Nature 501:321–322

Marchenko AV, Morozov EG (2013) Asymmetric tide in Lake Vallunden (Spitsbergen). Nonlin Proc Geophys 20:935–944. https://doi.org/10.5194/npg-20-935-2013
Marchenko AV, Morozov EG, Muzylev SV, Shestov AS (2010) Interaction of short internal waves with ice cover in an Arctic fjord. Oceanology 50(1):18–27
Marchenko AV, Morozov EG, Muzylev SV, Shestov AS (2011) Short-period internal waves under an ice cover in Van Mijen Fjord, Svalbard. Adv Meteorol 2011:573269. https://doi.org/10.1155/2011/573269
Marchenko A, Morozov E, Muzylev S (2013) Measurements of sea ice flexural stiffness by pressure characteristics of flexural-gravity waves. Ann Glaciol 54:51–60
Marchenko AV, Morozov EG, Kozlov IE, Frey DI (2021) High-amplitude internal waves southeast of Spitsbergen. Cont Shelf Res 227:104523. https://doi.org/10.1016/j.csr.2021.104523
Marchenko A, Kowalik Z (2017) Investigation of ocean currents in navigational straits of Spitsbergen. In: Proceedings of the 12th International conference on marine navigation and safety of sea transportation (TransNav 2017), 21–23 June 2017. Gdynia, Poland, pp 267–272
Marchenko AV, Morozov EG (2016) Surface manifestations of the waves in the ocean covered with ice. Russ J Earth Sci 16(1):ES1001. https://doi.org/10.2205/2016ES0005612016
Marchenko A, Kowalik Z, Brazhnikov D, et al (2015) Characteristics of sea currents in navigational strait Akselsundet in Spitsbergen. In: Proceedings of the 23rd International conference on port and ocean engineering under arctic conditions, June 14–18, 2015. Trondheim, Norway
Maxworthy T (1979) A note on the internal solitary waves produced by tidal flow over a three-dimensional ridge. J Geophys Res 84(C1):338–342
Morozov EG (1995) Semidiurnal internal wave global field. Deep-Sea Res 42(1):135–148
Morozov EG, Marchenko AV (2012) Short-period internal waves in an Arctic fjord (Spitsbergen). Izv Atmos Oceanic Phys 48:401–408
Morozov EG, Paka VT (2010) Internal waves in a high-latitude region. Oceanology 50(5):668–674. https://doi.org/10.1134/S0001437010050048
Morozov EG, Pisarev SV (2002) Internal tides at the Arctic latitudes (numerical experiments). Oceanology 42(2):153–161
Morozov EG, Pisarev SV (2004) Internal waves and polynya formation in the Laptev Sea. Doklady Earth Sci 398(7):983–986
Morozov EG, Pisarev SV (2023) Internal waves in the region of the Akselsundet Strait of Western Spitsbergen Island. Izv Atmos Oceanic Phys 59(4):432–442
Morozov EG, Trulsen K, Velarde MG, Vlasenko VI (2002b) Internal tides in the Strait of Gibraltar. J Phys Oceanogr 32:3193–3206
Morozov EG, Neiman VG, Shcherbinin AD (2003a) Internal tide in the Kara Strait. Dokl Earth Sci 393(9):1312–1314
Morozov EG, Parrilla-Barrera G, Velarde MG, Scherbinin AD (2003b) The straits of Gibraltar and Kara Gates: a comparison of internal tides. Oceanol Acta 26(3):231–241
Morozov EG, Pisarev SV, Neiman VG, Erofeeva SY (2003c) Internal tidal waves in the Barents Sea. Doklady Earth Sci 392(5):686–688
Morozov EG, Paka VT, Bakhanov VV (2008) Strong internal tides in the Kara Gates Strait. Geophys Res Let 9:L16603
Morozov EG, Kozlov IE, Shchuka SA, Frey DI (2017) Internal tide in the Kara Gates Strait. Oceanology 57(1):8–18. https://doi.org/10.1134/S0001437017010106
Morozov EG, Frey DI, Gladyshev SV, Klyuvitkin AA, Novigatsky AN (2019a) Internal tides in the Strait of Denmark. Izv Atmos Oceanic Phys 55(3):295–302. https://doi.org/10.1134/S0001433819030034
Morozov EG, Marchenko AV, Filchuk KV, Kowalik Z, Marchenko NA, Ryzhov IV (2019b) Sea ice evolution and internal wave generation due to a tidal jet in a frozen sea. Applied Ocean Res 87:179–191
Morozov EG, Pisarev SV, Erofeeva SY (2002a) Internal tidal waves in the Arctic seas of Russia, In: Lavrenov IV, Morozov EG (eds) Surface and internal waves in the arctic seas Gidrometeoizdat, pp. 217–230. [in Russian]

Morozov EG (2006) Internal tides. Global field of internal tides and mixing caused by internal tides. In: Waves in geophysical fluids. Springer, Wein, New York, pp 271–332

Morozov EG (2018) Oceanic internal waves. Observations, analysis and modeling. a global view. Springer, p 316. https://doi.org/10.1007/978-3-319-73159-9

Nash JD, Moum JN (2005) River plumes as a source of large-amplitude internal waves in the coastal ocean. Nature 437:400–403. https://doi.org/10.1038/nature03936

Osadchiev AA (2018) Small mountainous rivers generate high-frequency internal waves in coastal ocean. Sci Rep 8:16609. https://doi.org/10.1038/s41598-018-35070-7

Osborne AR, Burch TL (1980) Internal solitons in the Andaman Sea. Science 208(4443):451–459

Padman L, Dillon TM (1991) Turbulent mixing near the Yermak Plateau during the coordinated Eastern Arctic Experiment. J Geophys Res 96:4769–4782

Padman L, Plueddemann A, Muench R, Pinkel R (1992) Diurnal tides near the Yermak Plateau. J Geophys Res 97:12639–12652

Parsons AR, Bourke RH, Muench RD et al (1996) The Barents sea polar front in summer. J Geophys Res Oceans 101(C6):14201–14221

Pisarev SV (1988) Experimental frequency spectra of internal waves in an ice-covered high-latitude basin. Oceanology 28(5):577–580

Pisarev SV (1991) Some measurements of the spatial and temporal characteristics of internal waves in an ice-covered high-latitude basin. Oceanology 31(1):42–46

Pisarev SV (1992) Spatial and temporal characteristics of internal waves at the edge of the continental shelf in the Arctic basin. Oceanology 32(5):579–583

Pisarev SV (1996) Low-frequency internal waves near the shelf edge of the Arctic basin. Oceanology 36(6):771–778

Plueddemann AJ (1992) Internal wave observations from the arctic environmental drifting buoy. J Geophys Res 97:12619–12638

Plueddemann AJ, Krishfield R, Takizawa T et al (1998) Upper ocean velocities in the Beaufort Gyre. Geophys Res Lett 25(2):183–186

Prinsenberg SJ, Wilmot WL, Rattray M (1974) Generation and dissipation of coastal internal tides. Deep-Sea Res 21(4):263–281

Rippeth TP, Vlasenko V, Stashchuk N et al (2017) Tidal conversion and mixing poleward of the critical latitude (an Arctic case study). Geophys Res Lett 44:12349–12357. https://doi.org/10.1002/2017GL075310

Sabinin K, Serebryany A (2005) Intense short-period internal waves in the ocean. J Marine Res 63:227–261

Shestov A, Wrangborg D, Marchenko A (2015) Hydrology of Braganzavagen under ice-covered conditions. In: Proceedings of the 23rd International conference on port and ocean engineering under arctic conditions POAC15–00181, Trondheim, Norway, p 11

Stashchuk N, Vlasenko V (2009) Generation of internal waves by a supercritical stratified plume. J Geophys Res Oceans 114:C01004. https://doi.org/10.1029/2008JC004851

Sugiyama Y, Niwa Y, Hibiya T (2009) Numerically reproduced internal wave spectra in the deep ocean. Geophys Res Lett 36:L07601

Svalbardposten (2007) Dramatikk på Van Mijenfjorden, Publisert: 16.03.2007

Svalbardposten (2011) Fenomenet Murdoch-hullet, Publisert: 16.04.2011

Svalbardposten (2017) Russisk turfølge gjennom isen. Publisert: 27.04.2017

Tareev BA (1966) On the dynamics of internal gravity waves in a constantly stratified ocean. Izv Akad Nauk USSR Fiz Atmos Okeana 2(10):1064–1075

The Norwegian Pilot (2016) Sailing DIrections, vol. 7. Svalbard and Jan Mayen, 3rd edn. The Norwegian Hydrographic Service and Norwegian Polar Institute

Timmermans M-L, Melling H, Rainville L (2007) Dynamics in the deep Canada Basin, Arctic Ocean, inferred by thermistor-chain time series. J Phys Oceanogr 37(4):1066–1076

Torgrimson GM, Hickey BM (1979) Barotropic and baroclinic tides over the continental slope and shelf off Oregon. J Phys Oceanogr 9:945–961

Vlasenko VI (1992) Nonlinear model for the generation of baroclinic tides over extensive inhomogeneities of bottom topography. Phys Oceanogr (Morskoy Gidrofizicheskiy Zhurnal) 3:417–424

Vlasenko V, Hutter K (2002) Numerical experiments on the breaking of solitary internal waves over a slope-shelf topography. J Phys Oceanogr 32:1779–1793

Vlasenko V, Stashchuk N, Hutter K (2002) Water exchange in fjords induced by tidally generated internal lee waves. Dyn Atmos Oceans 35(1):63–83

Vlasenko V, Stashchuk N, Hutter K, Sabinin K (2003) Nonlinear internal waves forced by tides near the critical latitude. Deep-Sea Res 50:317–338

Vlasenko V, Stashchuk N, Hutter K (2005) Baroclinic tides: theoretical modeling and observational evidence. Cambridge University Press, Cambridge, p 351

Watson G, Robinson IS (1990) A study of internal wave propagation in the Strait of Gibraltar using shore-based marine radar images. J Phys Oceanogr 20:374–395

Wood D, Grue J (2003) Modelling of high amplitude internal waves integrating the primitive (Navier-Stokes) equations. Appl Ocean Res 24(6):331–340

Worthington LV (1969) An attempt to measure the volume transport of Norwegian Sea overflow water through the Denmark Strait. Deep-Sea Res 16:421–432

Wrangborg D, Grady R, Marchenko A, Karulin E (2013) Analysis of tidal deformations of land fast ice in shallow Arctic fjord. In: Proceedings of the International conference on port and ocean engineering under arctic conditions (POAC), POAC13–130, Espoo, Finland, p 11

Wright CJ, Scott RB, Ailliot P, Furnival D (2014) Lee wave generation rates in the deep ocean. Geophys Res Let 41:2434–2440. https://doi.org/10.1002/2013GL059087

Chapter 4
Processes in Svalbard Fjords

4.1 General Notes

The chapter presents oceanographic research in the Arctic related to the study of various processes in the Arctic fjords of Spitsbergen.

A tsunami wave generated by the motion of the Tuna Glacier in Svalbard was studied. The properties of this tsunami wave were similar to a landslide tsunami. The wave amplitude was 30 cm, the period was 93 s, and the period of disturbances after the wave was about one hour. The wavelength was 1850 m. Comparison of the spectra of ice oscillations, which are caused by various wave processes: tide, seiche oscillations, wind waves and swell, internal waves, and tsunami wave, is presented.

A method to estimate the flexural stiffness and effective elastic modulus of floating ice was suggested. The method is based on the analysis of water pressure records at two or three locations below the bottom of floating ice when flexural-gravity waves propagate through the ice. The relative errors in the calculations of the ice flexural stiffness and the water depth were analyzed. The method was tested using data from field measurements in Spitsbergen fjords, where flexural-gravity waves were generated by motion of a glacier.

We observed strongly asymmetric tide in a channel connecting the Van Mijen Fjord and Lake Vallunden in Spitsbergen. The channel is approximately 100 m long, 10 m wide, and 1–2 m deep. Asymmetric tide was also observed in the lake. The form of the semidiurnal tide is described by a combination of only three harmonics (M_2, S_2, and M_4) with different amplitudes and phases. The flood dominant form of the surface elevation in the channel (unlike the ebb dominant tide in the fjord) is formed over a horizontal distance of 50 m over a shallow bottom at the entrance to the channel. The tide in shallow places becomes asymmetric due to nonlinear effects caused by the influence of the bottom and especially by choking of the tidal flow over a sill that is located between the fjord and channel.

We summarized winter oceanographic measurements in the Spitsbergen fjords. The measurements allow us to analyze different processes that occur in the fjords

A. Marchenko et al., *Geophysical Hydrodynamics of Currents, Waves, and Ice in the Atlantic Arctic*, https://doi.org/10.1007/978-3-032-14122-4_4

covered with ice: semidiurnal tides, tsunami wave, wind waves, seiches, and high-frequency internal waves. Space and time scales of these processes were compared and the frequency spectra of the surface manifestations of these waves have been plotted on one graph.

4.2 Estimates of Sea Ice Bending Stiffness Based on the Characteristics of Flexural-Gravity Waves

This section is written together with S. Muzylev

Surface waves influence the structure of sea ice when they penetrate under ice. Continuous sea ice cover can be destroyed by spreading of surface waves into small fragments over sufficiently large areas during several hours or faster. Penetration of swell and low frequency waves into the ice over distances of several tens of kilometres and further was observed in numerous expeditions in Antarctica (Keliher 1976; Liu and Mollo-Christensen 1988), Central Arctic (Wadhams and Wells 1995; Smirnov 1996), Bering Sea (Squire and Moore 1980), Beaufort Sea (Bates and Shapiro 1980; Prinsenberg and Peterson 2011), Greenland Sea (Squire et al. 1995) and Okhotsk Sea (Marko 2003), and Van Mijen Fjord in Svalbard (Marchenko et al. 2011). A fast decrease of the representative sizes of floes caused by wave action can influence heat exchange processes between the atmosphere and ocean, sea ice dynamic, and ice loads on the offshore and coastal structures. In this section we review publication (Marchenko et al. 2013).

Field measurements show that periods of surface waves recorded in ice covered areas of seas exceed 10 s. For example, Liu and Mollo-Christensen (1988) describe surface waves in the sea covered by continuous 0.8 m thick ice with a period of 18 s and amplitude of 1 m in Antarctica. Marko (2003) reports about 14 s surface waves in continuous ice with thickness from 1.35 m to 2.5 m and wave amplitude below 1 m on the East Sakhalin shelf. Wadhams and Well (1995) describe an event of surface wave propagation with a period of 16 s over the Central Arctic. Bates and Shapiro (1980) describe waves with 600 s period recorded before each event when the land fast ice was pushed out on the beach near Point Barrow.

A criterion of ice destruction by waves is based on the conception of flexural strength (Timco and Weeks 2010). Flexural strength of sea ice is measured in standard experiments with cantilever beams. Bending failure of elastic and brittle ice plates occurs at a crest or a trough of bended plate after the initiation of a crack. Since compressive strength of sea ice is greater than the tensile strength the crack initiation occurs due to the tensile deformations of surface or bottom ice layers. Thus, the flexural strength of sea ice is close to its tensile strength. The influence of fatigue effects on sea ice flexural strength was discussed in Langhorne et al. (2001).

Calculation of critical strains is performed using rheological models of sea ice (Squire et al. 1996). Viscosity influences the formation of ice stresses when representative time of ice deforming exceeds 10 min. Pure elastic model or the model of floating elastic plate can be used for modelling ice response to the penetration of surface waves with periods shorter than 1 min. The basic equation describing bending deformations of floating ice plate is constructed using averaging of the equations of momentum balance and angular momentum balance of the ice over the plate thickness and the assumption about the existence of a mean surface where the bending stresses are zero. Vertical heterogeneity of the temperature and brine content of the ice influences the mechanical properties of the ice and coefficients in the basic equation. Kerr and Palmer (1972) have shown that a modified flexural stiffness may be defined, which allows us using a fully homogeneous plate theory with the effective elastic modulus.

Rheological properties of isotropic elastic materials are characterized by two constants: Young's modulus and Poisson's ratio. The Poisson's ratio of sea ice is very stable and varies near 0.33 (Bogorodskii and Gavrilo 1980; Weeks and Hibler 2010). The Young's modulus of ice can be determined by static and dynamic methods. The static method is based on the mechanical tests with ice samples. The dynamic method is based on the measurement of the speed of elastic waves in an ice sample. The dynamic method shows more stable results and larger values of the Young's modulus than the static method. The Young's modulus of sea ice varies from 1 to 10 GPa depending on the temperature and brine volume. Mechanical bending tests were used for the measurements of the effective elastic modulus of sea ice by Tabata et al. (1967). The tests with sea ice beams demonstrated that the effective elastic modulus increases with the increasing loading rate and decreasing temperature.

A new method for the calculation of the effective ice bending stiffness is described and analyzed. The method is based on the analysis of water pressure records at two or three locations below the bottom of floating ice when flexural-gravity waves propagate through the ice. Analysis of the relative errors in the definition of the effective ice bending stiffness and water depth is performed. The method is tested using the data from the field measurements in the Tempel Fjord in Svalbard. Flexural-gravity waves were excited by the ice destruction at the front of the Tuna Glacier in February 2010.

A water layer of the finite depth H beneath an elastic ice sheet of the thickness h is considered (Fig. 4.1). Linearized equations describing potential motions of small amplitude of the water layer beneath the floating elastic ice sheet are formulated as follows

$$\left(\Delta + \frac{\partial^2}{\partial z^2}\right)\varphi = 0, \quad -H < z < 0; \quad \Delta = \frac{\partial^2}{\partial x^2} + \frac{\partial^2}{\partial y^2}, \tag{4.1}$$

$$\frac{\partial \varphi}{\partial z} = 0, \quad z = -H, \tag{4.2}$$

$$\frac{\partial \eta}{\partial t} = \frac{\partial \varphi}{\partial z}, \quad z = 0, \tag{4.3}$$

$$\frac{\partial \varphi}{\partial t} + \frac{p_{iw}}{\rho_w} + g\eta = f(t), \quad z = 0, \tag{4.4}$$

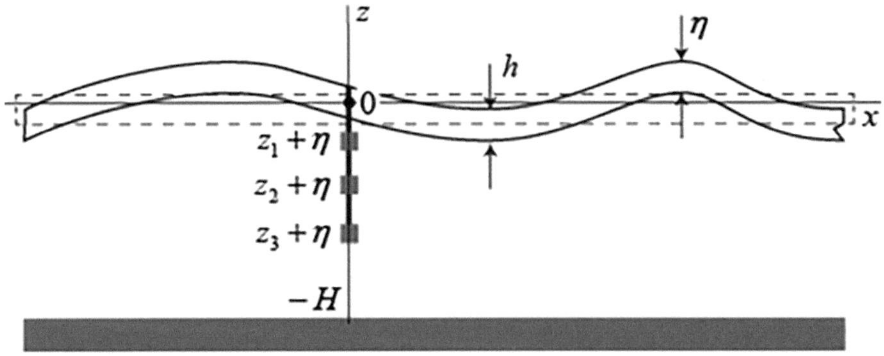

Fig. 4.1 Water layer beneath an elastic ice sheet. Scheme of deployment of three pressure sensors mounted on the floating ice

where, φ is the water velocity potential, η is the water surface elevation, H is the water depth, p_{iw} is the water pressure at the ice-water interface, ρ_w is the water density, t is the time, x, y and z are the horizontal and vertical coordinates.

Water pressure p_{iw} and the atmosphere pressure p_a satisfy the following equation (Landau and Lifshits 1987)

$$p_{iw} - p_a = D\Delta^2\eta, \ D = \frac{Eh^3}{12(1-\nu^2)}, \tag{4.5}$$

where, E and ν are the Young's modulus and Poisson ratio of the ice, respectively, h is the ice thickness, and D is the flexural stiffness of the ice. Representative values of E and ν are assumed equal to

$$E = 2\,\text{GPa}, \ \nu = 0.33. \tag{4.6}$$

In the natural conditions the Young's modulus varies over the ice column. Kerr and Palmer (1972) have shown that modified flexural stiffness D may be introduced in this case, which allows us to use a fully homogeneous plate theory with the effective elastic modulus E.

It is assumed that the atmosphere pressure is constant $p_a = const$ and $f(t) = p_a/\rho_w$. In this case, kinematic and dynamic boundary conditions (4.3) and (4.4) at the ice-water interface are reduced to one boundary condition for the water velocity potential

$$\frac{\partial^2\varphi}{\partial t^2} + \left(g + \frac{D}{\rho_w}\Delta^2\right)\frac{\partial\varphi}{\partial z} = 0, \ z = 0. \tag{4.7}$$

Water pressure (p) at an arbitrary point of the domain $z \in (-H, 0)$ is determined from the Cauchy-Lagrange integral as follows

$$\frac{\partial \varphi}{\partial t} + gz + \frac{p}{\rho_w} = \frac{p_a}{\rho_w}, \tag{4.8}$$

where, $\varphi(x, y, z, t)$ is the solution of Eqs. (4.1), (4.2), and (4.6).

Water pressure measured at the fixed distance z from the moving ice-water interface consists of the hydrostatic and dynamic parts

$$p = p_{gst} + p_{dyn}, \tag{4.9}$$

expressed by the formulas

$$p_{gst} = -\rho_w gz + p_a,\ p_{dyn} = -\rho_w \left(\frac{\partial \varphi}{\partial t} + g\eta \right). \tag{4.10}$$

It follows from formulas (4.10) that the dynamic part of the pressure becomes equal to zero, when water motion is absent. The solution of Eqs. (4.1), (4.2), and (4.7) describing periodic waves is expressed by the formulas

$$\varphi = \varphi_0 \frac{\cosh[k(z+H)]}{\cosh[kH]} e^{i\theta},\ \eta = \eta_0 e^{i\theta},\ \theta = \boldsymbol{k} \cdot \boldsymbol{x} - \omega t, \tag{4.11}$$

where, $\boldsymbol{k} = (k_x, k_y)$ and ω are the wave vector and wave frequency, respectively, $\boldsymbol{x} = (x, y)$. The amplitudes of the velocity potential ($2\varphi_0$) and the elevation of water–ice surface ($2\eta_0$) are related as follows

$$k\varphi_0 \tanh[kH] = -i\omega\eta_0. \tag{4.12}$$

The frequency and wave vector satisfy the dispersion equation

$$\omega^2 = gk \tanh[kH]\left(1 + dk^4\right),\ k = |\boldsymbol{k}|,\ d = D/(\rho_w g). \tag{4.13}$$

It follows from Formula (4.10) that the dynamic water pressure induced by the wave is expressed by the formula

$$p_{dyn} = p_{dyn,0} e^{i\theta},\ p_{dyn,0} = \rho_w g \eta_0 P(\omega, k, z),\ P(\omega, k, z) = \frac{\omega^2 \cosh[k(z+H)]}{gk \sinh[kH]} - 1. \tag{4.14}$$

Properties of water pressure distribution over the water column

Substituting dispersion Eq. (4.11) into Formula (4.12) we find

$$P(k, z) = \frac{(1 + dk^4) \cosh[k(z+H)]}{\cosh[kH]} - 1. \tag{4.15}$$

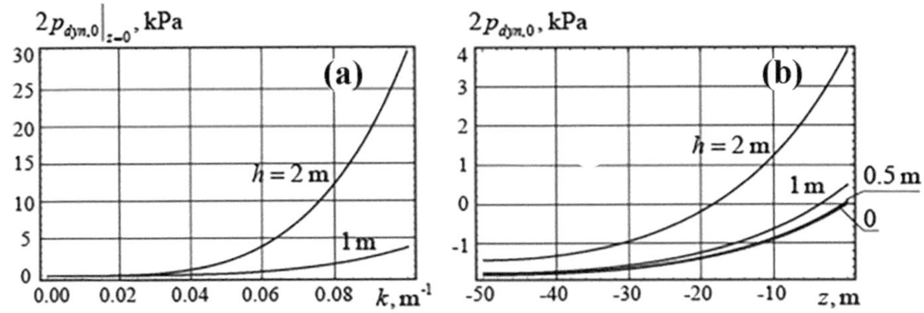

Fig. 4.2 Pressure amplitude at the ice-water interface versus wave number (**a**); profiles of pressure amplitude over the vertical water column constructed with $k = 0.06\ \text{m}^{-1}$ (**b**). Wave amplitude is $2\eta_0 = 0.2$ m

The pressure at the ice-water interface is equal to $P(k, 0) = dk^4$. The amplitude of the dynamic water pressure induced by the wave at the ice-water interface with the representative values introduced by Formula (4.6) is written as

$$2p_{dyn,0}\big|_{z=0} = 37421\rho_w g\eta_0 h^3 k^4, \tag{4.16}$$

where, all quantities are expressed in the SI units. One can see that $2p_{dyn,0}\big|_{z=0}$ is approximately several kPa, when $k \sim 0.1\ \text{m}^{-1}$, $\eta_0 \sim 0.1$ m and $h \sim 1$ m. The amplitude of water pressure at the ice-water interface is positive when $h > 0$ and it is equal to zero when the water surface is ice free. Figure 4.2a shows the dependence of $2p_{dyn,0}\big|_{z=0}$ on the wave number when ice thickness is equal to 1 and 2 m. The pressure below 2 m thick ice is 8 times greater than the pressure below 1 m thick ice since it is proportional to h^3.

The amplitude of water pressure versus depth z is shown in Fig. 4.2b at water depth $H = 50$ m and wave number $k = 0.06\ \text{m}^{-1}$ (wave length is about 100 m). Pressure amplitude $2p_{dyn,0}$ decreases monotonously from the positive values at the ice-water interface to negative values at the water bottom, so that there is a depth where it is equal to zero. The maximal value of $2|p_{dyn,0}|$ is reached either at the ice-water interface or at the bottom depending on the ice thickness, wave number, and water depth. For example, in Fig. 4.1b, $2|p_{dyn,0}|$ reaches the maximum at the ice-water interface when the ice thickness is 2 m and at the bottom when the ice thickness is 1 m.

Dependencies of $2|p_{dyn,0}|$ on the depth z and wave number k are shown in Fig. 4.3 when the water depth is equal to 50 m. The ice thickness is 1 m in Fig. 4.3a and the water surface is ice free in Fig. 4.3b. One can see that the influence of the ice on the water pressure is extended up to the certain depth depending on the wave number. For example, in Fig. 4.3a the amplitude $p_{dyn,0} = 0$ at a depth of 10.5 m when the wave number is $k = 0.1\ \text{m}^{-1}$, and also at a depth of 2 m when the wave number is $k = 0.05\ \text{m}^{-1}$. When the water surface is free the dependencies of $2|p_{dyn,0}|$ on the depth z and wave number k are monotonous (Fig. 4.3b).

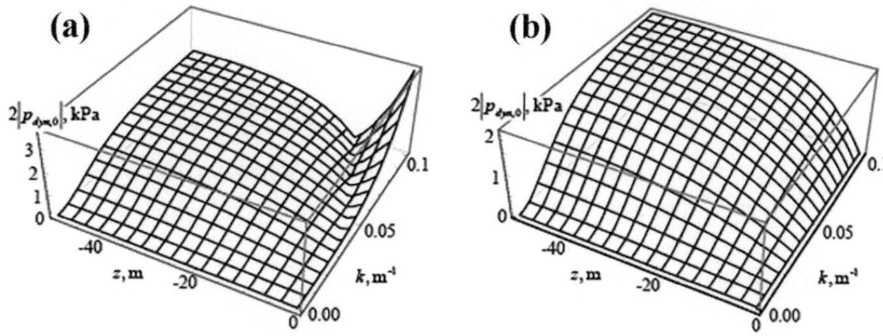

Fig. 4.3 Profiles of absolute values of water pressure amplitude over a vertical water column of 50 m depth constructed with different values of the wave numbers. Ice thickness $h = 1$ m (**a**) and $h = 0$ (**b**)

Figure 4.4 illustrates the dependence of vertical profiles of $2\left|p_{dyn,0}\right|$ on water depth H. Ice thickness is $h = 1$ m in Fig. 4.4a and $h = 2$ m in Fig. 4.4b. The graphs are constructed for three values of wave numbers: $k = 0.1$ m^{-1}, $k = 0.06$ m^{-1}, and $k = 0.02$ m^{-1}. For each of these three values the graphs are constructed for the water depths of 50, 60, 70, 80, 90, and 300 m. All graphs constructed with the same wave number begin from the same point at $z = 0$. They are very close to each other at relatively small depths. Their divergence increases with the increasing depth. The divergence is stronger at smaller wave numbers. It follows from formula (4.15) that

$$P(k, H) = \frac{1 + dk^4}{\cosh[kH]} - 1. \tag{4.17}$$

Therefore, the absolute value of the water pressure amplitude at the bottom increases with increasing water depth. A limit value of $P(k, H)$ is equal to –1 when $H \to \infty$.

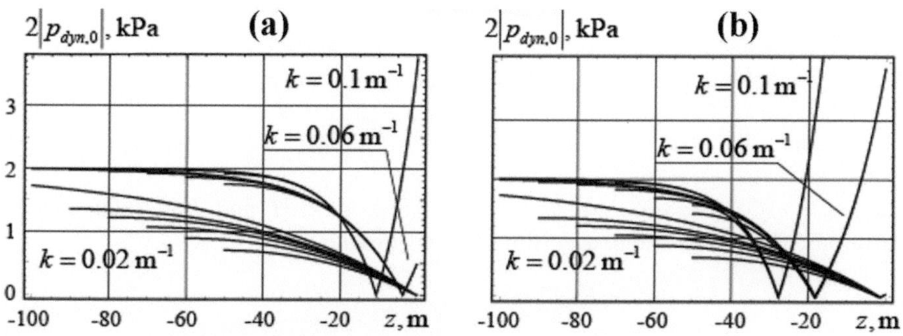

Fig. 4.4 Profiles of absolute values of the water pressure amplitude over the vertical water column constructed for water depths of 50, 60, 70, 80, 90, and 300 m. Ice thickness is equal to 1 m (**a**) and 2 m (**b**)

It means that the dimensional amplitude of the water pressure $2\rho_w g\eta_0$ at high depths depends only on the vertical motions of the instruments at the points of measurements.

4.2.1 Estimate of Ice Stiffness from Water Pressure Records Below Floating Ice

Let us assume that three pressure sensors are fixed on a wire under floating ice at distances z_1, z_2, and z_3 from the open water surface (Fig. 4.1). The open water surface is located at $z = 0$. Floating ice is deformed by surface waves with amplitudes much smaller than the wave lengths. The ice thickness is much smaller than the wave lengths and the water depth. Vertical positions of the sensors in the absolute frame of reference are $z_1 + \eta$, $z_2 + \eta$ and $z_3 + \eta$ since the sensors move together with the ice. Each of the sensors records water pressure at discrete time moments. Spectral analysis shows frequencies of waves with the maximal energy.

Let us consider one of these frequencies (ω), and assume that the spectral amplitudes of the pressures recorded by sensors at frequency ω are equal to $p_{1,dyn}$, $p_{2,dyn}$, and $p_{3,dyn}$. The ratios of these amplitudes are written according to formulas (4.14) as follows

$$\gamma_{12} \equiv \frac{p_{1,dyn}}{p_{2,dyn}} = \frac{\omega^2 \cosh[k(z_1 + H)] - gk \sinh[kH]}{\omega^2 \cosh[k(z_2 + H)] - gk \sinh[kH]}, \tag{4.18}$$

$$\gamma_{13} \equiv \frac{p_{1,dyn}}{p_{3,dyn}} = \frac{\omega^2 \cosh[k(z_1 + H)] - gk \sinh[kH]}{\omega^2 \cosh[k(z_3 + H)] - gk \sinh[kH]}, \tag{4.19}$$

where, γ_{12}, γ_{13}, and ω are known, and k, and H are unknown quantities, which can be calculated from Eqs. (4.18) and (4.19). Then, the dimensionless ice stiffness d is calculated from the dispersion Eq. (4.13).

The accuracy of this method depends on the errors in the calculations of γ_{12}, γ_{13}, and ω. It is assumed that γ_{12}, γ_{13}, and ω are determined with certain accuracy, and their variations near explicit values $\gamma_{12,0}$, $\gamma_{13,0}$, and ω_0 are equal to $\Delta\gamma_{12}$, $\Delta\gamma_{13}$, and $\Delta\omega$. Explicit solution of Eqs. (4.13), (4.18), and (4.19) is expressed by formulas

$$k = k_0, \; H = H_0, \; d = d_0, \tag{4.20}$$

when $\omega = \omega_0$, $\gamma_{12} = \gamma_{12,0}$, $\gamma_{13} = \gamma_{13,0}$. Variations of k, H, and d are related to the variations of γ_{12}, γ_{13}, and ω according to the formulas

$$\begin{aligned} \Delta\gamma_{12} - \frac{\partial\gamma_{12}}{\partial\omega}\Delta\omega &= \frac{\partial\gamma_{12}}{\partial k}\Delta k + \frac{\partial\gamma_{12}}{\partial H}\Delta H, \\ \Delta\gamma_{13} - \frac{\partial\gamma_{13}}{\partial\omega}\Delta\omega &= \frac{\partial\gamma_{13}}{\partial k}\Delta k + \frac{\partial\gamma_{13}}{\partial H}\Delta H, \end{aligned} \tag{4.21}$$

$$2\omega_0 \Delta\omega = \frac{\partial \omega}{\partial k}\Delta k + \frac{\partial \omega}{\partial H}\Delta H + \frac{\partial \omega}{\partial d}\Delta d, \tag{4.22}$$

where, all partial derivatives in Formulas (4.21) and (4.22) are calculated using Formulas (4.18) and (4.19) and dispersion Eq. (4.13) at the point of the explicit solution.

Quantities $m_\omega = \Delta\omega/\omega_0$, $m_{\gamma 12} = \Delta\gamma_{12}/\gamma_{12,0}$, and $m_{\gamma 13} = \Delta\gamma_{13}/\gamma_{13,0}$ are the relative errors of measurements of wave frequency $\Delta\omega$ and ratios γ_{12} and γ_{13}. These errors determine the errors of the calculation of the ice stiffness and water depth. Below, we analyse the relative errors of the calculation of the ice stiffness and water depth $m_d = \Delta d/d_0$ and $m_H = \Delta H/H_0$.

Numerical estimates were obtained with $m_{\gamma 12} = m_{\gamma 13} = m_\gamma = \pm 0.05$ when m_ω is equal to –0.05, 0.00, and 0.05. It means that all quantities determined from the field measurements are known with 95% of reliability. Figures 4.5 and 4.6 show relative errors m_d and m_H versus wave frequency ω_0. It is assumed that ice thickness is $h = 1$ m and pressure sensors 2 and 3 are located at depths $z_2 = -10$ m and $z_3 = -15$ m, pressure sensor 1 is located at depth $z_1 = -5$ m (Fig. 4.5) and $z_1 = -2$ m (Fig. 4.6). Water depth is given in the captions to the figures. Specification of curves shown in the figures is given in Table 4.1.

It follows from Fig. 4.5a, c that relative error m_d does not exceed 20% when wave frequency $\omega_0 > 0.6$ s^{-1}, i.e., the wave period is smaller than 10.5 s. Figure 4.6a, c

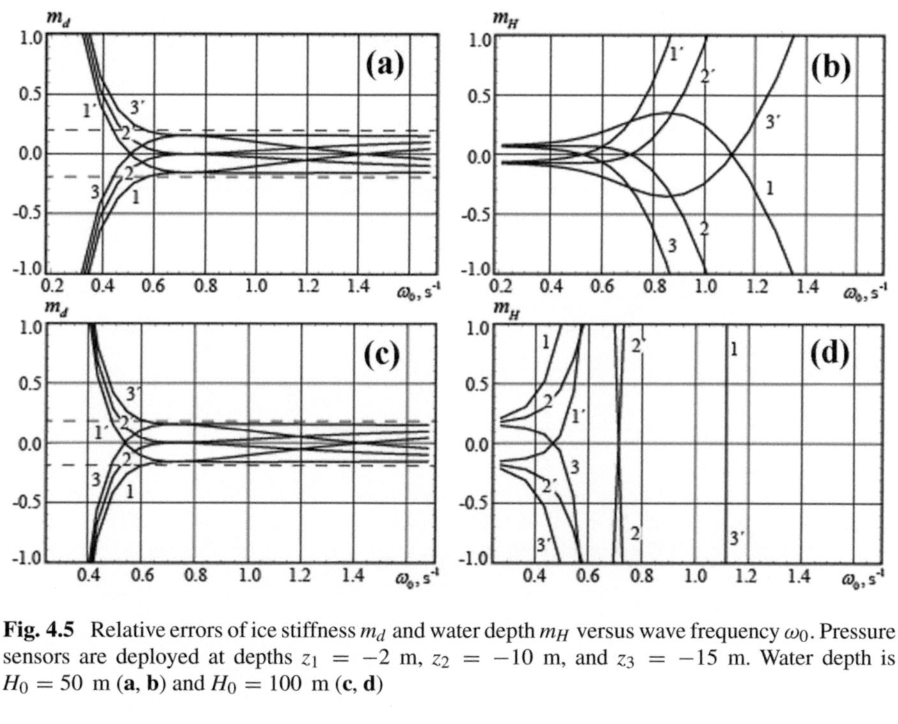

Fig. 4.5 Relative errors of ice stiffness m_d and water depth m_H versus wave frequency ω_0. Pressure sensors are deployed at depths $z_1 = -2$ m, $z_2 = -10$ m, and $z_3 = -15$ m. Water depth is $H_0 = 50$ m (**a**, **b**) and $H_0 = 100$ m (**c**, **d**)

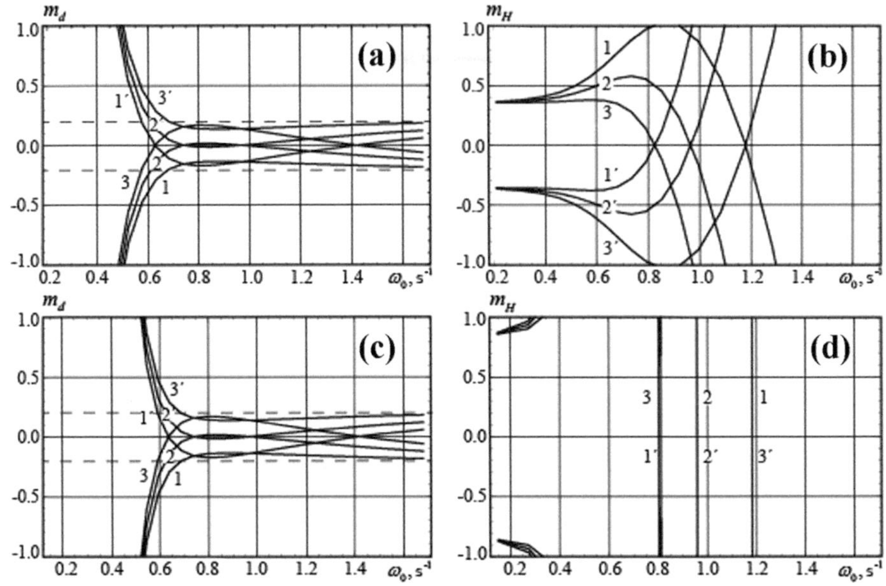

Fig. 4.6 Relative errors of ice stiffness m_d and water depth m_H versus wave frequency ω_0. Pressure sensors are deployed at depths $z_1 = -5$ m, $z_2 = -10$ m, and $z_3 = -15$ m. Water depth is $H_0 = 50$ m (**a**, **b**) and $H_0 = 100$ m (**c**, **d**)

Table 4.1 Specification of curves shown in Figs. 4.5, 4.6, 4.7 and 4.8

	$m_\omega = 0.05$	$m_\omega = 0$	$m_\omega = -0.05$
$m_\omega = 0.05$	1	2	3
$m_\omega = -0.05$	1′	2′	3′

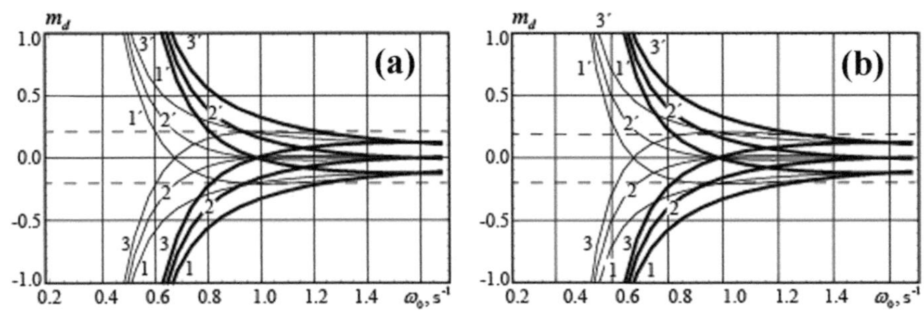

Fig. 4.7 Relative errors of ice stiffness m_d and water depth m_H versus wave frequency ω_0. Pressure sensors are deployed at the depths $z_1 = -10$ m, $z_2 = -15$ m (thick lines), and $z_1 = -5$ m, $z_2 = -10$ m (thin lines). Water depth is equal to $H_0 = 50$ m (**a**) and $H_0 = 100$ m (**b**)

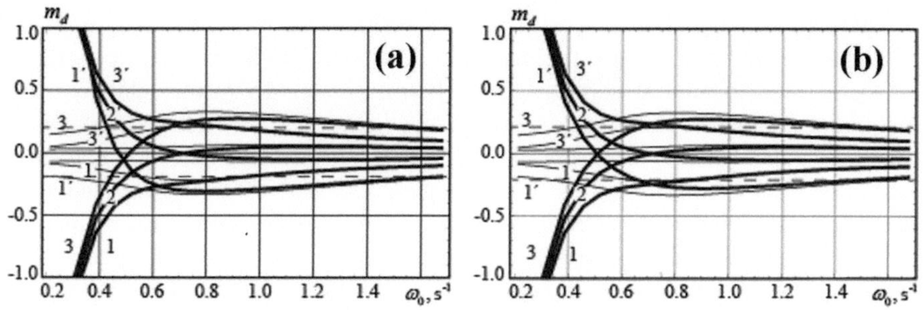

Fig. 4.8 Relative errors of ice stiffness m_d and water depth m_H versus wave frequency ω_0. Pressure sensors are deployed at depths $z_1 = -2$ m, $z_2 = -15$ m (thick lines) and $z_1 = 0$, $z_2 = -15$ m (thin lines). Water depth is equal to $H_0 = 50$ m (**a**) and $H_0 = 1000$ m (**b**)

show that that relative error m_d does not exceed 20% when wave frequency $\omega_0 >$ 0.7 s^{-1}, i.e., wave period is smaller than 9 s. Thus, approaching of sensor 1 to the ice bottom influences the increase in the accuracy in the definition of the ice stiffness. Figure 4.5b shows that relative error m_H is smaller than 10% when the wave frequency is smaller than 0.4 s^{-1}, i.e., the wave period is greater than 15.7 s. It follows from Fig. 4.5d that m_H is very high when the wave frequency is greater than 0.2 s^{-1}. Therefore, for the calculation of the water depth it is better to use the long wave part of the wave spectrum, and for the definition of the ice stiffness the shortwave part of wave spectrum is more helpful.

Formula (4.18) could be used for the calculation of the wave number and then the ice stiffness could be calculated from dispersion Eq. (4.13) when wave frequency ω_0, ratio $\gamma_{12,0}$, and water depth H_0 are measured. Now, variations of k and d are related to the variations of γ_{12}, γ_{13}, and ω according to the formulas

$$\Delta\gamma_{12} - \frac{\partial\gamma_{12}}{\partial\omega}\Delta\omega - \frac{\partial\gamma_{12}}{\partial H}\Delta H = \frac{\partial\gamma_{12}}{\partial k}\Delta k, \tag{4.23}$$

$$2\omega_0\Delta\omega = \frac{\partial\omega}{\partial k}\Delta k + \frac{\partial\omega}{\partial d}\Delta d, \tag{4.24}$$

where, all partial derivatives in Formulas (4.23) and (4.24) are calculated using Formula (4.18) and dispersion Eq. (4.13) at the point of the explicit solution.

Numerical simulations of the dependencies of relative error m_d on wave frequency ω_0 were performed with $m_{\gamma 12} = m_{\gamma 13} = m_\gamma = \pm 0.05$ when m_ω was equal to –0.05, 0.00, and 0.05. The results of the numerical simulations are shown in Figs. 4.7 and 4.8. Specification of curves shown in the figures is given in Table 4.1. Ice thickness is $h = 1$ m. Vertical coordinates z_1 and z_2 of the pressure sensors deployment and the water depth are given in the captions to the figures. Variations of the water depth within $m_H = \pm 0.1$ have very small influence on the shape of the curves shown in Figs. 4.7 and 4.8. Therefore, from hereinafter, the dependence of relative error m_d on m_H is not analyzed.

Figure 4.7 shows that relative error m_d does not exceed 20% when wave frequency $\omega_0 > 0.9\ \text{s}^{-1}$ and the sensors are located at depths $z_1 = -5$ m and $z_2 = -10$ m. For the sensors located at depths $z_1 = -10$ m, $z_2 = -15$ m the relative errors are smaller than 20% when the wave frequency is $\omega_0 > 1.2\ \text{s}^{-1}$. Figure 4.8 shows that the relative error m_d does not exceed 30% when pressure sensor 1 is located just below the ice bottom and the wave frequency is $\omega_0 > 0.2\ \text{s}^{-1}$, i.e., the wave period is smaller than 31.4 s. The displacement of the pressure sensor at 2 m distance from the ice shrinks the range of wave frequencies providing a 30% accuracy of the calculation of the ice stiffness to $\omega_0 > 0.6\ \text{s}^{-1}$. Figures 4.7 and 4.8 demonstrate small influence of water depth H_0 on the dependence of relative error m_d on wave frequency ω_0.

4.2.2 Determination of Sea Ice Stiffness Using the Data of Field Measurements

Field works were performed in 2011 and 2012 in the Tempel Fjord in Spitsbergen near the glacier front (Fig. 4.9a). Two profiles of the sea bottom along and across the fjord are shown in Fig. 4.9b, c. Two outflow glaciers Tuna Glacier and Von-Post Glacier form the glacier front of about 3 km wide in end of the fjord. The measurements with Total Station confirmed that the height of the glacier wall reached 37 m above the ice level near the place of the filed works. The thickness of sea ice near the glacier front was about 1 m in 2011 and 2012. The snow thickness did not exceed 15 cm. The temperature profiles through the ice were almost linear, and the mean salinity of the ice was 5–6 ppt (Marchenko et al. 2011).

Although the glaciers were stable since 2005, the visual observations have shown significant buckling deformations and folds of sea ice near the front of the Tuna Glacier in February of 2011 (Fig. 4.10a). The winter in 2012 was warmer and the

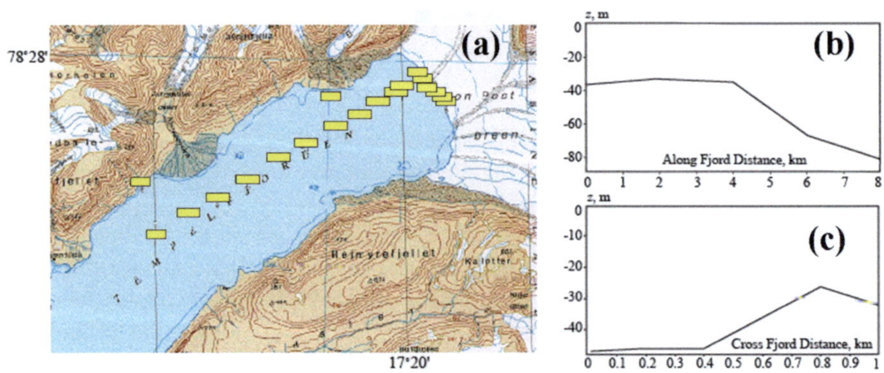

Fig. 4.9 Map of the Tempel Fjord (**a**) and sea bottom profiles along the fjord (**b**) and across the fjord (**c**). Locations of sea depth measurements are marked by yellow rectangles

glacier front was less stable. Fallings of ice blocks on the surface of sea ice near the glacier were observed in April. The surface waves excited by these events destroyed the sea ice into polygonal blocks near the glacier (Fig. 4.10b). At the same time the surface of sea ice near the glacier front in 2012 was flatter than in 2011. It is probably because of the smaller coverage of the Tempel Fjord by sea ice in 2012 in comparison with 2011. The confinement of sea ice in the fjord was much less in 2012 in comparison with 2011.

In February 2011, the sea ice thickness near the glacier front (~300 m from the front) was 94 cm. Two temperature and pressure sensors SBE-39 (sensors 1 and 2) were lowered from the ice at depths of $z_1 = -9.75$ m and $z_2 = -16.82$ m at a distance of ~300 m from the front of the Tuna Glacier. The sampling interval was 1 s at both of the sensors. An SBE-37 instrument (sensor 3) was deployed at the bottom at the same point with a sampling interval of 6 s. A scheme of the deployment is shown in Fig. 4.11. The weight at the bottom was used to keep sensor 3 on the bottom when the water level was changed by the tide. A record of the water pressure at the bottom by sensor 3 shows dominant semidiurnal variations of the water depth with an amplitude of about 1 m so that the water depth varied between 46 and 48 m (Fig. 4.12a). The time in Fig. 4.12a is accounted in hours from the beginning of the field measurements.

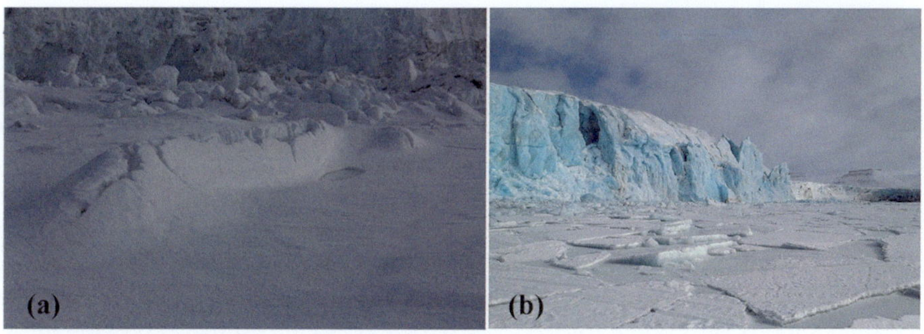

Fig. 4.10 Sea ice near the glacier front in February 2011 (**a**) and in April 2012 (**b**)

Fig. 4.11 Scheme of the deployment of pressure sensors 1, 2, and 3 near the glacier front

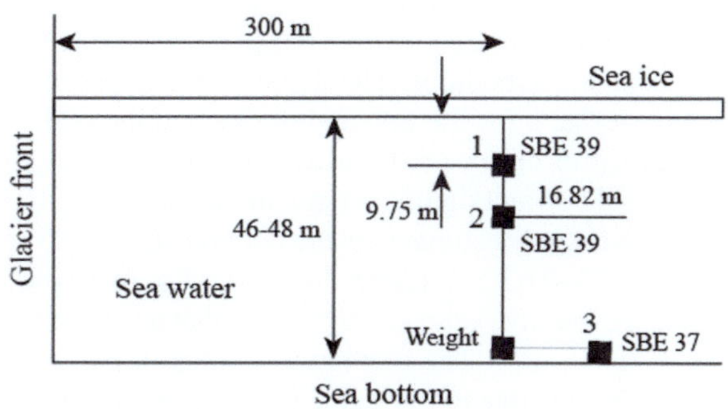

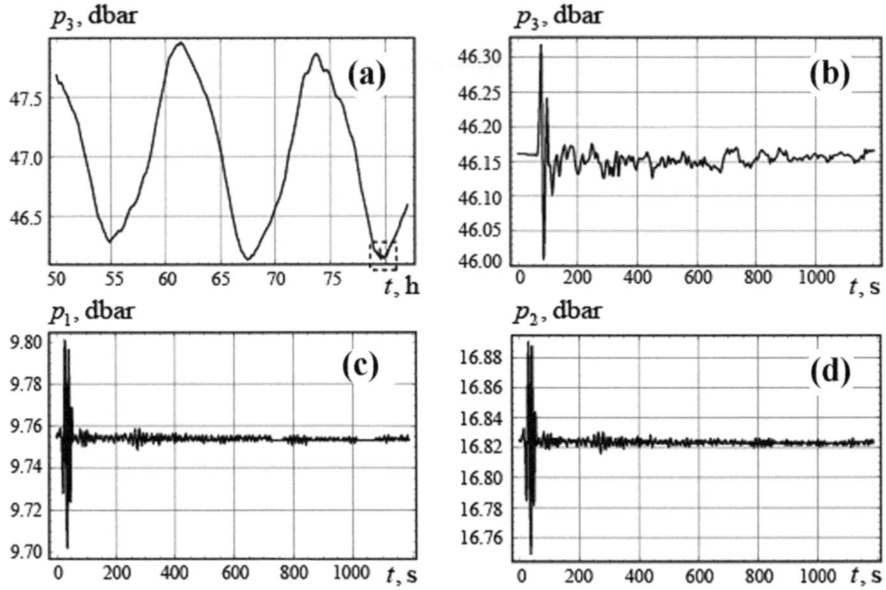

Fig. 4.12 Records of pressure sensors 3 (**a**, **b**), 1 (**c**), and 2 (**d**). The interval shown in Fig. 4.12b is marked with a dashed rectangle in Fig. 4.12a

Approximately at 07:40 on February 17, the instruments recorded the event of tsunami wave shown by the dotted square in Fig. 4.12a. The records of the pressure sensors are shown in Fig. 4.12b–d. The time is accounted in seconds from the beginning of the tsunami event. The analysis of spectral composition of the first wave pulse causing vertical displacement of the ice of about 30 cm is performed in (Marchenko et al. 2012). Here, we analyse the properties of the wave tail to obtain the estimates of the sea ice stiffness. Since the ice thickness was measured ($h = 94$ cm) and the Poisson's ratio is very stable for ice ($\nu = 0.31 - 0.34$) the estimates provide information about the effective elastic modulus E of sea ice in bending deformations. The records of sensor 3 deployed at the bottom show that the water depth at the moment of the tsunami event was $H_0 = 46.15$ m.

Two consecutive 5 min intervals of the wave tail record were used for the estimates of the Young's modulus. The dynamic water pressure caused by the wave is shown in Fig. 4.13a, b for the first and the second intervals. Red and blue lines are related to the records of sensors 1 and 2. Figure 4.13 shows about 4–5 oscillations during an interval of 50 s and significant modulations of their amplitude over 2–3 min. The wave period is roughly estimated in the interval 10.0–12.5 s and the wave frequency ranges within 0.50–0.63 s^{-1}. Using dispersion Eq. (4.13) with water depth $H_0 = 46.15$ m and representative values (4.6) we find that the wave number varied within the interval 0.026–0.045 m^{-1}. The theoretical curves in Fig. 4.4a and experimental records in Fig. 4.13 show that the dynamic pressure amplitude at depth $z_1 = -9.75$ m is higher than the dynamic pressure amplitude at depth $z_2 = -16.82$ m.

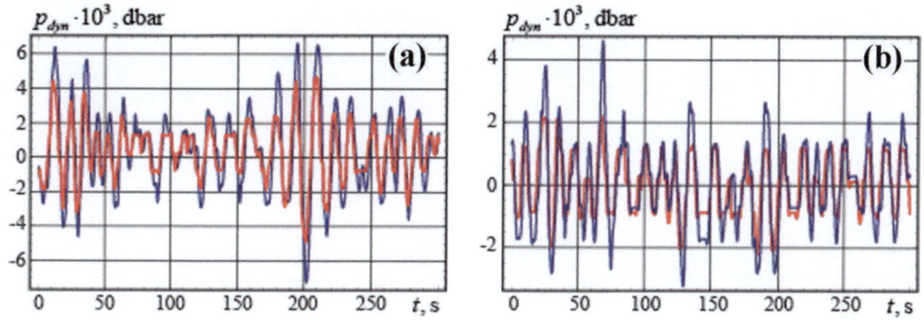

Fig. 4.13 Dynamic water pressure p_{dyn} versus time t recorded in two consecutive fragments of the wave tail at depths $z_1 = -9.75$ m (red lines) and $z_2 = -16.82$ m (blue lines)

Fourier transforms of the dynamic pressure records were calculated using the formula

$$p_F = \frac{1}{\sqrt{2\pi}} \int_{-\infty}^{\infty} p_{dyn} e^{i\omega t} dt. \tag{4.25}$$

The data processing consisted of the following steps for each time series of the dynamic pressure data shown in Fig. 4.13. Raw data performed as time series of the pressure values with 1 s time resolution were interpolated using the procedure Interpolation in Mathematica 8 with default setting: InterpolationOrder → 3. Then, a new time series with a time resolution of 0.1 s was formed for the numerical calculation of the Fourier integral (4.25). The integral was calculated as a sum of the integrand expression at points $t = t_j$ multiplied by $dt = 0.1$ s for the frequency values from the range $\omega \in (0, 3)$ s^{-1} with a resolution of 0.001 s^{-1}. The absolute values of the Fourier integrals shown in Fig. 4.14a, b versus frequency are smoothed using the MovingAverage procedure over 60 points. The red and blue spectral curves correspond to the pressure records at depths $z_1 = -9.75$ m and $z_2 = -16.82$ m. Spectral peaks of the dynamic pressure are seen in the range $\omega \in (0.2, 0.8)$ s^{-1}. Spectral maximum is located in the range $\omega \in (0.35, 0.5)$ s^{-1}.

In the next step, we calculated the wave numbers numerically from Eq. (4.18) using the FindRoot procedure for each value of the wave frequency in the range $\omega \in (0.4, 0.8)$ s^{-1} with a resolution of 0.001 s^{-1}. The values of the Yong's modulus calculated from the dispersion Eq. (4.13) are shown in Fig. 4.14c, d by continuous lines when $H_0 = 46.15$ m and by dashed lines when $H_0 = 30$ m and $H_0 = 50$ m. The ice thickness and Poisson's ratio are $h = 94$ cm and $\nu = 0.33$, respectively. The negative values of the calculated Young's modulus are not shown in Fig. 4.14c, d. The maximum value of the Young's modulus in the frequency range $\omega > 0.6$ s^{-1} is 1.5 GPa in Fig. 4.14c and 3.5 GPa in Fig. 4.14d. Both of them only slightly depend on the water depth variations. The maximal value of the Young's modulus in the frequency range $\omega \in (0.4, 0.55)$ s^{-1} is 2.3 GPa in Fig. 4.14c and 4.4.3 GPa in Fig. 4.14d.

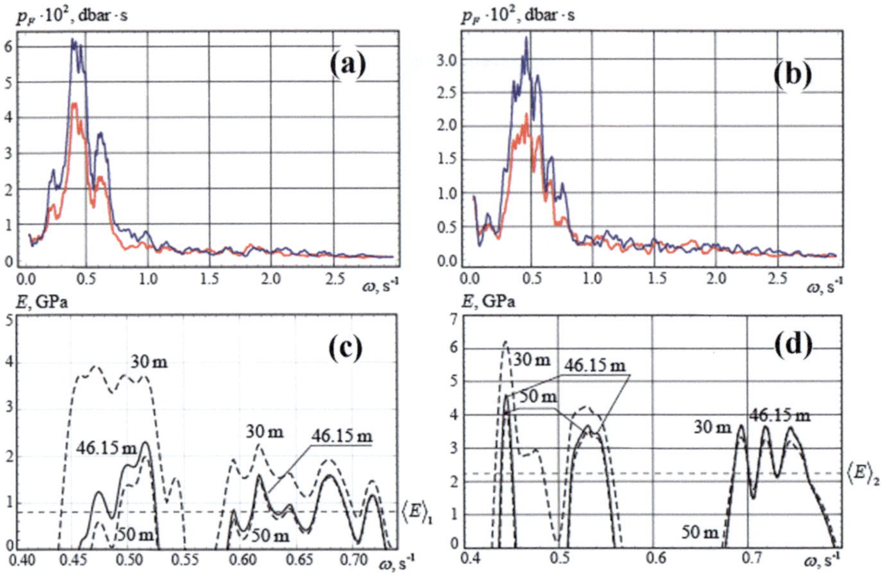

Fig. 4.14 Absolute value of Fourier transform of the dynamic pressure p_F (**a**, **b**) and calculated Young's modulus E (**c**, **d**) versus wave frequency ω in two consecutive fragments of the wave tail

Their dependence on the water depth variation is much stronger. The mean value of the Young's modulus shown in Fig. 4.14c in the range $\omega \in (0.59, 0.76)$ s^{-1} is $\langle E \rangle_1 = 0.77$ GPa. The mean value of the Young's modulus shown in Fig. 4.14d in the range $\omega \in (0.675, 0.795)$ s^{-1} is $\langle E \rangle_2 = 2.21$ GPa.

4.2.3 Estimates of the Loading Rate

In addition, we explain how the loading rate was estimated. Visual comparison of graphs in Fig. 4.13a and b shows that the quality of the data in Fig. 4.13a is better since the peak points of graphs in Fig. 4.13b are not smooth. Therefore we assume that the best estimate of the effective elastic modulus is given by the high frequency part of the wave signal with frequencies in the range $\omega \in (\omega_1, \omega_2)$, where $\omega_1 = 0.59$ s^{-1} and $\omega_1 = 0.76$ s^{-1}. The mean value of the effective elastic modulus is $\langle E \rangle_1 = 0.77$ GPa. Tabata et al. (1967) estimated the effective elastic modulus below 1 GPa in the experiments with in-situ beam tests conducted in 1964. In these tests, the mean temperature of the ice beams was –1.5 °C and the loading rate was below 100 kg/cm^2 s.

When flexural-gravity waves propagate through the ice the loading rate is estimated by the formula.

$$\dot{\sigma} < \left| \frac{\partial \sigma_s}{\partial t} \right|_{\max}, \tag{4.26}$$

where, σ_s is the surface stresses in the ice plate bended by the wave. Surface stresses in the bended plate are determined by the formula following from the theory of thin elastic plates (Timoshenko and Woinowsky-Krieger 1970)

$$\sigma_s = \frac{Eh}{2(1-\nu^2)} \frac{\partial^2 \eta}{\partial x^2}, \tag{4.27}$$

where, function $\eta(x)$ describes the shape of the mean surface of the plate. It follows from Formulas (4.11) and (4.14) that the stresses on the surface of the ice plate deformed by the periodic flexural-gravity wave are expressed by the formula

$$\sigma_s = -\frac{Eh\eta_0 k^2 e^{i\theta}}{2(1-\nu^2)}. \tag{4.28}$$

It follows from (4.28) that the maximal loading rate is determined as follows

$$\left| \frac{\partial \sigma_s}{\partial t} \right|_{\max} = \frac{Eh\eta_0 \omega k^2}{2(1-\nu^2)}. \tag{4.29}$$

The dynamic pressure amplitude $p_{dyn,0}$ at depth z below the ice is expressed through wave amplitude η_0 with Formula (4.14). Therefore,

$$\left| \frac{\partial \sigma_s}{\partial t} \right|_{\max} = \frac{Eh\omega k^2 \left| p_{dyn,0} \right|}{2(1-\nu^2)\rho_w g |P(\omega, k, z)|}. \tag{4.30}$$

For the estimates of the loading rate from the high frequency part of the wave signal we substitute in Formula (4.30) $z = z_1 = -9.75$ m, $\omega = \omega_1 + (\omega_2 - \omega_1)/2 = 0.63$ s^{-1} and

$$p_{dyn,0} = \frac{1}{\sqrt{2\pi}} \int_{\omega_1}^{\omega_2} p_F(\omega) d\omega, \tag{4.31}$$

where, function $p_F(\omega)$ is shown in Fig. 4.14a by the red line. Wave number $k = 0.41$ m^{-1} is found from the dispersion Eq. (4.13) with $H_0 = 46.15$ m, $h = 94$ cm and $E = \langle E \rangle_1 = 0.77$ GPa. Finally, we find that

$$\left| \frac{\partial \sigma_s}{\partial t} \right|_{\max} = 139.7 \text{ kg/(cm}^2 \text{ s)}. \tag{4.32}$$

Thus, the loading rate is similar to the loading rate in the experiments of Tabata et al. (1967).

4.2.4 Conclusions of This Section

A new method for calculation of the effective ice stiffness is described and analyzed. It is based on the analysis of water pressure records at two or three locations below the ice bottom when flexural-gravity waves propagate through the ice. The analysis of the relative errors in the definition of the effective stiffness shows their reduction with the increase of the wave frequency. The relative error is also smaller when one of the pressure sensors is deployed below the ice while the other sensors are deployed at a depth of about 10 ± 15 m.

The method is tested using the data from the field measurements in the Tempel Fjord in Svalbard. Flexural-gravity waves were excited by the ice destruction at the front of the outflow Tuna Glacier in February 2010. The water pressure was recorded by two sensors deployed at depths $z_1 = -9.75$ m and $z_2 = -16.82$ m. The water depth was measured by the pressure sensor deployed at the sea bottom. The ice thickness was measured by drilling. Several spectral peaks were found in the frequency range $\omega \in (0.2, 0.8)$ s^{-1}. The data were used for the calculation of the effective elastic modulus $E = 0.77$ GPa. Similar values of the effective elastic modulus were obtained in the experiments of Tabata et al. (1967) which were performed under similar loading rates.

4.3 Asymmetric Tide in Lake Vallunden (Spitsbergen)

The phenomenon of asymmetric tide is well known. Tidal asymmetry is usually developed in shallow water. If the duration of decreasing sea level due to the ebb exceeds that of the increasing tide (when flood tidal currents are stronger), the system is referred to as flood dominant or flood asymmetric. If the duration of the decreasing tide is shorter than that of the increasing tide (and consequently, the ebb currents exceed the flood currents), the tide is referred to as ebb dominant (Walton 2002). Usually, asymmetric tide is developed in channels and inlets (Hoitink et al. 2003; Mirfendersk and Tomlinson 2008). This phenomenon is analyzed in many papers (Gallagher 1973; Aubrey and Speer 1985).

The presence of higher harmonics is an additional mechanism for tidal velocity asymmetry in inlets (Boon and Byrne 1981; Aubrey and Speer 1985). If two tidal constituents (M_2 and M_4) are developed, we can express the tidal elevation of the surface as

$$h = A_{M2}\cos(\omega_{M2}t) + A_{M4}\cos(\omega_{M4}t - \varphi) \qquad (4.33)$$

where, A_{M2} is the amplitude of M_2 component of tide; and A_{M4} is the amplitude of M_4 component of tide; φ is the amplitude phase difference between M_2 and M_4 constituents of tide.

Boon and Byrne (1981) showed that flood dominance exists if $\pi \le \varphi \le 2\pi$, and ebb dominance exists if $0 \le \varphi \le \pi$. The greater the ratio A_{M4}/A_{M2}, the greater the flood or ebb dominance. If $\varphi = 0$, the crests are sharp and troughs are flat. If $\varphi = 2\pi$, the crests are flat and troughs are sharp.

We observed strongly asymmetric tide in a channel connecting the Van Mijen Fjord and Lake Vallunden (and also in Lake Vallunden) in Spitsbergen. The channel that connects Lake Vallunden with the fjord (Littrownesec) is a natural test area for investigating strong asymmetric tides. Tides in fjords were investigated by Stigebrandt (1980), Farmer and Freeland (1983). The studies in Van Mijen Fjord were conducted by Fer and Widell (2007) and Støylen and Weber (2010). Long-term measurements of tides in Van Mijen Fjord were performed by Caline (2009). The amplitudes of different tidal constituents based on his measurements are presented in Table 4.2. We see from the table that M_2 and S_2 constituents dominate in the tidal motion.

The motivation of this section is to show that strong tidal asymmetry develops in a small channel that connects the lake with the fjord. The structure of tides becomes flood-dominating in the channel and lake. This transition occurs over a small distance of 50 m.

Field works were performed in 2008–2012 in Van Mijen Fjord (Spitsbergen). Lake Vallunden, approximately 1 km long, is connected to the Van Mijen Fjord by a shallow channel 100 m long, 10 m wide, and 1–2 m deep (Littrownesec Channel). The lake is located almost near the summit of the fjord at a distance of 80 km from the open ocean. Strong tidal currents develop in this channel because the tidal displacement of the level in the Van Mijen Fjord reaches 3 m. Photos and chart of the region, in which points of measurements are indicated, are shown in Fig. 4.15.

We deployed bottom tide gauges (SBE-26, SBE-39, and SBE-37) in the lake, in the channel, and in the fjord near the channel. We also deployed a current meter (Aquadopp, operating frequency 2 MHz) in the channel and measured the time evolution of the level with graduated poles (vertical pole with marks of centimeters). The currents were measured in the fjord from ice near the channel using an RDI Monitor instrument (operating frequency 1.2 kHz). A summary of measurements is given in Table 4.3.

A down-looking current profiler ADCP RDI Sentinel was deployed in the fjord at point 2 (Fig. 4.15a) in the winter time (1200 kHz, bin size 5 cm, high resolution mode, bottom tracking, time/ping 3.52 min, average ensemble interval 2 min). It

Table 4.2 Semidiurnal harmonic constants based on the long-term measurements of tide near the Littrownesec channel

	M_2	S_2	K_1	K_2	N_2
A, m	0.5109	0.1867	0.0701	0.0541	0.0967
s^{-1}	14.0519×10^{-5}	14.5444×10^{-5}	7.29212×10^{-5}	14.5842×10^{-5}	13.788×10^{-5}
G (phase)	0.759393	1.6921	4.16697	1.65527	0.427606

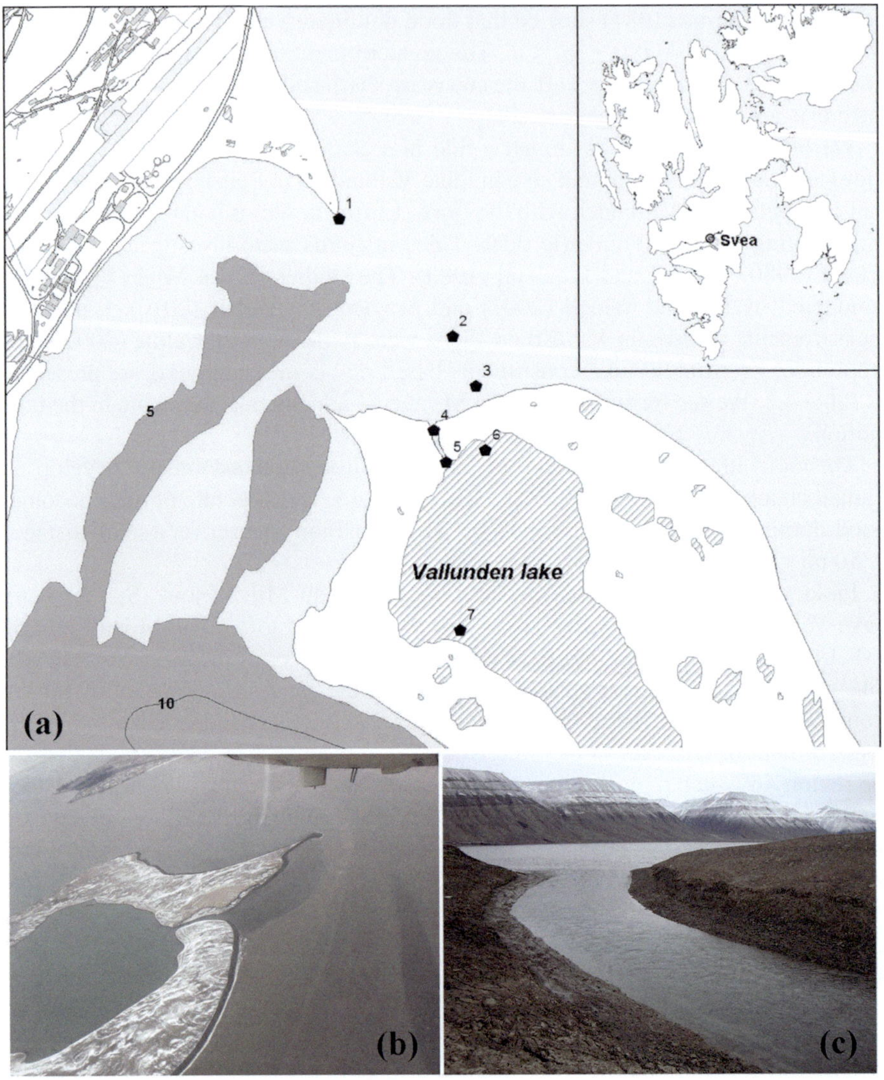

Fig. 4.15 **a** Chart of the region and locations of the instruments. Isobaths of 5 m and 10 m are indicated. The inset shows location of the region in Spitsbergen; **b** Aerial photograph of the lake and channel. **c** The channel near the entrance to the lake. Numerals 1–8 show locations of measurements presented in Table 4.3

was fixed in a hole in the ice as shown in Fig. 4.16a. The Aquadopp current meter (2 MHz, bin size 50 cm, time sampling 6 s) was located in the channel horizontally as shown in Fig. 4.16 at point 4 (Fig. 4.15a). It was tightly fixed at a distance of 20 cm from the bottom of the channel. The second and third bins were averaged so that the instrument measured integrated currents in the interval 0.5–1.5 from the transducer.

Table 4.3 Summary of measurements

Time	Points of measurements (see Fig. 4.15)							
	1	2	3	4	5	6	7	8
August, 2009	SBE-39			Graduated pole	Graduated pole	SBE-39	SBE-39	SBE-39
Sept., 2010	SBE-26		SBE-39	ADCP Aquadopp		SBE-39	SBE-39	
February, 2009	SBE-39					SBE-39		
March, 2012	SBE-37	ADCP RDI				SBE-39		

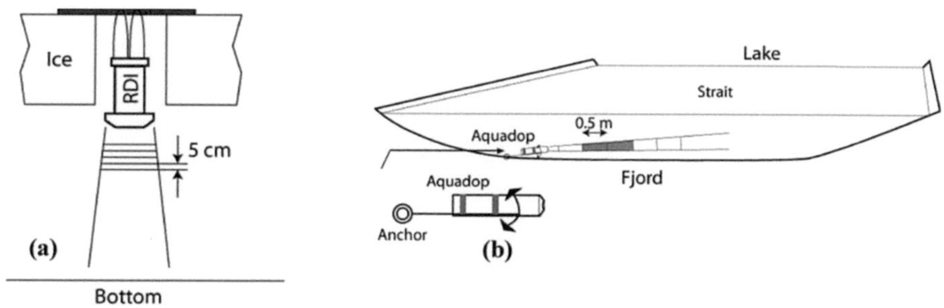

Fig. 4.16 Deployment schemes of current profilers ADCP: **a** RDI Sentinel and **b** Aquadopp

We placed two graduated poles at both ends of the channel (points 4 and 5 in Fig. 4.15a) and observed the tidal elevation of the level during half period of the semidiurnal tide (Fig. 4.17a). The displacement of the sea level at the fjord side of the channel reached 50 cm at the place where the pole was located in the channel, whereas at the lake side it was only 15 cm. We note that the surface displacement in the fjord reached ± 1.5 m (Fig. 4.19). The maximum tide in the lake occurs with a 90 min delay relative to the maximum tide in the fjord. When the currents in the fjord turn from tide to ebb, the current continues to flow into the lake for 90 min. This happens because when the surface elevation starts to decrease in the fjord it is still higher than in the lake, and the equilibrium occurs only after one and a half hour of level decreasing in the fjord (Fig. 4.17b).

In August 2009, two pressure gauges SBE-39 were deployed on the bottom of the lake at points 6 and 7 and pressure gauge SBE-26 was deployed on the fjord bottom at point 1 (Fig. 4.15a). The instruments in the lake recorded clear flood dominant time series of currents and pressure. The results of measurements are shown in Figs. 4.18 and 4.19. The tide amplitude in the lake is two times smaller than in the fjord.

The graphs of water pressure in the fjord on the lake side and on the opposite side of the fjord based on the data of pressure gauges SBE-26 and SBE-39 deployed at points 1 and 3 in August 2009 are shown in Fig. 4.19. The measurements demonstrate

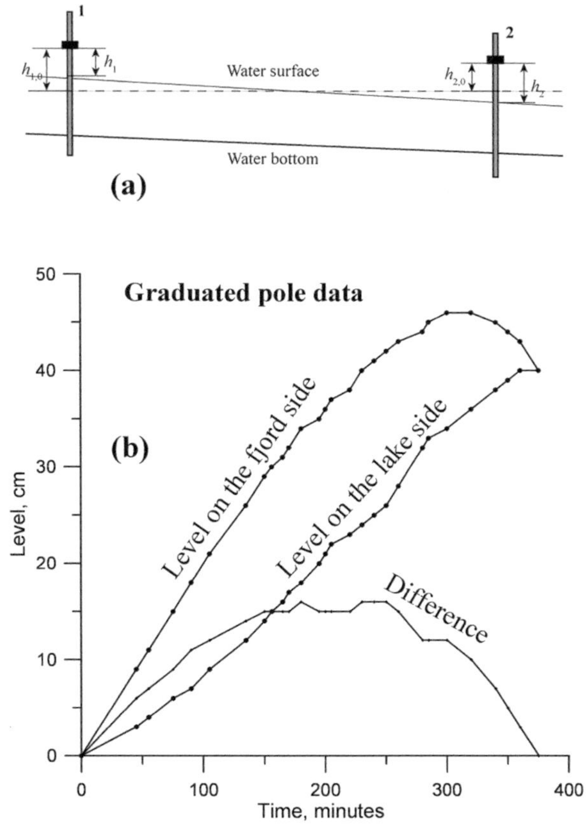

Fig. 4.17 **a** Scheme of water level measurements using two graduated poles in the channel; **b** Graphs of sea level variations at two graduated poles in the channel

that the amplitude of the tide is almost the same at both shores of the fjord. The records in the channel and in the lake show that the tide record is flood dominating.

The measurements in the lake were repeated in September 2010. In this experiment, one pressure gauge SBE-39 was deployed in the lake at point 6 and current meter (Aquadopp) was deployed in the channel at the point 4. The records of SBE-39 are shown in Fig. 4.18b and d. The record of currents in the channel at point 4 is shown in Fig. 4.20. The form of the pressure and current records is flood dominating. The maximum absolute speed of the current in the channel was about 1.5 m/s. The measurements were made in the ice-free season using the Aquadopp current meter.

The form of the tide changes owing to the presence of a shallow rocky bar where the channel connects with the lake. In summer, the depth of this bar in the low water phase is less than one meter. In winter the effect of the bar is enhanced by the ice freezing, which makes the bar even shallower. In winter, strong currents in the channel influence the reduction of ice thickness in comparison with the ice thickness in the fjord and in the lake. At the same time the snow thickness on the ice surface in the channel is greater than in the fjord. As a result, the tidal amplitude in the lake becomes smaller and the duration of the flood tidal phase in the lake becomes shorter.

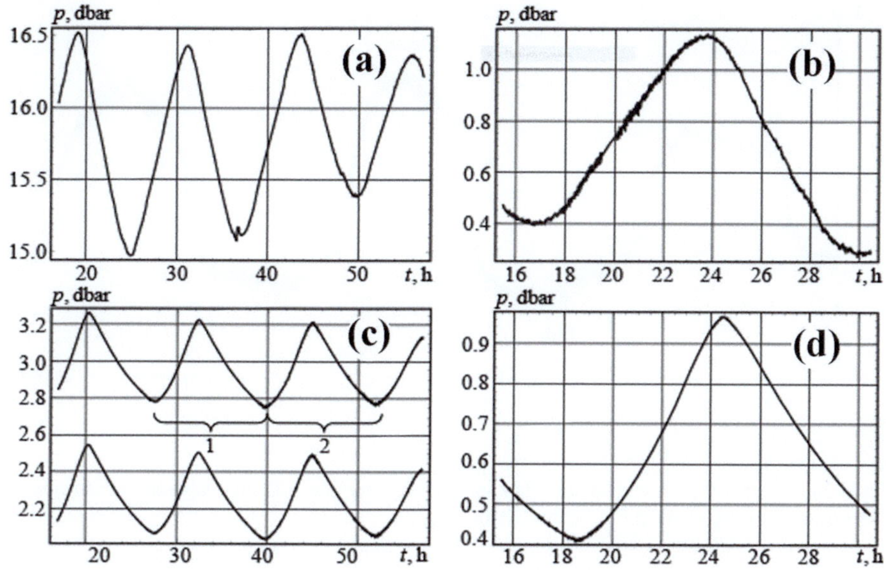

Fig. 4.18 Seabed water pressure measured in the fjord (**a**, **b**) and in the lake (**c**, **d**). Field works were carried out on August 26–28, 2009 (**a**, **c**) and September 4–5, 2010 (**b**, **d**). Braces with numbers 1 and 2 in this figure show the data used to compare with the model calculations

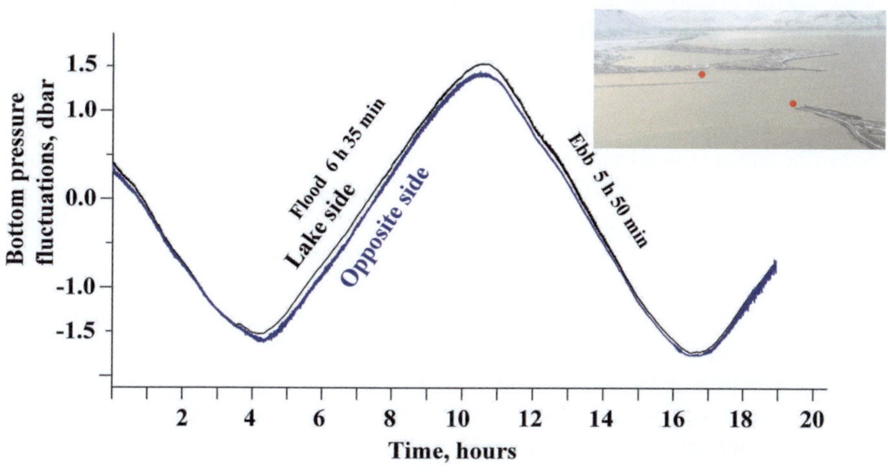

Fig. 4.19 Bottom pressure variation in the fjord near the channel and on the opposite side of the fjord versus time (red dots) recorded in August 2009. A photo of the place with red points indicating the locations of measurements is shown in the upper right angle

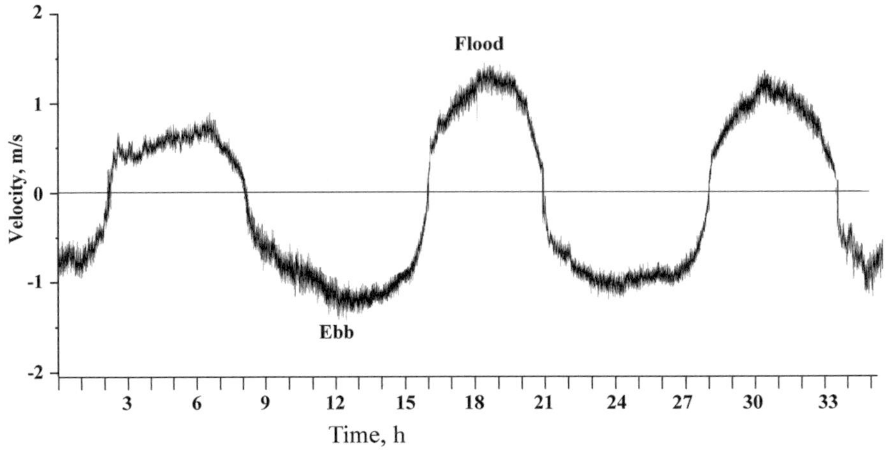

Fig. 4.20 Record of currents in the channel 20 cm over the bottom (Aquadopp instrument). Measurements were made in September 2010

This effect is shown in Fig. 4.21. In the winter expeditions in April 2009, 2010, and 2011, the tide in the lake was zero when the channel was completely locked due to freezing. The measurements of water velocities in the channel in the ice season were not successful because of the strong water flood on the ice through the drilled holes.

Sea current profile in the fjord was recorded from ice with ADCP RDI (1200 kHz) at point 2 on March 2012. The measurements were carried out in a shallow channel in the fjord where the tidal currents are maximal. Examples of the ADCP time series versus time for the velocity direction and velocity magnitude are shown in Fig. 4.22. The northeastern velocities (60°, blue color) are related to the flood cycle. The flood currents last approximately from 10:00 to 18:20 (8 h 20 min). The flood cycle based on the pressure measurements (increasing depth) lasts from 11:30 to 18:05 (6 h 35 min). The southwestern currents (210°, yellow color) are related to the ebb cycle. The ebb currents last approximately from 6:00 to 12:00 (6 h) or even shorter because from 9:30 to 12:00; the ebb currents are only in the upper layer, while the lower layer is occupied by the tidal currents. The ebb cycle based on the pressure measurements (decreasing depth) lasts from 5:40 to 11:30 (5 h 50 m). Table 4.4 summarizes all measurements in the fjord near the channel made in different periods of time.

The maximum absolute velocity up to 50 cm/s shown in Fig. 4.22b by red color was recorded during a very short time period in the beginning of the ebb phase. Generally, the absolute velocities were lower than 30 cm/s. The tidal current below the ice in the narrow part of the Van Mijen Fjord between points 1 and 3 is not symmetric over the semidiurnal cycle unlike the water pressure variations at the bottom (Fig. 4.19). However, the pressure cycle is also asymmetric. Properties of the velocity profile across the fjord depend on the locations were the ice is grounded in the low water phase and river discharge from the valley in the end of the fjord (Wrangborg et al. 2013).

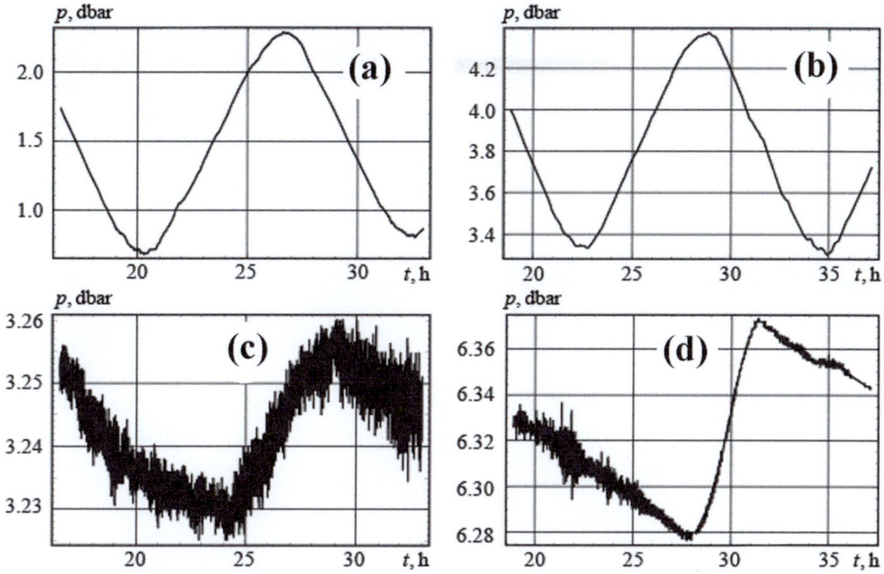

Fig. 4.21 Seabed water pressure measured in the fjord (**a**, **b**) and in the lake (**c**, **d**). Field works were performed under ice on February 8–9, 2009 (points 1, 3 in Fig. 4.15a) (**a**, **c**) and on March 8–9, 2012 (points 6, 7 in Fig. 4.15a) (**b**, **d**)

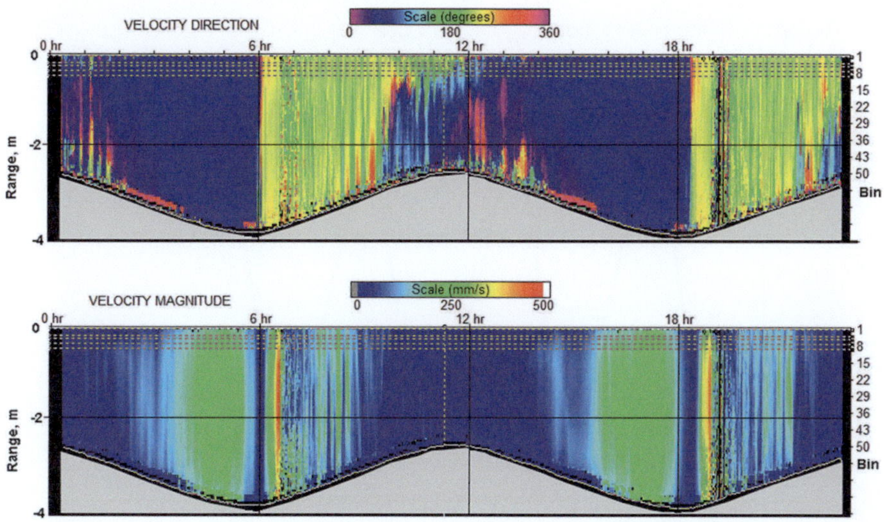

Fig. 4.22 Record of currents measured by the ADCP instrument at point 2 (see Fig. 4.15a for its location)

Table 4.4 Duration of flood and ebb periods at different points in the fjord near the channel

Point (Fig. 4.15) and time of measurements; figure, in which the graph is shown	Duration of periods	
	Flood period	Ebb period
Point 1, August 2009. Figure 4.19	6 h 35 min	5 h 50 min
Point 3, February 2009. Figure 4.21a	6 h 27 min	5 h 57 min
Point 8, August 2009. Figure 4.18a	6 h 49 min	5 h 35 min
Point 1, September 2010. Figure 4.18b	6 h 25 min	5 h 59 min
Point 1, March 2012. Figure 4.21b	6 h 28 min	5 h 56 min
Point 2, March 2012. Figure 4.22	6 h 35 min	5 h 50 min

We tried to find a fit for the measurements by a bottom pressure gauge at point 6 (see Fig. 4.15a). A record of the tide gauge in the lake is shown with a blue line with variable amplitude (Fig. 4.23).

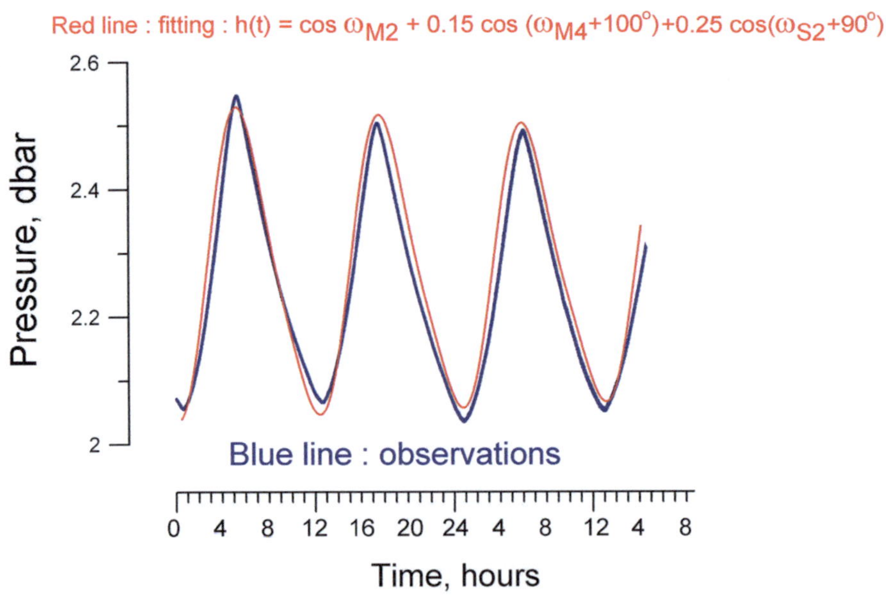

Fig. 4.23 Variations in the bottom pressure in the lake close to the channel (blue line), the measurements were made in September 2009 at point 6 (Fig. 4.15a); a fit with three sinusoids, relation written in the figure (red line)

We consider three types of variations:

(1) Decreasing or increasing amplitude
(2) Uneven periods of flood and ebb
(3) Sharp crests and flat troughs.

The decreasing amplitude of the tide is related to the spring-neap variability of tides that is caused by the presence of two main semidiurnal tidal constituents M_2 (period 12.42 h) and S_2 (period 12.00 h) (Caline 2009). In the mathematical sense, this variability is the beating caused by two sinusoids with close frequencies. Irregularity superimposed over this variation is caused by onshore and offshore wind currents that slightly change the regular tidal variation of the level.

Choking over the bar at the entrance to the channel causes generation of higher harmonics of the semidiurnal tide (Stigebrandt 1980; McClimans 1977); hence uneven periodicity of flood and ebb tides appears. Among them the most important is the M_4 harmonic (period 6.21 h). The phase difference between the M_2 and M_4 constituents of the tide causes asymmetry of the tide with uneven periods of flood and ebb. Asymmetry between crests and troughs of the record is caused by the phase difference between the M_2 and M_4 harmonics initially induced by tide choking over the bar.

We tried to find a simple fit of the tide amplitudes variation using three sinusoidal tidal harmonics: M_2 (12.42 h), S_2 (12.00 h), and M_4 (6.21 h). The curve that fits the measured amplitude variation is shown in Fig. 4.23 with a red line. The fit was found using several iterations. First, we selected the amplitude of the M_4 harmonic by sorting different amplitudes with a step of 0.05 of the M_2 harmonic. The selection was made by finding the minimum root-mean-square deviation between the fit and measured bottom pressure variation. Then, we selected a fit by changing the phase of the M_4 harmonic with a step of 5 degrees. A similar procedure was applied to the S_2 harmonic that changes the amplitude of the signal due to the spring-neap variability. Finally, we obtained the following fit:

$$h = A_{M2}\cos(\omega_{M2}t) + 0.15A_{M4}\cos(\omega_{M4}t + 100^{\circ}) + 0.25A_{S2}\cos(\omega_{S2}t + 90^{\circ}) \tag{4.34}$$

We note that no bottom friction and no bottom topography were introduced into the fit. They are included indirectly by the existence of M_4 constituent. A surprising thing is that the point at which the flood dominating tide was recorded exists in the channel at a distance of only 50 m from the beginning of the channel.

Tidal asymmetry problem was raised in several publications (Huang et al. 2008; Cheng et al. 2010; Dias and Sousa 2009), in which the authors investigated tidal asymmetry in narrow inlets. It was mentioned that the dependence of coefficient of resistance on the Reynolds number, bottom roughness and the form of the channel is most important for the simulations. In our case, the resistance is characterized by product κC, where C is the drag coefficient and κ characterizes the influence of variations of hydraulic radius and water speed over the channel on the resistance. We used average characteristic of the channel and assumed that $\kappa = 1$. Thus, we

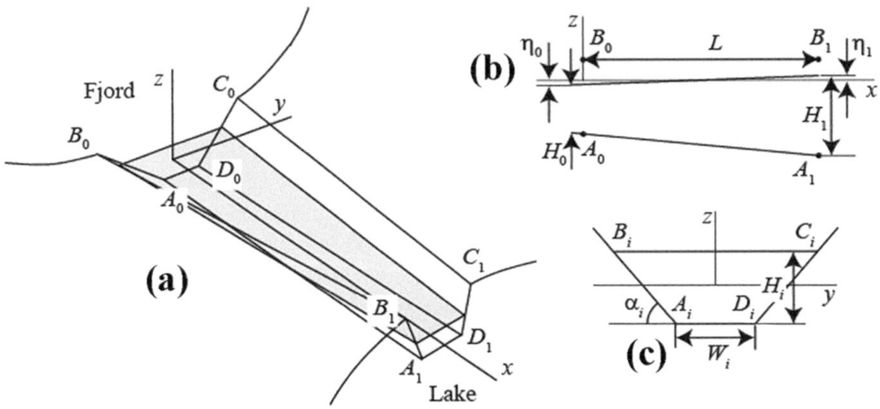

Fig. 4.24 Scheme of the channel: 3D (**a**) and projections (**b**, **c**)

obtained that drag coefficient C is most important for the numerical simulations. The value of coefficient $C = 0.01$ was chosen based on the paper (Mofield 1988) when representative water depth is 1 m and seabed roughness is slightly above 1 cm.

Numerical modeling was performed for a channel of the trapezoidal shape shown in Fig. 4.24a. The length of the channel in the x-direction is L, the bottom of the channel is slightly inclined, and the water level elevation changes along the channel (Fig. 4.24b). Vertical cross-sections of the channel at the fjord side and at the lake side have trapeze forms $A_0B_0C_0D_0$ and $A_1B_1C_1D_1$ (Fig. 4.24c). Bottom bases of the trapezes are equal to W_0 and W_1, and slope angles of their lateral sides are equal to α_0 and α_1. The thicknesses of the water levels above the bases are equal to H_0 at the fjord side and H_1 at the lake side, and the depths of undisturbed water in the channel are equal to $h_0 = H_0 - \eta_0$ and $h_1 = H_1 - \eta_1$, respectively.

The squares of areas A_i and hydraulic radii R_i were calculated using the following formulas

$$A_i = \frac{H_i^2}{\tan\alpha_i} + H_i W_i,\ R_i = \frac{A_i}{P_i},\ P_i = \frac{2H_i}{\sin\alpha_i} + W_i,\ i = 0, 1. \tag{4.35}$$

Numerical simulations were performed with the following numerical values, which are close to those in the natural conditions

$$\begin{aligned} &S_l = 537800\ \text{m}^2,\ L = 100\ \text{m},\ W_0 = 7\ \text{m},\ W_1 = 3\ \text{m},\ \alpha_{0,1} = 30^\circ, \\ &h_0 = 1.2\ \text{m}\ h_1 = 1.5\ \text{m},\ C = 0.01,\ \kappa = 1. \end{aligned} \tag{4.36}$$

S_l is the square of the lake. The values in Formula (4.36) correspond to geometrical characteristics of the channel. The drag coefficient corresponds to very shallow water conditions and high water velocities, therefore it is relatively high.

Water level variations at the fjord side of the channel (forcing) are determined by function

$$\eta_0 = A_{M2}\cos(\omega_{M2}t + (v_0 + u)_{M2} - G_{M2}) + A_{S2}\cos(\omega_{S2}t(v_0 + u)_{S2} - G_{S2}) + \\ A_{K1}\cos(\omega_{K1}t + (v_0 + u)_{K1} - G_{K1}) + A_{K2}\cos(\omega_{K2}t + (v_0 + u)_{K2} - G_{K2}) + \\ A_{N2}\cos(\omega_{N2}t + (v_0 + u)_{N2} - G_{N2}), \tag{4.37}$$

where numerical values of harmonic constants A, ω and G are given in Table 4.5; the values of initial arguments $(v_0 + u)_i$ are calculated using standard astronomical tables, and values of reduction coefficients are assumed equal to 1 (Vladimirsky 2010).

The results of the numerical simulations are illustrated in Fig. 4.25. Figure 4.25a shows that the simulated variations in the water level at the lake side are very similar to the measured values shown in Fig. 4.18. Simulated water velocities at the fjord side shown in Fig. 4.25c are very similar to the velocities measured using the Aquadopp current meter (Fig. 4.20). The maximum velocities are gained at the lake side of the channel when the water level in the fjord is close to the maximum and at the fjord side of the channel when the water level in the fjord is close to the minimum. Their maximum absolute values exceed 2 m/s. Model calculations demonstrate that the tidal flow in the channel is flood dominating.

Figure 4.25b shows that a constriction of the channel due to the ice formation in the winter time explains the decrease in the tidal amplitude and the duration of flood phase of the tide in the lake measured in the ice season (Fig. 4.21). Figure 4.25d demonstrates the decrease of the current velocities in the channel due to the constriction.

Figure 4.26a shows water pressure variation at the bottom of the lake measured during two consecutive periods of semidiurnal tide (12.42 h) versus time (see Fig. 4.18c), and Fig. 4.26b shows an example of water level variations in the lake calculated in three consecutive periods of the semidiurnal tide.

These graphs were used for the analysis of the change in the form of tide elevation. Curves 1 and 2 in Fig. 4.26a correspond to intervals 1 and 2 in Fig. 4.18c. Curves 1, 2 and 3 in Fig. 4.26b correspond to intervals 1, 2 and 3 in Fig. 4.25a. Each of the curves shown in Fig. 4.26 was approximated by formula

$$F = a_1 + a_2\cos(\omega_{M2}t + a_3) + a_4\cos(2\omega_{M2}t + a_5). \tag{4.38}$$

Table 4.5 Absolute values of the amplitude ratios of M_2 and M_4 tides in the lake

Ratio of amplitudes	Measurements related to intervals 1 and 2 shown in Fig. 4.18c		Three consecutive intervals of model calculation. The graphs are shown in Fig. 4.25b		
	1	2	1	2	3
$\lvert a_4/a_2\rvert$	0.196	0.198	0.232	0.178	0.177

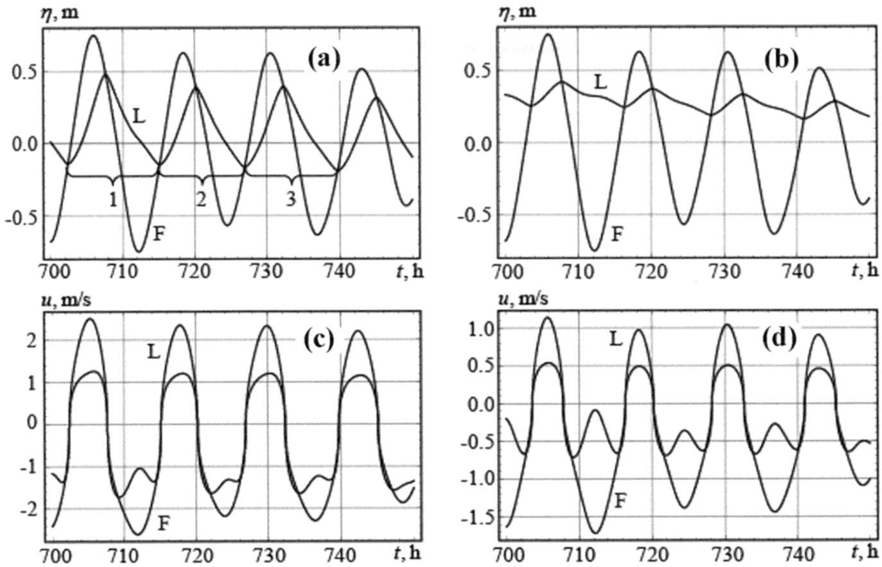

Fig. 4.25 Model calculations of water level that demonstrate tidal asymmetry. Variations in the water level at the lake side (L) and at the fjord side (F) at low tide (**a**) and high tide (**b**). Variations in the water velocity at the lake side (L) and at the fjord side (F) at low tide (**c**) and high tide (**d**). Braces in this figure a show the time intervals used for comparison in Fig. 4.26b

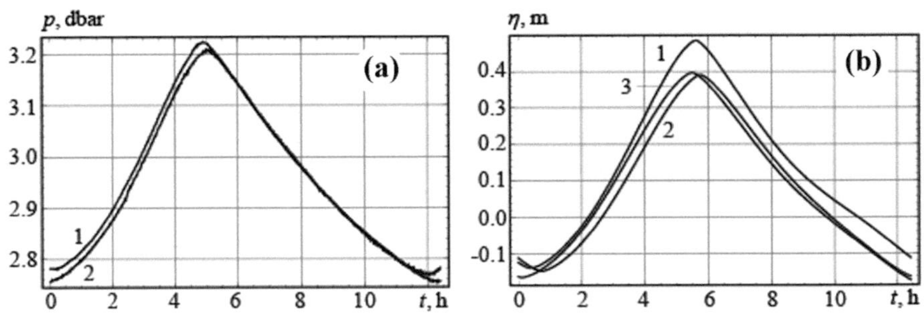

Fig. 4.26 Water pressure at the bottom of the lake measured in two consecutive periods of semidiurnal tide versus time (**a**) (see Fig. 4.18c). Water level in the lake calculated in three consecutive periods of semidiurnal tide versus time (**b**)

Constants a_i were calculated using operations FindFit in "Mathematica 8". The absolute values of the calculated ratios $|a_4/a_2|$ of M_2 and M_4 tide amplitudes in the lake are shown in Table 4.5. The ratios $|a_4/a_2|$ calculated with field data are close to the ratios calculated for simulated tides in the lake.

The physical interpretation of the change in the form of the tide is explained by the presence of a shallow rocky bar where the channel connects with the fjord. The depth of this bar in the low water phase is less than one meter. This is the tidal choking effect considered in Stigebrandt (1980), McClimans (1977). The presence of the bar leads to the generation of the second harmonic of semidiurnal tide; hence the form of the tide elevation changes. Different harmonics of the tide lose different parts of their energy when passing over a sill before flowing into the channel. Ice conditions in winter enhance this effect.

We attribute this rapid change in the asymmetric form of the tide to a very small scale of the experiment. Usually, such transition occurs over distances of the order of one kilometer (Dronkers 1986; Nidzieko 2010). The horizontal scales of the channel are of the order of a few kilometers and the depth scale is of the order of a few tens of meters. In our case, the Littrownesec Channel is only 10 m wide and 1–2 m deep, which is two orders of magnitude smaller than the usual scales in straits and estuaries where tidal asymmetry was observed (Friedrichs and Aubrey 1994; Blanton et al. 2002; Lanzoni and Seminara 1998; Savenije and Veling 2005). We can suppose that such change in the scales can lead to a decrease in the horizontal scale over which tidal asymmetry forms.

The tide in shallow places becomes asymmetric due to nonlinear effects caused by the influence of the bottom and especially by choking of the tidal flow over a sill that is located between the fjord and channel. The form of asymmetric tide in the channel can be satisfactory approximated using the M_2 harmonic and only two additional tidal components S_2 and M_4.

Measurements in the channel (Littrownesec) demonstrate that strong asymmetric effects develop over a distance of 50 m very close to the sill that chokes the tidal flow.

The causes of such rapid development can be related to the small scale of the channel, which is two orders of magnitude smaller than the coastal channels connecting the open sea and lagoons and in the estuaries, where tidal asymmetry is usually observed.

4.4 Seiche Oscillations in Lake Valunden (Spitsbergen)

Seiches are standing waves in enclosed or semi-enclosed water reservoirs. In general, seiches are generated by external forcing mostly by wind and atmospheric pressure variations. The gravity force restores the horizontal surface of water, which is in hydrostatic equilibrium. Pulses of harmonic motion are reflected from the boundaries of the reservoir. Repeated reflections result in standing waves. Oscillation nodes appear at points that experience no vertical motion. The frequencies of seiches depend on the fundamental features of a specific reservoir and do not depend on the mechanism of external forcing. Oscillation frequency (eigen period of seiches) is determined by the geometry and size of the basin, its depth, and configuration of the coastline (Proudman 1953; Rabinovich 2009). Irregularities of the basin geometry do not prevent development of seiches and basins with irregular shapes oscillate with very stable frequencies.

However, the amplitudes of oscillation strongly depend on the intensity of external forcing. Hence, the time variability can be strong. The natural period of a seiche can be estimated from the Merian's formula:

$$T_{seich} = {2L}\Big/{n\sqrt{gH_{mean}}}$$

where, T is the period of seiche, L is the length of the basin, H_{mean} is the average depth of the basin, g is the acceleration due to gravity, and n is the mode number (Merian 1828). The seiches occur as natural oscillations in a basin. The mode with the lowest frequency ($n = 1$) is referred to as the fundamental mode.

Usually seiches are unnoticeable, except during periods of very smooth water surface. The amplitudes of seiches can be as high as a few centimeters, while their periods are of a few minutes. Seiches can be induced by wind, changes in the atmospheric pressure, tsunamis or tides. In this paper we study the seiches under an ice cover in a lake connected to a fjord in Spitsbergen. In this section we review publication (Marchenko and Morozov 2016b).

We studied seiches in Lake Vallunden, which is located in Spitsbergen. At its northwestern coast the lake is connected with the Van Mijen Fjord by a narrow (10 m) 100 m long channel (Littrowneset) with strong tidal currents (Marchenko and Morozov 2016b). A chart of the lake is shown in Fig. 4.27. The lake extends from the southeast to the northwest. The lake is approximately 1300 m long and 600–650 m wide. The mean depth of the lake is 10–11 m. Starting from 2008, we made repeated measurements of seiches at different points of the lake in winter and summer using bottom based pressure gauges SBE-39. Seiche oscillations almost always exist in the lake. In March 2015, we used four bottom pressure gauges SBE-39 to measure the pressure variation during six days. The lake and the fjord were covered with ice, but the Littrowneset channel was ice free due to strong tidal currents.

In 2015, we made measurements at the points shown in the Table 4.6 and in Fig. 4.27.

Two spectral peaks usually exist on the spectra of bottom pressure fluctuations in Lake Vallunden. The frequencies of peaks are inverse values to periods of 240 and 128 s. These periods correspond to the longitudinal and transversal natural oscillations of the basin. According to the Merian formula, the periods of spectral peaks correspond to the first modes of seiche oscillations, while the oscillation nodes are located in the middle of the lake.

A typical spectrum of bottom pressure fluctuations is shown in Fig. 4.28. It was calculated from the bottom pressure time series from March 02 04:00 to 07:00 with a time sampling of 1 s. Two major reliable spectral peaks corresponding to the longitudinal and transversal seiches with periods of 240 s (0.0041667 cps) and 128 s (0.0078125 cps) are seen.

Usually, the seiches appear at strong winds and changes in the atmospheric pressure, and also due to tidal forcing through the Littrowneset channel. If the lake is covered with ice, wind forcing at the surface decreases or even vanishes. In this case, wind forcing is applied to the ice cover plate as a whole. Hence, strong winds do not

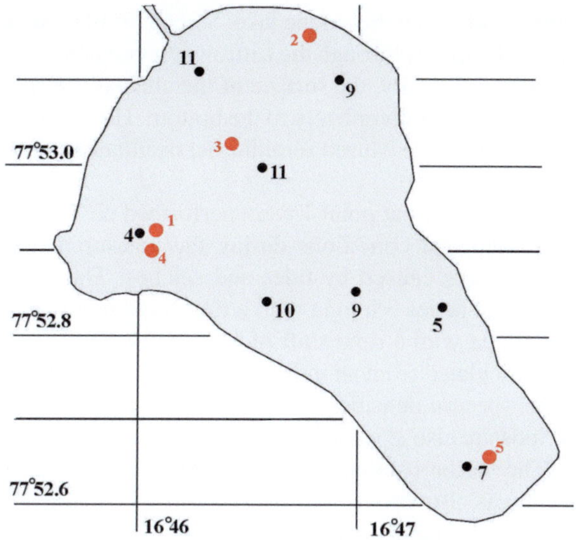

Fig. 4.27 Map of the Lake. Locations of pressure gauges with numbers of stations are shown with large red dots. Smaller black dots indicate the locations of depth measurements with a rope. Depths at black dots are in meters. The amplitude of tidal variations in depth is 0.5 m

Table 4.6 Locations and dates of measurements

Beginning of measurements	Point	Latitude	Longitude
March 01	1	77°52.881′ N	16°46.098′ E
March 05	2	77°53.160′ N	16°46.625′ E
March 05	3	77°53.017′ N	16°46.323′ E
March 05	4	77°52.898′ N	16°46.019′ E
March 05	5	77°52.634′ N	16°47.550′ E

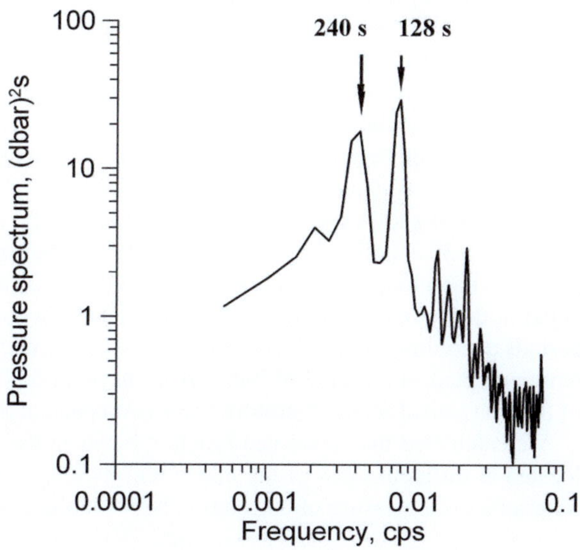

Fig. 4.28 A typical spectrum of bottom pressure fluctuations. Measurements at point 1 on March 2, 2016

always induce seiches in the lake. Seiche oscillations in the lake can be generated by the tidal currents through the Littrowneset channel. The depth of the channel is about one meter. Usually, the surface of the channel freezes in winter, and sometimes the channel freezes completely to the bottom. However, in March 2016, the channel was ice free. The tide caused semidiurnal oscillations of the ice cover with an amplitude up to 50 cm.

Observations at point 1 were performed on March 2–3, 2016. Figure 4.29 shows meteorological conditions during the measurements. Pressure fluctuations at the bottom were caused by tides and seiches. The spectral densities were calculated from time series 3 h long with a time sampling of 1 s. Calculations of spectra were performed with a time shift of each time series equal to 3 h. Thus, each spectrum was calculated from an independent time series of bottom pressure. Time variations in the spectral densities of bottom pressure at inverse frequencies of 240 and 128 s periods are also shown in Fig. 4.29.

The seiche oscillations were intensified at night on March 3. On this day, the wind was also strong, and fluctuations of the atmospheric pressure were recorded, which could have been the causes of intensification of the seiches. However, in the beginning of night on March 2, the wind was even stronger, and the atmospheric pressure decreased rapidly. However, the seiches were not intensified. Hence, strong winds and jumps in the atmospheric pressure do not always lead to intensification of the seiches if the surface is covered with ice.

Intensification of seiches in the lake was observed during the period of spring tides on March 9. This time coincided with the new moon and intensification of tide. Figure 4.30 shows time variations of bottom pressure at point 4. Seiches intensify during the period of low water. This can be seen from the pressure graph. The pressure curve in the figure seems thicker due to the fluctuations related to seiches. We used a band filter to separate the seiches with a period of 266 s. The band-pass of the filter was within 235–278 s. This filtering removes the tidal oscillations. An envelope curve of the fluctuations after filtering is shown in Fig. 4.30. The vertical lines in the figure connect the maxima of longitudinal seiches with the curve of pressure variations. It is seen that seiches intensify during the ebb phase. Thus, tidal oscillation force fluctuations of the ice cover in the lake and as a response to this forcing the ice cover fluctuates with its natural frequencies of seiches.

Figure 4.31 shows two spectra calculated from the time series on March 10 from 11:30 to 13:30 at points 3 and 5. This is the period of spring tides. Transversal mode at point 5 has a longer period (145 s) than at points 2 and 4 (128 s) because the mean depth in the southeastern part of the lake is shallower, which corresponds to longer periods according to the Merian formula despite a decrease of the lake width there. Only one peak at a period of 266 s exists at point 3. This point is located at a node of the transversal mode; therefore there is no peak at a period of 128 s.

We calculated the coherence function between the oscillations at points 3 and 5 located at different sides of the lake along its longer axis. Coherence between two time series is a measure of correlation between them. It is determined as

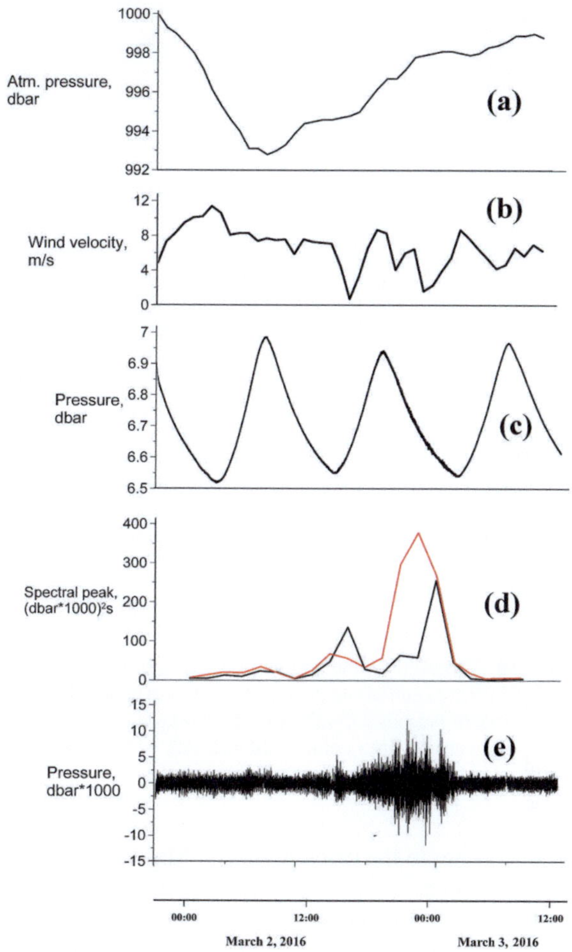

Fig. 4.29 Time variation of different values: **a** atmospheric pressure; **b** wind velocity; **c** bottom pressure; **d** spectral peaks at frequencies of 0.0078125 cps (128 s) (thick line) and 0.0041667 cps (240 s) (thin line); **e** bottom pressure after filtering out the tidal signal

$$Ch = \sqrt{\frac{Co^2(\omega) + Q^2(\omega)}{S_1(\omega) \cdot S_2(\omega)}},$$

where, Co is the Co-spectrum, Q is the quadrature spectrum, and S_1 and S_2 are the frequency spectra of fluctuations calculated from the two time series (Blackman and Tukey 1958). Coherence is a function of frequency and we consider the frequencies of the seiche oscillations. The coherence varies between zero (no correlation) to unity (full correlation). The phase difference between the two time series is determined as

$$\Delta\varphi = \arctan \frac{Co(\omega)}{Q(\omega)}$$

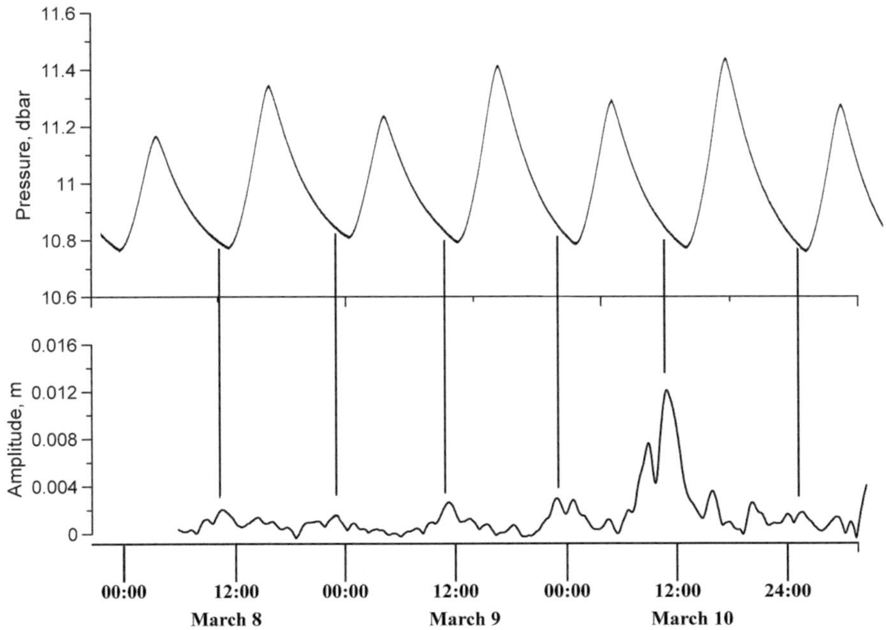

Fig. 4.30 Variations in the bottom pressure at point 4 on March 8–10, 2016. Superposition of the tidal and seiche oscillations is shown in the top panel; the envelope curve of the seiche oscillation in the range of periods 235–278 s after the tidal oscillations were filtered out is shown in the bottom panel. The vertical lines connect the maxima of the amplitudes of seiche oscillations and periods of the ebb tidal phases

According to the calculations on March 9, coherence between the oscillations at these two points (3 and 5) was 0.98, while the phase difference was 182°.

We calculated the spectra of bottom pressure fluctuations at point 2 and 4 to compare the amplitudes of the transversal mode at the opposite sides of the lake. The amplitudes of the fluctuations with a period of 128 s are almost equal (Fig. 4.32). Coherence between the fluctuations at these two points (2 and 4) was 0.92 (Blackman and Tukey 1958), while the phase difference was 179°.

We studied seiche oscillations in Lake Vallunden, which is covered with ice in winter. It is likely that forcing of seiche oscillations occurs due to the tides. Tidal currents flow into the lake through a narrow (10 m) and shallow (1 m) Littrowneset channel. Tidal currents raise and lower the ice cover plate in the lake and the ice cover oscillates with its natural seiche frequencies. Since the lake has an elongated form, two modes of seiche oscillations existed in the lake. The longitudinal mode has a period of 240–266 s and the transversal mode has a period of 128–145 s.

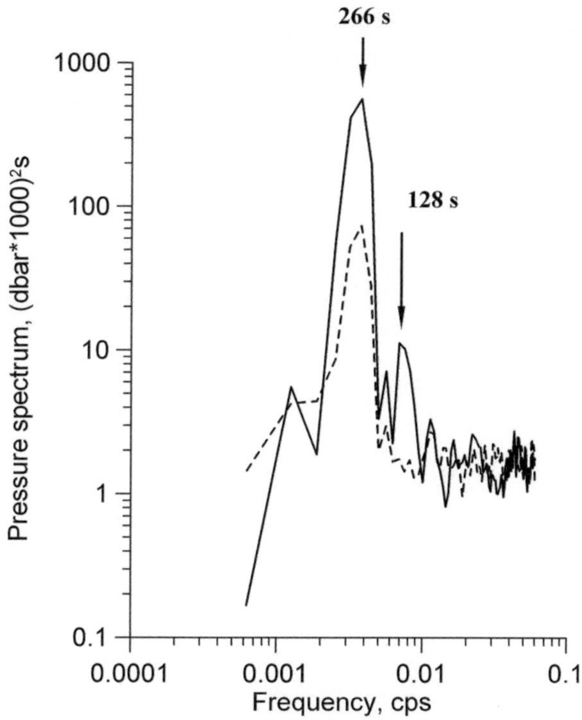

Fig. 4.31 Spectra of bottom pressure fluctuations at points 3 (dashed line) and 5 (solid line). Peaks are at frequency 0.00375 cps (period of longitudinal fluctuations is 266 s). There is no peak in the frequency range 0.0068–0.0078 cps at point 3 because it is located at a node of the transversal mode. There is such a peak at point 5 located in the southeastern part of the lake, which is narrow and shallow

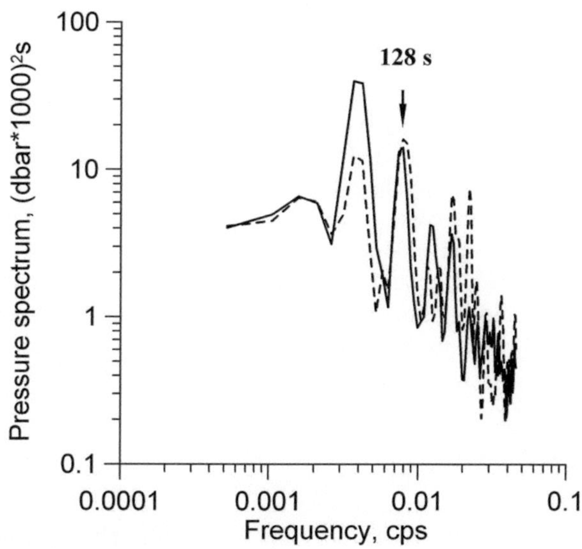

Fig. 4.32 Spectra of bottom pressure fluctuations at points 2 (dashed line) and 4 (solid line). The spectral peaks are related to the transversal mode of seiche oscillations at a frequency of 0.0078125 cps (128 s)

4.5 Surface Manifestations of the Waves in the Ocean Covered with Ice

In this section, we overview our winter oceanographic measurements in the Spitsbergen fjords and our publication (Marchenko and Morozov 2016a). We studied different wave processes in the ocean covered with ice. Our measurements allow us to analyze several processes that occur in the fjords covered with ice that include: semidiurnal tides, tsunami wave, wind waves, seiches, and high-frequency internal waves. We conducted our measurements in the Tempel Fjord and Van Mijen Fjord near the Paula and Tuna glaciers. These classes of waves occupy different space and time scales; however, the frequency spectra of the surface manifestations of these waves can be plotted on one graph. The spectra occupy 14 orders of vertical displacement fluctuations versus five frequency orders of magnitude.

Here, we summarize winter oceanographic measurements in the Spitsbergen fjords. The measurements allow us to analyze different processes that occur in the fjords covered with ice: semidiurnal tides, tsunami wave, wind waves, seiches, and high-frequency internal waves. Space and time scales of these processes were compared and the frequency spectra of the surface manifestations of these waves have been plotted on one graph.

4.5.1 *Tides*

We performed measurements of tides every winter from 2008 to 2015. Usually, the vertical displacements of the ocean surface covered with ice were recorded using SBE-39 pressure gauges deployed on the ocean bottom at the depths ranging from two meters to 50 m (Marchenko et al. 2015). The maximum vertical displacements of the ice caused by the tidal motions are slightly less than two meters. The spring-neap differences are within 50 cm. The M_2 tidal wave with a period of 12.4 h dominates near the coasts of Spitsbergen. The other tidal constituents are several times smaller. The spectrum was calculated from the data of two months duration in the Van Mijen Fjord. The sampling time was 10 s. The dominating spectral peak at the tidal frequency is clearly seen in Fig. 4.33.

4.5.2 *Seiches*

We observed seiches in Lake Vallunden. This is a natural lagoon formed by the glacier. It is about 1.3 km long and 0.7 km wide. The maximum depth of the lagoon is 12 m. The lagoon is connected to the Van Mijen fjord by a shallow passage 10 m wide. Seiche oscillations in this lagoon are generated by meteorological forcing and tides. Usually, the duration of the time series was one day. The sampling time was

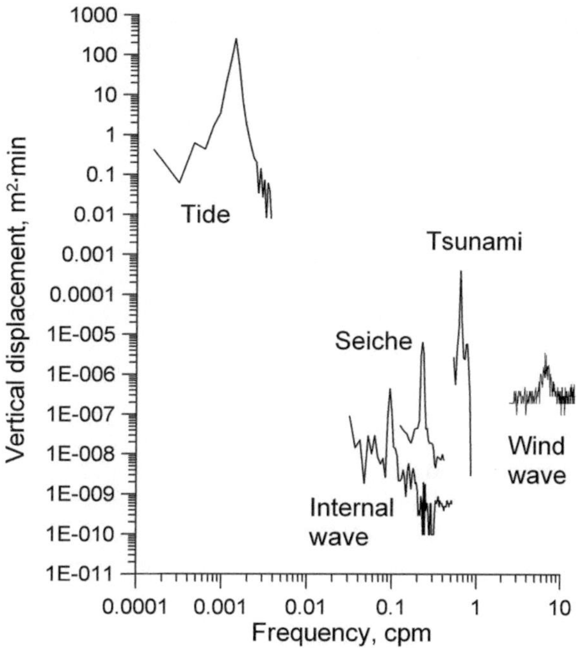

Fig. 4.33 Combined spectra of the surface manifestation of five different wave processes measured under the ice cover in Spitsbergen fjords. Adopted from (Marchenko and Morozov 2016a) with permission from the Editor of Russian Journal of Earth Sciences

1 s. The ice fluctuations were measured by SBE-39 pressure gauges located at the bottom at a depth of 3–10 m. The vertical displacements of the ice caused by seiches were approximately 2 cm, and the spectral peak is very sharp. The period of seiches is about 5 min, which corresponds to the theoretical period of these oscillations in a basin of the given size (Rabinovich 2009).

4.5.3 Tsunami

We observed a landslide tsunami near the glacier front in the Tempel Fjord (Spitsbergen) caused by the motion of the Tuna Glacier (Marchenko et al. 2012). An SBE-37 pressure recorder was deployed at the bottom approximately 300 m from the glacier front at a depth of 46 m. The sampling time was 10 s. The vertical displacement of the ice was approximately 30 cm and the period of the tsunami wave was 90 s. The calculated wave length was 1850 m. The tsunami wave was generated at the low tide phase so that the water resistance to the glacier motion was minimal. This phase of the tide usually coincides with the landslide motions observed in the nature. We interpret the tsunami wave generation caused by the horizontal displacement of the glacier or probably a landslide of the moraine similarly to the landslide tsunami generated by the motion of a block of rocks down the sloping bottom. We consider that this was a glacier motion without ice fall. No new fallen ice blocks were found, but the ice cover was deformed. The initial pulse of the tsunami wave was followed by oscillations of smaller amplitude.

4.5.4 Wind Waves and Swell

The surface elevation of ice caused by wind waves and swell were measured in the Tempel Fjord at a point located approximately at a distance of 1000 m from the ice edge. The measurements were made at a depth of 7 m. An SBE-39 pressure gauge was located on a sloping bottom at a distance of 300 m from the shore. The measurements were continuing for three days. The weather was very frosty and calm during the first two days. The wind increased on the third day reaching 15 m/s causing waves in the ice-free part of the Tempel Fjord. The ice fluctuations recorded by the instrument were approximately 2 cm, while the spectral peak was wide. The frequency of the spectral maximum corresponds to a period of 9 s. Compared to the ice displacements during the calm weather the spectral peak increased by two orders of magnitude. The frequency of the maximum spectral peak during the calm period corresponded to a period of 13 s.

4.5.5 Internal Waves

Usually, the "rigid lid" approximation is assumed in the theoretical studies of internal waves. The vertical velocity at the surface is assumed to be equal to zero; therefore, due to the kinematic condition, the internal waves cannot result in any vertical displacements of the ice cover. However, this is not absolutely true because this approximation filters off the surface mode. Even for the high frequency internal waves whose frequency is close to the Brunt-Väisälä frequency, the ice cover cannot be considered a rigid lid. Measurements of internal waves in a shallow fjord in Spitsbergen showed that fluctuations of temperature and velocity with a period of approximately 10 min and amplitude of internal waves about 1 m correlate with the fluctuations of the ice cover of the same period with an amplitude of 5–8 mm (Marchenko et al. 2011). These results agree with the theory (Muzylev 2008). One can distinguish a spectral peak at a frequency corresponding to a period of 10 min on the spectrum of bottom pressure measured with an SBE-39 pressure gauge on a submarine slope at a depth of 8 m.

Thus, we consider that pressure measurements at the bottom reflect the vertical displacements of the ice surface in the frequency range from the semidiurnal frequency to a frequency corresponding to the periods of a few seconds. Figure 4.33 presents combined spectra of ice displacement caused by different wave classes. Surface manifestations of different waves in the ocean cover a spectral range of 14 orders of vertical displacement fluctuations versus five frequency orders of magnitude. Tidal displacements of the surface are, of course, the strongest among the wave classes considered in this study. Their frequency is the lowest in the spectral band of fluctuations presented in Fig. 4.33. Tsunami wave generated by the motion of the Tuna Glacier has the second amplitude after the tide. Seiches, surface waves, and surface displacements induced by high-frequency internal waves are several orders of magnitude smaller. They cause vertical displacement of the ice of the order of one centimeter.

4.6 Collisional Interaction of Floating Ice Blocks

This section was written together with P. Chistyakov, E. Karulin, V. Markov, M. Karulina, and A. Sakharov

Numerous publications are devoted in the investigation of collisional interaction of immersed particles with each other and with solid walls. Theoretical consideration of a sphere near a solid wall in ideal incompressible fluid leads to the conclusion that direct contact between the sphere and wall is not possible due to the action of the water pressure (Lamb 1975). Starovoitov (2003) investigated theoretically this problem in the case when the motion occurs in incompressible viscous fluid and concluded that the body approaches the wall with zero velocity or cannot touch it at all. Tchieu et al. (2010) investigated numerically the interaction of two cylinders submerged in ideal incompressible fluid. They found that central collision is not realized, and cylinders can change trajectories to avoid direct contact in case if they approach each other not along the line connecting their centers.

Davis et al. (1986) investigated the influence of elastic properties on the collisional interaction of bodies. They concluded that rebound of the approaching bodies can be realized without their direct contact. The rebound velocity and restitution coefficient are controlled by the Stokes ($\sim\rho_p Re/\rho$) and Reynolds ($\sim\rho v_0 d/\mu$) numbers, where ρ_p and ρ are the densities of particles and water, v_0 is the relative velocity, d is the diameter of particles, and μ is the fluid viscosity. Joseph et al. (2001) investigated particle-wall collisions in viscous fluid experimentally with the particles made of different materials. It was revealed that rebound velocity decreases with the decreasing of the Stokes number. It controls the diameter of particles allowing the direct contact by the interaction. Large effective viscosity of grease ice (De Carolis et al. 2005) influences large sizes of particles interacting without collisions in cold waters.

Wave induced ice floes collisions were investigated using the data of in-situ (Martin and Becker 1987) and laboratory (Li and Lubbad 2018; Li et al. 2021) experiments. Martin and Becker (1987) concluded that the mean square velocity associated with collisions was 1 mm/s, and ice floe collisions are not strong enough to contribute strongly to the ice deformation. Li and Lubbad (2018) concluded that wave energy dissipated by collisions is significant. Marchenko et al. (2019) investigated dry and wet collisions of ice blocks with floating ice in the field experiments. They found that representative time of dry collision of ice cube with flat floating ice is of the order of 0.01 s with the maximal relative acceleration greater than 40 g. The representative time of wet collision of the ice pyramid with floating ice is estimated at ~0.15 s with the maximal acceleration of the pyramid of 5 g. Numerical modeling confirmed that in the wet collision the energy of ice pyramid was transferred into the kinetic energy of water volume forming a spray due to the collision.

Here, we describe preliminary results of the field tests where ice-ice collisions were realized both by the pendulum method and by the direct puling of floe by a rope. In both experiments the collisions occurred in the water. The sizes of colliding ice masses were large enough to demonstrate importance of the ice shape and sizes on the collision process. The influence of water was also important for energy dissipation and damping. In this section we review publication (Marchenko et al. 2022).

Field experiments were performed to investigate collisional interaction of ice floes and characteristics of water currents between interacting floes. The experiments were organized in the Van-Mijen Fjord in Spitsbergen in the winter season of 2019. In the first experiment an ice block was hanged on an iron rig by a chain as a pendulum. In the rest position the ice block was submerged in the water in artificially made pool in the ice and had a contact with ice edge of land fast ice. The accelerations of the ice block were recorded after it was replaced from the rest position and released. In the second experiment a floe with sizes 5 m × 10 m was made in a pool with sizes 5.2 m × 12 m. The floe motion along the pool back and forward was initiated by mechanical pooling by a rope. During the experiments we recorded the floe acceleration and water velocity below the ice. The experimental data were used to describe properties of floe-floe collisions and characteristics of water motion initiated by floe motion and floe-floe interactions.

Let us consider inelastic collision of a floe with a solid wall. Floes interact with obstacles (other floes and constructions) by collisions. Energy dissipation during collisions is associated both with fracture and viscous deformations of ice and with water-ice interaction. Energy losses during the collisions make them inelastic. Figure 4.34 shows qualitative description of the dependencies of velocities and accelerations of point O of circular floe from time during the collision. Any other point of the floe has similar motion because of the symmetry. The collision happens if the floe velocity changes sign. It is assumed that the floe velocity equals v_0 before the collision, and velocity approaches $v_1 \leq 0$ after the collision. Thin line in Fig. 4.34b corresponds to ideal elastic collision when $v_1 = -v_0$, and viscous-elastic collision when $v_1 = v_{1,0} > -v_0$. Thick line in Fig. 4.34b corresponds to the collision in the water. The water action influences a reduction of the floe speed immediately after the collision.

Figure 4.34c shows accelerations versus time. The integral from acceleration with respect to time equals to the change of the velocity. The thin line around dark gray region corresponds to the acceleration at the thin line in Fig. 4.34b. The area of dark gray region equals to $v_0 – v_{1,0}$. Thick line in Fig. 4.34c corresponds to the acceleration at the thick line in Fig. 4.34b. The area of light gray region equal to $v_0 – v_1$ is smaller than the area of the dark gray region because light gray areas above and below axis t are accounted with different signs.

Figure 4.35 illustrates the time dependence of the velocity and acceleration of arbitrary point O of rectangular floe when it collides with a wall. This interpretation

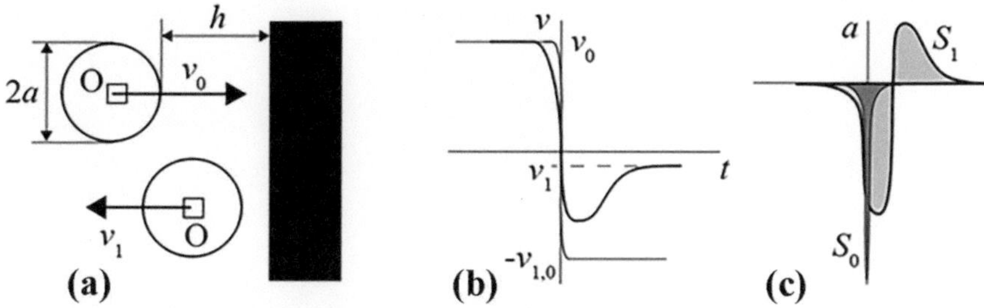

Fig. 4.34 Inelastic collision of disk with a wall (**a**). Qualitative dependencies of the disk velocity (**b**) and acceleration (**c**) on time during the collision

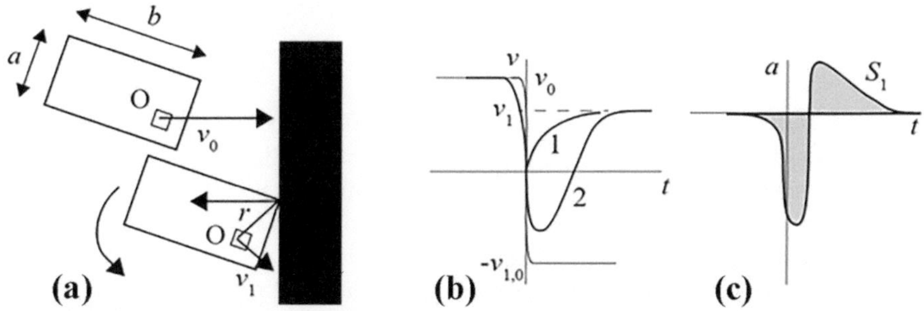

Fig. 4.35 Inelastic collision of a rectangular body with a wall (**a**). Qualitative dependencies of the rectangle velocity (**b**) and acceleration (**c**) on time during the collision

is important for the understanding of accelerometer records in the field experiments. It is assumed that the accelerometer is installed at point O which does not coincide with the point of the first contact of the floe with the wall. After the first contact the floe stops and begins to rotate relative to the contact point.

The speed of point O can be estimated using the assumption that the kinetic energy of translational floe motion is equal to the kinetic energy of the rotational motion. It leads to the formula ${v_0}^2 = \omega^2(a^2 + b^2)/12$, where v is the floe velocity before the collision, ω is the angular floe velocity after the collision, and a and b are the floe sizes (Fig. 4.35). The speed v_1 of point O after the collision equals ωr, where r is the distance between the contact point and point O. Thus, we have:

$$v_1 = 2v_0\sqrt{3r^2/\left(a^2 + b^2\right)}. \tag{4.39}$$

In case if $r = a = b$ we find $v_1 = 2.5v_0$, i.e., rotational velocity of the floe points can be higher than the velocity of their translational motion before the collision. Qualitative dependence of velocity v_1 on time after the collision is shown by line 1 in Fig. 4.35b. Line 2 shows the dependence of velocity v_1 on time in the case when rebound occurs, the floe velocity changes the sign, and the influence of water drag

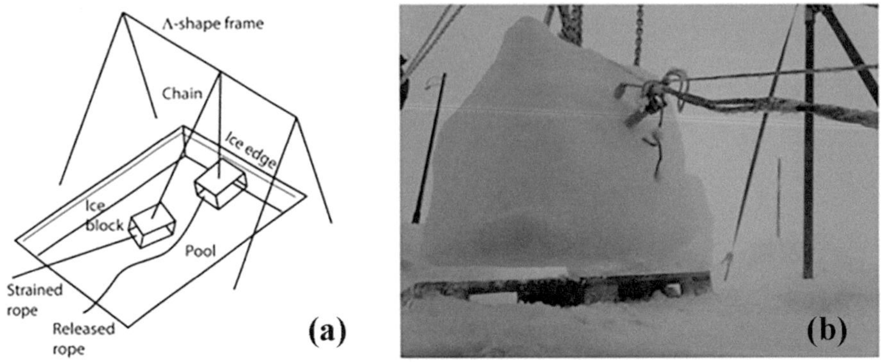

Fig. 4.36 Schematic of pendulum experiment with ice block (left panel). Ice block with rope used in the experiment (right panel)

influences floe rotation. Figure 4.35c shows qualitative dependence of the acceleration at point O in the case when the floe rotates after the collision. It looks similar to Fig. 4.34c, but the area of the gray region above the axis t is larger than in Fig. 4.34c. Total area of the gray region in Fig. 4.35c is equal to zero when $v_1 = v_0$.

We performed pendulum experiment with an ice block. An ice block of rectangular shape with a size of about 0.5 m was cut from sea ice and mounted on an L-shape frame by ice screws and a chain (Fig. 4.36). During the experiment the block was submerged in a pool with a size of 2 m × 5 m and collided with the vertical edge of floating sea ice. The collision was initiated by initial displacement of the block to the side from the ice edge by a rope and consequent release of the rope. The motion of the submerged block continued until the block collided with the ice edge. The block motion was recorded by video and the block acceleration was measured by the uniaxial accelerometer Bruel & Kjær DeltaTron Type 8344 with a sampling interval of 0.2 ms.

Four video frames in Fig. 4.37 show the beginning of the test ($t = 0$) and the collision process. The water surface was covered by slush in the pool. The motion of block influences cleaning of the water surface from slush behind the block, splashing of water on the ice edge, and formation of surface waves. The collisional interaction extended over about 0.2 s. Wave reflection from the ice edge influenced the motion of ice block in the opposite direction after the collision. Separation of the block from the ice edge is seen in Fig. 4.34 at time moment $t = 2$ s.

The horizontal velocity of the ice block reconstructed by consequent analysis of the video frames is shown by the blue line in Fig. 4.38a. One can see that after the rope allows the free block motion, the block velocity increases during 0.3 s and then decreases. Just before the collision the block velocity is estimated lower than 0.8 m/s. The velocity dropped to zero over 0.1 s, and after that the block was "glued" to the wall over 0.5 s. Rebounds were not observed. The inverse motion of the block was initiated by reflected wave by $t > 1.5$ s. The blue line in Fig. 4.38b shows the recorded acceleration versus time. The black line shows running mean values of the

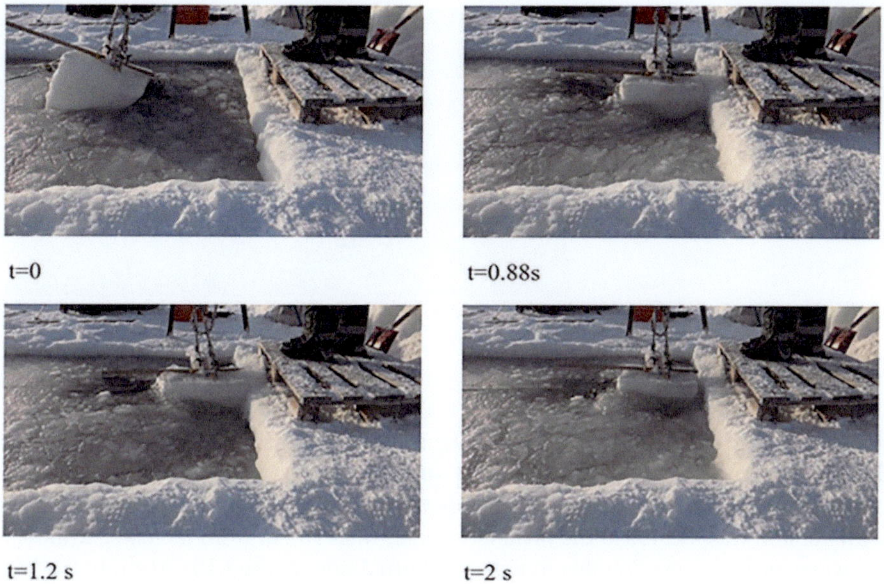

Fig. 4.37 Fragments of the block position reconstructed from video file

recorded acceleration. We think that positive accelerations recorded by $t > 0.9$ s are related to the rotational motion of the block relatively the contact points of the block with ice edge. The oscillations with a period of about 50 Hz in Fig. 4.38b can be explained by vibrations of the wooden block with the accelerometer mounted to the ice block by screws.

Next, we performed a floe drift experiment. A floe of rectangular shape was cut by Ditch Witch trencher in land fast ice of Vallunden Lake in the Van Mijen Fjord (Spitsbergen) on March 11, 2019 (Fig. 4.39). The ice thickness was 80 cm, and the floe size was 5 m × 10 m. The size of a pool where the floe was floating was 12 m × 5.2 m. Lake Vallunden is connected to the Van Mijen Fjord by a strait of 100 m length

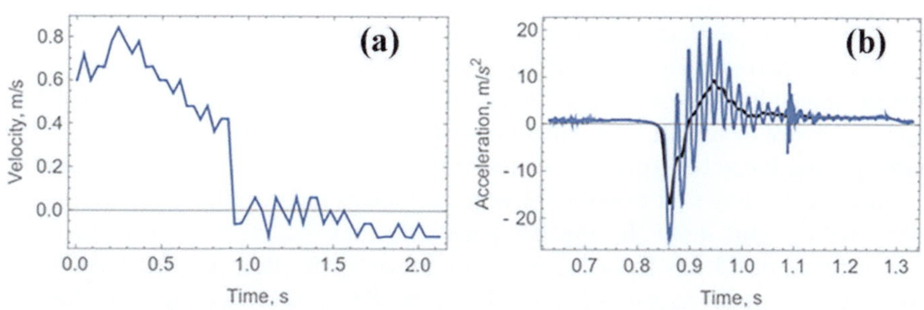

Fig. 4.38 Horizontal velocity of the ice block versus time reconstructed from the video (**a**). Measured (blue line) and averaged (black line) accelerations versus time (**b**)

Fig. 4.39 Photograph of drift floe experiment. Lake Vallunden, Van-Mijen Fjord, Spitsbergen, March 12, 2019

and 10–20 m width. Tidal jet penetrating through the strait provides strong mixing of fjord water and lake water (Morozov et al. 2019). Since there are no regular fresh water supply to the lake by rivers, the lake is filled with sea water with a salinity of 34 ppt. Sea ice in the lake has columnar structure and belongs to the S2 type. The range of the influence of the tidal jet is approximately 200 m from the strait. The pool with the floe was located at the distance greater than 300 m from the lake. Although the water surface elevation due to the semidiurnal tide reached 10–20 cm in the lake the currents were not recorded in the lake during the ice season at distances greater than 200 m from the strait.

A sketch of the drift floe experiment is shown in Fig. 4.40. The experiment was focused on the investigation of water motion caused by the floe motion in the pool. The floe motion was initiated by consequent manual pooling of two ropes fixed at the floe edges. The floe motion was recorded by two accelerometers (Bruel & Kjær DeltaTron Type 8344) mounted to the floe at two locations marked by A1 and A2 in Fig. 4.40. The sampling frequency of accelerometers was set to 100 Hz. Three components of the water velocity were measured by ocean probe ADV SonTek Hydra 5 MHz with sampling frequency 10 Hz at the three locations marked by ADV1,2,3 in Fig. 4.40.

Figure 4.41 shows an example of acceleration record in the drift floe experiment during four collisional interactions between the floe and the ice edges. The interaction events are marked as 1, 2, 3, 4 (Fig. 4.41a). The duration of each event was estimated at 1 s. Each event started from the floe deceleration during approaching to the ice edge, and consequent acceleration. The deceleration periods were shorter than the acceleration periods (Fig. 4.41b, c). Integrals of the accelerations taken from $t = 0$ to t are shown in Fig. 4.41d versus time t. One can see that all integrals are almost equal to zero at $t = 3$ s. It means that floe-ice interaction follows line 1 in Fig. 4.35b with $v_1 \approx v_0$. It can be explained by the floe rotation during the interaction events.

Figure 4.42 shows an example of water velocities recorded by ADV during the experiment. Figure 4.42a, b show the water velocities relative to the floe measured at a distance of 35 cm from the bottom surface of the floe (Location ADV1 in Fig. 4.40). Figure 4.43a shows zoomed fragment of this record together with the floe accelerations. The maximal absolute values of the velocities reached a few seconds after the collisions. The amplitudes of the horizontal velocities varied in the range

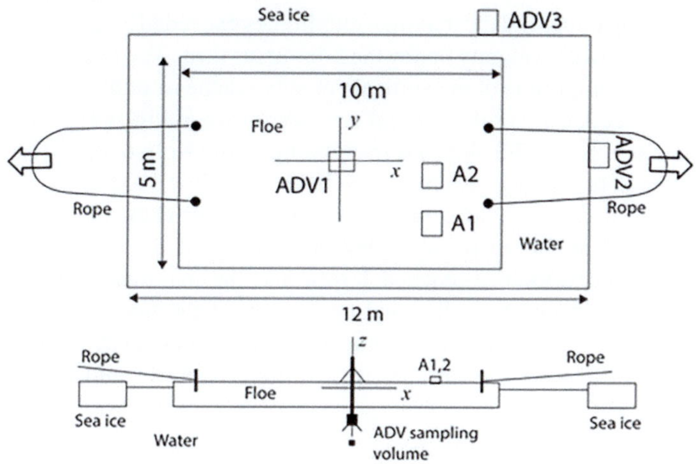

Fig. 4.40 Sketch of the drift floe experiment. A1,2 are locations of accelerometers, ADV1,2,3 are locations of ADV velocity measurements

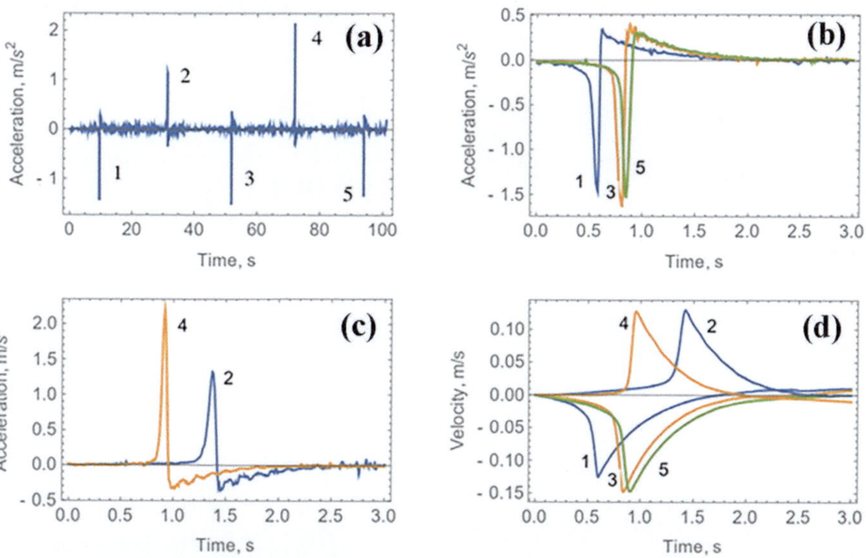

Fig. 4.41 Example of acceleration record in the drift floe experiment. Accelerations recorded versus time after four collisions (**a**). Zoomed dependencies of the accelerations versus time (**b**, **c**), and integrated accelerations versus time (**d**)

between 20 and 30 cm/s, and the maximal amplitudes exceeded 30 cm/s. Figure 4.43b shows the vertical water velocity below the floe versus time. The floe starts to move at $t = 880$ s, and the amplitude of the vertical velocity reaches a stable value already at $t = 1000$ s. Spectral density of the vertical velocity shown in Fig. 4.43 b demonstrated high correlation with the 5/3 slope corresponding to the Kolmogorov-Obukhov law for the developed turbulence (Kolmogorov 1941; Obukhov 1941).

Figure 4.42c, d show the horizontal and vertical water velocities recorded at the location of ADV2 at 50 cm from the bottom surface of the ice (Fig. 4.41). The maximal horizontal velocities reached 1 m/s, and the maximal vertical velocities exceeded 20 cm/s during the collisional interaction of the floe with the ice edges. One can see that each collision was accompanied by consequent events of downward (negative vertical velocity), upward (positive vertical velocity), and again downward

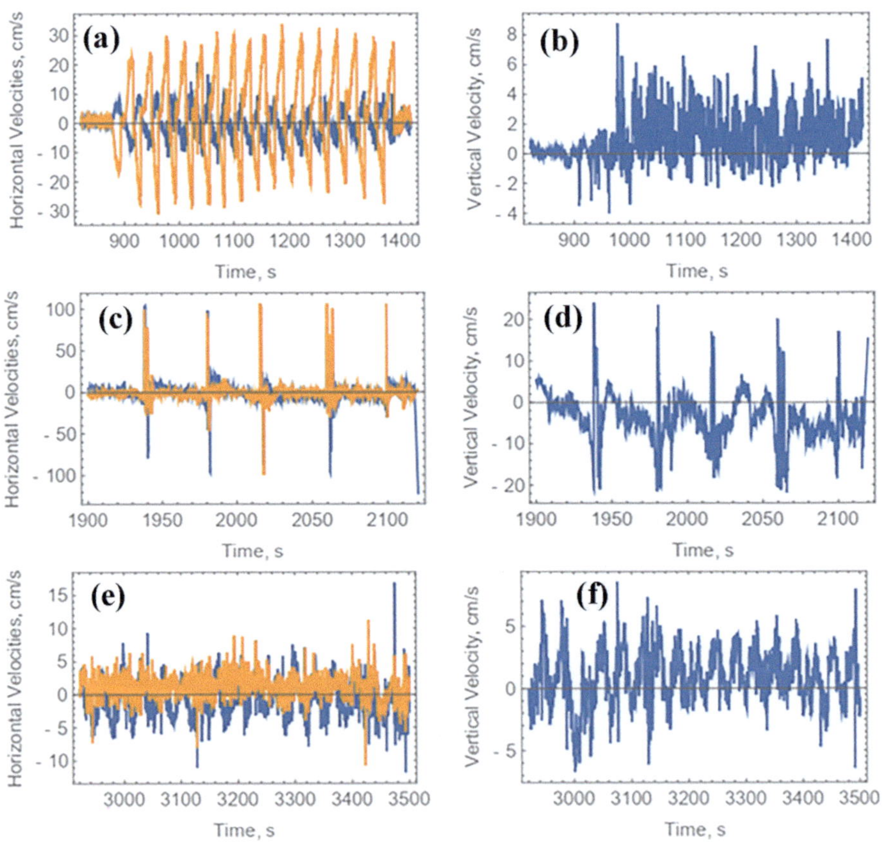

Fig. 4.42 Example of the horizontal (**a**, **c**, **d**) and vertical (**b**, **d**, **f**) water velocities recorded by ADV versus time. The blue and yellow lines show respectively the eastern and northern components of velocity

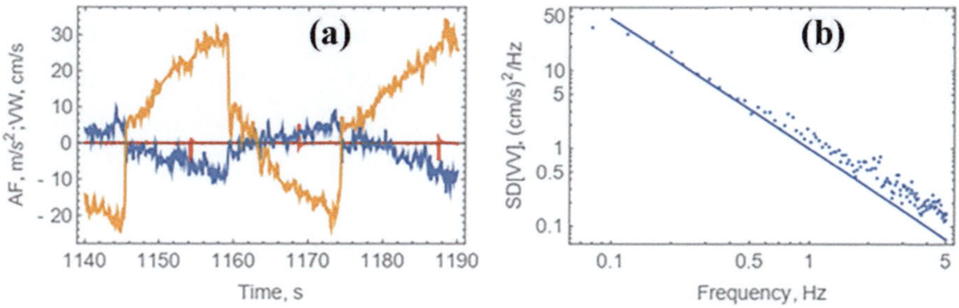

Fig. 4.43 ADV1 measurements. Floe acceleration (AF, red line) and the eastern (blue line) and northern (yellow line) components of water velocities (VW) relative to the floe (VW) versus time (**a**). Spectral density of the vertical velocity (dotted line) and 5/3 slope (solid line)

motions of the water. Duration of each event was approximately 2 s. The water jets were seen on the ice surface during the collisions.

Figure 4.42e, f show the horizontal and vertical water velocities recorded at the location of ADV3 at 65 cm from the bottom surface of the ice (Fig. 4.41). At this location the amplitude of the horizontal and vertical velocities of the water were of the same order of 5 cm/s. This measurement shows the velocity of water motion in the shear zone between the floe and the ice edges extended along the direction of the floe motion.

The mean kinetic energy (KE) and kinetic energy of fluctuations (KEF) of the water motion are calculated from the formulas:

$$2\text{KE} = \langle u^2 + v^2 + w^2 \rangle,\ 2\text{KEF} = \langle du^2 + dv^2 + dw^2 \rangle, \tag{4.40}$$

where, $\langle\ \rangle$ means averaging over the measurement time, u, v, and w are the velocities measured by ADV, and δu, δv, and δw are the velocity fluctuations. The calculated velocity fluctuations are equal to the differences between the actual and averaged velocities. The averaged velocities were calculated using the operation MovingAverage over 1 s applied to the actual data.

There are two time scales in the experiment. One time scale (~10 s) equals to the time interval of the floe motion between two consequent collisions. The other time scale equals to the ADV sampling interval set to 0.1 s. Averaging over 1 s extracts "laminar" velocities initiated by the floe motion. The fluctuation velocities are related to the turbulent motion at smaller spatial scales initiated by the "laminar" velocities.

Table 4.7 shows the mean kinetic energy (KE) and the kinetic energy of fluctuations (KEF) of the water motion calculated from the measurements at locations ADV1, ADV2, and ADV3. KE is largest at the location of ADV1, while KEF/KE is largest at the location of ADV2. The minimal value of KE is reached at the location of ADV3, while KEF/KE reaches the minimum at the location of ADV1. This energy distribution is explained by the regular water motion below the central part of the floe, not regular water motion with strong fluctuations near the frontal edges of the

Table 4.7 The mean kinetic energy (KE) and kinetic energy of fluctuations (KEF) of the water motion calculated from the ADV measurements

	ADV1	ADV2	ADV3
KE · 10^{-2}, $(cm/s)^2$	1.1	0.63	0.065
KEF · 10^{-2}, $(cm/s)^2$	0.025	0.68	0.01
KEF/KE, %	2.2	100	17.1

floe colliding with the ice edges, and less intense, but still regular motion of the water near lateral edges of the floe.

Thus, two field experiments were performed to investigate ice-ice collisions and water motion initiated by the floe motion and collisions. We did not observe significant ice breaking of the contacting surfaces of ice during each test, although gradual accumulation of slush in the water and some adjustment of the shapes of the contacting surfaces were seen after several tests in each experiment. We found that water pressure and floe rotation are important components of the collisional interaction. The action of water pressure increases the interaction time up to 1–2 s in comparison to the interaction time of about 0.01 s measured in the dry collision tests (Marchenko et al. 2019). The rotation of the floes influences the floe acceleration after the deceleration caused by the first contact and increases the interaction time in addition to the action of water pressure.

Cyclic motion of the floe influences currents and turbulence in the water. We observed the formation of developed turbulent motion of the water by cyclic motion of the floe (5 m × 10 m) with a period of 20 s in a pool (5.2 m × 12 m) during 120 s. The integration of the floe acceleration over the time (Fig. 4.41d) showed that the floe velocity amplitudes were similar the water velocity amplitudes measured at 35 cm distance from the bottom surface of the floe (Fig. 4.42a). It means that the mean kinetic energy of the floe was of the order of the mean kinetic energy of water. Therefore, relative motion of floes should have strong damping caused by the generation of turbulence in the upper ocean layer.

The dependence of floe acceleration on time shown in Fig. 4.43 is similar to the dependence of floe accelerations over time obtained in the field experiment of (Martin and Becker 1987; Fig. 4.37). In both cases floe accelerations were recorded immediately after the decelerations. It confirms the rotation of interacting floes caused by wave propagation below the ice. The acceleration amplitudes measured by Martin and Becker (1987) were below 5 mm/s^2, that is 200–400 times smaller than in our tests. Therefore, energy dissipation during floe-floe collisions in the experiment of Martin and Becker (1987) was very small and indistinguishable in comparison with the energy of drift ice and water current below the ice.

4.7 Ice Thickening Caused by Freezing of Tidal Jet

This section was written in cooperation with A.V. Ivanov and T.G. Elizarova

Lake Vallunden is a lake in Spitsbergen approximately 1300 m long, its width is 600–650 m. The mean depth of the lake is 10–11 m. The lake is located near the shallow summit of the fjord at a distance of 55 km from the open ocean. The lake is connected to the Van Mijen Fjord by a channel 100 long and 10 m wide (Marchenko and Morozov 2013). Strong tidal currents develop in this channel because the sea level in the fjord changes with the tidal periods. The amplitude of sea surface height in the fjord is approximately ± 1 m. Due to strong tides, periodical tidal current in the channel exceeds 1 m/s. The lake and the fjord are covered with ice in winter, but the channel usually remains ice-free even in very cold winters because of the strong tidal flow in the channel. In winter, water temperature in the ice-free channel is close to freezing. Water strongly cools while propagating along the channel. The jet of high velocity from the channel continues into the lake and its velocity decreases in the lake approximately over a distance of 100 m. Field works were performed in the shallow area located in the end of the Van Mijen Fjord (Marchenko and Morozov 2013, 2016a). A chart of the region is shown in Fig. 4.44. We study the continuation of the strong flow from the channel between the Van Mijen Fjord and Lake Vallunden near the Svea settlement (77°53′ N, 16°46′ E). In our previous studies (Marchenko and Morozov 2013; Morozov et al. 2019) we observed a strong tidal flow in the channel connecting the Van Mijen Fjord and Lake Vallunden (and in Lake Vallunden). This tidal flow generates short period internal waves (Morozov and Pisarev 2002; Marchenko and Morozov 2016b; Morozov et al., 2019). In this section we review publication (Marchenko et al. 2021).

The goal of our research is to investigate the influence of very cold-water current on sea ice, which can cause changes in the ice thickness. The strong tidal currents in the channel continue into the lake preventing ice freezing along a narrow strip during relatively warm weather.

Field works were performed in 2019 in Lake Vallunden (Spitsbergen) near the channel connecting the lake with the fjord. A narrow (~1–2 m wide) strip of water in the continuation of the flow from the strip does not freeze (Fig. 4.45). We found an interesting phenomenon: ice thickness at the end of the ice-free strip appeared notably greater than in the lake. The mean ice thickness in the lake was approximately 70 cm.

To make a detailed map of the ice thickness near the channel in the northern part of the lake we performed a survey of ice thickness by drilling holes in the ice and measuring ice thickness with a graduated rod. The survey showed that the characteristic size of the region, in which thicker ice was found, is ~100 m. The region of thicker ice is located immediately near the end of the non-freezing tidal stream and around it. The maximum ice thickness in this region reaches 120 cm. At the border of this region, a significant gradient of ice thickness was found. Ice thickness sharply decreases over several meters to 80–90 cm and then gradually decreases to 70 cm, which is typical for the ice thickness throughout the lake (Fig. 4.46).

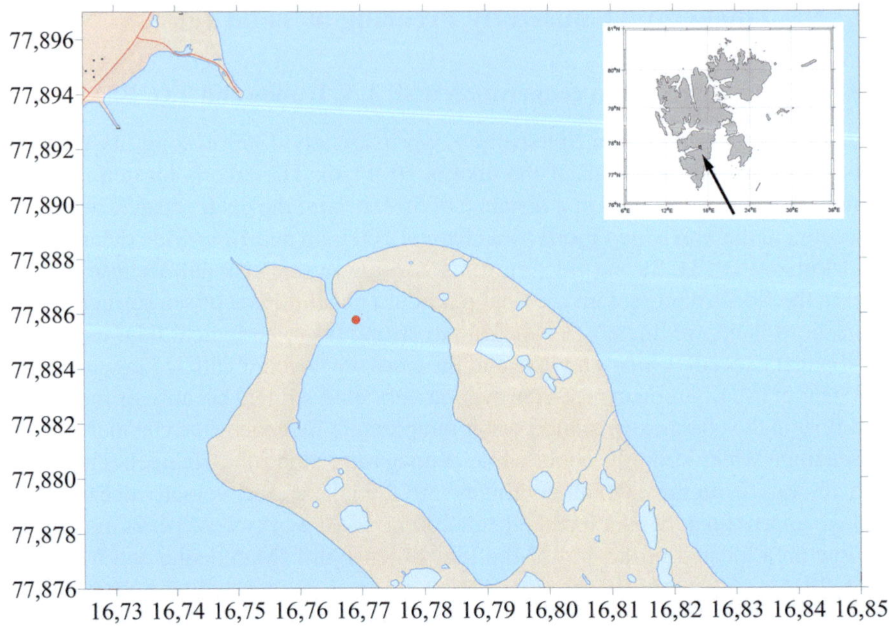

Fig. 4.44 Chart of Lake Vallunden. The red dot indicates location of the experiment. Location of the lake in Spitsbergen is shown in the inset

Fig. 4.45 Photo of the ice-free strip (left) and the flow of cold water under ice at the end of the strip (right). Photos by E. Morozov

We interpret this phenomenon as follows. When water flows into the lake along the non-freezing channel, the water temperature drops almost to the freezing point, and ice crystals and larger slash formations appear at the surface of the stream. This near-freezing water flows under the ice and ice crystals and slash join the ice sheet in the lake.

Strongly varying ice thickness near the strip was revealed by drilling measurements (Morozov et al. 2019). Ice thickness near the ice-free strip sharply increased

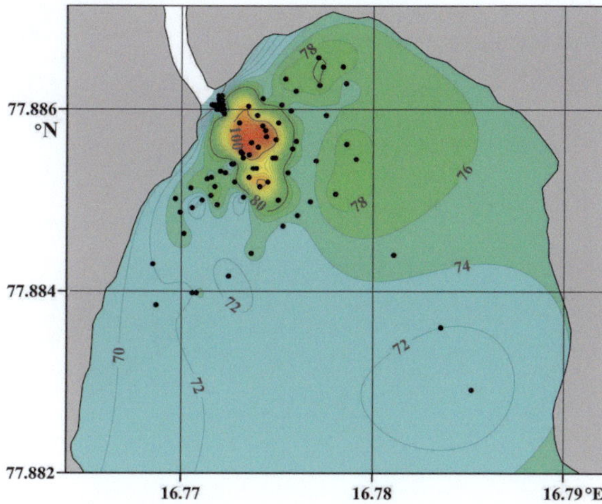

Fig. 4.46 Map of ice thickness in the lake. Black dots show the locations of ice thickness measurements. Gray color shows land. Light cyan color shows ice in the fjord

from 25 cm at the ice edge to 40–50 cm at about 1 m from the ice edge. At a distance of 5–8 m from the ice edge, the ice was as thick as 60 cm. In the continuation of the narrow strip, the ice thickness was almost 120 cm and then the thickness gradually decreased to about 70 cm in the entire lake. We explain the growth of ice thickness by the transport of small ice pieces and freezing water under the ice cover by the strong current and their accumulation under ice where the current speed decreases (Morozov et al. 2019).

Let us consider a mathematical model. A system of regularized shallow water equations (RSWE) described in Bulatov and Elizarova (2011) has been used for numerical simulation of hydrodynamic processes in Lake Vallunden:

$$\frac{\partial h}{\partial t} + \operatorname{div} \vec{j}_m = 0, \tag{4.41}$$

$$\frac{\partial h \vec{u}}{\partial t} + \operatorname{div}\left(\vec{j}_m \otimes \vec{u}\right) + \vec{\nabla} \frac{gh^2}{2} = h^*\left(\vec{f} - g \vec{\nabla} b\right) + div\Pi, \tag{4.42}$$

$$h^* = h - \tau \operatorname{div}\left(h \vec{u}\right), \tag{4.43}$$

$$\vec{j}_m = h\left(\vec{u} - \vec{w}\right), \tag{4.44}$$

$$\vec{w} = \frac{\tau}{h}\left[\operatorname{div}(h\vec{u} \otimes \vec{u}) + gh\vec{\nabla}(b + h) - h\vec{f}\right], \tag{4.45}$$

$$\Pi = \Pi_{NS} + \tau \vec{u} \otimes \left[h\left(\vec{u} \cdot \vec{\nabla}\right)\vec{u} + gh\vec{\nabla}(b + h) - h\vec{f}\right] + \tau I\left[gh \operatorname{div}(h\vec{u})\right], \tag{4.46}$$

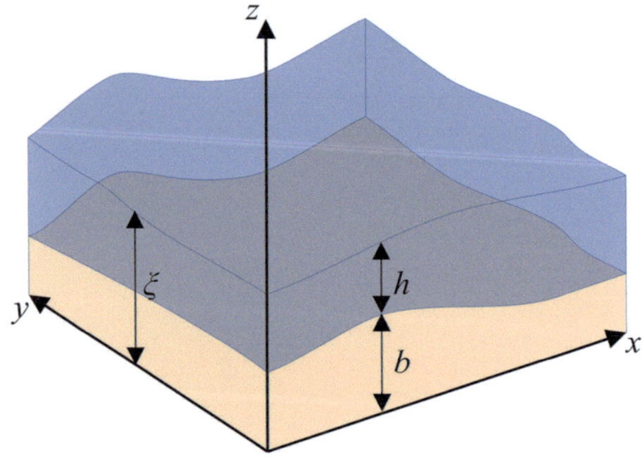

Fig. 4.47 Schematic view of the RSWE model designations

where, $h(x, y, t)$ is the thickness of the water layer measured from the bottom, $b(x, y)$ is the bathymetry function; therefore, $\xi(x, y, t) = h(x, y, t) + b(x, y)$ is the level of water surface (Fig. 4.47), $\vec{u} = \{u_x, u_y\}$ is the vector of horizontal velocity, $g = 9.81$ m/s^2 is the acceleration due to gravity, $\vec{f}(x, y, t)$ is the vector of external volume force; $\tau > 0$ is regularization parameter, whose dimension is time. Besides, Π_{NS} is the Navier-Stokes viscous stress tensor, which, in a number of problems is considered as an additional regularizer and can be included or dropped, see, for example, (Bulatov and Elizarova 2011; Sheretov 2016).

The coefficient of kinematic viscosity of fluid μ is considered artificial; it is calculated from parameter τ:

$$\Pi_{\mathrm{NS}} = \mu h\left[\left(\vec{\nabla} \otimes \vec{u}\right) + \left(\vec{\nabla} \otimes \vec{u}\right)^{\mathrm{T}}\right], \mu = \tau g h. \tag{4.47}$$

Besides, a simplified model of the passive scalar transport is used to calculate the distribution of the inflowing cold water over the lake. Thus, the inflowing cold water is considered as an "impurity", where its concentration $C(x, y, t)$ could be proportional to the ice thickness. Concentration $C(x, y, t)$ is specified in dimensionless units. This model is based on the solution of the regularized transport equation:

$$\frac{\partial Ch}{\partial t} + \mathrm{div}\left(\vec{j}_m C\right) = \mathrm{div}\left(Dh\vec{\nabla} C + \tau \vec{u} h\left(\vec{u} \cdot \vec{\nabla} C\right)\right), \tag{4.48}$$

together with the RSWE system. The methods of deriving and solving the system of Eqs. (4.41–4.48) are described in (Elizarova and Ivanov 2020).

External volume force is represented by the Coriolis force

$$\vec{f} = \{f_x, f_y\}, f_x = f^c u_y, f_y = -f^c u_x.$$

The Coriolis force may be important because circular currents are formed in the lake and the Coriolis force can influence their formation. The Coriolis parameter is $f^c = 2\Omega \sin\varphi$, where $\Omega = 7.2921 \times 10^{-5}$ s^{-1} is the angular velocity of the Earth's rotation, which was assumed constant in the entire domain of simulations at latitude $\varphi = 77.87°$.

Before considering the model of bathymetry, let us describe the following designations: $\eta(x, y, t) = \xi(x, y, t) - \xi_0$ is the deviation of water level, where ξ_0 is the mean water level; $B(x, y) = b(x, y) - \xi_0$ is the depth of the lake measured from the mean level ξ_0.

Since the exact bathymetry data of Lake Vallunden has not been measured, an approximate bathymetry model, which is consistent with observations, has been developed, Fig. 4.48. The shape of the lake was reproduced by transforming a satellite image, and the bathymetry was built on the assumption that the bottom 10-m deep is approximately flat, and the slopes at the coasts are very steep. In the northern part, there is an inlet channel, whose depth is 1.5 m, but there is a stone bar at the end of the channel where the depth reaches 1 m, Fig. 4.48b. The width of the channel is approximately 10–15 m, the length is approximately 100 m. The height of the shores above the lake is 3–5 m.

Since the y-axis coincides with the direction of the normal to the boundary of the computational domain, the following boundary conditions were set to simulate the flow of cold water into the lake, at the boundary of the channel flowing into the lake ($x = 1250$ m, Fig. 4.48b)

$$\frac{\partial h}{\partial y} = 0,\ u_y = -\sin\left(\frac{\pi t}{6}\right)(1 - e^{-t/6})\ \text{m/s},\ u_x = 0\ \text{m/s},\ C = 1,$$

where, t is given in hours. The profile of the normal velocity component at the channel boundary is shown in Fig. 4.49.

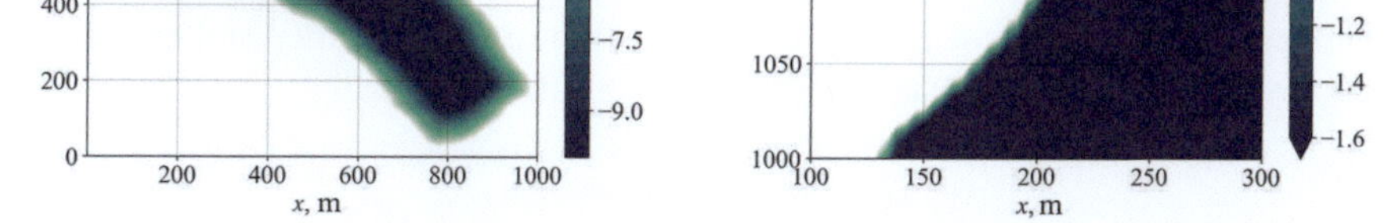

Fig. 4.48 Model of lake bathymetry. Entire lake (**a**); location of the flow from the channel into the lake (**b**)

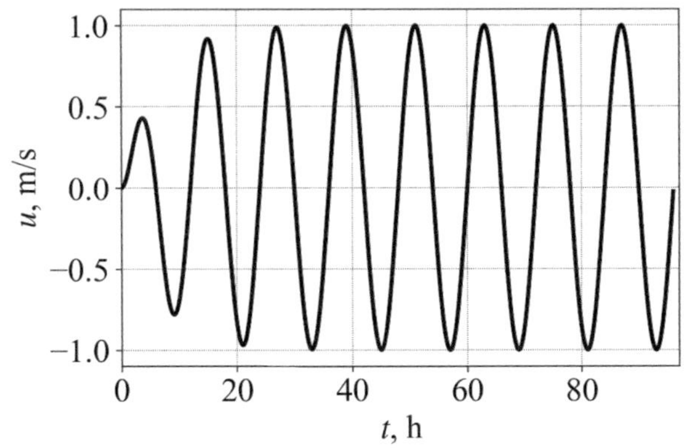

Fig. 4.49 Variation of velocity $u_n = -u_y$ at the inflow into the lake

The "dry bottom" condition was used at the other boundaries, as the boundary condition. The cutoff parameter ε represents the minimum water level. Below this level, i.e., in the regions of the dry bottom, the liquid is at rest; therefore, a constraint is imposed on τ and $\overrightarrow{u}$ in the form

$$\tau = \begin{cases} \frac{\alpha\sqrt{\Delta x \Delta y}}{\sqrt{gh}}, & h > \varepsilon; \\ 0 & h \leq \varepsilon, \end{cases} \tag{4.49}$$

$$\text{if } h < \varepsilon, \text{ then } h = \varepsilon \text{and } \overrightarrow{u} = \overrightarrow{0}, \tag{4.50}$$

where Δx, Δy are spatial grid steps. A similar problem without taking into account the propagation of impurities was considered using the system of regularized shallow water equations; its solution, as well as a description of the method for calculating the zones of flooding/drying are given in (Bulatov and Elizarova 2016).

Since we consider cold water transport, it can be assumed that function C is the temperature, and D is the coefficient of thermal diffusivity of water. This coefficient is very small (at 0 °C, $D \approx 13.2 \times 10^{-8}$ m^2/s); therefore, it can be neglected. The currents in the lake are low (about 10 cm/s); therefore, the regularizer on the right-hand side of Eq. (4.48), represented by the dissipative term of the order of $\sim \tau h u^2$, is insufficient to ensure the stability of the numerical solution. That is why, in addition to the Navier-Stokes regularizer (4.47) in Eq. (4.46), an artificial term $\delta\mu$ was introduced to coefficient D in the transport Eq. (4.48) of the model, where δ is a dimensionless coefficient chosen from the conditions of stability and calculation accuracy, and μ is the viscosity coefficient taken from (4.47):

$$D \rightarrow D + \delta\mu, \, \mu = \tau g h.$$

The first-order Euler scheme for time derivatives and second-order scheme for spatial derivatives were used for the numerical solution of the system of Eqs. (4.41–4.48). A uniform rectangular grid with the following parameters was used:

$$N_x = 247,\ \Delta x = 4.1\,\text{m};$$

$$N_y = 234,\ \Delta y = 5.4\,\text{m};$$

$$\alpha = 0.3,\ \delta = 0.1,\ \varepsilon = 0.01\,\text{m}.$$

The time step was chosen in accordance with the Courant–Friedrichs–Lewy condition:

$$\Delta t = \beta \frac{(\Delta x + \Delta y)}{2\sqrt{g h_{\max}}},\ \beta = 0.2.$$

We start modeling when the water is at rest. The inflow velocity is set (Fig. 4.49) to gradually establish the water circulation in the lake. However, the periodical variations in the inflow velocity cause sea level fluctuations in the lake, which are established only from the 40-th model hour (Fig. 4.50), and the mean water level increases up to about 20 cm. This can be explained by many factors. First, it should be noted that there are no drains in the lake, and its slopes are very steep. The narrow and shallow channel does not enable the water to completely flow from the lake during low tide. Part of the water remains in the lake due to which the total water level gradually increases during the first 50–60 model hours. Besides, during the numerical experiments, it was revealed that the shape and depth of the channel also play an important role. Various variants of the model approximations of the channel were considered in the simulations, for example, if we increased the width of the deep part of the channel, the water level in the lake changed with an amplitude of 1 m. The current view of the channel model agrees well with satellite images and experimental observations, and provides an amplitude of the water level change in the lake equal to ±20 cm, which is consistent with experimental observations.

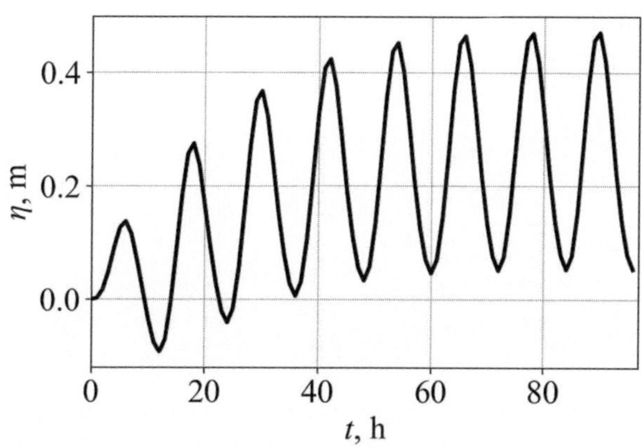

Fig. 4.50 Change in water level at point $x = 406.5$, $y = 804.7$ (the coordinate system is shown in Fig. 4.48a)

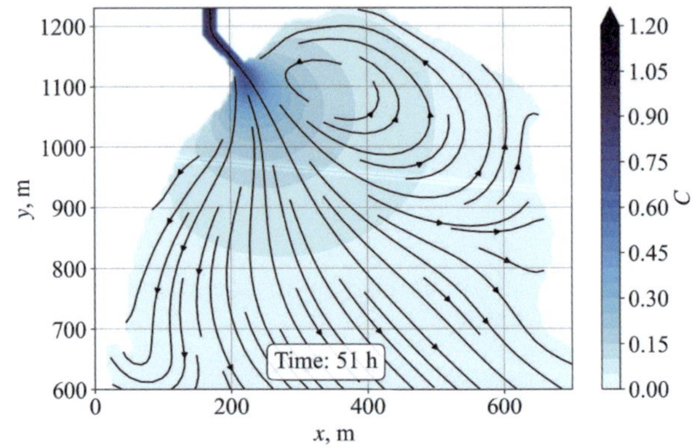

Fig. 4.51 Distribution of concentration C in the lake at time moment $t = 51$ h, black arrows indicate the streamlines

As a result of several numerical experiments, it was found that the change in the mean water level almost did not have any effect on the circulation in the lake. The distribution of the concentration of coldest water is repeated with a periodicity of 12 h, which corresponds to the period of changes in the boundary conditions. A typical example of cold water propagation is shown in Fig. 4.51. One can see a region with a radius of about 100 m and a larger region with a radius of about 300 m.

Figure 4.52 shows the pattern of the flow and streamlines of the magnitude of velocity $|u| = \sqrt{u_x^2 + u_y^2}$ for the inflow (Fig. 4.52a) and outflow (Fig. 4.52b). The distribution of the velocity (of the order of several centimeters per second) far from the channel is consistent with the field measurements (Morozov et al. 2019). A pattern of the formation of a circular cyclonic eddy with a diameter of 200–300 m is seen, which is also consistent with observations.

The concentration of the inflowing cold water determines the formation and distribution of ice thickness when the lake freezes. An eddy is formed in the solution. However, the form of the eddy and its relative location to the channel depends on small variations in the location of the channel and bottom topography near the channel. These variations almost do not influence the size of the eddy. The numerical experiments have shown that the presence of the Coriolis force does not affect the

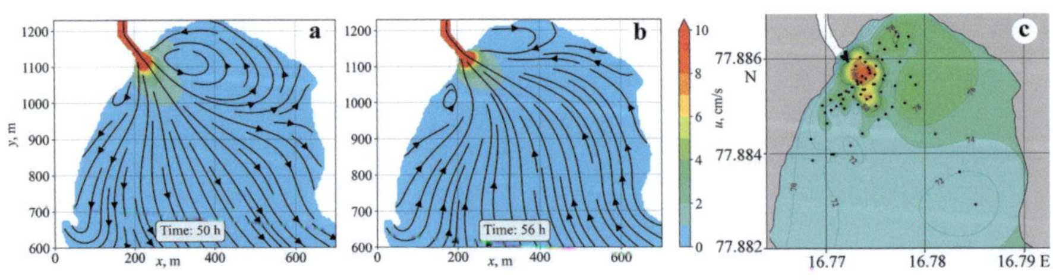

Fig. 4.52 Circulation in the lake and streamlines of the absolute value of velocity $|u|$ during the inflow at $t = 50$ h (**a**); during the outflow at $t = 56$ h (**b**); map of ice thickness on the same scale (**c**). Color scale of velocity is shown

formation of circulation, i.e., circulation is caused solely due to the shape of the lake and the position of the inflow. The low influence of the Coriolis force is explained by the small spatial scale of the phenomenon. The effect of the Earth's rotation becomes important when the time scales of the motion are comparable with the inertial period. The size of the eddy in the lake is of the order of ~100 m and the velocity scale of motion in the lake is of the order of ~0.1 m/s. Hence, the time scale is of the order of ~1000 s, which is much smaller than the inertial period at latitude 78° N (~12 h). Cold water is concentrated near the outflow from the channel to the lake (Fig. 4.51). This corresponds to the real pattern of ice thickness distribution: ice is much thicker in the region of the eddy near the channel.

Let us compare the map of ice thickness in the inflow region with the circulation pattern. Figure 4.52 combines maps of ice thickness and currents in the lake, calculated using a numerical model. In the region of the lake, which is located close to the channel, the ice thickness is thicker, and a cyclonic eddy is formed by the inflowing tidal flow here. The size of the region of thick ice is the same as the region of high velocities near the channel as well as the size of the region of impurity (very cold water) in Fig. 4.52a, c.

Ice thickness in the region of the tidal jet flowing from the Van Mijen Fjord to Lake Vallunden was investigated in the winter season. The ice in the continuation of the ice-free strip is much thicker than in the lake. The ice-free continuation of the flow in the lake was about 2 m wide and 100 m long. Strong currents of almost freezing water penetrate under ice transporting ice crystals and slash, which attach to the ice cover from below, and thus ice thickness increases in the continuation of the flow. A numerical simulation allowed us to model the structure of currents in the lake when the tidal current flows into the lake. The flow initially forms an eddy which expands over the basin of the lake with decreasing velocities.

4.8 Freezing of Tidal Current Under Ice in a Shallow Channel

This section was written together with K.V. Filchuk

We studied tidal currents under ice in a shallow tip of the Van Mijen Fjord in Spitsbergen. The location is called Braganzavägen. A creek of freshwater is flowing down from melting mountain snow and glaciers along the Kjellstromdalen valley to the southwest approximately along the middle line of the valley. This geological structure continues in a shallow channel (~1 m depth) in the Braganzavägen basin. This basin is subjected to strong tides in the Van Mijen Fjord. The open ocean is located at a distance of 56 km. The basin is covered with ice in winter. The whole bulk of ice in the basin is generally aground because most of the basin area freezes to the bottom. However, continuations of the flow from the mountains along shallow channels do not freeze to the bottom in February leaving a shallow channel that allows a tidal flow.

A summer aerial photo in Fig. 4.53 shows the stream from the mountains in the valley, which flows into the fjord in the shallow Braganzavägen basin. Another aerial photo of this region is also shown in Fig. 4.53. A chart of the region is shown in Fig. 4.54. In this section, we review publication (Morozov et al. 2022).

Tidal currents flow into these channels, which leads to the periodical elevations and depressions of the ice cover along these channels. Tidal currents generally follow the middle channel in the basin. Previous researches in this region are reported in (Bogorodskii et al. 2020; Marchenko et al. 2010, 2011, 2013, 2021; Marchenko and Morozov 2016a; Morozov et al. 2019). The works on ice in the Van Mijen Fjord were reported in Skarðhamar and Svendsen (2010), Høyland (2009). Ice regime and tides were studied by Høyland and Løset (1999), Liferov and Høyland (2004), Høyland and Liferov (2005) and Strub-Klein and Høyland (2012). Moslet (2008) performed medium scale ice-structure interaction experiment in the Van Mijen Fjord. Ice thermodynamics was studied in Launiainen and Cheng (1998), Leppäranta (1993). The work by Cottier et al. (2010) is a good review of processes in many fjords.

Our measurements were conducted in the beginning of March, 2016. We found locations of these channels by drilling holes in the ice. Generally, the drill hit land after

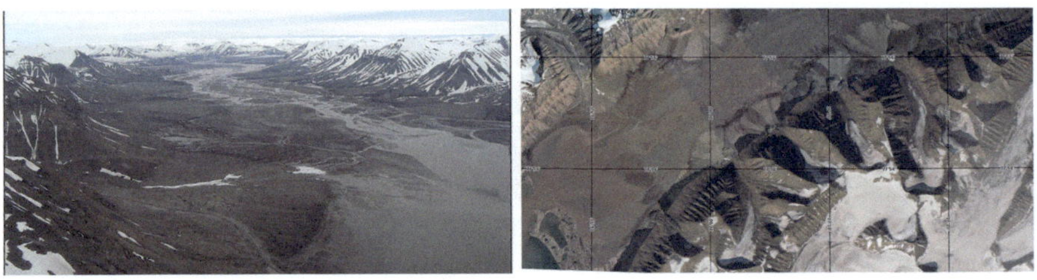

Fig. 4.53 Photos of the region

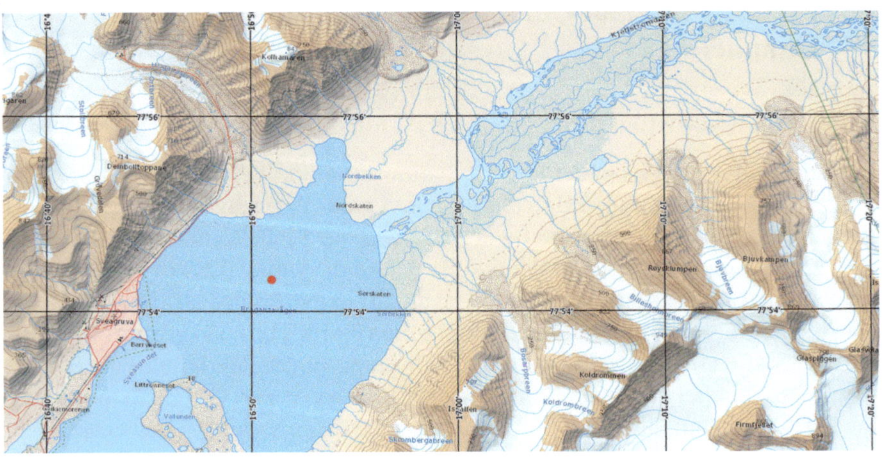

Fig. 4.54 Chart of the region. The red dot indicates location of measurements

drilling the ice. During the ebb tide we saw depressions in the ice cover. After drilling ice at these locations, we found water under ice. During high water periods (flood) of tides these depressions disappeared and water under ice became deeper. One of these locations is marked with a red dot in Fig. 4.54. Its coordinates are 77°54.322′ N, 16°50.922′ E. The thickness of the water layer under ice in this channel ranges from 50 to 70 cm. The width of the channel is approximately 35–40 m. The channel is 1.2 m deep relatively to the upper boundary of the ice cover (water level).

We made a section of 10 holes across this channel to reveal the horizontal structure of the channel. This section was repeated several times during the high and low water tide. Then, we deployed an SBE-37 instrument in the deepest place on the bottom. The instrument measured pressure, temperature, and salinity. Ice thickness was measured with a graduated rod. A few temperature and salinity meters (Star-Oddi) were frozen in the ice at the surface. During high tide, water seeped to the surface and we could record salinity time series. Possible locations of the channels in Braganzavägen were reported in (Shestov et al. 2015) based on field works (Fig. 4.55).

In March 2016, we made a section in the BV-1 region. A section of the channel bottom and ice cover during high and low water are shown in Fig. 4.56.

One can see from Fig. 4.56 that depression of ice cover during low water is well pronounced. The depth of the ice depression is almost 20 cm. Hence it is clearly visible on the ice surface. Records of pressure and salinity measured by SBE-37 instrument at the bottom are shown in Fig. 4.57 together with the records of surface salinity by the Star-Oddi sensors.

It is clearly seen from the records that during the flood period of the tide (increasing pressure), salinity of the inflowing water decreases because water is transported from the open regions of the fjord. During the ebb period of tide (decreasing pressure), water flows from the channel to the fjord. Part of the water was frozen under ice and brine was released into water, thus increasing salinity of the outflowing stream. During the high-water period, water seeped to the surfaces that caused increased salinity at the ice surface. Star-Oddi sensor recorded salinities within 10–12 PSU. Let us estimate the increment of ice thickness during a tidal period, which is caused by partial freezing of seawater flowing into the channel. After freezing, salinity of the outflowing becomes greater. Stefan (1891) recommended the following formula for practical calculations of ice thickness in time:

$$\sqrt{2 * 86400 * \lambda * \Sigma Tj/(L * \rho)}$$

where, 86400 is the number of seconds during 24 h, $\lambda = 2.2$ W (m °C) is coefficient of thermal conductivity of ice; $\rho = 920$ kg/m^3 is ice density; ΣT_j is the sum of degrees of frost multiplied by the number of days; $L = 3.3\ 10^5$ J/kg is heat of fusion of ice. If we approximately assume that the ice thickness has been formed during 50 days at a temperature of –10 °C, we get 79 cm, which is reasonable. The daily increment at a temperature of –10 °C is 1 cm.

This channel continues to the northeast from the location of the section. We also placed an SBE-37 instrument there at a distance of approximately 1000 m from the section. The water layer under ice was quite thin, approximately 10–15 cm. After

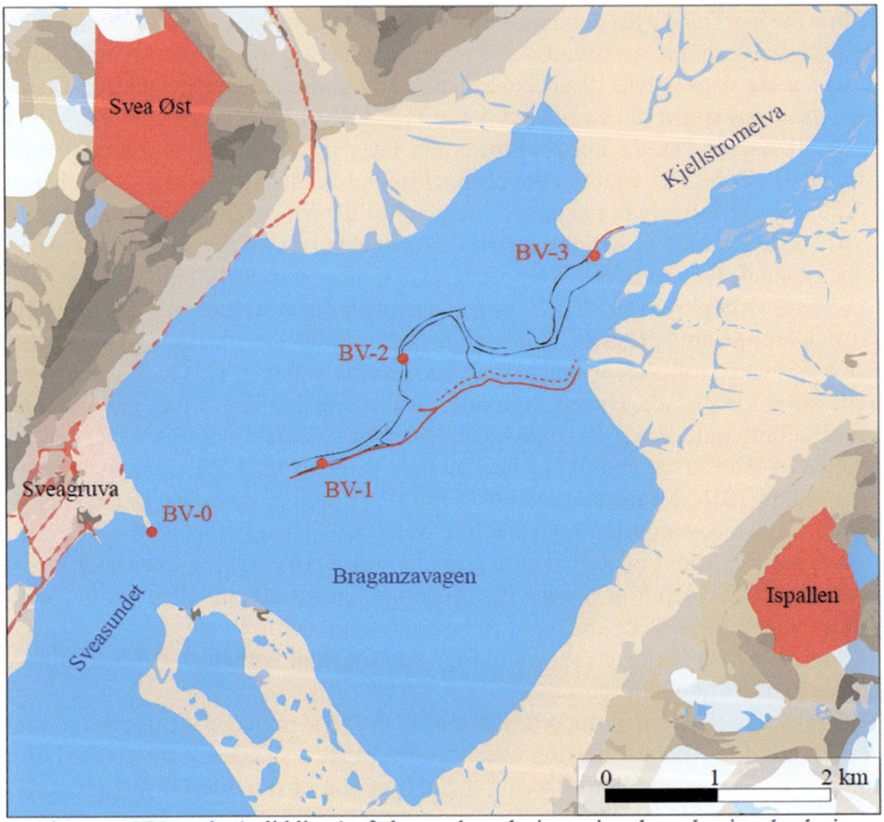

Fig. 4.55 Locations of studies in 2014–2015. GPS recorded tracks (solid lines) of observed cracks in the ice in the Braganzavägen basin are shown. BV0, BV1, BV2, and BV3 are points of CTD measurements in 2014–2015 (Shestov et al. 2015)

12 days we tried to recover it, but it was already covered with aground ice. The instrument was damaged during recovery.

Thus, we can calculate the amount of water transported by the tidal flow across our section during 6 h. The cross-section of the water during high-water tide is 10 m^2. It decreases by 3 m^2 at the low-water tide. The square of the channel surface (upper boundary of water) is approximately (40 m × 1000 m)/2 = 20000 m^2. Water thickness difference between high and low water is 0.15 m. We assume that the channel becomes narrower in the northeastern direction (divide by 2). Thus, the amount of water transported to the northwest across the section during 6 h is 3000 m^3. Knowing that the square of the cross-section is 10 m^2, the velocity of the current is estimated at 3000 $m^3/(21600\ s \times 10\ m^2) = 0.014$ m/s, which is a reasonable velocity.

The ice becomes thicker by 0.005 m during 12 h of the tidal cycle, which makes 20000 $m^2 \times 0.005\ m = 100\ m^3$ of new ice.

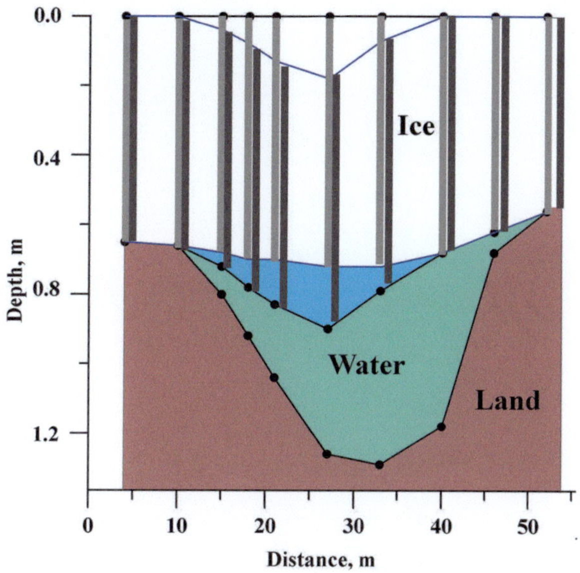

Fig. 4.56 Section across the channel. Gray vertical lines show drillings through ice during high water and those of black color are related to low water. Ice is shown in two positions: high water (flat) and low water (depression). Land is shown with brown color

Thus, 3000 m^3 of water with salinity 34.5 is transported to the northeast during the flood tide (6 h). The amount of transported salt is 3000 $m^3 \times 34.2$ kg/m^3 = 102600 kg.

We assume that ice salinity is 5. The amount of salt frozen in the ice (5 kg/m^3) is 100 $m^3 \times 5$ kg/m^3 = 500 kg.

The volume of outflowing water is 100 m^3 smaller and amounts to 2900 m^3.

The amount of salt transported out of the channel with salinity 35.2 is

2900 m^3 * 35.2 kg/m^3 = 102080 kg. Thus, the salt transport is approximately balanced.

Tidal currents flow under ice in a shallow basin near the tip of the Van Mijen Fjord in Spitsbergen. The ice is almost aground but a small channel exists under ice. This channel was formed in summer time by a flow of freshwater from the mountains. During the flood cycle of tide seawater flows under ice approximately over a distance of one kilometer. Small amounts of seawater in the channel strongly freeze. Salinity of water increases due to salt rejection during freezing. During the ebb cycle, salinity of the outflowing water is greater than that of the inflowing water. Salt and mass balances are estimated. During high-water tide, salt water seeps through the cracks in the ice to the surface. This increase has been recorded by salinity sensors frozen in the ice near the surface.

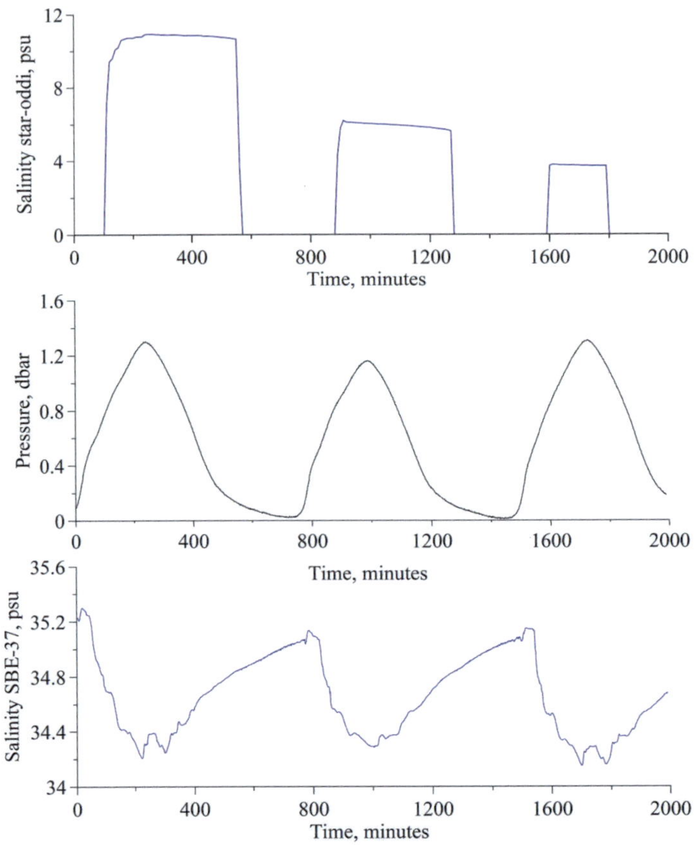

Fig. 4.57 Time series of salinity at the surface measured by a Star-Oddi sensor (top panel); pressure at the bottom (middle panel) and salinity (bottom panel) measured by SBE-37

4.9 Field Studies of Sea Water and Ice Properties in Svalbard Fjords

This section was written together with A. Shestov, E. Karulin, M. Karulina, P. Bogorodsky, S. Muzylev, D. Onishchenko, and A. Makshtas

Multidisciplinary field studies were performed on land fast ice in the Advent Fjord, Tempel Fjord, and Van-Mijen Fjord in 2008–2010 (Figs. 4.58 and 4.59). The goal of our research is to study local peculiarities of oceanographic and ice processes, which can potentially influence coastal infrastructure and navigation in Svalbard and other coastal Arctic regions. Local coastal infrastructure includes quays and transportation routes by ships, small boats, and other vehicles (snow scooters, hovercraft, belt wagon etc.). The influence of environment on local infrastructure is very important for organizing life and communications between settlements and industrial objects in the Arctic. The understanding of local ice and water processes can help us to avoid catastrophic events because of ice impact on coastal structures and can give new ideas for optimal design solutions. Regions of field studies were located near Longyearbyen and Svea, where industrial and tourist activity in Svalbard is concentrated. Longyearbyen is a fast-growing town with local population about 2000 people. Tourist activity in Svalbard is intense in summer and winter seasons. Tourist groups arrive to Longyearbyen mostly by ships in ice free season. In winter time tourist groups and local people use snow scooters to travel over sea ice to cottages, sightseeing places and to other local settlements in Svea, Barentsburg, and New Alesund.

Coal boats up to 70000 t deadweight pass the narrow strait between Akseløya Island and coast to sail in Van Mijen Fjord and Kapp Amsterdam coal quay. Although main activity in harbors occurs in ice free season all quay structures should survive over many winter seasons to be profitable. Field studies near Akseløya were

Fig. 4.58 Satellite image of Svalbard on March 23, 2010

Fig. 4.59 Photos of study sites. Paula Glacier in summer (**a**); Aerial image of Akseløya (**b**); Paula Glacier in winter (**c**); Paula Glacier from a drone (**d**), Tuna Glacier in winter (**e**); Tempel fjord between sail boat *Noorderlicht* and Tuna Glacier (photograph of W. Gurter) (**f**), Tuna Glacier in summer (**g**); Tuna Glacier from a drone (**h**)

performed in May 2010 when the RV *Lance* was moored to sea ice in the Van Mijen Fjord close to the island. Field studies were also performed near the glaciers Tuna Glacier (Fig. 4.59e–h) and Paula Glacier (Fig. 4.59a, c, d) located in the Tempel Fjord and Van Mijen Fjord, respectively. The expedition based in 2010 and 2011 on the Dutch sail boat *Noorderlicht* frozen in the sea ice in winter time in 10 km from the Tuna Glacier. This location made available the access for field studies on the ice between the boat and the Tuna Glacier. The main goal of our research was to understand the influence of glaciers on sea ice conditions near their fronts. Here, we describe some results of field works in relation to tide measurements, ice and water profiling, flexural strength, and surface wave phenomena.

Tide measurements were performed with SBE-39 sensors deployed at the sea bottom. They measured water pressure and temperature with sampling frequency 1 Hz to record surface waves. In this case, the battery capacity allowed us to record data during 2 days, and then the battery should be replaced. In the ice season, the sensors were deployed on a wire through the hole in the ice. Additional weight on the wire was used to avoid the lift up of sensors from the bottom due to water level changes if the wire would be frozen into the ice. Measurements were performed since 2008 in the Advent Fjord, Tempel Fjord, Sveabukta, near Colesbukta, in Lake Vallunden, and near Akseløya Island. Water level in New Alesund is available from http://vannstand.no.

Figure 4.60 shows that the shape of water pressure at a fixed point of sea bottom over semidiurnal tidal cycle is not symmetric in the Ice Fjord in comparison with tides in New Alesund and Van Mijen Fjord. Water level in New Alesund (NA) is elevated by 9.5 m for better visual comparison in the figures. Tidal measurements near Akseløya were performed by two sensors deployed from the opposite beaches of the island at about 1 km from each other. Sensor B was deployed from Bellsund side (B) and sensors VMF was deployed at the inner side in the Van Mijen Fjord. Records of sensor B shown in Fig. 4.60c are damaged because the wire was moved two times by drifting ice. Nevertheless, it is possible to distinguish the times of minimal and maximal tides and compare it with water level records in New Alesund (Fig. 4.61). The phase shift between sensors B and VMF is about 30 min at low and high tides both due to the Akseløya influence. Phase shift between the NA records and VMF sensor is almost zero at low tide and 30 min at high tide. The tidal wave arrives to New Alesund later than to Akseløya.

Lake Vallunden is located on the other side of the fjord opposite to Svea (Fig. 4.62a). The lake is connected with the fjord by a narrow strait 10 m wide and 100 m long. The depth in the strait varies from 0.5 m in low tide to 1.5 m in high tide. During winter time the strait freezes completely. In summer time the water velocity can reach 2–3 m/s in the strait. Water depth in the lake is shallower than 12 m, while water depth in the fjord is shallower than 4 m at high tide. Figure 4.62b shows synchronous records of water pressure at the bottom of fjord (F) and at two locations on the lake bottom close to the strait (S) and near the opposite coast of the lake (L) in August 2009. One can see that tide in the lake is not symmetric over the semidiurnal cycle. Phase shift between the minimal water levels in the fjord and in the lake is 1.5 h, and between the maximal water levels it is about 0.5 h. Tidal

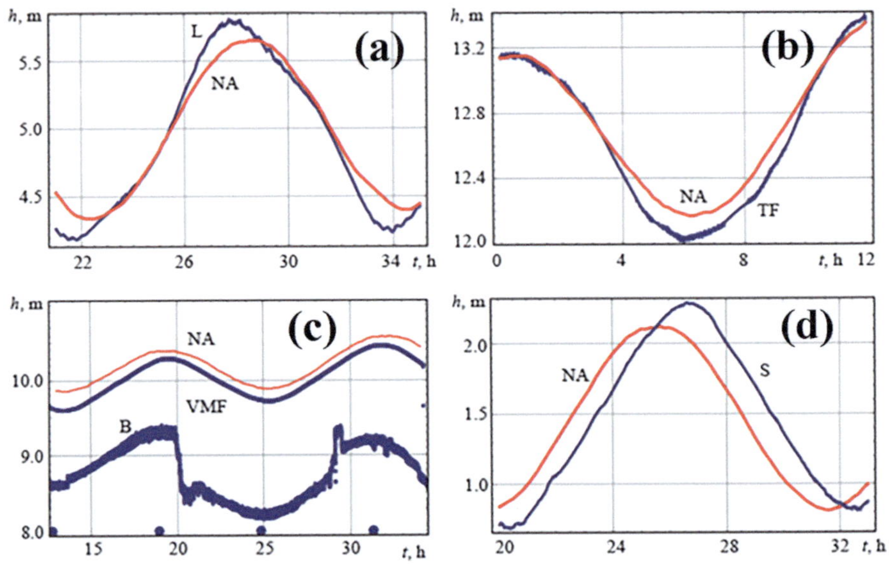

Fig. 4.60 Comparison of curves showing water level changes over semidiurnal tide cycle recorded in New-Alesund (NA) and Longyearbyen (L) (**a**), NA and Tempel Fjord (TF) (**b**), NA and near Akseløya from Bellsund (B) and Van Mijen Fjord (VMF) sides (**c**), and NA and Svea (S) (**d**)

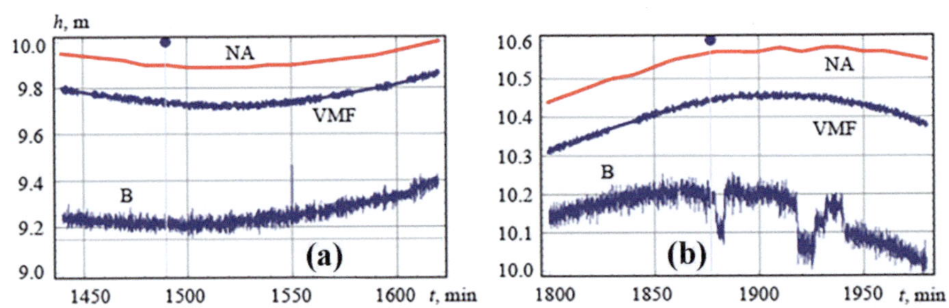

Fig. 4.61 Comparison of curves showing water level changes over the low tide (**a**) and high tide (**b**) recorded in New-Alesund (NA), from Bellsund (B) and Van Mijen Fjord (VMF) sides of Akseløya

amplitude in the lake is two–three times smaller than in the fjord. These effects can be explained by the influence of local topography of the strait and its carrying capacity. The bottom of the strait is inclined to the lake. Therefore, it creates potential barrier for the water outflow from the lake. As a result, the rise-up time of the water in the lake is shorter than the time to reduce the water level in the lake. Probably similar processes influence the difference of water pressure curves in Fig. 4.60.

Figure 4.63 shows the amplitude and phase difference of water level variations recorded in New Alesund and Longyearbyen (a, b) and in New Alesund and in Svea

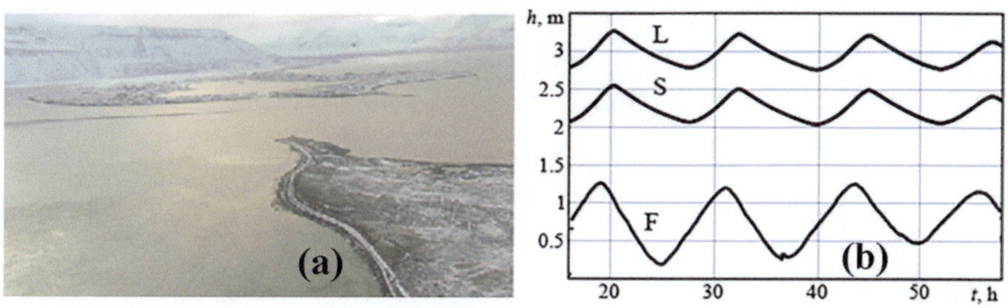

Fig. 4.62 Lake Vallunden on the other side of the VMF from Svea (**a**). Tidal variations of water level recorded in the lake near the strait (S) and near the beach opposite to the strait (L) in comparison with tide in the fjord (F)

(c, d). H_{NA}, H_L, and H_S denote water level difference between high and low tide measured in the same semidiurnal cycle. Time difference ΔT_i is equal to time shift between maximal (or minimal) water levels measured in Longyearbyen and New Alesund (b), and in Svea and New Alesund (d) at certain half of the semidiurnal cycle. The time differences of the first half of the semidiurnal cycle between the minimal and maximal water levels measured in Longyearbyen and New Alesund and in Svea and New Alesund are the same; ΔT_{i+1} shows the same shift between minimal (or maximal) water levels on the next half of the semidiurnal cycle. Blue dots in Fig. 4.63a, b are related to the data collected from December 16, 2010, to January 24, 2011, in the Advent Fjord when the ice was extended over the whole fjord and the West Coast of Svalbard. Red dots in Fig. 4.63a, b are related to the data collected from March 28 to April 7, 2010. In this period the ice was almost absent in the Ice Fjord and on the West Coast of Svalbard from the Ice Fjord to New Alesund. Green dots are related to the data collected in the Tempel Fjord on March 4–6, 2010. Blue dots in Fig. 4.63b, c are related to the data collected in Sveabukta in February–April, 2008–2010, and red dots in Fig. 4.63b, c are related to the data collected in Sveabukta in August–September, 2008–2010. The Van Mijen Fjord is typically covered by fast ice from February to June due to the locking effect by Akseløya Island.

Blue dots in Fig. 4.63a, c are located on straight lines, while red dots have higher dispersion. Blue straight lines fitting blue dots in Fig. 4.63a, c are described by equations $H_{NA} = 0.07 + 0.77H_L$ and $H_{NA} = 0.01 + 0.8H_L$. A possible explanation is as follows. The height of semidiurnal tide (A) depends on the Moon position characterized by angle φ and other parameters α_i describing the influence of local topography of seabed as follows $A = F(\varphi)G(\alpha_i)$. The values of H in Fig. 4.63a, c are equal to $H = F(\varphi)G(\alpha_i) - F(\varphi + \Delta\varphi)G(\alpha_i)$, where angle $\Delta\varphi$ shows the displacement of the Moon over a half of the semidiurnal cycle. Using asymptotic series, we can write $H = \partial F(\varphi, \alpha_i)/\partial\varphi G(\alpha_i)\Delta\varphi$. Then we find

$$\begin{aligned} H_{NA}/H_L &= \partial F(\varphi_{NA})/\partial\varphi G(\alpha_{i,NA})\big(\partial F(\varphi_L)/\partial\varphi G(\alpha_{i,L})\big)^{-1}, \\ H_{NA}/H_S &= \partial F(\varphi_{NA})/\partial\varphi G(\alpha_{i,NA})\big(\partial F(\varphi_S)/\partial\varphi G(\alpha_{i,S})\big)^{-1}. \end{aligned} \tag{4.51}$$

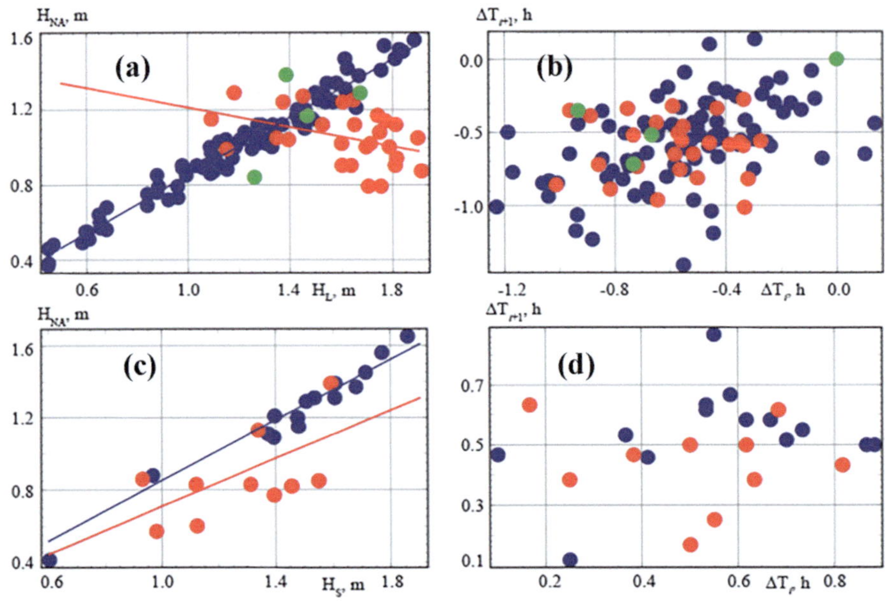

Fig. 4.63 Water level variations over the same semidiurnal tidal cycle recorded in New Alesund and Longyearbyen (**a**), and in New Alesund and Svea (**c**). Shifts between times of extreme water levels recorded in Longyearbyen and New Alesund (**b**), and in Svea and New Alesund (**d**). Blue and red dots are related to ice- and ice-free seasons respectively. Green dots are related to the measurements in Tempel Fjord in 2010

Since all locations of measurements are very close to each other it is possible to assume that $\partial F(\varphi_{NA})/\partial\varphi = \partial F(\varphi_L)/\partial\varphi = \partial F(\varphi_S)/\partial\varphi$. Substituting these formulas into (4.51) we find that the ratios of tidal heights depend only on the local characteristics

$$H_{NA}/H_L = G(\alpha_{i,NA})/G(\alpha_{i,L}), \; H_{NA}/H_S = G(\alpha_{i,NA})/G(\alpha_{i,S}). \tag{4.52}$$

In the ice season local characteristics α_i depend on water level A since the sea surface can easily extend onto the shallow places. In the ice season, this dependence is reduced significantly because of the ice effect. Experiments on flexural strength were performed in February–March in 2009–2011 in the Tempel Fjord, Sveabukta, and Advent Fjord, and on May 5–7, 2010, in the Van Mijen Fjord near Akseloya. All experiments were performed with the same rig anchored to the ice. In 2009 and 2010 the load was applied by manual jack and measured with oil pressure sensors. In 2010 the load was applied by hydraulic cylinder with 15 t power capacity powered from electric pump produced by Enerpac. The load was measured with load cell NTT with a maximal capacity of 10 t. Parameters of the ice beams used in the tests are shown in Table 4.8.

Table 4.8 Parameters of the ice beams in flexural strength tests

Beam #	Date	Position	Length, m	Width, m	Thickness, m	Salinity mean, ppt	Temperature mean, degree C	Flexural strength, kPa
Place		Tempel Fjord						
1	05.03.2010	78°24.31′ N	1.97	0.64	0.23	5.57	–8.00	346.7
2		16°59.95′ E	1.27	0.58	0.22	5.57	–8.00	258.3
3			1.40	0.56	0.22	5.57	–8.00	302.6
4			1.38	0.56	0.22		–8.00	256.7
5	06.03.2010	78°25.72′ N	1.90	0.74	0.35	5.57	–5.18	310.5
6		17°11.80′ E	1.98	0.68	0.35	5.57	–5.18	281.8
7	07.03.2010	78°26.18′ N	2.50	0.51	0.51	6.03	–3.77	213.2
8		17°17.50′ E	2.28	0.56	0.51	6.03	–3.77	232.1
Place		Svea						
9	10.03.2010	77°52.56′ N	3.40	0.69	0.75	3.63	–5.92	334.3
10	11.03.2010	16°43.91′ E	3.36	0.69	0.75	3.63	–5.92	368.9
Place		Van Mijen Fjord						
11	01.05.2010		2.15	0.47	0.50	3.0	–2.27	126.7
12	03.05.2010		3.39	0.50	0.45	3.5	–2.10	169.0
13			2.13	0.48	0.40	2.7	–1.40	149.8
14			2.30	0.46	0.39	2.7	–1.60	186.1
15			2.27	0.45	0.37	5.7	–2.20	131.6
16			2.15	0.47	0.50	5.9	–1.78	106.7
17			3.02	0.60	0.49	3.9	–2.10	128.1

Temperature effect on the flexural strength is shown in Fig. 4.64a based on the tests performed in 2009 and 2010. Measured values of flexural strength are shown in Fig. 4.64b versus square root of brine volume $\sqrt{\nu_b}$ calculated from Frankenstein and Garner formula (1967) using measured values of mean ice salinity (S) and temperature (T)

$$\nu_b = S\left(\frac{48.185}{|T|} + 0.532\right), \quad -22.9\ ^\circ\mathrm{C} < T < -0.5\ ^\circ\mathrm{C}. \tag{4.53}$$

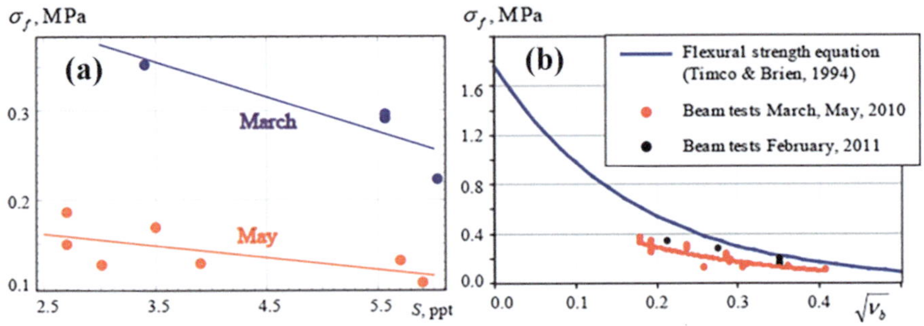

Fig. 4.64 Flexural strength versus ice salinity (**a**) and versus square root of brine volume (**b**)

One can see that constructed experimental points can be approximated by the solid blue line in Fig. 4.64 described by formula (Timco and O'Brien 1994)

$$\sigma_f = 0.829 \exp(-5.245\sqrt{\nu_b}). \tag{4.54}$$

Surface waves and swell with periods from 7 s and higher are always observed in different locations of the Ice Fjord. At the same time similar swell was never observed in Sveabukta. Wave influence on the ice is realized by the formation of alongshore bands with broken ice (Fig. 4.65a). In February 2011 several snow scooters were trapped on broken ice (Fig. 4.65b). The effect of ice destruction near shoreline is explained by the increase of swell amplitude when it propagates to the smaller depths. Let us consider the propagation of flexural gravity waves with small amplitude in shallow water. Ice surface elevation is described by the formula $\eta = A\cos(kx - \omega t)$, where A, k and ω are the wave amplitude, wave number and wave frequency. Wave frequency is related to wave number by the dispersion equation

$$\omega^2 = k^2 H(g + Dk^4), \ D = \frac{Eh_i^3}{12\rho_w(1-\nu^2)}, \tag{4.55}$$

Fig. 4.65 Band of broken ice near shore line (**a**), snow scooter on broken ice near shoreline (photographs from Svalbardposten) (**b**)

where, h_i, E, and ν are the ice thickness, Young's modulus, and Poisson's ratio, H and ρ_w are the water depth and density, and g is gravity acceleration. Variations of wave number caused by slow changes of water depth are calculated from Eq. (4.55), where the wave frequency is assumed constant.

Variations of wave amplitude caused by slow changes of water depth are calculated under the assumption that the mean flux of wave energy $\langle F \rangle$ is ~~a~~ constant. It means that wave reflection from the bottom is ignored. Mean flux of wave energy is calculated with the formula (Marchenko and Voliak 1997)

$$\langle F \rangle = \frac{c_g}{2} \frac{\omega^2 A^2}{k^2 H}, \tag{4.56}$$

where, $c_g = \partial\omega/\partial k$ is the group velocity. Ratio of wave amplitudes A_1 and A_2 at different depths H_1 and H_2 is calculated using the formula

$$r = A_1 A_2^{-1} = k_1 k_2^{-1} \sqrt{c_{g,2}/c_{g,1}}, \tag{4.57}$$

where, wave numbers k_1 and k_2, and group velocities $c_{g,1}$ and $c_{g,2}$ are calculated from Eq. (4.55) with $H = H_1$ and $H = H_2$ respectively. The ice is destroyed by wave when the maximal tensile stresses ($\sigma_{\max}$) on the bended ice surface reach critical value σ_f:

$$\sigma_{\max} = \frac{E h_i k^2 A}{2(1-\nu^2)} = \sigma_f. \tag{4.58}$$

Figure 4.66a, b show records of water pressure at the bottom when the incoming swell destroyed the ice 0.5 m thick around Longyearbyen harbor where the water depth was $H = 1.75$ m. Wave period was approximately 15 s, and wave amplitude reached $A = 0.1$ m. For numerical simulations it is assumed that $E = 30$ MPa and $\nu = 0.33$. Figure 4.66c shows the reduction of wave amplitude r and wave number k with the increase in water depth. Figure 4.66d shows maximal tensile stress in the ice plate when the wave propagates over the regions with different water depths. One can see that at depth $H \approx 8$ m tensile stress $\sigma_{\max}$ reaches 200 kPa. This value is close to the flexural strength of the ice measured at the same place several days before this event. Therefore, the ice floating at depths shallower than 8 m was destroyed by the waves. The same waves at greater depths have smaller amplitudes and cannot destroy the ice.

Tide measurements demonstrated proportional changes of the water level amplitudes in New Alesund, Longyearbyen, and Svea during the ice season. In the ice-free season, water level amplitudes in these locations have higher dispersion around their mean values. Phase shifts between the tides in New Alesund and Longyearbyen vary within 1 h with high dispersion around the mean value 0.55 h in the ice season and 0.59 h in the ice-free season, so that the highest and lowest water levels in Longyearbyen occur earlier than in New Alesund over the semidiurnal cycle. Phase shifts between tides in New Alesund and Svea vary within 1 h with high dispersion around

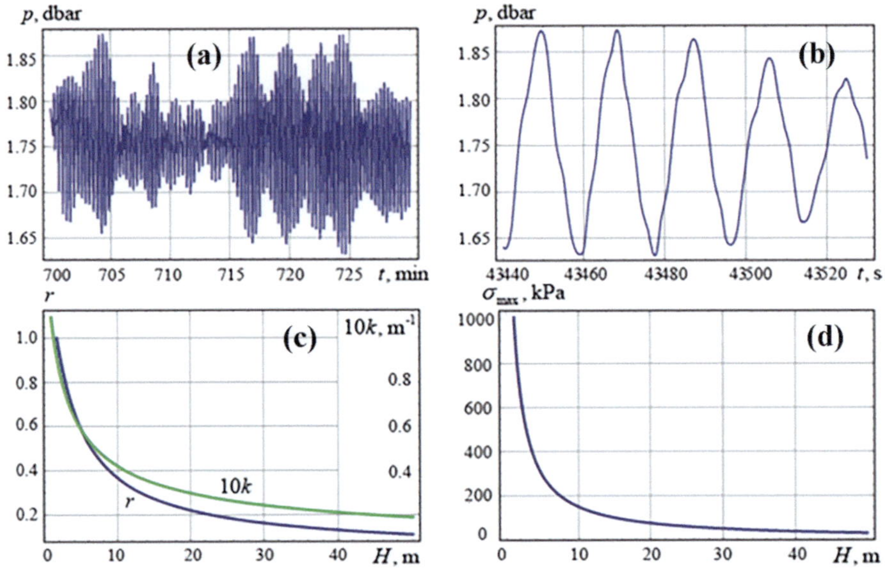

Fig. 4.66 Wave records near Longyearbyen on March 1, 2011 (**a**, **b**). Reduction of wave amplitude (r) and wave number (k) versus sea depth H (**c**). Maximal tensile stress in elastic ice plate bended by the wave versus sea depth H (**d**)

the mean value 0.55 h in the ice season and 0.45 h in the ice-free season, so that the highest and lowest water levels in Svea occur later than in New Alesund over the semidiurnal cycle. Akseløya Island influences phase shift about 0.5 h between the tides on different sides of the island.

Salinity of sea water near glaciers is lower and water temperature is close to the freezing point. Water temperature near the Paula Glacier is higher than water temperature in Sveabukta and adjacent regions of the Van Mijen Fjord. Water temperature in the Tempel Fjord depends on the inflow of sea water from the ocean. In 2010, the water temperature at a distance of 10 km from the glacier varied from –1.8 to –1.1 °C and was much higher than the water temperature –1.9 °C measured near the Tuna Glacier. In 2011, water temperature at a distance of 10 km from the glacier varied from –1.86 to –1.82 °C and was almost the same as near the Tuna Glacier. Ice thickness near glaciers is well interpolated by the Zubov's (1944) formula based on the calculation of degree days of the frost. Heat fluxes from the water can significantly reduce the ice thickness at a distance of 10 km from the Tuna Glacier.

Flexural strength measured in Svalbard fjords on ice beam of different thickness corresponds well to the formula of Timco and O'Brien (1994) showing the dependence of flexural strength from the brine volume. Values of flexural strength of the ice in Svalbard fjords vary from 0.1 to 0.38 MPa. Effect of ice destruction along the shore line in Ice Fjord is explained by the impact of swell on the ice over shallow depth. The flexural strength was used in the criterion of ice destruction by swell.

References

Aubrey DG, Speer PE (1985) A study of non-linear tidal propagation in shallow inlet/estuarine systems. Part i: Observ Estuarine Coast Shelf Sci 21:185–205

Bates HF, Shapiro LH (1980) Long-period gravity waves in ice-covered sea. J Geophys Res 85:1095–1100

Blackman RB, Tukey JW (1958) The measurement of power spectra from the point of view of communications engineering. Dover, New York, p 190

Blanton JO, Lin GQ, Elston SA (2002) Tidal current asymmetry in shallow estuaries and tidal creeks. Cont Shelf Res 22(11–13):1731–1743

Bogorodskii PV, Demidov NE, Filchuk KV et al (2020) Growth of landfast ice and its thermal interaction with bottom sediments in the Braganzavågen Gulf (West Spitsbergen). Russ J Earth Sci 20:ES6005. https://doi.org/10.2205/2020ES000718

Bogorodskii VV, Gavrilo VP (1980) Ice: its physical properties: contemporary methods of glaciology. Leningrad, Gidrometeoizdat, 384 p. [in Russian]

Boon JD, Byrne RJ (1981) On basin hypsometry and the morphodynamic response of coastal inlets. Mar Geol 40:27–48

Bulatov OV, Elizarova TG (2011) Regularized shallow water equations and an efficient method for numerical simulation of shallow water flow. Comp Math Math Phys 51(1):160–173

Bulatov OV, Elizarova TG (2016) Regularized shallow water equations for numerical simulation of flows with a moving shoreline. Comp Math Math Phys 56(4):661–679

Caline F (2009) Coastal sea ice conditions around a breakwater in a microtidal, inner-fjord river delta in Svalbard. Ph.D. thesis. University Center in Svalbard

Cheng P, Valle-Levinson A, de Swart HE (2010) Residual currents induced by asymmetric tidal mixing in weakly stratified narrow estuaries. J Phys Oceanogr 40:2135–2147

Cottier FR, Nilsen F, Skogseth R et al (2010) Arctic fjords: a review of the oceanographic environment and dominant physical processes. In: Howe JA, Austin WEN, Forwick M, Paetzel M (eds) Fjord systems and archives. Geological Society, vol 344, London, Special Publications, pp 35–50. https://doi.org/10.1144/SP344.4

Davis RH, Serayssol J-M, Hinch EJ (1986) The elastohydrodynamic collision of two spheres. J Fluid Mech 163:479–497

De Carolis G, Olla P, Pignagnoli L (2005) Effective viscosity of grease ice in linearized gravity waves. J Fluid Mech 535:369–381

Dias JM, Sousa MC (2009) Numerical modeling of Ria Formosa tidal dynamics. J Coastal Res 56:1345–1349

Dronkers J (1986) Tidal asymmetry and estuarine morphology. Netherlands J Sea Res 20(2/3):117–131

Elizarova TG, Ivanov AV (2020) Numerical modeling of passive scalar transport in shallow water based on the quasi-gasdynamic approach. Comp Math Math Phys 60(7):1208–1227

Farmer DM, Freeland HJ (1983) The physical oceanography of fjords. Prog Oceanogr 12(2):147–219

Fer I, Widell K (2007) Early spring turbulent mixing in an ice-covered Arctic fjord during transition to melting. Cont Shelf Res 27:1980–1999

Frankenstein GE, Garner R (1967) Equations for determining the brine volume of sea ice from –0.5° to –22.9 °C. J Glaciol 6(48):943–944

Friedrichs CT, Aubrey DG (1994) Tidal propagation in strongly convergent channels. J Geophys Res 99(C2):3321–3336

Gallagher B (1973) Model for nonlinear tides in small basins with openings of restricted depth. J Geophys Res 78(27):6395–6400

Hoitink AJF, Hoekstra P, van Maren DS (2003) Flow asymmetry associated with astronomical tides: implications for the residual transport of sediment. J Geophys Res 108(C10):3315–3323. https://doi.org/10.1029/2002JC001539

Høyland KV, Løset S (1999) Measurements of temperature distribution, consolidation and morphology of a first-year sea ice ridge. Cold Regions Sci Technol 29(1):59–74

Høyland KV, Liferov P (2005) On the initial phase of consolidation. Cold Regions Sci Technol 41(1):49–59

Høyland KV (2009) Ice thickness, growth and salinity in Van Mijenfjorden, Svalbard, Norway. Polar Res 28(3):339–352. https://doi.org/10.1111/j.1751-8369.2009.00133

Huang H, Chen C, Blanton JO, Andrade FA (2008) A numerical study of tidal asymmetry in Okatee Creek, South Carolina. Estuarine Coastal Shelf Sci 78:190–202

Joseph GG, Zenit R, Hunt ML, Rosenwinkel AM (2001) Particle-wall collisions in a viscous fluid. J Fluid Mech 433:329–346

Keliher TE (1976) An investigation of the effect of large-amplitude ocean waves on Antarctic pack ice. AIDJEX Bull 34:114–136

Kerr AD, Palmer WT (1972) The deformations and stresses in floating ice plates. Acta Mech 15:57–72

Kolmogorov AN (1941) The local structure of turbulence in incompressible viscous fluid for very large Reynolds number. Dokl Akad Nauk SSSR 30:9–13

Lamb H (1975) Hydrodynamics, 6th edn. Cambridge University Press

Landau LD, Lifshitz EM (1987) Course of theoretical physics: fluid mechanics, vol 6. Pergamon Press

Langhorne PJ, Squire VA, Fox C, Haskell TG (2001) Lifetime estimation for a land-fast ice sheet subjected to ocean swell. Ann Glaciol 33:333–338

Lanzoni S, Seminara G (1998) On tide propagation in convergent estuaries. J Geophys Res 103(C13):30793–30812

Launiainen J, Cheng B (1998) Modelling of ice thermodynamics in natural water bodies. Cold Regions Sci Tech 27:153–178

Leppäranta M (1993) A review of analytical models of sea-ice growth. Atmos Ocean 31(1):123–138

Li H, Lubbad R (2018) Laboratory study of ice floes collisions under wave action. In: Proceedings of the 28th international ocean and polar engineering conference (ISOPE), Sapporo, Japan, pp 1516–1524

Li H, Gedikli ED, Lubbad R (2021) Laboratory study of wave-induced flexural motion of ice floes. Cold Regions Sci Technol 182:103208. https://doi.org/10.1016/j.coldregions.2020.103208

Liferov P, Høyland KV (2004) In-situ ice ridge scour tests: experimental set up and basic results. Cold Regions Sci Technol 40(1–2):97–110

Liu AK, Mollo-Christensen E (1988) Wave propagation in a solid ice pack. J Phys Oceanogr 18(11):1720–1712

Marchenko AV, Voliak KI (1997) Surface wave propagation in shallow water beneath an inhomogeneous ice cover. J Phys Oceanogr 27:1602–1613

Marchenko AV, Morozov EG, Muzylev SV, Shestov AS (2010) Interaction of short internal waves with the ice cover in an Arctic fjord. Oceanology 50(1):18–27

Marchenko AV, Morozov EG, Muzylev SV, Shestov AS (2011) Short-period internal waves under an ice cover in Van Mijen Fjord, Svalbard. Adv Meteor, Article ID 573269. https://doi.org/10.1155/2011/573269

Marchenko AV, Morozov EG, Muzylev SV (2012) A tsunami wave recorded near a glacier front. Nat Hazards Earth System Sci 12:415–419

Marchenko AV, Morozov EG (2013) Asymmetric tide in Lake Vallunden (Spitsbergen). Nonlin Process Geophys 20:935–944

Marchenko AV, Morozov EG, Muzylev SV (2013) Measurements of sea ice flexural stiffness by pressure characteristics of flexural-gravity waves. Ann Glaciol 54:51–60. https://doi.org/10.3189/2013AoG64A075

Marchenko AV, Kowalik Z, Brazhnikov D et al (2015) Characteristics of sea currents in navigational strait Akselsundet in Spitsbergen. In: Proceedings of the 23rd International conference on port and ocean engineering under arctic conditions, Trondheim, Norway

Marchenko AV, Morozov EG (2016a) Surface manifestations of the waves in the ocean covered with ice. Russ J Earth Sci 16:ES1001. https://doi.org/10.2205/2016ES000561
Marchenko AV, Morozov EG (2016b) Seiche oscillations in Lake Vallunden (Spitsbergen). Russ J Earth Sci 16:ES2003. https://doi.org/10.2205/2016ES000567
Marchenko A, Markov V, Taylor R (2019) Influence of water on collisions of floating ice blocks. ISOPE 19TPC-0906: 757–764
Marchenko AV, Morozov EG, Ivanov AV et al (2021) Ice thickening caused by freezing of tidal jet. Russ J Earth Sci 21:ES2004. https://doi.org/10.2205/2021ES000761
Marchenko AV, Chistyakov PV, Karulin EB et al (2022) Field experiments on collisional interaction of floating ice blocks. In: Proceedings of the 26th international conference on port and ocean engineering under arctic conditions (POAC 2021)
Marko JR (2003) Observations and analyses of an intense wave-in-ice event in the Sea of Okhotsk. J Geophys Res 108(C9):3926. https://doi.org/10.1029/2001JC001214
Martin S, Becker P (1987) High-frequency ice floe collisions in the Greenland Sea during the 1984 marginal ice zone experiment. J Geoph Res 92(C7):7071–7084
McClimans TA (1977) On the energetics of tidal inlets to land-locked fjords. Mar Sci Commun 4:121–137
Merian JR (1828) Ueber die Bewegung tropfbarer Flüssigkeiten in Gefässen [On the motion of drippable liquids in containers] (thesis) (in German). Schweighauser, Basel
Mirfendersk H, Tomlinson R (2008) Living with sea-level rise and climate change: observation and analysis of hydrodynamic parameters in tidal inlets in a predominantly semidiurnal regime. J Coastal Res 24(5):1229–1239
Mofjeld HO (1988) Depth dependence of bottom stress and quadratic drag coefficient for barotropic pressure-driven currents. J Phys Oceanogr 18:1658–1669
Morozov EG, Pisarev SV (2002) Internal tides at the Arctic latitudes (numerical experiments). Oceanology 42(2):153–161
Morozov EG, Marchenko AV, Filchuk KV et al (2019) Sea ice evolution and internal wave generation due to a tidal jet in a frozen sea. Appl Ocean Res 87:179–191
Morozov EG, Marchenko AV, Filchuk KV (2022) Freezing of tidal current under ice in a shallow channel. Russ J Earth Sci 22(2):ES2003. https://doi.org/10.2205/2022ES000793
Moslet PO (2008) Medium scale ice–structure interaction. Cold Regions Sci Tech 54(2):143–152
Muzylev SV (2008) Internal waves under an ice cover. Doklady Earth Sci 418:145–148
Nidzieko NJ (2010) Tidal asymmetry in estuaries with mixed semidiurnal/diurnal tides. J Geophys Res 115:C08006. https://doi.org/10.1029/2009JC005864
Obukhov AM (1941) On the distribution of energy in the spectrum of turbulent flow. Dokl Akad Nauk SSSR 32:19–22
Prinsenberg SJ, Peterson IK (2011) Observing regional-scale pack-ice decay processes with helicopter-borne sensors and moored upward-looking sonars. Ann Glaciol 52(57):35–42
Proudman J (1953) Dynamical oceanography. Vol 117. Methuen, London, 225p
Rabinovich AB (2009) Handbook on coastal and ocean engineering. In: Kim YC (ed) Chapter 9, Seiches and harbor oscillations: 193–236. World Scientific Publ., Singapore
Savenije HHG, Veling EJM (2005) Relation between tidal damping and wave celerity in estuaries. J Geophys Res 110:C04007. https://doi.org/10.1029/2004JC002278
Sheretov YV (2016) Regularized hydrodynamic equations. Tver State Univ. 222 p. [in Russian]
Shestov A, Wrangborg D, Marchenko A (2015) Hydrology of Braganzavågen under ice-covered conditions. Proceedings of the 23rd international conference on port and ocean engineering under arctic conditions, Trondheim, Norway
Skarðhamar J, Svendsen H (2010) Short-term hydrographic variability in a stratified Arctic fjord. In: Howe JA, Austin WEN, Forwick M, Paetzel M (eds) Fjord systems and archives. Geological society, vol 344, London, Special Publications, pp 51–60
Smirnov VN (1996) Dynamic processes in sea ice. Gidrometeoizdat, St.-Petersburg 162 p. [in Russian]

Squire VA, Hosking RG, Kerr AD, Langhorne PJ (1996) Moving load on ice plates. Kluwer Academic Publishers, Netherlands, p 230
Squire VA, Dugan JP, Wadhams P et al (1995) Of ocean waves and sea ice. Ann Rev Fluid Mech 27:115–168
Squire VA, Moore SC (1980) Direct measurement of the attenuation of ocean waves by pack ice. Nature 283:365–368
Starovoitov VN (2003) Behavior of rigid body in an incompressible viscous fluid near a boundary. Int Ser Numer Math 147:313–327
Stefan J (1891) Über die Theorie der Eisbildung, insbesondere über Eisbildung im Polarmeere. Ann Phys 3rd Ser 42:269–286
Stigebrandt A (1980) Some aspects of tidal interactions with fjord constrictions. Estuarine Coastal Mar Sci 11(2):151–166
Støylen E, Weber E (2010) Mass transport induced by internal Kelvin waves beneath shore-fast ice. J Geophys Res 115(C3):C03022. https://doi.org/1029/2009JC005298
Strub-Klein L, Høyland KV (2012) Spatial and temporal distributions of level ice properties: experiments and thermo-mechanical analysis. Cold Regions Sci Tech 71:11–22
Tabata T, Fujino K, Aota M (1967) Studies on the mechanical properties of sea ice: the flexural strength of sea ice in situ. In: Physics of snow and ice, vol 1. Hokkaido University, Japan, pp 539–550
Tchieu AA, Crowdy D, Leonard A (2010) Fluid-structure interaction of two bodies in an inviscid fluid. Phys Fluids 22:107101
Timco GMW, Weeks WF (2010) A review of the engineering properties of sea ice. Cold Reg Sci Technol 60:107–129
Timco GW, O'Brien S (1994) Flexural strength equation for sea ice. Cold Reg Sci Tech 22:285–298
Timoshenko S, Woinowsky-Krieger S (1970) Theory of plates and shells, 2nd edn. McGraw-Hill, Singapore
Vladimirsky NP (ed) (2010) Rukovodstvo po obrabotke i predskazaniu prilivov (1941), Gidrograficheskoe upravlenie VMF USSR, Leningrad, 347 pp. (Handbook for the processing and prediction of tides)
Wadhams P, Wells SCS (1995) Ice surface oscillation measurements on SIMI using strain, heave, and tilt sensors. In: Proceedings of the sea ice mechanics and arctic modeling workshop, Anchorage, AK, Office of Naval Research, pp 176–189
Walton TL (2002) Tidal velocity asymmetry at inlets, ERDC/CHL CHETN IV-47. U.S. Army Engineer Research and Development Center, Vicksburg, MS. http://chl.erdc.usace.army.mil
Weeks WF, Hibler WD (III) (2010) On sea ice. University Alaska Press, Fairbanks, Alaska, 664 p.
Wrangborg D, Grady R, Marchenko A, Karulin E (2013) Analysis of tidal deformations of land fast ice in shallow Arctic fjord. In: Proceedings of the 22 international conference on port and ocean engineering under arctic conditions, Espoo, Finland
Zubov NN (1944) Arctic ice. Moscow, Izdatel'stvo Glavsevmorputi, 360 pp. [in Russian]

Chapter 5
Field Experiments Near Glaciers

5.1 General Notes and Introduction

Glaciers are huge bodies of dense ice. They are formed from snow precipitation. The glaciers compress under the weight of the snow layer. If glaciers are formed on a sloping terrain they slowly descend downhill moving under their own weight. Generally, the glaciers are formed in the regions where snow accumulation exceeds its ablation (ablation is related to all processes that remove snow, ice, or water from a glacier or snowfield) over centuries. During the motion of glaciers, many large cracks (crevasses) appear in the ice and sometimes high columns of ice (seracs) are formed during ice deformation under stresses induced by the glacier weight. During the motion over land, glaciers transport soil and rocks from the land surface and may change the surrounding terrain. Some of the glaciers melt at low heights above the sea, the others reach water in polar regions. Accumulation of glacial ice is considered the largest reserve of freshwater on the planet. The amount of ice is estimated as 69% of the Earth's freshwater.

The modern glaciation on the Spitsbergen archipelago covers an area of approximately 36,600 km^2 and includes glaciers of different morphological types (Hagen et al. 1993). About 60% of this area is occupied by tidewater glaciers, the fronts of which terminate in the sea (Błaszczyk et al. 2009). The volume of ice in Spitsbergen is about 7000 km^3. Evaporation from the Barents Sea provides snow precipitation. The balance between summer temperatures and winter precipitation determines whether glaciers grow or shrink.

The largest glaciers of Spitsbergen are in the eastern part of the islands. The dominating type of glaciers in Spitsbergen are large contiguous ice masses divided into many smaller streams of ice. Many of glaciers terminate in the ocean, where they calve.

Analysis of ASTER 2000 space images from 2006 showed that the total length of ice fronts of 163 tidewater glaciers of the archipelago is currently 860 km (Muzylev et al. 2013). Over the past 30–40 years, some of the ice fronts have retreated onto land

A. Marchenko et al., *Geophysical Hydrodynamics of Currents, Waves, and Ice in the Atlantic Arctic*, https://doi.org/10.1007/978-3-032-14122-4_5

(14 of them), while 11 glaciers that previously terminated on land, advanced into the sea. Tidewater glaciers are divided into four groups based on the dynamic state and the pattern of fracturing of the frontal part: very slow, slow-moving, fast-moving, and surging glaciers, which are in a quiet or active phase. According to estimates (Błaszczyk et al. 2009), the loss of ice mass by these glaciers due to iceberg calving occurs throughout the archipelago (except for Kvitøya Island) and amounts to 5.0–8.4 km^3/year (on average 6.75 km^3/year in water equivalent).

Their contribution to the global ocean level rise is about 0.03 mm/year. When icebergs melt in the summer, a significant amount of fresh water is formed, which has a significant effect on the temperature, density and salinity of sea water, its circulation, thermal and strength properties of sea ice, and its distribution and thickness in coastal areas and in fjords where tidewater glaciers end. Meltwater flowing from the surface of these glaciers into the sea during the melting period has the same effect on the state of sea level.

An additional, but difficult to quantify, source of fresh water entering the sea may also be melting in the ice columns of glaciers on the bed. Internal and subglacial melting is most likely in polythermal glaciers, consisting of an upper layer of cold ice and a lower layer of warm, water-containing ice. According to radio sounding data (Macheret and Zhuravlev 1982), such glaciers are widely distributed in Spitsbergen including those terminating in the sea. It was also established that the mean water content in the lower warm layer of such glaciers may range from 0.1% in spring to 2.4% in summer. Fresh water accumulated in the bottom layer freezes at the boundary of the cold and warm ice, and some part flowing out into the sea, including in winter (Macheret and Glazovsky 2000). Currently, 50 polythermal (two-layer) glaciers have been identified on Spitsbergen using radio sounding data; most of them end in the sea and are classified as surging glaciers (Jiskoot et al. 2000).

5.2 A Tsunami Wave Recorded Near a Glacier Front

In this section we analyze a tsunami wave observed near the glacier front in the Tempel Fjord (Spitsbergen). Several different recorders were deployed on a wire from ice and on the sea bottom approximately 300 m from the glacier front. The measurements allowed us to reveal an ice-slide tsunami wave generated by the motion of the glacier. In our description we will follow our publication (Marchenko et al. 2012).

Landslide tsunamis are generated by the gravity flow of the rocks, land, or sediments on an inclined bottom. Usually, horizontal scales of landslide tsunamis are smaller than those of the tsunamis generated by earthquakes. These tsunamis decay quickly due to the dispersion effects (Gonzalez and Kulikov 1993). Theoretical investigations and laboratory modeling performed by (Fine et al. 2003) indicate that submarine landslides are less effective at tsunami generation compared with subaerial slides. A rigid-body slide produces much higher tsunami waves than a viscous (liquid) slide. This result is confirmed in the review of tsunami mechanisms by Harbitz et al. (2006). Only a limited part of the potential energy released by the landslide is transferred to the wave energy. The critical parameter determining the generation of surface waves is the Froude number (*Fr* is the ratio between slide and wave speeds). The most efficient generation occurs near resonance when $Fr = 1.0$. Subaerial slides displace a considerable volume of water at relatively high speed as they slide into the water from the slope. One of the well-known examples of subaerial landslides is the event in Nice on October 16, 1979, when a volume of ground slumped into the Mediterranean Sea during landfilling operations (Assier-Rzadkiewicz et al. 2000). Other examples are the landslide of November 3, 1994 in Skagway (Kulikov et al. 1996) during the collapse of a cruise ship wharf undergoing construction and the landslide of July 17, 1998 in Papua New Guinea caused by a moderate earthquake (Tappin et al. 1999).

Here, we consider a tsunami wave generated by the displacement of a glacier (Tuna Glacier) in the Tempel Fjord (Spitsbergen). We will discuss the results of the research carried out in February 2011 near the front of the Tuna Glacier. The Tuna Glacier is located in the interior of the Tempel Fjord approximately 60 km from the Longyearbyen town, its geographic coordinates are 78°26′ N, 17°18′ E. (Fig. 5.1a). It is 35 km long; its area is ~203 km^2. The glacier experienced strong displacements in 1930, 1970, and 2003–2005 (Hodgkins and Dowdeswell 1994). At present, it is in a quiet phase and merges with the Von Post Glacier, forming an ice front broken into blocks, 3260 m long and up to 37 m high.

According to aerial radio sounding data (Macheret et al. 1992), the bed of the glacier, like all tidewater glaciers of Spitsbergen, is located below the sea level near their fronts and the ice at the front does not float in the sea but rests on the bottom. During the second advance in 1970 and in the spring of 1971, which was better provided with observations, the end part of the tongue of the Tuna Glacier moved at a speed of about 1 m/day and by August had advanced 150–550 m compared to July 1970 and about 1.5 km compared to 1966 (Hodgkins and Dowdeswell 1994). In February 2011, that is, 7–9 years after the last advance in 2003–2005, the glacier was probably already in a quiet phase, but could have experienced micro-motions, as evidenced by the accumulation of ice blocks in front of its front and tension cracks on the surface of the sea ice.

Field works were performed on February 16–19, 2011 in the Tempel Fjord in Spitsbergen near the glacier front. The air temperature was quite low (~–20 °C). Two outflow glaciers Tuna Glacier and Von-Post Glacier form the glacier front of about 3.5 km wide (Fig. 5.1). Although the glaciers were stable since 2005, the visual observations show significant buckling deformations and folds of sea ice near the front of the Tuna Glacier (Fig. 5.2). The buckling deformations can be excited by the ice creep (Collins and McCrae 1985). At the same time, ice compression can induce sudden destruction of the ice due to the buckling instability (Sodhi et al. 1983).

The thickness of sea ice near the glacier was 94 cm. Two SBE 39 temperature and pressure sensors were lowered from the ice covering the fjord at depths $z_1 \approx 9.7$ m and $z_2 \approx 16.8$ m at a distance of ~300 m from the front of the Tuna Glacier (Figs. 5.1 and 5.3). The sampling time interval was 1 s. An SBE 37 instrument was deployed at the bottom under them with a sampling interval of 6 s (Fig. 5.4).

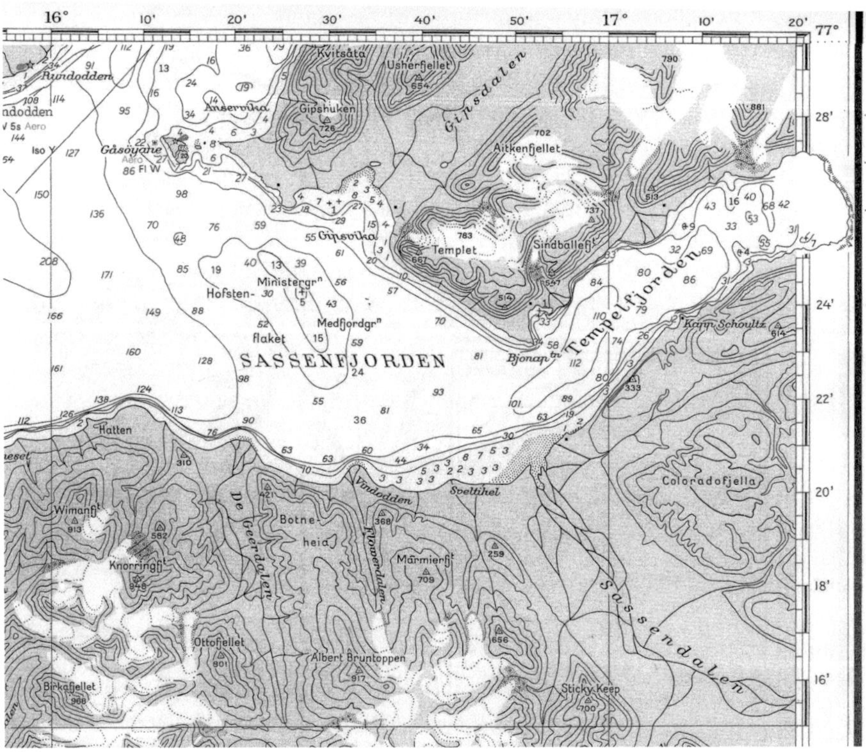

Fig. 5.1 Map of the Tempel Fjord

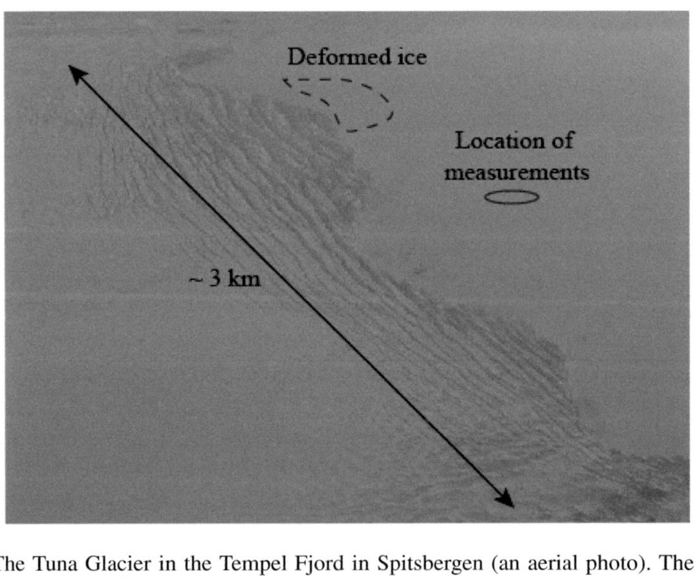

Fig. 5.2 The Tuna Glacier in the Tempel Fjord in Spitsbergen (an aerial photo). The horizontal scales on the photo are distorted

Fig. 5.3 Deformed sea ice near the glacier front: buckling (**a**) and fold with a small puddle of seawater (**b**)

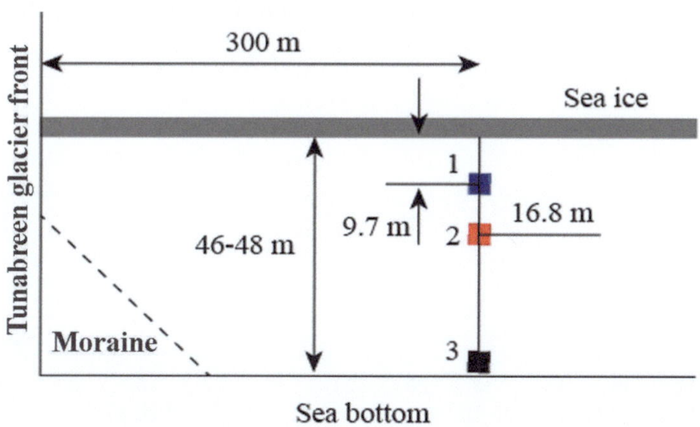

Fig. 5.4 Scheme of the deployment of pressure recorders SBE 39 (1 and 2) and SBE 37 (3)

The bottom gauge generally records the hydrostatic fluctuations of the water column (the frequency is of the order of 0.01 Hz and lower). The pressure gauges fixed on a wire from the ice surface do not record the hydrostatic fluctuations but record high frequency (of the order of 1 Hz) non-hydrostatic vibrations of the ice cover.

The depth of the location was 46–48 m depending on the tide phase. The height of the glacier over the ice cover was 37–40 m. Then, on February 17 the instruments were recovered and one SBE 39 was lowered into the same hole in the ice together with the Aquadopp current meter. These instruments operated for one day.

Approximately at 07:40 on February 17, 2011, the instruments recorded a tsunami wave. A record of the pressure tide gauge at the bottom is shown in Fig. 5.5. The time in hours from the beginning of the deployment is laid off as the abscissa. The tidal signal dominated in the record. A pressure perturbation is seen at the time period approximately corresponding to 80 h. A detailed record of this pressure perturbation is shown in Fig. 5.5. The zero-time moment corresponds to 79.17 h after the deployment. We assume that this perturbation was caused by a tsunami wave. The time interval is shown in detail in the inset of Fig. 5.5. One can see that the leading pulse displaced the surface by more than 30 cm. It was followed by 10 cm fluctuations that later decreased to a few centimeters. It is worth noting that the event occurred at the low tide, which seems a characteristic feature of landslides (Bjerrum 1971; Johns et al. 1986; Kulikov et al. 1996). Records of the pressure gauges on the wire are shown in Fig. 5.6, where the zero-time moment is the same as in Fig. 5.5. These records also include the leading pulse and the trailing wave tail.

The spectrum of the bottom pressure gauge reveals a peak of high confidence at a period of 93 s. The spectrum is shown in Fig. 5.7. A landslide tsunami of a close period (3 min) was reported in (Rabinovich et al. 1999). We estimate the tsunami wave velocity as $c = \sqrt{gH}$, where H is the water depth; then the velocity is approximately equal to 20 m/s and the tsunami wave length over the depth of 46–48 m is approximately 1850 m.

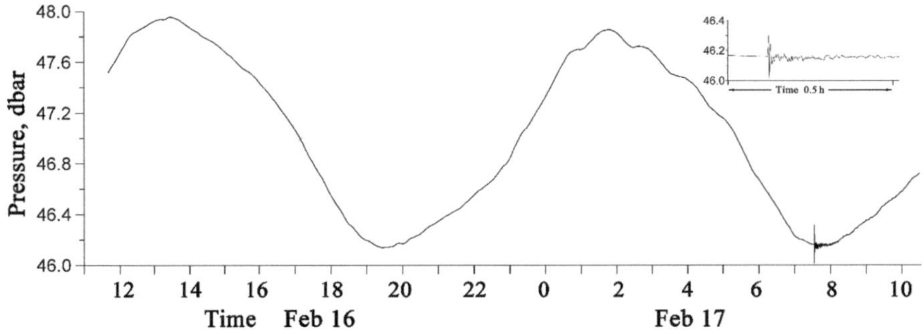

Fig. 5.5 Record of the pressure tide gauge at the bottom, which recorded a high amplitude tidal signal and fluctuations of pressure caused by the tsunami wave. A detailed record of the tsunami perturbations is shown in the inset

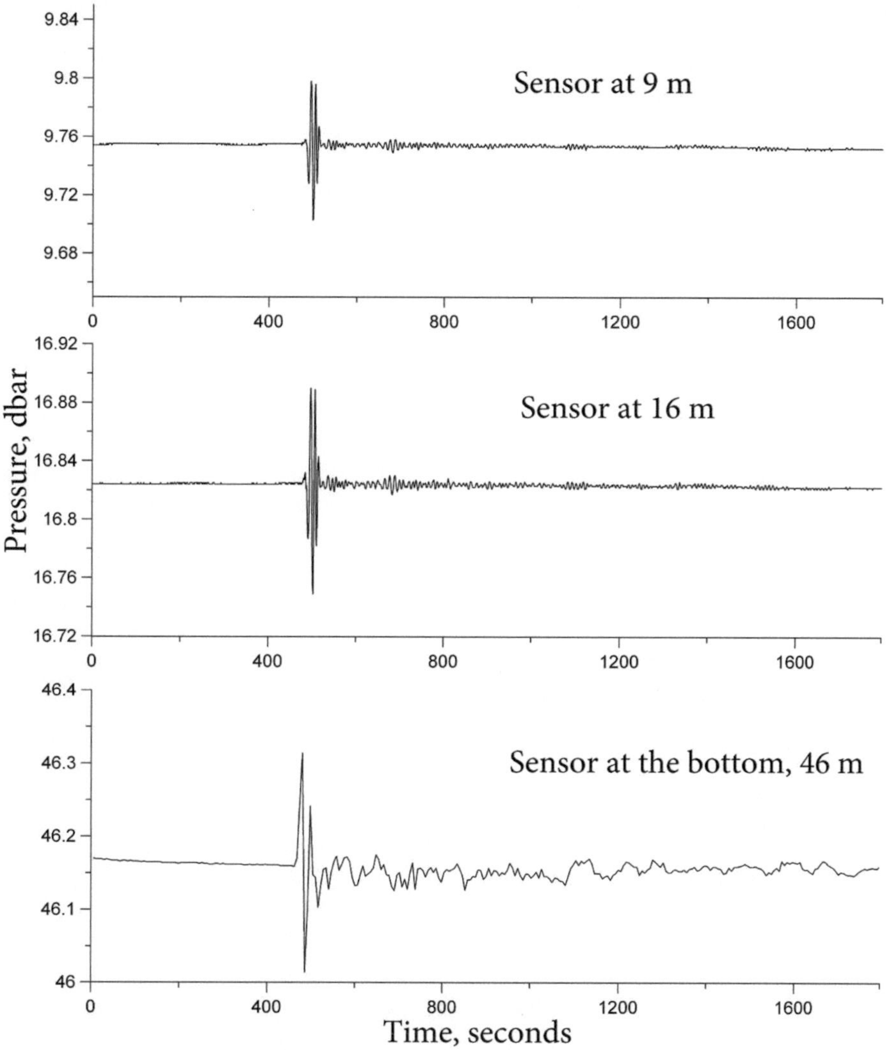

Fig. 5.6 Records of pressure gauges at 9, 16 m, and at the bottom (46 m)

Not knowing that the instruments recorded a tsunami wave we recovered them and replaced by one SBE 39 and Aquadopp current meter. The event recorded by the SBE 39 instrument on February 19 was weaker than the previous one. Again, the tsunami wave was recorded during the low water tide phase. The current meter recorded horizontal velocities reaching 80 cm/s.

The sampling frequency of gauge 3 at the bottom was not sufficient to resolve the high-frequency surface waves produced by the event. The waves are clearly seen in the detailed records. The leading signal and wave tails of water pressure perturbations

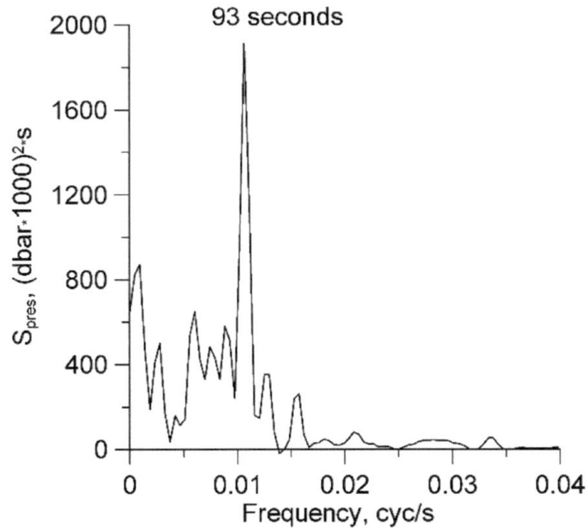

Fig. 5.7 Spectrum of the pressure gauge record at the bottom during the time interval of the tsunami event. The main peak corresponds to 93 s

created by the event are shown in Fig. 5.8a and b by blue lines for gauge 1 (9 m) and red lines for gauge 2 (16 m). The Fourier transforms of the first signals and wave tails were calculated from relation

$$p_F = \frac{1}{\sqrt{2\pi}} \int_{-\infty}^{\infty} p e^{i\omega t} dt. \tag{5.1}$$

The Fourier transforms of records by pressure gauges on the wire are shown in Figs. 5.8c and d with blue lines for gauge 1 and red lines for gauge 2. One can see that the leading signals of both gauges have spectral maxima at a frequency of $\omega_0 \approx 0.5\ \mathrm{s}^{-1}$. Wave tails in the records of both gauges have two spectral maxima at frequencies $\omega_1 \approx 0.435\ \mathrm{s}^{-1}$ and $\omega_2 \approx 0.625\ \mathrm{s}^{-1}$. In the linear theory of surface periodical waves, the perturbations of water pressure at depth z are described by the following relation (Mei 1983):

$$p_z = \rho_w a \left(\frac{\omega^2 \cosh k(z+H)}{k \sinh kH} - g \right) \cos(kx - \omega t), \tag{5.2}$$

where, a, ω, and k are the wave amplitude, frequency, and wave number, respectively; H and ρ_w are the water depth and density, and g is the gravity acceleration. In the water covered by an elastic plate, the wave frequency is related to the wave number by the dispersion equation (Marchenko et al. 2010; Squire et al. 1996):

$$\omega^2 = k \tanh kH (g + Dk^4), \quad D = \frac{Eh^3}{12\rho_w (1 - \nu^2)}, \tag{5.3}$$

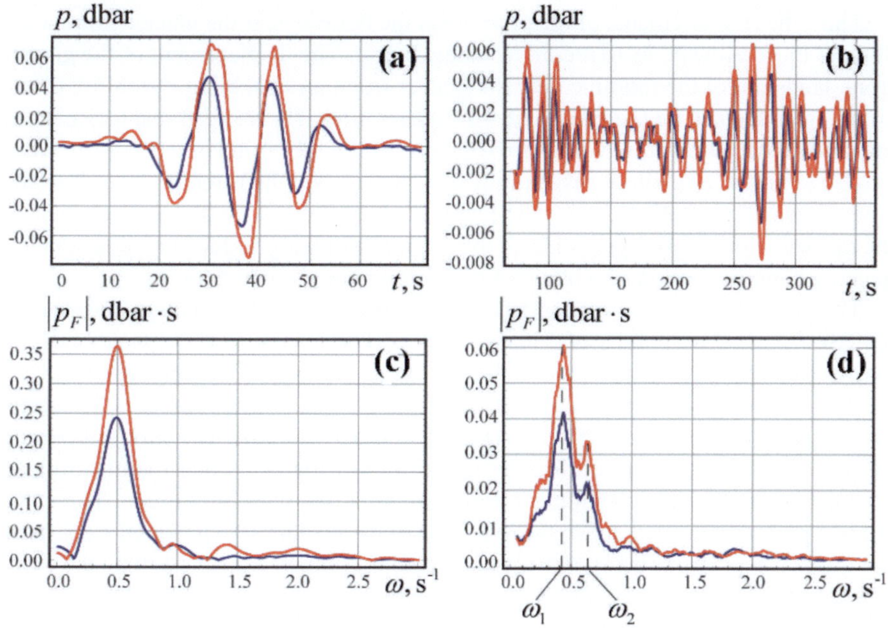

Fig. 5.8 Water pressure perturbations by the leading pulse (**a**) and wave tails (**b**) and absolute values of their Fourier transforms (**c**, **d**). Blue and red lines correspond to gauges 1 and 2, respectively

where E and ν are the Young's modulus and Poisson ratio of the ice, and h is the ice thickness. The ratio of the water pressures at depths z_1 and z_2 is determined as follows

$$\Delta_{12} = \frac{p_{z1}}{p_{z2}} = \frac{\omega^2 \cosh k(z_1 + H) - gk \sinh kH}{\omega^2 \cosh k(z_2 + H) - gk \sinh kH}. \tag{5.4}$$

Equation (5.4) can be used for the calculation of wavenumber k when the water pressures δp_{z1} and δp_{z2} are known. Then, the dispersion equation is used for the calculation of the ice rigidity D. Since ice thickness $h = 0.94$ m was measured and the Poisson ratio of ice $\nu \approx 0.34$ is very stable, the dispersion equation can be used for the estimates of the Young's modulus E of ice.

Ratio Δ_{12} is estimated as a ratio of the Fourier transform maxima at frequencies ω_1 and ω_2. It results in wave numbers $k_1 = 0.0246\ \text{m}^{-1}$ and $k_2 = 0.0405\ \text{m}^{-1}$, respectively. The wave lengths are equal to 255.4 m and 155.1 m. The amplitude of the wave tail shown in Fig. 5.7b is estimated by formula (5.2) is smaller (2.2 cm). It corresponds to the amplitude of the wave tail recorded by gauge 3 at the bottom. The estimate of the Young's modulus performed for the shortest wave with frequency ω_2 gives a value of $E = 1.7$ GPa. It is close to the values measured by mechanical and acoustic methods. The measured values of the Young's modulus are within 1–5 GPa (Bogorodskii and Gavrilo 1980).

Thus, the measurements of the pressure at the bottom near the glacier front in the Tempel Fjord allowed us to record a tsunami wave. The displacement of the glacier front, landslide of the moraine near the glacier front, and buckling instability of the compressed ice can be the causes of wave generation. The vertical displacement of the ice by the leading pulse was approximately 30 cm, the period of the tsunami wave was 90 s and the estimated wave length was 1850 m. We interpret the records as the tsunami wave generation during the low tide phase caused by the displacement of the glacier or probably a landslide of the moraine similarly to the landslide tsunami generated by the motion of a block of rocks down the sloping bottom.

The glacier motion also generated a short-period (12 s) deformation wave in the ice cover with wavelengths of 150 and 250 m. The response of the water column was strongly non-hydrostatic that allowed the pressure gauges hanging on a wire to record this signal. The duration of the leading signal was about 40 s. Then, the trailing wave tail was recorded during approximately 20 min. The wave amplitude in the tail reached 2.2 cm during the first 5 min of the event and then it dropped gradually to zero with insignificant modulations of the wave amplitude. The wave tail was probably formed due to the reflection of the leading signal from the coasts of the fjord. The pressure measurements at two depths allowed us to estimate the wave number of these waves and the Young's modulus of the ice $E = 1.7$ GPa, which is close to the measurements based on other methods (1–3 GPa).

5.3 Supercooling of Water Near the Glacier Front in a Fjord

We will analyze measurements of seawater temperature and salinity in the immediate vicinity of the Paula Glacier front in Spitsbergen. Our analysis is based on the thermohaline properties of seawater under the conditions close to the freezing point. The theory of supercooled water was considered in Creyts and Clarke (2010). The processes that occur in the water near glaciers frequently lead to water supercooling. In this relation a special term glaciohydraulic supercooling was introduced in (Röthlisberger 1972). Glaciohydraulic supercooling is a process that occurs when the pressure melting point of water ascending the adverse slope of a subglacial overdeepening rises faster than the water is heated by viscous dissipation resulting in water remaining liquid below 0 °C. In other words, glaciohydraulic supercooling occurs when water rapidly ascends to smaller depths and cools to the temperature needed for freezing. Upwelling and cooling of water with lesser salinity than the ambient waters is a particular case of this phenomenon. Transport of water from the region of high pressure to the region of low pressure without equalizing of its internal energy can lead to freezing (Röthlisberger 1972; Creyts and Clarke 2010). The authors of (Cook et al. 2006) analyze the processes in the basement of glaciers and state that glaciohydraulic supercooling allows the water not to freeze at temperatures lower than the freezing point. They write: glaciohydraulic supercooling is a process that

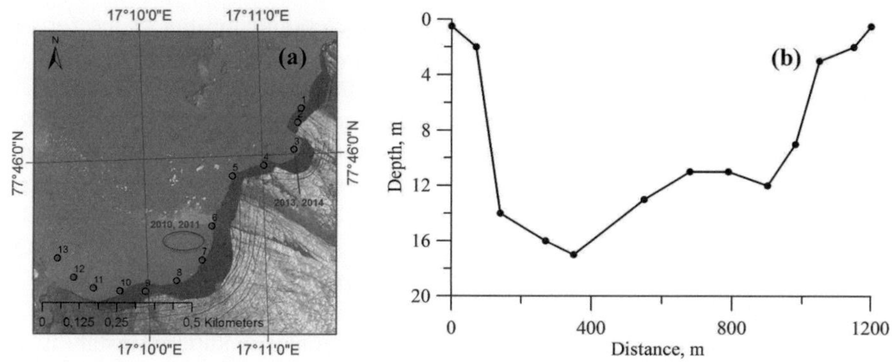

Fig. 5.9 Chart of the Paula Glacier front. Red points show the locations of depth measurement (High resolution satellite image of the Paula Glacier made in 2011 provided by the Norwegian Polar Institute). White circles show positions of CTD-casts in 2011 and 2013 (**a**); Depth profile along the glacier front. Thirteen points of measurements are related to the locations shown in Fig. 5.9 with red dots in the direction from northeast to southwest (**b**)

allows water at the base of a glacier to remain liquid at a temperature below its freezing point in response to the geometry of water flow and subglacial pressure.

The conditions for seawater supercooling appear when water is cooled to the freezing point in situ due to the contact with ice in deep water. If the water then ascends to the depths where the local freezing temperature is higher, the water appears at a temperature below the freezing point (Stevens et al. 2010) because the freezing temperature decreases when pressure decreases. It is shown in (Debenham 1965; Jeffries and Weeks 1992) that floating glaciers over the shelf can be the sources of supercooled water. Such a phenomenon was found in the data of CTD-casts (Stevens et al. 2010, Fig. 5.9) in seawater close to a glacier. They found that seawater is cooler than the freezing point by 0.01 °C under floating ice of the glacier tongue in the ocean in Antarctica. We think that in this case two mechanisms are responsible: fast cooling and fast upward motion to lower pressure.

If there are no nucleation centers, the supercooled water remains in the form of platelets and frazil ice crystals (Dmitrenko et al. 2010). The observations of this phenomenon are described in Robinson et al. (2010). Brewster and Gebhart (1994) report on the laboratory studies of strong supercooling of water with oceanic salinity 35. They found supercooling up to 5.0 °C before water began to freeze. Supercooled water was found far from the cooling surface. A model was suggested in Lawson et al. (1998) and field measurements were described that were made near the Matanuska glacier in Alaska. The authors state that glaciohydraulic supercooling is an important mechanism of the glacier ice growth due to accretion of ice crystals from supercooled water, which flows from the drainage system of the glacier. Such water contains suspended filaments: ice spicules.

Supercooling was frequently observed in polynyas (Skogseth et al. 2009; Dmitrenko et al. 2010). The water supercooling was as high as 0.01–0.02 °C with the formation of frazil ice from ice needles. Supercooling results in ice formation that occurs initially throughout the surface water layer in the form of small millimeter-scale crystals, called frazil ice crystals, that float slowly to the surface.

Laboratory experiments on frazil ice formation in the upper low saline water layer cooled from below by a more saline cold-water layer are described in (Zatsepin and Golovin 2001). The authors report that frazil ice production increases when both layers are in the turbulent motion. A similar effect of ice freezing below the water surface was described by Nansen (1897) in the sea when he observed a freshwater layer in the sea over cold saline sea waters at a temperature close to freezing.

Several papers are dedicated to the water properties near a vertical ice wall. Foldvik and Kvinge (1974) considered a thermohaline convection of conditional instability in the ocean near the ice shelf released by the formation of ice crystals in the ascending water. A paper by Huppert and Turner (1980) considered a laboratory experiment when water originating from ice melting in stratified fluid forms a step structure. A similar structure is formed when water with a vertical gradient of salinity is cooled at a vertical wall. Laboratory experiments reported in (Josberger and Martin 1981) study the convection generated in polar oceans when a fresh-water wall melts in salt water of different uniform far field temperature and salinity. Field measurements near the Erebus Glacier tongue in Antarctica were considered in Jacobs et al. (1981). A vertically stable step structure was observed in the water column 400 m thick from the surface to the bottom.

The measurements were carried out in Spitsbergen in the Tempel Fjord near the Tuna Glacier in March and September of 2010, in March 2011, in February 2012, and in Rinders Bay near the Paula Glacier in February 2011 and March 2013. In the winter season, the measurements were carried out at a minimum distance of 15 m from the glacier front. In the summer season, this distance increased to 50 m. The measurements of temperature and salinity profiles were made with CTD-profilers (temperature, conductivity, depth) SBE 19. In winter, the measurements were made from ice. The instrument was lowered into a hole drilled in the ice. In summer, the measurements were made from a zodiac boat. The first analysis of measurements was published in Morozov et al. (2015); in this section we partially review this publication.

A chart of the Paula Glacier front is shown in Fig. 5.9. The red dots show the points of depth measurements along the front line. The lower panel shows the depth profile along the front line.

Measurements in the winter time of 2010 and 2011 were carried out at a distance of 100–200 m from the glacier front without fixing the GPS positions of the points where the measurements were performed. Their locations are within the blue ellipse. The GPS coordinates of the points where CTD profiling was performed in 2013 and CTD and ADCP profiling was performed in 2014 are the same (Fig. 5.9). The glacier wall was located at a distance of 10–15 m from the point of measurements in 2013. The glacier retreated by 100 m from the same point of measurements in 2014.

On February 23–24, 2011, the ice was 60 cm thick and covered with a 15 cm snow layer. The air temperature was –15 °C. In March 2013 and 2014 the ice was 100 cm thick and covered with a 10 cm snow layer. The air temperature was –25 °C on March 12, 2013, and –10 °C on March 30, 2014. The ice surface was flat and not deformed in all field campaigns in 2010–2014. In 2013 and 2014 the ice thickness near the entrance to Rinders Bay at the distance of about 15 km from the glacier front was 50–60 cm, i.e., the sea ice near the Paula Glacier was thicker by 40–50%. Similar effect of substantial increase in the ice thickness by 40–50% over 10 km distance to the glacier front was also observed near the Tuna Glacier in Spitsbergen in 2010 and 2011 (Marchenko et al. 2011).

Water temperature near the glacier front was measured with thermistor string (EBA Engineering, Edmonton, Canada) on March 10–11, 2010. The accuracy is 0.01 °C and the resolution is 0.001 °C. The distance between thermistors was 1 m and 1.5 m between three end thermistors. Results of measurements performed with a 2 min sampling time interval are shown in Fig. 5.10a (depths 1–5 m) and in Fig. 5.10b (depths 6–10 m). Synchronous variations in the temperature within 0.05 °C are seen on the graphs. Water temperature at the bottom at a depth of 10 m is slightly higher than the surface temperature of the water. Temperature measurements performed using the SBE 39 recorder (the accuracy is 0.002 °C and resolution is 0.0001 °C) on February 23, 2011 show smaller water temperatures in the range of shallower sea depth (~5.4 m) (Fig. 5.10c).

In winter, the background distributions of temperature and salinity with depth (made at a large distance from the glacier were almost vertically homogeneous (Fig. 5.11). Almost vertically homogeneous water facilitates the upward motion of fresh water from the glacier.

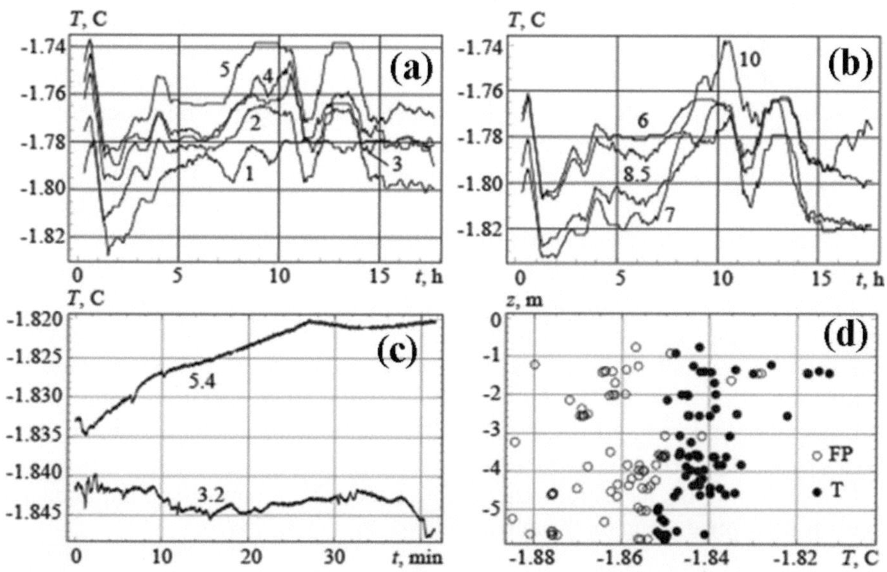

Fig. 5.10 Water temperature versus time on March 10–11, 2010 (**a**, **b**) and on February 23, 2011 (**c**). Location depth of thermistors is shown by numbers. Temperature profile (T) and water freezing points (FP) calculated from the CTD data were collected on February 24, 2011 (**d**)

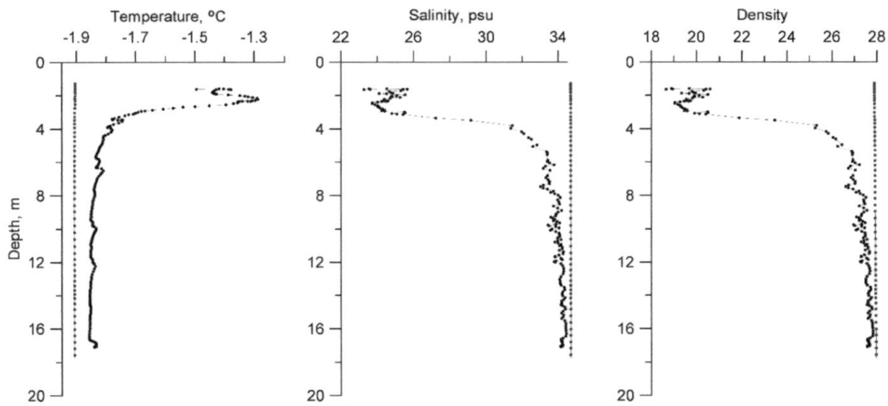

Fig. 5.11 Vertical distributions of temperature, salinity, and density based on the CTD-data near the front of the Paula Glacier in March 2013. Almost vertical blue dotted lines correspond to background vertical distributions measured at a distance of 15 km from the glacier

We made several attempts to measure the water parameters near the glacier. Initially the instrument was kept in a warm box not to freeze in the air. Then it was lowered into the cold water and kept there for approximately 10 min to equalize the temperature. Only after this, the vertical profiles were measured. The black square in Fig. 5.11 shows the background temperatures and salinities in the fjord at a distance of 15 km from the glacier. (T = –1.8 °C, S = 34.4; salinity is given in practical units). Water parameters close to supercooling were found only at a distance less than 100 m from the glacier front. In winter of 2011, the measurements were made at a distance of approximately 50 m from the front. The minimum difference between the measured temperature and freezing temperature at given salinity was close to 0.01 °C. In winter (March) of 2013, the measurements were made near the Paula Glacier at a distance of 15 m from the front at a depth of 17 m. The CTD-cast was made in the deepest place of the fjord near the front. Five sequential CTD-casts were made. They gave close distributions of temperature and salinity with depth.

The vertical distributions of temperature, salinity, and density are shown in Fig. 5.11. There is a peculiarity of their distribution near the glacier at a depth of 2–4 m over the background distribution. Here, we find higher temperatures up to –1.29 °C and a salinity decrease to 23.7. This is accompanied by unstable vertical distribution of density. The background distributions of temperature, salinity, and density in the Rinders Fjord at a distance of 15 km from the glacier are shown in the graphs almost by vertical dotted lines.

Comparison of the measured water temperature with the freezing point temperature was performed using the formula (Millero and Leung 1976)

$$T_f = S\left(-57.5 + 1.710523\sqrt{S} - 0.2154996S\right), \tag{5.5}$$

where, the freezing point T_f is in mK, and water salinity S is in psu. Deviation of the actual temperature from the freezing point is calculated using the following formula:

$$\Delta T = T - T_f(S), \tag{5.6}$$

where, T and S are the actual values of the water temperature and salinity measured during CTD profiling. The influence of pressure on the freezing point is ignored because of small water depths.

The temperature-salinity dependence near the front of the glacier based on these data is shown in Fig. 5.12 with small circles. The freezing temperature dependence on salinity based on (Millero and Leung 1976) is also shown in this figure as almost a straight line. Large circles mark the measurements at depths of 2, 3, and 4 m. Upwelling of melting fresh water in the environment of cooler seawater leads to its glaciohydraulic supercooling up to 0.4 °C. The core of supercooled water is located at a depth of 3 m. We believe that it ascended from the deeper depths. The temperature-salinity characteristics of water deeper than 5–6 m at the point of measurements almost correspond to its homogeneous distribution. A jet of melting fresh water admittedly flows from the basement of the glacier or from its middle part and ascends to the surface being cooled.

In summer, a strong temperature, salinity, and density interface exists at a depth of a few meters due to warming at the surface and freshwater flow from the coast. This layer prevents vertical water motions. Glaciohydraulic supercooling cannot develop in summer also because the temperature of seawater is greater than 0 °C (+1°–5 °C).

Temperature and salinity profiles were measured using CTD-profiler SBE 19 near the glacier front. In winter, supercooled water was found at a distance of 15 m from

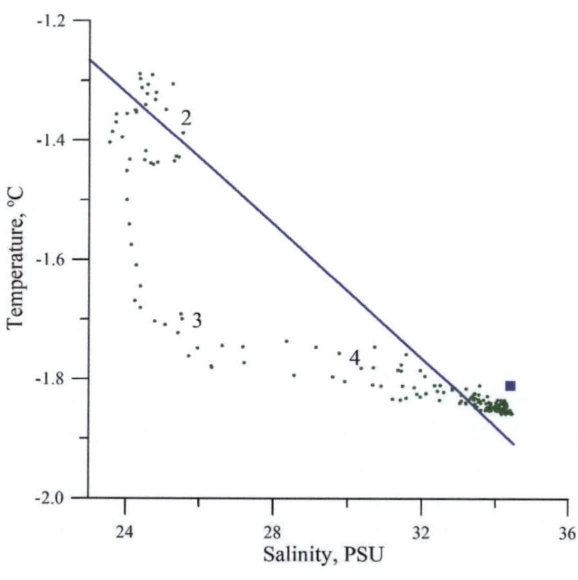

Fig. 5.12 θ,S-diagram of measurements near the glacier front and at a long distance from it. The blue solid line shows freezing temperature versus salinity (Millero and Leung 1976). The blue square corresponds to the background conditions at a large distance from the glacier. Small green circles denote measurements in 2013 at a distance of 15 m from the front at a depth of 17 m. Numerals 2, 3, and 4 near the circles denote measurements at depths of 2, 3, and 4 m, respectively in 2013

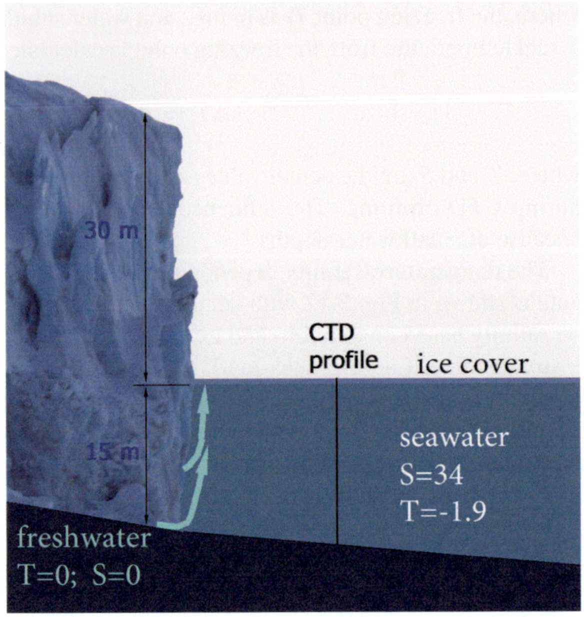

Fig. 5.13 Scheme of freshwater flow from the glacier and measurements of water properties near the glacier front (Morozov et al. 2015). Used with permission of the Pleiades Publishing, Ltd

the glacier front at a depth of 2–4 m, while the depth of the sea near the glacier is 17 m. We assume that melting freshwater flows from the glacier at a temperature of 0 °C. This freshwater appears in the environment with a temperature close to –1.8 °C. While flowing to the surface the freshwater cools to the freezing temperature and lower due to the contact with the surrounding seawater. This temperature is close to the point of seawater freezing at a salinity of 33 and knowingly lower than the freezing temperature of freshwater. Salinity of the water flowing from the glacier is close to zero. A scheme of measurements near the glacier is shown in Fig. 5.13.

Freshwater salinity increases due to mixing with seawater in the fjord. As the freshwater flows from the glacier at the depths with relatively high pressure and ascends to lesser depths with smaller pressure its internal energy does not reach the equilibrium state at the freezing point, which also depends on pressure.

Internal melting is observed in polythermal glaciers. These glaciers consist of the upper layer with cold ice and lower layer of warm water containing ice. In Spitsbergen such glaciers that descend to sea are widely spread (Macheret 2006; Macheret and Zhuravlev 1982; Kristensen and Benn 2012). The mean water content in the lower warmer layer of such glaciers can exceed 0.1%. Part of accumulated fresh water in this layer can freeze at the boundary between the cold and warm ice, and the other part can flow into the sea even in the winter period (Macheret and Glazovsky 2000). Schemes and explanation of freshwater flows from polythermal glaciers are given in (Hambrey and Glasser 2012) and (Moorman 2003).

The greatest concavities of the glacier wall are found in the deepest places along the front of the Paula Glacier. This makes us think that the glacier is most subjected to the loss of its mass in these places. Water flux from under the glacier can influence the sediment transport from the glacier bed, make the cohesion between the ice and soil weaker and contribute to the formation of local ice stresses and crevasses.

Mixing of freshened water at the freezing point with colder sea water can influence the formation of supercooled water if the sea water temperature is close enough to its freezing point. The weather in February and March 2013 was very cold with air temperatures between –25 and –30 °C, and therefore, the entire water column below the ice was almost at the freezing point in relatively shallow places in Spitsbergen fjords. This weather factor is important for the explanation of the observations in March 2013.

Freshwater freezing in the seawater develops in the form of ice crystals formation, which flow to the surface (Tweed et al. 2005). In the natural conditions such crystals reach the lower surface of the ice cover and increase the ice thickness. The measurements of ice thickness near the glacier show that ice thickness increases closer to the glacier.

In addition, when we observe the water surface in the holes, which we drill in the ice to lower our instruments, ice crystals very quickly appear at the water surface in the holes drilled near the glacier (provided that the hole is initially cleared of snow and remains of the drilled ice). In 15–20 min the water surface in the hole can be covered with a layer of wet crystals approximately one centimeter thick. This phenomenon is not observed far from the glacier.

Thus, investigation of vertical profiles of seawater temperature and salinity close to the front of the Paula Glacier in the Rinders Fjord in Spitsbergen demonstrated the presence of water with smaller salinities than the surrounding water in the fjord. Such water was found at a depth range of 2 to 5 m, while the total depth near the glacier was 17 m. The water was supercooled up to 0.35 °C at 3 m depth, an effect related to glaciohydraulic supercooling. We attribute this to the fact that meltwater that contains in the polythermal glacier flows from the glacier in the places of depressions of glacier bed into the seawater and cools by the surrounding water of lower temperature to the temperatures lower than the freezing temperature.

5.4 Ice and Water Structure Near Glaciers

Ice thickness varies significantly in Svalbard fjords. Fast ice thickness in Sveabukta and near the Paula Glacier varies typically within 0.8–0.9 m in March–April. Ice thickness in the Advent Fjord reached 0.5 to the beginning of March 2011, while in 2010 the ice was absent in the Advent Fjord. Ice thickness from the inner side of Akseløya Island was 0.6 m in the beginning of May 2010. Figure 5.14a shows the distribution of ice thickness from the Tuna Glacier to the *Noorderlicht* vessel over 10 km distance in the beginning of March 2010 (red dots) and in mid-February 2011 (blue dots). The *Noorderlicht* is a schooner frozen in the ice approximately 10 km from the glacier front, which was used for living of the expedition participants. Since the weather conditions over 10 km distance are the same the reduction of the ice thickness with the distance from the glacier is related to heat fluxes from the sea. A decrease of sea depth from 80–90 m near the *Noorderlicht* to 40 m near the glacier in combination with the ice action on the upper water layer can lock warm water in the deeper part of the fjord. It will decrease heat fluxes from the sea to the ice in the shallow part of the fjord near the glacier and influence the ice growth. In ice free season, wind and wave actions induce higher water exchange within the fjord (Fig. 5.15).

Ice thickness (h_i) without accounting for the heat fluxes from the sea can be estimated with the Zubov's formula (Zubov 1944): $h_i^2 + 50h_i = 80R$, where R is the

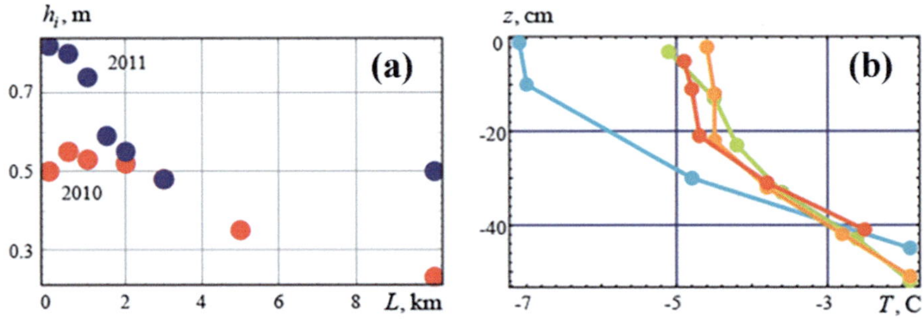

Fig. 5.14 Ice thickness variation over Tempel Fjord from the Tuna Glacier to the *Noorderlicht* (**a**). Typical temperature profiles through the ice (**b**)

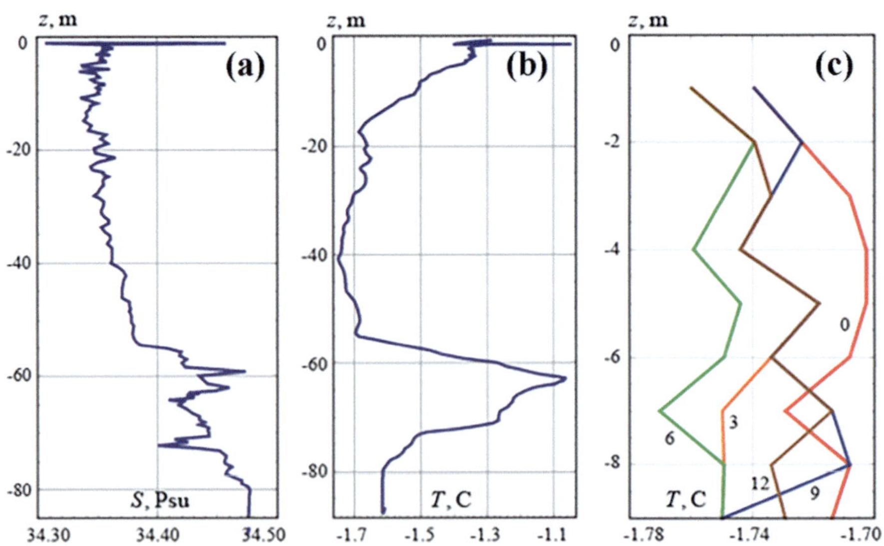

Fig. 5.15 Vertical profiles of salinity (**a**) and temperature (**b**) in the Tempel Fjord near the *Noorderlicht*. Temporal evolution of vertical temperature profile near the Paula Glacier on March 3–11, 2010 (**c**)

amount of degree days (degree days is degrees multiplied by days) of the frost, and h_i is the ice thickness in cm. Amount of degree days of the frost estimated with the data from Svalbard Lufthavn (www.eklima.met.no) is equal to 753 from December 1 to March 8 in 2010, and 1458 from November 1 to February 28 in 2011. According to the Zubov's formula the ice thickness is $h_i = 56$ cm in March 2010 and $h_i = 85$ cm to the end of February 2011. Figure 5.14a demonstrates that the ice thickness near the glacier was close to these values in 2010 and 2011. The measured salinity of sea ice was between 5 and 6 ppt in the Tempel Fjord. Temperature profiles through the ice thickness are close to linear (Fig. 5.14b). Snow thickness on sea ice was typically smaller than 10 cm.

Sea water temperature and salinity changes significantly with the distance from the Tuna Glacier in the ice season. Figure 5.16a and b show vertical salinity and temperature profiles measured with SBE 19 near the *Noorderlicht* in the beginning of March 2010. The temperature over the entire water column below the ice is higher than the freezing point which should be close to –1.85 °C according to the measured water salinity. It explains small ice thickness 23 cm. Red lines in Fig. 5.16c and d show vertical profiles at the same point in February 2011. One can see that the temperature of almost the entire water column is close to the freezing point. Water temperature near the Tuna Glacier was higher than water temperature near the *Noorderlicht*, but the water near glacier was fresher. The same situation was observed near the Paula Glacier. Water temperature near the Paula Glacier was higher than in Sveabukta, but the water was fresher. Figure 5.16c shows water temperature profiles measured over tidal cycle after each 3 h with thermistor string EBA on March 11, 2010. One can see that the water temperature was much warmer than on February 24, 2011, shown in Fig. 5.16d near the Paula Glacier by the blue curve.

The CTD profiling performed with SBE 19 on September 1, 2010, demonstrated uniform water properties with the distance from the Tuna Glacier. Figures 5.16a and b show examples of salinity and temperature profiles taken at different distances from the glacier wall. In all profiles maximal salinity was measured near the bottom, and maximal temperature near the surface. The stratification was stable. Variation in the ice thickness with the distance from the glacier is shown in Fig. 5.17 based on the measurements in March 2011.

5.5 Temperature Gradient Inside the Paula Glacier

In 2011 we performed measurements of temperature gradient in the interior of the Paula Glacier. Ice temperature measurements were carried out at the front of the Paula Glacier on February 23, 2011. We drilled a 4-m long hole using an auger of 2 inch diameter and fuel drilling mashine. The stainless drill consisted of buildable sections each 1 m long. The drilling process is shown in the photo in Fig. 5.18. Dust particles in the glacial ice damaged the blade and did not allow further drilling. The hole was made at approximately 1 m distance from the surface of sea ice near the glacier wall. A temperature string was placed inside the hole to measure ice temperature during

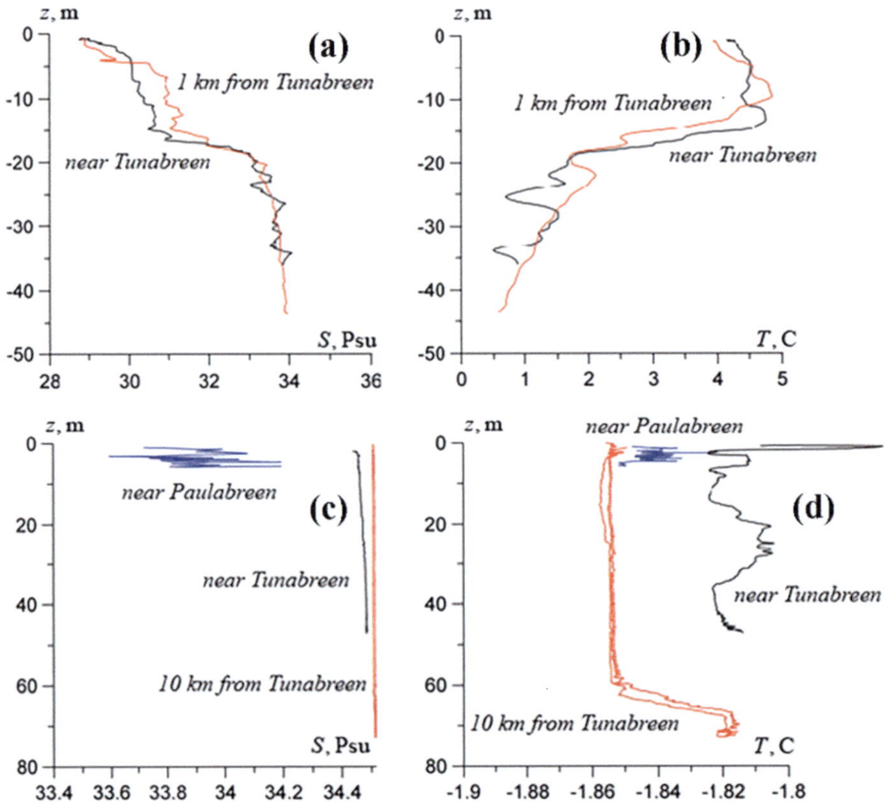

Fig. 5.16 Vertical profiles of salinity (**a**) and temperature (**b**) in the Tempel Fjord in summer on September 1, 2010 near the Tuna Glacier and 1 km from the glacier. Vertical profiles of salinity (**c**) and temperature (**d**) in Tempel Fjord and near the Paula Glacier on February 17–22, 2011

approximately 1 h from 13:00 to 14:00 UTC at low phase tide. The readings were recorded by the unit located near the glacier.

Water level during the day of measurements is shown in Fig. 5.19. Sea water level was calculated using the function TideData in Wolfram Mathematica software applied to the point with the coordinates 77.7922° N and 17.02551° E in the Rindersbukta of the Van Mijen Fjord. Then, we removed the temperature string since the sea ice was rising because of the tide.

Figure 5.20 shows the water level changes depending on time during one month before the date of the measurements. Solid line marks the level of temperature measurements relative to the water level. It is clear that this location was mostly above the ice surface. During the spring tide tide the location could be below the the ice surface by approximately 0.2 m. The ice thickness in the February exceeded 1 m near Paulabreen. Therefore, it is possible to conclude that the conducted temperature measurements can be used to estimate the heat flux from the glacier into the air in winter time.

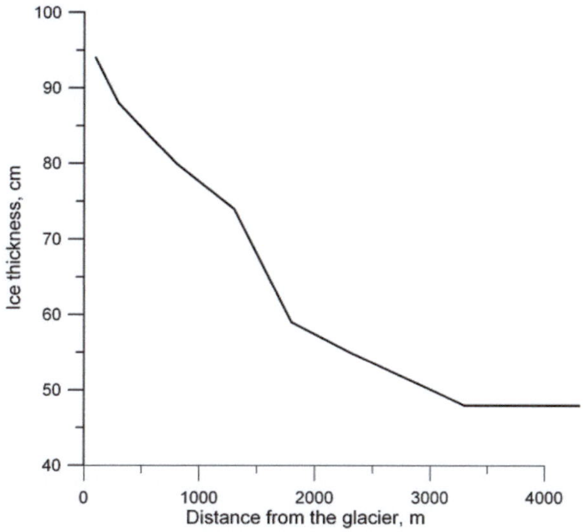

Fig. 5.17 Ice thickness versus distance from the glacier front. The measurements were made in 2010 near the Tuna Glacier

Fig. 5.18 Drilling the Paula Glacier

The temperature profile is shown in Fig. 5.21. Assuming the thermal conductivity of ice of 2.24 W/(m K) we find that the heat flux from the glacier ice into the air is 3.2 W/m^2. The heat flux was calculated from the temperature measurements at points located at 2, 3, and 4 m from the glacier front. The Paula Glacier is a polythermal glacier which basal temperature is close to 0 °C (Hagen et al. 2003).

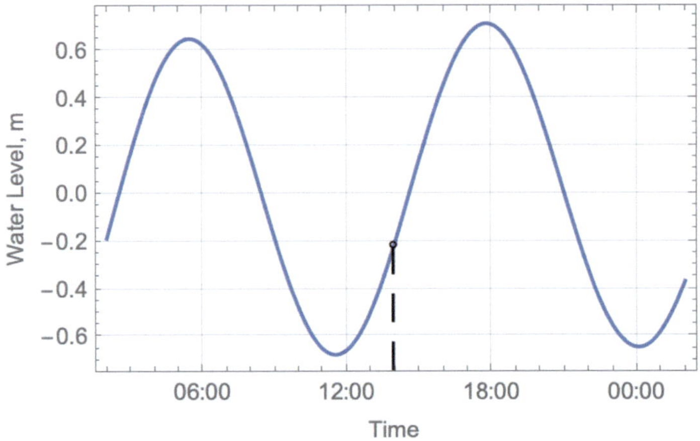

Fig. 5.19 Water level depending on time on February 23, 2011. Dushed line mark the time of measurements of the temperature profile

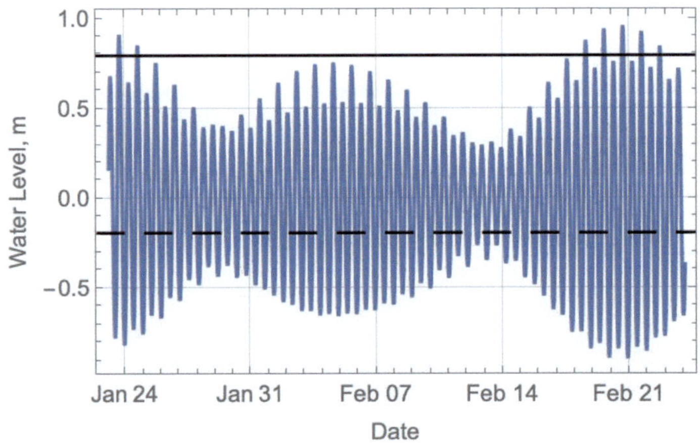

Fig. 5.20 Water level depending on time from January 23 to February 23, 2011. Dushed line marks the water level at 14.00 UTC on February 23, 2011. Solid line marks the level where the temperature profile was measured

The outside temperature was −13.5 °C. The temperature gradually increased in the glacier. The Paula Glacier is a polythermal glacier, which is a glacier with a complex thermal structure, having both frozen, or "cold-based," and unfrozen, or "warm-based," areas. This mixed thermal regime leads to complex ice flow and subglacial drainage systems. The outer parts of the glacier and surface ice are typically frozen. The thicker ice in the accumulation area or interior can be at or near the melting point. The cause of heat may be geothermal flux, which is common in Spitsbergen.

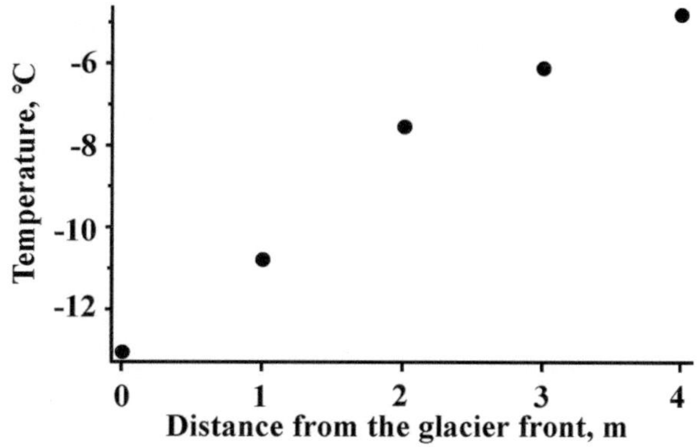

Fig. 5.21 Temperature profile in the interior of the glacier in the horizontal direction at approximately 1 m distance from sea ice surface in the front of Paula Glacier measured at 14.00 UTC on February 23, 2011

Sea water temperature near the glacier front is at the freezing point of −1.9 °C in ice season and close to 0 °C in ice free time. This warm environment below Paulabreen influences the heat flux into the air. The geothermal heat flux estimated from lithology logs of ten boreholes is significantl and ranges from 60 to 147 mW/m^2 (Nuus et al., 2025). The oceanic heat flux through sea ice in the ice season estimated at a temperature gradient of 10 °K in 1 m thick ice, is about 20 W/m^2. In case of 2 m thick ice the ocean heat flux decreases in two times. Thus, dominant influence of the ocean heat flux on ice processes near the glacier fronts is expected and the heat flux from glacier can make 30% contribution to the driving forces influencing the heat fluxes in the air when the thickness of sea ice near the glacier front is about 2 m and greater.

References

Assier-Rzadkiewicz S, Heinrich P, Sabatier PC et al (2000) Numerical modelling of a landslide-generated tsunami: the 1979 Nice event. Pure Appl Geophys 157:1707–1727

Bjerrum L (1971) Sub-aqueous slope failures in Norwegian fjords, vol 88. Norwegian Geotechnical Institute Publication, Oslo, p 8

Błaszczyk M, Jania JA, Hagen JO (2009) Tidewater glaciers of Svalbard: recent changes and estimates of calving fluxes. Polish Polar Res 30(2):85–142

Bogorodskii VV, Gavrilo VP (1980) Ice, physical properties. Modern methods of glaciology. Leningrad. Gidrometeoizdat, pp 384. [in Russian]

Brewster RA, Gebhart B (1994) The effects of supercooling and freezing on natural convection in seawater. Inter. J Heat Mass Transf 37(4):543–552

Collins IF, McCrae IR (1985) Creep buckling of ice shelves and the formation of pressure rollers. J. Glaciology 31(109):242–252

Cook SJ, Waller RI, Knight PG (2006) Glaciohydraulic supercooling: the process and its significance. Prog Phys Geography 30(5):1–12

Creyts TT, Clarke GKC (2010) Hydraulics of subglacial supercooling: theory and simulations for clear water flows. J Geophys Res 115:F03021. https://doi.org/10.1029/2009JF001417

Debenham F (1965) The glacier tongues of McMurdo Sound. Geogr J 131:369–371

Dmitrenko IA, Wegner C, Kassens H, Kirillov SA et al (2010) Observations of supercooling and frazil ice formation in the Laptev Sea coastal polynya. J Geophys Res 115:C05015. https://doi.org/10.1029/2009JC005798
Fine IV, Rabinovich AB, Thomson RE, Kulikov EA (2003) Numerical modeling of tsunami generation by submarine and subaerial landslides. In: Yalciner AC, Pelinovsky E, Synolakis CE, Okal E (eds) Submarine landslides and tsunamis. Kluwer Academic Publishers, pp 73–93
Foldvik A, Kvinge T (1974) Conditional instability of sea water at the freezing point. Deep-Sea Res 21:169–174
Gonzalez FI, Kulikov YA (1993) Tsunami dispersion observed in the deep ocean. In: Tinti S (ed) Tsunamis in the World. Kluwer Acad. Norwell. Mass, pp 7–16
Hagen JO, Liestøl O, Roland E, Jørgensen T (1993) Glacier Atlas of Svalbard and Jan Mayen. Norsk Polarinstitutts Meddelelser. № 129, pp 160
Hagen JO, Kohler J, Melvold K, Winter J-G (2003) Glaciers in Svalbard: mass balance, runoff and freshwater flux. Polar Res 22(2):145–159
Harbitz CB, Løvholt F, Pedersen G, Masson DG (2006) Mechanisms of tsunami generation by submarine landslides: a short review. Norwegian J Geology 86:255–264
Hambrey MJ, Glasser NF (2012) Discriminating glacier thermal and dynamic regimes in the sedimentary record. Sed Geol 251–252:1–33
Hodgkins R, Dowdeswell JA (1994) Tectonic processes in Svalbard tide-water glacier surges: evidence from structural glaciology. J Glaciology 40(136):553–560
Huppert HE, Turner JS (1980) Ice blocks melting into a salinity gradient. J Fluid Mech 100:367–384
Jacobs SS, Huppert HE, Holdsworth G, Drewry DJ (1981) Thermohaline steps induced by melting of the Erebus Glacier Tongue. J Geophys Res 86(C7):6547–6555
Jeffries MO, Weeks WF (1992) Structural characteristics and development of sea ice in the 15 Western Ross Sea. Antarct Sci 5(1):63–75
Jiskoot H, Murray T, Boyle P (2000) Controls on the distribution of surge-type glaciers in Svalbard. J. Glaciology 46(154):412–422
Johns MW, Prior DB, Bornhold BD et al (1986) Geotechnical aspects of a submarine slope failure, Kitimat Fjord, British Columbia. Mar Geotechnol 6:243–279
Josberger EG, Martin S (1981) Convection generated by vertical icewalls. J Fluid Mech 111:439–473
Kristensen L, Benn DI (2012) A surge of the glaciers Skobreen-Paulabreen, Svalbard, observed by time-lapse photographs and remote sensing data. Polar Res 31:11106. https://doi.org/10.3402/polar.v31i0.11106
Kulikov EA, Rabinovich AB, Thomson RE, Bornhold BD (1996) The landslide tsunami of November 3, 1994, Skagway Harbor, Alaska. J Geophys Res Oceans 101(C3):6609–6615
Lawson DE, Strasser JC, Evenson ED et al (1998) Glaciohydraulic supercooling: a freeze-on mechanism to create stratified, debris-rich basal ice: I. Field evidence. J Glaciology 44(148):547–562
Macheret YY, Glazovsky AF, Ignatyeva IY et al (1992) Structure, hydrothermal state and regime of subpolar glaciers, Regime and evolution of polar ice sheets. Gidrometeoizdat, St. Petersburg, pp 48–115
Macheret YY (2006) Radiosounding of Glaciers. Moscow. Nauchnyi Mir, p 400. [in Russian]
Macheret YY, Glazovsky AF (2000) Estimation of absolute water content in Spitsbergen glaciers from radar sounding data. Polar Res 19(2):205–216
Macheret YY, Zhuravlev AB (1982) Radio-echo sounding of Svalbard glaciers. J Glaciology 28(99):295–314
Marchenko AV, Morozov EG, Muzylev SV, Shestov AS (2010) Interaction of short internal waves with the ice cover in an Arctic fjord. Oceanology 50(1):18–27
Marchenko A, Shestov A, Karulin E et al (2011) Field studies of sea water and ice properties in Svalbard fjords. In: Proceedings 21st internal conference port and ocean engineering arctic conditions July 10–14, 2011 Montréal, Canada
Marchenko AV, Morozov EG, Muzylev SV (2012) A tsunami wave recorded near a glacier front. Nat Hazard 12:415–419

Mei CC (1983) The applied dynamics of ocean surface waves. Wiley, New York, p 740
Millero FJ, Leung WH (1976) Thermodynamics of seawater at one atmosphere. Am J Sci 276(9):1035–1077
Moorman B (2003) Glacier-permafrost hydrology interactions, Bylot Island, Canada. In: Phillips M, Springman SM, Arenson L (eds) Proceedings of 8th internal permafrost conference Zurich, Switzerland. A.A. Balkema Publishers, pp 783–788
Morozov EG, Marchenko AV, Fomin YV (2015) Supercooled water near the glacier front in Spitsbergen. Izv Atmosph Oceanic Phys 51(2):203–207
Muzylev SV, Macheret YY, Morozov EG, Lavrentiev II, Marchenko AV (2013) Fluctuations of ice cover and sea water pressure nearby the Tunabreen Glacier front at Spitsbergen. Ice Snow 4(124):119–124
Nansen F (1897) Fram over Polhavet. Den norske polarfærd 1893–1896, Aschehoug, Kristiania
Nuus MH, Fuchs S, Smyrak-Sikora AA, Kubutat T, Senger K (2025) Thermal conductivity and heat flow modeling of petroleum exploration and research wells onshore Svalbard. SSRN: https://ssrn.com/abstract=5396091 or https://doi.org/10.2139/ssrn.5396091
Rabinovich AB, Thomson RE, Kulikov EA et al (1999) The landslide-generated tsunami of November 3, 1994 in Skagway Harbor, Alaska: a case study. Geophys Res Let 26(19):3009–3012
Robinson NJ, Williams MJM, Barrett PJ, Pyne AR (2010) Observations of flow and ice–ocean interaction beneath the McMurdo Ice Shelf. Antarctica J Geophys Res 115:C03025. https://doi.org/10.1029/2008JC005255
Röthlisberger H (1972) Water pressure in intra- and subglacial channels. J Glaciol 11(62):177–203
Skogseth R, Nilsen F, Smedsrud LH (2009) Supercooled water in an Arctic polynya: observations and modeling. J Glaciology 55(189):43–52
Sodhi DS, Haynes FD, Kato K, Hirayama K (1983) Experimental determination of the buckling loads of floating ice sheet. Ann. Glaciology 4:260–265
Squire VA, Hosking JR, Kerr AD, Langhorne PJ (1996) Moving loads on ice plates. Kluwer Academy Publishers, pp 230
Stevens CL, Stewart CL, Robinson NJ et al (2010) Flow and mixing around a glacier tongue. Ocean Sci 7:293–304
Tappin D, Matsumoto T, Watts P et al (1999) Sediment slump likely caused 1998 Papua New Guinea tsunami. Eos 80:329–334
Tweed FS, Roberts MJ, Russell AJ (2005) Hydrologic monitoring of supercooled discharge from Icelandic glaciers. Quatern Sci Rev 24:2308–2318
Zubov NN (1944) Arctic ice. Moscow, Izdatel'stvo Glavsevmorputi, pp 360. [in Russian]
Zatsepin AG, Golovin PN (2001) Laboratory experiments on frazil ice formation with application to the Arctic marine environment. Memoirs of National Institute of Polar Research (Tokyo), Special Issue, No. 54, pp 417–421

MIX
Papier aus verantwortungsvollen Quellen
Paper from responsible sources
FSC® C105338

If you have any concerns about our products,
you can contact us on
ProductSafety@springernature.com

In case Publisher is established outside the EU,
the EU authorized representative is:
Springer Nature Customer Service Center GmbH
Europaplatz 3, 69115 Heidelberg, Germany

Printed by Libri Plureos GmbH
in Hamburg, Germany